THEODORE ROOSEVELT

Books by Nathan Miller

Theodore Roosevelt:
A Life
Stealing from America: Corruption in American Politics
from Jamestown to Reagan
The U.S. Navy: A History
Spying for America:
The Hidden History of U.S. Intelligence
FDR: An Intimate History
The Naval Air War: 1939–1945
The Roosevelt Chronicles
The U.S. Navy: An Illustrated History
The Founding Finaglers:
A History of Corruption in American Politics
Sea of Glory:
The Naval History of the American Revolution
The Belarus Secret (with John Loftus)

Theodore ROOSEVELT

★ *A Life* ★

Nathan Miller

Perennial
An Imprint of HarperCollinsPublishers

A hardcover edition of this book was published in 1992 by William Morrow.

THEODORE ROOSEVELT. Copyright © 1992 by Nathan Miller. All rights reserved. Printed in the United States of America. No part of this book may be used or reproduced in any manner whatsoever without written permission except in the case of brief quotations embodied in critical articles and reviews. For information address HarperCollins Publishers Inc., 10 East 53rd Street, New York, NY 10022.

HarperCollins books may be purchased for educational, business, or sales promotional use. For information please write: Special Markets Department, HarperCollins Publishers Inc., 10 East 53rd Street, New York, NY 10022.

First Quill edition published 1994.

Reprinted in Perennial 2003.

Designed by Paul Chevannes

Library of Congress Cataloging-in-Publication Data

Miller, Nathan.
    Theodore Roosevelt : a life / Nathan Miller.
       p.   cm.
    ISBN 0-688-13220-0
    1. Roosevelt, Theodore, 1858–1919. 2. Presidents—United States—
Biography.   I. Title.
    E757.M67   1992
973.91'1'092—dc20
[B]   91-44155
                                                                                                                    CIP

06 07 QW 20 19 18 17 16 15 14

Happily once again,
to
Jeanette

Preface

A flash of teeth . . .
The big stick . . .
The strenuous life . . .

These are the things that immediately come to mind when Theodore Roosevelt's name is mentioned. Probably no president has more successfully captured the imagination of the American people than the Rough Rider. Three quarters of a century after his death, what H. G. Wells called Roosevelt's "friendly, peering snarl of a face" is still immediately recognizable to most Americans. In every ranking of our presidents, he usually comes out in the top half-dozen or so, with the rank of "Great" or "Near Great." If Americans were polled on which past president they would like to have in the White House today, Roosevelt would probably be the winner.

The sheer delight of the man was overwhelming. "The special mark of Theodore Roosevelt was joy—joy in everything he did," as John Morton Blum has observed. But his appeal goes far beyond that. In an age that offers few genuine heroes, Roosevelt continues to cast a magic spell over the collective consciousness. Even when the veneer of mythology is scraped away, unlike many

9

past heroes, he survives scrutiny. What is the reason for a phenomenon so at variance with the usual? Although he could be as
ruthless as any politician when circumstances demanded, Roosevelt's greatness lies in the fact that he was essentially a moral
man in a world that has increasingly regarded morality as super-
fluous.

Prodigiously endowed with gifts of mind and spirit, Roosevelt
was in many ways ahead of his time. He was the first American
president to use government as a force to meliorate the harsh
social and economic conditions of the time. He was the first to
grant recognition to the wave of immigrants flocking to the United
States at the turn of the century. He put a brake on the reckless
exhaustion of the nation's natural resources and preserved wilderness areas for future generations of Americans. He not only
won the Nobel Peace Prize but understood that the United States
could neither shield itself against involvement in global affairs,
nor police the world alone. Moreover, he realized, as few of our
leaders since have done, that the most important task facing any
political leader is to educate the public—to arouse the nation's
consciousness about existing evils and create a demand for
change.

Roosevelt was a walking bundle of contradictions. On the surface, he seemed a sunny and open personality and almost per-
petually in motion. Often it is assumed that his perceptions of
people were limited by a tendency to reduce situations to stark
black and white. But he was a far more elusive figure than usually
credited. He not only made history but wrote it. He led the charge
up San Juan Hill and could sit all afternoon, silent and alone, to
observe an unfamiliar bird. He wrote the definitive analysis of a
species of North American deer. Often, he was troubled by melancholy and premonitions of impending disaster.

The public appetite for information about Roosevelt has been
whetted by two excellent books, David McCullough's *Mornings
on Horseback* and *The Rise of Theodore Roosevelt* by Edmund
Morris, but both works cover only his early years. This book, the
first full-scale, one-volume biography in more than three decades,
is intended for those readers who wish to know the full story of
his life. Moreover, I have had access to the letters of Roosevelt's
courtship and marriage to his first wife, Alice Lee, that were

unavailable to earlier writers. Although I have placed ample emphasis upon Roosevelt's public career, I was attracted by the man himself and by his relations with his close associates and family, his children, and particularly his two wives. I have tried to portray a three-dimensional figure of flesh and blood who confronted failures as well as triumphs. In creating this portrait, I have relied to a large extent on Roosevelt's own words as expressed in his letters and other voluminous writings.

A great many people helped in the preparation of this book. I would particularly like to express my appreciation to the staffs of the Manuscript Division of the Library of Congress, the Nimitz Library of the U.S. Naval Academy at Annapolis, and the Houghton Library at Harvard University. Wallace F. Dailey, the curator of the Theodore Roosevelt Collection at Harvard, not only responded promptly and fully to my queries and supplied many of the pictures printed in the book, but also made numerous suggestions about research material that should be consulted. I owe a special debt of gratitude to Dr. John A. Gable, the able executive director of the Theodore Roosevelt Association, who not only fielded numerous questions and suggested new material, but took time out from a hectic schedule to give the manuscript a thorough reading and made countless suggestions that improved it. Connie Roosevelt—Mrs. Theodore Roosevelt IV—read the manuscript and lent encouragement to the author. In Lisa Drew, my editor, and her assistant, Bob Shuman, I was gifted with an editorial team that, in a perfect world, every writer would have with every book. Lastly, I cannot sufficiently thank my good friends Cassie and Pat Furgurson for their continuing enthusiasm, encouragement, and numerous kindnesses throughout the long process of researching and writing this book.

Nathan Miller
Washington, D.C.

Contents

Life is a great adventure, and the worst
of all fears is the fear of living.

—THEODORE ROOSEVELT

Prologue

Inauguration Day—March 4, 1905.

Promptly at ten o'clock that morning, the heavy glass doors of the White House abruptly swung open and President Theodore Roosevelt, tombstone-like teeth bared in a broad grin, strode out onto the North Portico. Roosevelt never walked anywhere: He bustled, he hurried, he raced. This was a special day—the day on which he would become president in his own right. Ignoring blustery winds and the threat of rain in the air, he bounded into an open landau drawn by a matched team of four horses and set off at a brisk pace for the Capitol. In the distance, the majestic dome appeared etched against the steel-gray sky.

Periodically doffing his top hat, the president bowed again and again to the cheering crowds lining both sides of Pennsylvania Avenue. It seemed as if some mysterious force had tipped the continent in the direction of Washington and suddenly filled the streets with people. Shouts of "Hooray for Teddy! Hooray for Teddy!" rolled in upon him like the waves of the sea. Hats were in the air; handkerchiefs waved. Forty Rough Riders, members of Roosevelt's Spanish-American War regiment, trotted beside his carriage as an honor guard. Well lubricated by three days of celebration, the veterans had trouble maintaining formation. Red-faced, their Colonel stood in his carriage to shout orders to keep lined up. One moment the throb of bands playing the wartime song "There'll Be a Hot Time in the Old Town Tonight" would be heard above the cheering and then the disembodied tune was lost, only to sound out again above the din.

Roosevelt was in his glory. Over the last four years, detractors had derided him as an accidental president—"that damned cowboy"—who had been elevated from the obscurity of the vice presidency by the anarchist's bullet that killed William McKinley. Outwardly boisterous, he inwardly suffered from bouts of pessimism and melancholy and had feared that he would not be elected on his own. But he had overwhelmed the Democratic ticket headed by Judge Alton B. Parker and now had his own popular mandate. And he had made it clear that he intended to flex his muscles in his new term. "Tomorrow I shall come into office in my own right," he had reportedly told a visitor the previous evening. "Then watch out for me!"

The first lady's carriage soon followed that of the president. Edith Kermit Roosevelt was wearing a dark blue cloth coat and matching hat and was accompanied by the couple's gravely serious eldest son, Ted, and daughter Alice. Resplendent in a white cashmere outfit, "Princess Alice," as always the center of attention, struggled with a cartwheel-sized hat with an ostrich plume that was almost carried away by a sudden gust. In the next carriage rode the rest of the Roosevelt brood: Ethel, Kermit, Archie, and Quentin, who were being admonished by their old nurse to be still. Trailing behind were the president's sisters, Mrs. Corinne Robinson and Mrs. Anna Cowles, the latter riding with the engaged cousins, Eleanor and Franklin D. Roosevelt. Eleanor, the orphaned daughter of the president's brother Elliott, was his favorite niece and he had promised to come up to New York City in two weeks to give the bride away at her wedding.

Roosevelt embodied the overwhelming confidence of the nation as it entered a limitless new century—the American Century. Industry was booming; the unemployment, breadlines, and savage labor strife of the previous decade had faded away. Life bubbled over in him like laughter from a healthy child. He had punched cattle, led the charge up San Juan Hill, hunted big game, and waved his fist under J. P. Morgan's rubicund nose. A man of contradictions and barely contained energies, he exercised strenuously, yet read a book a day. Western rancher and society clubman, writer and soldier, conservationist and big-game hunter, he was, as Lord Morley said, a cross between Saint Vitus and Saint Paul. "Do you know the two most wonderful things I have seen in your country?" the visiting English statesman observed to an American. "Niagara Falls and the President of the United States . . . both great wonders of nature."

With a fierce joy, Roosevelt brandished a "Big Stick" abroad and promised a "Square Deal" at home. In his first term, he had personally settled a bitter anthracite coal strike, challenged the trusts, and as the first American president to play a role in world affairs, "took" Panama and began construction of a long-dreamed-of canal across the Isthmus. When a supposed American citizen was kidnapped by a Moroccan bandit, he sent the U.S. Navy to the rescue with the resounding demand: "We want Perdicaris alive or Raisuli dead!" By a happy "coincidence," his words became public as the Republican national convention was meeting to select the party's nominee for president. And even as the inaugural ceremonies were taking place, he was secretly trying to negotiate an end to the Russo-Japanese War.

Shortly before noon, Roosevelt, having witnessed the swearing-in of the new vice president, Charles W. Fairbanks, in the Senate chamber, stepped out onto the inaugural platform on the East Portico of the Capitol. A blaze of sunlight suddenly broke through the clouds. Forty acres of spectators, standing shoulder to shoulder in the surrounding plaza and filling every window and niche of the nearby buildings, saw it as an omen and let out a lusty cheer. "Roosevelt weather," said observers.

Placing his left hand on an open Bible and with right hand raised, Roosevelt slowly intoned the words of the oath of office in a high-pitched voice as it was administered by the white-haired Chief Justice Melville W. Fuller: "I, Theodore Roosevelt, do solemnly swear . . ." On the president's upraised hand was a curious ring. The gift of Secretary of State John Hay, it contained strands of Abraham Lincoln's hair. Hay, who had been one of Lincoln's private secretaries, asked Roosevelt to wear it at his inauguration because he "was one of the men who most thoroughly understand and appreciate Lincoln. . . ."

Ramrod-straight, Roosevelt carried himself like a soldier on parade, chin in, chest out, the line from the back of his close-cropped head falling straight as a plumb line to his heels. The originator of the strenuous life stood only about five feet, eight inches tall and weighed about two hundred pounds, but with his large head and heavily muscled body he looked monumental. When he spoke, his right hand shot out in a stabbing thrust, rose almost to the level of his shoulder, and came down with a powerful stroke on the other hand. Friends noted, however, that the presidency had taken its toll. Even though he was only forty-six, lines that had appeared about Roosevelt's eyes only when he smiled or laughed were now permanently incised. The drooping moustache was grizzled, the face fuller, the shoulders broader, the neck thicker, and the waistline larger.

The crowd strained to hear the inaugural address before the president's words were carried away on the wind. Roosevelt spoke of the good fortune enjoyed by the American people and the duties and responsibilities stemming from this fortune. "We are the heirs of the ages, and yet we have to pay few of the penalties which in old countries are exacted by the dead hand of a bygone civilization," he declared. Yet America must avoid complacency and vainglorious pride, and recognize that with great fortune went great responsibility. ". . . Much has been given us and much will rightfully be expected from us.* We have duties to others and duties to ourselves and we cannot shirk either."

Following lunch, the president was driven to the Peace Monument at the west entrance to the Capitol grounds. The four melancholy female figures marked the traditional starting point for inaugural parades. A cavalry trooper's bugle sounded and with a detachment of mounted policemen in the lead the parade proceeded back up Pennsylvania Avenue, where the president mounted a reviewing stand in front of the White House.

The cobblestone streets resounded with the clatter of marching feet and horses' hooves as various military units passed in review. Civic organizations and mechanics' clubs, some brave with colored sashes and other regalia, followed, among them a band of coal miners carrying a banner that read: We Honor The Man Who

---

*John F. Kennedy, whose speechwriters had combed texts of previous inaugural addresses, used the same words in his inaugural speech in 1961.

Settled Our Strike. Union and Confederate veterans, lean and aging, saluted the president, the latter undoubtedly remembering his Bulloch uncles from Georgia who had fought for the Lost Cause. Bad Lands cowboys and war-bonneted Indians, brandishing lances and tomahawks and riding green-, blue-, and red-painted ponies, were reminders of his ranching days in the West. "Everybody and everything" was there, the president later wrote an English friend.

Keeping up a spirited chatter with his guests, Roosevelt jumped to attention to applaud each new unit as it passed. He stamped his feet and swayed to the irresistible rhythm of "There'll Be a Hot Time in the Old Town Tonight." He laughed uproariously when the Rough Riders lassoed a spectator and carried him along in the march. Eight-year-old Quentin Roosevelt had trouble seeing, so a Supreme Court justice hoisted the boy up on his shoulders. The other Roosevelt children mingled with the crowds and snapped pictures with their Brownie cameras. To be perfect, thought Princess Alice, the parade needed only Judge Parker and other prominent Democrats marching along in chains as in a Roman triumph.

Joking to a friend about critics who complained that he had habitually usurped congressional prerogatives, Roosevelt declared: "I really shuddered today as I swore to obey the Constitution." Mindful of criticism of his foreign policy as imperialistic, he turned to Senator Augustus O. Bacon of Georgia, a prominent anti-expansionist, as a detachment of Puerto Rican militia passed in review and chuckled: "They look pretty well for an oppressed people, eh Senator?"

That evening there was a family dinner and attendance at the Inaugural Ball in the cavernous Pension Building on Fourth Street. The hall was decorated to resemble a Venetian garden with forty-foot-tall palms and blue draperies. The younger Roosevelts joined the hundreds of guests in dancing the night away while the president and first lady watched from a private box. From time to time, they moved to the front to be seen and applauded by the dancers. "I never expected to see another inauguration in the family," said Eleanor Roosevelt.

Upon their return to the White House, the Roosevelts were too excited for bed. They sat up with their guests chatting about all that had happened that day. Just as everyone was going off to sleep, the president remembered that an admirer had sent him a bottle of cherry liqueur and insisted that everyone join him for a nightcap. In various stages of undress, the party reassembled in Edith Roosevelt's bedroom to drink the president's health from water tumblers. They talked of old times, of childhood memories, laughing and feeling "so gay, so full of life and fun," the first lady recalled, they "could hardly bear to say 'goodnight.' "

Yet there was one person whose presence the president had sorely missed on this, the day of his greatest triumph. "How I wished Father could have been here to see it too!" he wistfully declared.

Chapter 1

"Teedie"

Theodore Roosevelt, Sr., was the center of his children's world. As soon as they heard his key rattle in the lock of the front door of the family brownstone at 28 East Twentieth Street in New York City,* the three youngest children—Theodore, Jr., Elliott, and Corinne—raced to be the first to greet him. Usually, he brought fruit or trinkets as special treats and they clung to him as he pressed on into the house. For an hour or so, he talked and played with them. While he dressed for dinner, recalled the future president, "we would troop into his room . . . to stay there as long as we were permitted, eagerly examining anything which came out of his pockets which could be regarded as an attractive novelty."

The elder Roosevelt was often away on business, so his children vied among each other for his attention when he was at home. "We all craved him as our most desired companion," said Cor-

---

*The house, just off Broadway, was numbered 33 East Twentieth Street until 1867, but for clarity's sake it will be called 28 during the Roosevelt family's entire stay there. The original structure was torn down, but the current building, a museum operated by the National Park Service, is a faithful reconstruction of the home in which the Roosevelts lived. The fourth floor was added later.

inne, indicating an active if unacknowledged rivalry among the children. Every morning, they lined up at the foot of the stairs eagerly waiting for Father to come down for the ritual of daily prayers. As soon as he appeared, they ran into the library to contend for the "cubby-hole," the space between him and the arm of the sofa where he sat to read from the Bible. This was considered the seat of favor while the other children felt momentarily like outsiders.

Roosevelt never laughed at his children when they asked questions—he laughed with them often—and no question seemed too troublesome or unimportant to answer. A firm believer in the benefits of outdoor life, he taught them to swim, to sail, to ride, and to climb trees. One morning he surprised the children with a new pony. "Who will get on first?" he asked. As her two brothers hung back, four-year-old Corinne jumped on the animal's back, and he declared that from then on, the pony was to be hers. "I think I did it," she later wrote, "to see the light in Father's eyes."

In the years just before the Civil War, New York was already a metropolis of more than a half million people where the rumble of loaded wagons, the stamping of big dray horses, and the hooting of tugboats on the rivers were never stilled. Ugly, overcrowded, and dirty, the city was sharply divided between rich and poor. While the Renaissance palaces of the wealthy were beginning to march uptown along Fifth Avenue, the poor huddled in foul, rat-infested slums such as the notorious Five Points in lower Manhattan. "All that is loathsome and decayed is here," observed the touring Charles Dickens. Only a few miles away, what Edith Wharton described as the fashionable "pioneers of the younger set" were taking over rows of stolid brownstones newly erected in the tree-shrouded streets adjoining Gramercy Park. "Thee" Roosevelt and his wife, Georgia-born Martha Bulloch Roosevelt, were among the most engaging couples in this society.

Tall, handsome, and with a leonine head, Roosevelt had a glossy chestnut beard and the natural grace of an athlete. He was the youngest of the five surviving sons of old Cornelius Van Schaack Roosevelt who, along with William B. Astor, Cornelius Vanderbilt, and A. T. Stewart, was one of the city's wealthiest men. Although active in Roosevelt & Son, the family glass-importing business in Maiden Lane, Thee was more interested in charitable

work and philanthropy, which eventually became his main interest. "Mittie" Roosevelt was a reigning beauty of Knickerbocker
society. Fine-featured and with lustrous, jet-black hair, she had
a tiny waist and coral-tinged skin that was, according to her daughter Anna, "the purest and most delicate white, more moonlight-
white than cream-white." Every day she took two baths, and she
wore white both summer and winter.

The Roosevelt family had lived in New York City for more than
two centuries, but after Theodore Roosevelt, Jr., became president, he insisted that his background was quite ordinary. For
example, in his *Autobiography* he described Claes Martenszen
van Rosenvelt, the family's founding father in America, as "our
very common ancestor" and observed he had come over from the
Netherlands in the seventeenth-century version of steerage. In
reality, however, this was part of a conscious effort to skew accounts of his life, to downplay the advantages of a privileged
background to bridge the chasm between himself and most Americans. Roosevelt was not a common man, and he knew it. Nevertheless, he wished to share the thoughts and aspirations of the
average American—and if this created a somewhat synthetic personality, it was an intellectual rather than a moral failing.

Claes settled in New Amsterdam, the struggling Dutch settlement at the tip of Manhattan, sometime before 1648. Like that
of many immigrants to America before and since, his past is a
recordless blank. One of the few clues is his name, which rendered
into English is Nicholas, son of Martin, of the Field of Roses. In
those days, Dutchmen took in addition to their baptismal name
that of their father and the locality from which they came. The
family name, spelled a dozen different ways in the old records—
Roosinffelt, Rosewelt, and Rosvelt among others—indicates that
the ancestral home of the Roosevelts was the island of Tholen in
the province of Zeeland, where a Van Rosevelt family was once
prominent. Their family motto, *Qui Plantavit Curabit*—He Who
Has Planted Will Preserve—was adopted by the American Roosevelts. The family name is pronounced by all its branches as
"Rose-uh-velt."

When Theodore, Jr., was a small boy, Dutch was still spoken
at the Saturday dinners at his grandfather's mansion on the southwest corner of Broadway and Fourteenth Street, just across from

Union Square. But the only Dutch he knew, taught by his grandmother, Margaret Barnhill Roosevelt, was a "baby song of which the first line ran, *'Trippe troppa tronjes.'* "* Over the years, the Roosevelts' original Dutch strain had been diluted by infusions of French, English, Scotch, German, Irish, and Welsh blood, a mixed ancestry shrewdly exploited by the rising politician. The first American president to play ethnic politics, Roosevelt struck up a kinship with voters with foreign-sounding names by eagerly shaking hands and proclaiming, "I'm partly Scotch, too!" or Welsh or Dutch or French or whatever the circumstances demanded. In fact, his ability to find a common ancestry with voters earned him the nickname "Old Fifty-seven Varieties."

Through a series of respectable marriages—which, as his outspoken daughter Alice tartly observed, "few of them . . . have made since"—the Roosevelts rose from the mercantile class into the upper rungs of society. Over the years, they served as New York City aldermen, in the state senate, and in Congress. The future president's great-grandfather, James I. Roosevelt, served honorably if without distinction in the Continental Army during the Revolution and when the war was over, opened a hardware business, Roosevelt & Son. The "son" was Cornelius Van Schaack Roosevelt, a curious-looking man—short, red-haired, and with a head that seemed too large for his body—but a shrewd judge of property values. When the real estate market collapsed following the Panic of 1837, he purchased property all over Manhattan at bargain prices and sat back to wait until it shot up in value. The richest of the Roosevelts, he amassed a fortune estimated at $7 million, worth more than ten times that at current values.

C.V.S. and his wife, Margaret, whose Quaker ancestors had come to Pennsylvania with William Penn, had six sons. Theodore, the youngest surviving child, was born in 1831 and was doted on by all the family. The boys did not attend school but were taught at home by tutors who must have been hard-pressed to control their boisterous charges. "There goes that lovely Mrs. Roosevelt with those five *horrid* boys," said the neighbors. C.V.S. thought

---

*"I always remembered this," he later recalled, "and when I was in East Africa it proved a bond of union between me and the Boer settlers, not a few of whom knew it, although at first they always had difficulty in understanding my pronunciation—at which I do not wonder." Roosevelt, *Autobiography*, p. 2.

college a waste of time and after his youngest son completed a tour of Europe, he went directly into the family business.

One evening in 1850, Thee Roosevelt attended a dinner at the home of his oldest brother, Silas Weir Roosevelt. Over cigars and port, the young man chatted with a brother-in-law, Dr. Hilborne West, who was married to Susan Elliott, a member of the Bulloch family of Georgia. West had just returned from a visit to his in-laws at Roswell in the Georgia uplands and entertained the company with stories about the plantation. Fascinated by a feudal way of life that seemed to have sprung from the novels of Sir Walter Scott—in a land of moonlight and magnolias and of chivalrous men and beautiful women—Thee resolved to see the South for himself.

He was captivated by life at Bulloch Hall. The white-columned house was crowded by a sprawling extended family presided over by Mrs. James S. Bulloch, whose late husband had been a planter and Savannah businessman before moving to Roswell, about nineteen miles north of what was to become the city of Atlanta.* The Bullochs ranked among Georgia's first families and had provided the state and nation with judges, lawmakers, and military officers. To a traveler fresh from the snow and slush of New York, the very air of the South was distinctive. White blossoms hung from the dogwoods and the azalea bushes were covered with flowers of a dozen hues. Majestic live oaks dripped with mournful moss and the air was fragrant with the scent of purple wisteria.

Parties, dinners, and dances were held in honor of the guest from the North. As the weeks passed, Thee found himself paying more and more attention to Mittie Bulloch, a vivacious fifteen-year-old. Quiet strolls were followed by long rides together, for she was a spirited horsewoman. In the evenings, they read aloud to each other. When the young man returned to New York, he left his heart behind him. Over the next three years, the couple conducted a courtship, largely by mail, that ended with their

---

*The Bullochs were touched by scandal. James Bulloch had courted Martha Stewart, but she had rebuffed his offer of marriage and he married Hester Elliott, the daughter of U.S. Senator John Elliott. Not long afterward, Martha married the senator—making her the mother-in-law of her former suitor. When both the senator and Hester died, Bulloch and Martha Elliott were finally married—but the union offended the staid society of Savannah, and the couple moved to Roswell to escape gossip.

marriage at Roswell three days before Christmas, 1853. It was the social event of the season and ice was hauled from Savannah— more than two hundred miles away—to make ice cream.

Thee brought his bride to New York and the house on Twentieth Street, a gift from his father. With its heavy brocade draperies, dark furniture, and horsehair sofas, it had an air of serene gentility. C.V.S. Roosevelt had presented each of his sons a similar home when they married, and Theodore's brother Robert Barnwell Roosevelt and his wife, Lizzie, lived next door in a house that was a duplicate of his own. Robert's middle name was actually "Barnhill" but he is said to have changed it to Barnwell when he went into politics because he thought his real name made people think of a dunghill. Robert was also a ladies' man and presented each of his conquests with a pair of green gloves. A descendant has observed that he bought them by the gross from Stewart's nearby department store.

For a well-connected young couple like Theodore and Mittie Roosevelt, life in New York was a pleasant round of calls, dinner parties, and Assembly balls. Still, Mittie was new to this tightly structured world and the adjustment may have been difficult after the free and easy atmosphere of Roswell. The family dinners at her father-in-law's home at which only Dutch was spoken were undoubtedly a special trial. But girls of her generation were expected to accept the unexpected without too much fuss. Besides, she was soon pregnant with her first child.

Anna Roosevelt was born on January 7, 1855, and was known as Bamie, short for *bambina*. Almost from birth she suffered from "spinal trouble," which family tradition ascribed to her being dropped in the bath by a careless nurse. In reality, she appears to have been the victim of Pott's disease, a form of bone tuberculosis that causes severe curvature of the spine. For the first four years of her life, Bamie spent long days lying on a sofa while harnessed in a "terrible instrument" intended to strengthen her spine. The child's father sought help everywhere and eventually heard of a young physician named Dr. Charles F. Taylor, who had radical theories about orthopedics. Taylor prescribed special exercises for Bamie and devised a lighter harness for her. Soon, she was up and romping about, and lived a remarkably active life. She gave the impression of being slightly hunchbacked, but her zest for living made everyone forget her handicap. "Energy—thy name is 'Bamie' " became a family saying.

Theodore Roosevelt, Jr., was born in the Twentieth Street house on October 27, 1858. For late October, it had been unseasonably cold. Record high tides were reported in the East River and the river ferries were coated with ice. Mittie Roosevelt was bored at being cooped up at home and even though her second child was due within a few days, she ordered a carriage brought around to the front door when the weather moderated at last. She had some cloaks to order and other last-minute shopping to do. But upon returning home, she felt unwell and retired to rest in the large bed in the second-floor front bedroom.

Late in the afternoon, Mittie's mother, who was staying with the Roosevelts, looked in and found her in the beginning of labor. The family doctor was hurriedly sent for, but he was unable to come. With her son-in-law away on business, Mrs. Bulloch was anxious and worried. Servants were sent scurrying about the neighborhood to find another physician or a midwife while Bamie was hastily dispatched to a relative. A Dr. Marko, who lived nearby, finally responded.

The flickering lamps cast long shadows on the walls as Mittie continued "to get worse and worse." But at a quarter to eight that evening she gave birth to a boy. He weighed eight and a half pounds and was described by his grandmother "as sweet and pretty a young baby as I have ever seen. . . . No chloroform or any such thing was used, no instrument was necessary, consequently the dear little thing has no cuts or bruises." But the child's mother insisted, probably in jest, that he was "hideous" and looked like "a terrapin."

Rapidly recovering from her ordeal, Mittie nursed the baby without difficulty. "All quite well," Mrs. Bulloch reported on November 3. "Little Theodore is a week old. . . . Mittie is quite motherly, likes to have him lying quite near her." Ten months later his grandmother reported that the boy had acquired a nickname—"Teedie"*—and was "one of the brightest little fellows I ever knew." At eighteen months, he was a happy, chattering child

---

*Pronounced as in T.D.

and "almost a little beauty." In 1860, Bamie and Teedie were joined by another brother, Elliott, and the following year by a sister, Corinne. Like the other members of the Roosevelt family, they quickly received nicknames: Elliott became "Ellie" or "Nell" while Corinne was called "Conie."

When he was about three, Teedie began to suffer from severe bronchial asthma. For all of his life, he was haunted by a half-dream, half-nightmare of suddenly awakening in the dark and gasping for breath. Only the victim can know the terrors of an asthma attack. You feel as if you are drowning. You open your mouth for breath, but no air seems to come in. You breathe faster and gasp out a wheezing, whistling sound. Air can get in, but not out. The chest contracts with the effort of trying to breathe and the heart pounds, sometimes at twice the normal rate. Asthma can kill—and children are among its most vulnerable victims.

One of Roosevelt's earliest memories was "of my father walking up and down the room with me in his arms" while he struggled for breath. Sometimes, in desperation, a horse and rig was hastily summoned from a nearby livery stable, and the child, bundled in blankets and cradled in his father's lap, was driven at a breakneck pace through the empty streets of the city to force air into his straining lungs. Often he found it so difficult to breathe that he could only sleep propped up in bed or in a chair.

The threat of death and the frightening experience of watching their child struggle for breath had a profound effect on the boy's parents. Overprotective and nervously alert to the possibility of an oncoming episode, they centered their lives around the vagaries of Teedie's health. Trips were taken to the country, to the sea-shore, to the mountains, to help the boy breathe. Instead of improving, his illness became worse. Physicians examined foods, plants, household dust, particles of dog and cat hair, odors, air, and climate in an effort to determine the root of Teedie's asthma. Every effort to find the cause was unsuccessful, inspiring some observers to speculate that the illness may have been more psychosomatic than physical.

One recent biographer, David McCullough, argues that some of Teedie's asthma attacks were largely self-induced in order to avoid attending the rather forbidding church services of the day. He points out that Roosevelt's boyhood diaries during 1869 to

1870 reveal that most of his asthma attacks occurred on Saturday evenings and Sundays, and suggests he timed the attacks to occur on the weekends when his beloved father was present so he would have his undivided attention. "The pattern is too pronounced to be coincidental," according to McCullough.* But Paul Russell Cutright, a longtime Roosevelt scholar, contends that the boy suffered more asthma attacks on weekdays than weekends. "If it was the fear of churches that triggered Theodore's weekend asthmatic attacks, what sparked his attacks on weekdays, and what inhibited them on so many weekends—twenty-seven of fifty-two—when he did attend Sunday church services?"†

Treatment for asthma in the mid-nineteenth century was primitive. Black coffee, tobacco, and ipecac, an emetic used to promote vomiting, were often prescribed for victims, and Teedie's father used coffee as "a trump card" in dealing with his son's attacks when all else failed. Teedie also developed a form of nervous diarrhea, which he called "cholera morbus." He also suffered from a recurring dream of a werewolf waiting to spring upon him from the foot of his bed. The family correspondence of the period is filled with accounts of his illnesses. "Teedie has a very bad cold," Mittie wrote her husband during one of his absences. "Teedie was very unwell last night. . . . I was up with him six or seven times during the night," she reported a few weeks later. Letter after letter continues in the same vein. "I think Teedie looks badly and his appetite is poor . . . Teedie has a very warm fever . . . I think Teedie does not mend fast. . . ." And his father was depressed because "Teedie is too much sick. It worries me. . . ."

The elder Roosevelt was either founder or early supporter of almost every humanitarian endeavor in the city. In extending a hand to those less fortunate than himself, he acted from a disciplined sense of obligation as a Christian and a gentleman. Good fortune, he believed, must be balanced with productive work and service. He was a founder of the New York Orthopedic Hospital, of the Children's Aid Society, and was a guiding force in the

*Mornings on Horseback, Chapter IV.
†Cutright, Theodore Roosevelt: The Making of a Conservationist, pp. 20–23.

YMCA. The founding meeting of the board of the American Museum of Natural History was held in the front parlor on Twentieth Street, and he was a founder of the Metropolitan Museum of Art. Every Sunday, he taught a mission class at Madison Square Presbyterian Church.

When Thee Roosevelt realized the streets of New York were filled with homeless and abandoned children, he helped found the Newsboys' Lodging House on West Eighteenth Street. He not only provided financial support and persuaded wealthy friends to contribute—"All right, Theodore, how much is it this time?" they would say as he approached with a certain look in his eye— but gave his time to counseling the boys and listening to their problems. He found foster homes for some and dug into his own pocket to provide railroad fares home for runaways or for others to get a fresh start in the West. "He literally went about doing good," said one friend.

Unlike many high-minded men, however, Theodore, Sr., possessed a zest for living. He and his wife were known for their lavish and openhanded hospitality. Gay and fun-loving, he dressed with flair, loved flowers—especially yellow roses—was attractive to women, and danced all night at parties. "I can see him now," wrote a friend, "in full evening dress, serving a most generous supper to his newsboys in the Lodging House, and later dashing off to an evening party in Fifth Avenue." And his son Theodore remembered him as a "sport" who liked to drive a four-in-hand team with a verve approaching recklessness as the long tail of his linen duster bagged out behind him like a balloon.

No one had a greater influence upon his namesake. By example and instruction, he imbued Theodore, Jr., with a strong sense of moral values and remained an almost palpable presence at his side long after his death at the age of forty-six. These values were both a curb and a spur. In time of trial and triumph, the younger Roosevelt worried whether or not his conduct would have merited his father's approval. "My father . . .was the best man I ever knew," he often declared. Significantly, Roosevelt, who regarded surrender to fear as a deadly sin, added that "he was the only man of whom I was ever really afraid."

The coming of the Civil War cast a dark shadow over this happy family circle. Like the nation itself, the residents of the Twentieth

Street house were divided by the struggle. Both Mittie's mother and her sister Anna, having suffered financial reverses that forced them to give up Bulloch Hall, had come to live with her, with Anna earning her board as the children's governess. The three Georgia-born women formed a united front of staunch Confederate sympathizers among the Roosevelts, who supported the Union. Along with most New York businessmen with commercial contacts with the South, Theodore, Sr., had deplored the drift toward war, but once it came, he and his family were strong Lincoln Republicans.

In contrast, Mittie's brother James D. Bulloch, a onetime U.S. Navy officer, headed the Confederate Secret Service in England, and directed the building of the Confederate commerce raiders *Florida* and *Alabama,* while her brother Irvine Bulloch went to sea in the *Alabama* as a midshipman. A stepbrother, Stewart Elliott, served in the Confederate Army. Influenced by his wife's protestations that it would kill her if he joined the Union Army and fought against her brothers, Thee agreed, against his better judgment, not to volunteer for military service. "Always afterward," reported Bamie, "he felt he had done a wrong thing in not having put every other feeling aside to join the fighting forces."

In later years, young Theodore obviously felt that in this one instance his beloved father had let him down. Corinne believed his determination to make a military reputation was "in part compensation for an unspoken disappointment in his father's course in 1861." Robert B. Roosevelt served as a ninety-day volunteer in the Union Army at the outbreak of the war, but none of the Roosevelt brothers saw active service, although all were of military age. Like many prosperous men, they paid substitutes the going rate of $1,000 to shoulder muskets in their place.

Through his relief work, Theodore, Sr. discovered that the rapid mobilization of the Union Army had left many of the soldiers' families destitute because no arrangements had been made for them to receive part of the servicemen's pay. Quickly devising a scheme for voluntary allotments, he went to Washington to lobby for passage of enabling legislation. Through the influence of a friend, John Hay, one of President Lincoln's secretaries, he secured the support of the administration and was introduced to the president and Mrs. Lincoln. The first lady often invited him to go riding with her in the countryside around Washington and sought his opinion on new bonnets.

Obtaining congressional support for the allotment plan was far more difficult. Considerable skill was required to persuade the dubious legislators that he sought no personal gain from the proposal. Following several months of lobbying, the law was finally passed and Roosevelt was appointed one of the allotment commissioners to administer the plan and to persuade the soldiers to join it. This proved to be far more difficult than it sounded. The soldiers were suspicious and there was strong opposition from the sutlers, who had purchased licenses to operate the equivalent of modern post exchanges—which included saloons and sometimes brothels.

Over the next year, Theodore was away from his family most of the time, visiting troops, mostly on horseback and in the worst weather—all without pay. He probably saw more of army life than many men actually in the service. Through constant practice, he developed a sales pitch designed to persuade men to sign up. "I had three companies formed into three sides of a square and used all my eloquence," he wrote Mittie of one appeal. "When I had finished, they cheered me vociferously. I told them I would be better able to judge who meant their cheers by seeing which company made the most allotments, and raised a spirit of competition in this way that made the rolls the best we had taken during the day."

Theodore referred only rarely to the differences between himself and his wife. On New Year's Day 1862, he wrote: "I do not want you not to miss me but remember that I would never have felt satisfied with myself after this war is over if I had done nothing and that I do feel now that I am doing my duty. I know you will not regret having me do what is right, and I do not believe you will love me any the less for it." For her part, Mittie kept her anxieties to herself. She refused, however, to continue to attend the Saturday dinners at the home of her father-in-law, where anti-Southern feelings were strong. "Something occurred there which has made my blood boil," she wrote her husband upon one occasion. "I could not touch another mouthful. . . . Thee, I wish I could see you tonight."

The younger children were oblivious to their mother's anxieties.

In fact, Teedie paraded about the nursery in the uniform of the New York Zouaves, in triumphant celebration of Union victories. Resentful at being chastised for some sin by his mother, he openly prayed for the Lord to "grind the Southern troops to powder."* Bamie, however, was old enough to sense that something was amiss and later told a friend how unhappy those years had been for her mother. "I know the Roosevelts and I should hate to have married into them at that time unless I had been one with them in thought," she said. "They think they are just, but they are hard in a way."

Most of Mittie's letters to her husband are filled with details of life at home. Bamie and Elliott were sick, the latter complaining of frequent head pains, while little Corinne was showing signs of developing asthma. Teedie's own asthma attacks were alarming, disrupting family outings to the Jersey shore and Saratoga. "Teedie came downstairs this morning looking rather sad and said 'I feel badly—I have a tooth ache in my stomach. . . .' He is the most affectionate and endearing little creature." But his mother also noted that the boy was becoming a handful. "He begins to require his Papa's discipline badly. He is brimming full of mischief and has to be watched all the time."

Wisely, Mittie did not tell him that she and her mother and sister, although surrounded by Yankees, covertly kept up their contacts with the Confederacy. Periodically, they prepared parcels of medicine, clothing, and money that were smuggled to friends in the South. Bamie and Teedie would be taken on a picnic in Central Park, where the parcels were turned over to couriers who ran the blockade. Meetings were also arranged through innocent-appearing newspaper advertisements, with mysterious strangers who brought letters from James Bulloch and other relatives serving the Confederacy.

"One of my most vivid memories were [sic] of the days of hushed and thrilling excitement, which only occurred when Father was

---

*TR's uncle, Irvine Bulloch, is said to have been in charge of the gun that fired the *Alabama*'s last defiant shots before she was sunk by the U.S.S. *Kearsarge* in 1864. Ignoring his childhood pro-Union sympathies, Roosevelt turned the incident to his own political advantage when he campaigned in the South forty years later. "One would suppose that the President, himself, fired the last two shots from the *Alabama* instead of his uncle," wrote a *Washington Star* reporter.

away," Bamie recalled. "Grandmother, Mother and Aunt Annie would pack a box, while . . . Theodore and I helped, not knowing at all what it was about, except that it was a mystery and the box was going to run the blockade. Our favorite game for years afterward, needless to say, instigated by Theodore, was one of 'running the blockade' over the bridge in Central Park, in which I was the blockade runner, and he was the Government boat that caught me."

And then the event for which Teedie had waited so long came at last. Having completed his work as an allotment commissioner—and receiving the thanks of the New York Legislature for his efforts—Father came home for good. "Handsome dandy that he was, the thought of him now and always has been a sense of comfort," the president recalled nearly four decades later. "I could breathe, I could sleep, when he held me in his arms. My father—he got me breath, he got me lungs, strength—life."

Prolonged illness is a humiliating experience that carries with it the hazard of narcissistic self-absorption. Children with severe cases of asthma suffer psychologically as well as physically. Some become so frightened and depressed by the violence of the attacks that they despair of every enjoying good health. A pale child with a shock of unruly fair hair, prominent teeth, and nearsighted blue eyes, Teedie was almost a caricature of the pampered, protected child. It would have been easy for him to abdicate leadership to Ellie, the handsomest and strongest of the Roosevelt children. But he was intellectually precocious and already a skilled storyteller who held his own in his own way.

"We used to sit, Elliott and I, on two little chairs, near the higher chair which was his, and drink in these tales of endless variety," recalled Corinne. Usually they were serials, which invariably included a small boy—like Kipling's Mowgli—who understood the language of the animals and "never flagged in interest for us, though sometimes [the stories] continued from week to week, or even from month to month."

Teedie, Ellie, and Conie banded together into a group they called "We Three," from which Bamie, three years older than Teedie and classified among the "Big People," was excluded. "We Three" were joined by Edith Kermit Carow, the daughter of the Charles Carows, who lived near Grandfather Roosevelt's house on Union Square. The Carows were originally French Huguenots and their name was Quereau before it was anglicized. Having become one of the city's wealthiest shipping families, they moved up in society as well. Blue-eyed, fair-haired Edie was Conie's friend, but the little girl soon formed a special relationship with Teedie. On April 25, 1865, Teedie, Ellie, and Edie gathered on the second floor of the Roosevelt mansion to watch Abraham Lincoln's funeral procession slowly pass up Broadway. Frightened by the tolling of the church bells and the relentless blackness all about her, the four-year-old girl began to cry. Irritated, the boys shunted her into a back room.

Even before Teedie could read, he discovered a copy of David Livingston's *Missionary Travels and Researches in Southern Africa* in the family library and was fascinated by the illustrations of exotic birds and animals. Although the book was nearly as big as the boy himself, he dragged it about with him, begging his elders to tell him the stories behind these pictures. His formal education was undertaken by his aunt, Anna Bulloch,* a talented and energetic teacher. She not only taught him his numbers and to read and write, but also imbued him with a love of history. Roosevelt never learned to spell well, however, and was abysmal at any sort of mathematical reasoning.

Gathering "We Three" and Edith Carow in the second-floor nursery on Twentieth Street, Aunt Annie entertained them with tales of the daring exploits of the Bullochs during the American Revolution and of life on a Georgia plantation. They were enchanted by her stories of Br'er Rabbit and "the queer goings-on in the Negro quarters." From these tales, the children graduated to the McGuffey Readers, and as president, Roosevelt would often reach back to McGuffey to make a point. The phrase "Meddlesome Matties," which he used to describe political opponents, came from this source. Anna Bulloch's teaching was supplemented by tutors, including a French mademoiselle.

Like many semi-invalids, Teedie was an omnivorous reader with

---

*Mrs. Bulloch had died in October 1864, at about the time the Roswell plantation was sacked by Sherman's army on its march through Georgia. Bulloch Hall survived, however, and in 1978, the restored mansion, now owned by the city of Roswell, was opened to the public. Local tradition has it that Margaret Mitchell, who wrote about the house as an Atlanta newspaper reporter in the 1920s, used it as the model for Tara in *Gone With the Wind*. TR visited the house for the first time during a presidential tour of the South in 1905.

a bent toward stories of adventure. By the hissing gaslight, he pored over the books in the family library, a windowless room "of gloomy respectability" between the front parlor and dining room. He had free access to all the books except one—a novel by Ouida called *Under Two Flags,* which was considered somewhat racy—and of course he read that, too. The ribald parts made no impression upon him and he "simply enjoyed in a rather confused way the general adventures."

Just as contemporary small boys are excited by the challenge of outer space, children of the Victorian age, inspired by the discoveries of Darwin, were fascinated by natural history and zoology. Teedie was particularly attracted to the tales of Captain Mayne Reid with their simplified natural history and outdoor adventure. Reid was an Englishman who had lived in America, served in a New York regiment during the Mexican War, and had returned to Britain to write about his adventures. One book, *The Scalp Hunters,* opened with these words: "Unroll the world's map, and look upon the great northern continent of America. Away to the wild West—away towards the setting sun—away beyond many a far meridian. . . ."

Away to the wild West! These words set Teedie's vivid imagination afire. In his dreams he escaped to the prairies and mountain ranges beyond the rolling Mississippi. When his father saw how taken the boy was with these books, he provided him with two volumes by J. G. Wood, a popular natural history writer. Teedie was also enthralled by such works as Captain Frederick Marryat's *Mr. Midshipman Easy* and J. Fenimore Cooper's books on the American frontier. He loved Longfellow's *Saga of King Olaf* and memorized it, although he was shocked when he learned the Norsemen killed women and children. He did not care for the Greek epics because he felt the gods used loaded dice and did not play fair with ordinary mortals. *Our Young Folks,* a magazine for boys that taught "manliness, decency and good conduct," was also among his favorite reading.

Overcoming fear was one of Roosevelt's major concerns throughout his life. "I was nervous and timid," he recalled. "Yet from reading of the people I admired, ranging from the soldiers at Valley Forge and Morgan's riflemen, to the heroes of my favorite stories, and from hearing of the feats performed by my

Southern forebears and kinfolk, and from knowing my father, I had a great admiration for men who were fearless and who could hold their own in the world, and I had a great desire to be like them."*

One day when he was about seven, Teedie was sent to a market on nearby Broadway to buy strawberries for breakfast and there discovered a seal that had been killed in New York harbor laid out for display. "The seal filled me with every possible feeling of romance and adventure," he recalled. Teedie haunted the market and made careful measurements of the carcass, which were dutifully recorded in a copybook in simplified spelling, the first of a series of natural histories of his own. When the seal was removed from display, he was given the skull, which became the centerpiece of the "Roosevelt Museum of Natural History." The collection was first kept in his room, but following a chambermaid's rebellion, was banished to a bookcase in a rear hall.

Teedie's enthusiasm for natural history—and the vitality and exuberance that were to be facets of his adult personality—is clearly evident from the earliest of his 150,000 letters to survive. Written to his mother, who was visiting relatives in Savannah, it is dated April 28, 1868:

Dearest Mamma:

I have just received your letter! What an excitement. What long letters you write. I don't see how you can write them. My mouth opened wide with astonish[ment] when I heard how many flowers were sent to you. I could revel in the buggie ones. I jumped with delight when I found you heard the mocking-bird, get some of its feathers if you can. . . . Tell me how many curiousities and living things you have got for me. . . .

Yours loveingly,

Theodore Roosevelt. Jr.

---

*Roosevelt also told Hermann Hagedorn that he had been greatly helped in overcoming his timidity by the comments of an officer in Marryat's *Mr. Midshipman Easy.* When asked before going into battle if he was afraid, the man replied:

"Afraid? Of course, I'm afraid. But what you've got to do is look and act as though you weren't. Then by and by you won't be." Hagedorn, *Talk with TR,* in Theodore Roosevelt Collection at Harvard.

Unwary adults were often startled by Teedie's interest in zoology. One day, meeting Mrs. Hamilton Fish on a streetcar, he
politely lifted his hat and several frogs leaped to the floor. House-
guests learned to take their seats warily when they visited the
Roosevelts and to check their water pitchers for snakes. At dinner
one evening, one of his mice poked its head from a Dutch cheese
being passed about the table. The Roosevelts' cook was horrified
to find a dead mouse in the icebox, a specimen placed there by
Teedie for preservation until needed for one of his experiments.
"Oh, the loss to science," he mourned after the mouse was thrown
out. "Oh, the loss." Nevertheless, the boy's parents encouraged
this unsettling new interest—"as they always did in anything that
could give me wholesome pleasure or help develop me."

The asthma attacks suffered by Teedie and Corinne convinced
the elder Roosevelt that the nursery was unhealthy and he converted the back room overlooking the garden into an open-air
piazza where the children could play all year long. In the summer
of 1868, the family was packed off to Tarrytown on the Hudson,
where Teedie recorded continuing attacks of "asmer" in his diary,
but also did a great deal of riding, walking, and swimming. On
August 18, he recorded a ride of six miles. Two days later, he
repeated the feat and declared: "I will always have a ride of six
miles before breakfast now."

The world out on the teeming streets of Manhattan was
undergoing sweeping changes in the wake of the Civil War, but
the children were as carefully isolated from the brutalities of the
industrial age as if they lived inside one of the bell jars covering
the dried plants in the parlor. Although business was important
to Thee Roosevelt, he carefully compartmentalized his life and
never allowed the grubbier aspects of moneymaking to absorb his
attention completely. The Roosevelts never discussed financial
matters before the children; there had always been money and
servants and this was accepted as the norm.

Following the war, the elder Roosevelt helped organize the
Protective War Claims Association, which saved the families of
veterans with claims against the government millions of dollars in
fees that grasping agents tried to extract from them. He found
work for crippled veterans through a Soldiers' Employment Bu-

reau, which, according to one observer, did "more, and vastly better work than all the Soldiers' Homes combined." Appalled by the conditions he found during a tour of the city's insane asylum on Blackwell's Island in the East River,* he lobbied for construction of new asylums where "the patients could work out of doors, and gain strength and mental and physical health in useful occupations."

To instill his children with a sense of stewardship and responsibility, of noblesse oblige, toward those less fortunate than themselves the elder Roosevelt took them on his visits to the Newsboys' Lodging Home, to Miss Slattery's Night School for Italian Children, and various hospitals and other charities in which he had an interest. "Generally . . . Saturdays commenced by a ride on horseback in the Park, followed instantly . . . by a visit of inspection to both the Art Museum and the Museum of Natural History, and then to some one of the Children's Aid Society schools," according to Bamie. "We would get home for lunch very late, and as a rule would find whoever was most interesting of the moment in New York lunching with us. By the time that was over, we either drove in the park or visited a hospital."

Late in 1868, the Roosevelts began looking beyond their immediate horizons for new adventures. Mittie wished to see her brothers James and Irvine Bulloch, who were among the few southerners not granted amnesty following the war and had taken refuge in England. She persuaded her husband to take the family abroad, and with the glass business almost running itself, he readily agreed. The trip was also designed as a change of scene for Teedie with the hope of alleviating his asthma, to find a suitable finishing school for fifteen-year-old Bamie, and as an educational experience for all the children, none of whom had any formal schooling.

Plunging in enthusiastically, Theodore spent all winter studying maps and guidebooks and arranged a yearlong grand tour that began in England, proceeded to Scotland, the Low Countries, Paris, and then on to Switzerland, Germany, and Italy, where they were to spend Christmas in Rome. Obviously, Teedie's later devotion to the strenuous life owed much to his father's example.

---

*Now Roosevelt Island

But the children reacted with dismay—especially Teedie. He had hoped to return to Barrytown and his "buggie" things and experiments. A year seemed like forever to be away from his playmates, especially Edith Carow. "It was verry hard parting from our friend," he tearfully recorded in his diary as he boarded the Cunard steamer *Scotia* along with the rest of the Roosevelt brood on May 12, 1869. Nine days later, the ship docked at Liverpool and Teedie set foot in "Briten" for the first time.

Looking back in later years, Theodore Roosevelt said he "cordially hated" his first trip to Europe. "I do not think I gained anything from it." But his diary of the trip makes it clear that he enjoyed himself immensely. Although he was often homesick or suffering from stomach troubles, asthma attacks, and toothaches, entry after entry relates that he is having "fine fun" or had a "great play." The list of sights sometimes runs to dozens of items in a single day. He went on long hikes despite his recurrent illnesses and squirreled away maps, pictures, and other souvenirs for his collections.

While Mittie enjoyed a reunion with her brothers, the elder Roosevelt and the children toured the West Country. Castles with "dreary Dungens" where "prisoners of yore were kept," and all kinds of ruins and arms and armor fascinated Teedie, especially a "Leathern jacket" seen at Chatsworth, the seat of the dukes of Devonshire, "in which a lord received his death blow." At the Tower of London, "I put my head on the block where so maney had been beheaded." In London, early in July, his wheezing and coughing flared up again. The boy's lungs were perfect, reported an English physician, who prescribed a dose of fresh air. Following the doctor's suggestion, the elder Roosevelt bundled Teedie off to the seashore at Hastings.

For the next two days, father and son walked along the beach together and scrambled up the nearby cliffs to inspect the Saxon ruins. From the crest they looked down upon the town, glistening in the sunshine, and the beach and the surf stretching out before them. The following day, they attended church and went walking while the senior Roosevelt conducted an impromptu Sunday school, and they made "a feeble attempt" to cheer the Fourth of July. That night, Teedie put himself to bed, but unable to summon up enough breath to blow out the candle, doused the lighted end in a tumbler of water, and blissfully fell asleep. Overjoyed at having had his father all to himself, he wrote in his diary: "This is the happiest day I have ever spent."

In the Low Countries and on the Rhine, Teedie suffered sporadic asthma attacks. Nevertheless, by the time the Roosevelts reached Switzerland, he was hiking prodigious distances, matching his father stride for stride. On August 6, he climbed an eight-thousand-foot mountain at Chamonix, even though "some of it was pretty dangerous work." The following day, "I went all alone to a castle. It is a 3 mile walk—and 3 miles back—and I went and came in 1 hour." On August 11, the family crossed La Tête-Noire. "Papa walked 22 miles, I 19. . . ." And on August 21, they hiked through the Grimsel Pass. "I walked 20 and Papa 22 miles." Although these distances appear at first glance to be boyish exaggerations, they are confirmed by his father's letters.

Teedie was entranced by the beauty of Lake Como, where there was "no sound save the waterfall and the Italian breeze on my cheek. I all alone am writing my Journal." And in Venice, "we saw the moonlight on the waters and I contrasted it with the black gondolas darting about like water goblins." Teedie's asthma and diarrhea were incessant, and his mother anxiously reported they kept "us constantly uneasy and on the stretch."

"On Lake Maggiore," she wrote her sister Anna,* "it came to a point where he had to sit up in bed to breathe. After taking a strong cup of black coffee the spasmotic part of the attacks ceased and he slept. . . . Had the coffee not taken effect he would have gone on struggling through the night. . . . Thee has warded off one or two attacks with this coffee but likes to keep it as our trump card." But Teedie's round of sight-seeing hardly suggests a listless child. In one day in Venice, he commented on twenty-three separate sights, ranging from paintings to a "candleslarbra."

Visits to Austria, Germany, France, and Belgium followed before the Roosevelts settled in at Rome for the winter. In Salzburg, Teedie had a nightmare in which he dreamed that "the devil was

*Anna Bulloch had married James K. Gracie, a banker and member of an old New York family that gave its name to Gracie Mansion, now the official home of the city's mayor.

carrying me away." And in Munich, "I was verry sick. . . . Mama was so kind telling me storrys and rubing me with her delicate fingers." He observed his twelfth birthday in Cologne—"the first of my birthdays that it snowed on"—and the family celebrated by dressing for dinner. Looking over his gifts, which included a pocket looking glass, a portable writing stand, a compass, and a box of miniature "domminnoes," he pronounced the day, "Splendid!"

Perhaps because of the mild weather, Teedie's health improved during the family's two-month stay in southern Italy. Temples, tombs, churches, catacombs, and the Colosseum were methodically inspected by the Roosevelts. During a side trip to Naples, Teedie described his ascent of Vesuvius in a letter to Edie. "We Three" had a snowball fight and then "went on until we came to a small hole through which we saw a red flame inside the mountain. . . . I put my alpine stock in and it caught fire right away. The smoke nearly suffacated us." His father gave him a special gift—a Roman vase and coin—"what in my wildest dreams I never thought to have. . . . Just think of it!!!"

In March 1870, after careful inquiry, the senior Roosevelt left Bamie at Les Ruches, a school for girls operated by Mlle. Marie Souvestre at Fontainebleau, outside Paris. Mlle. Souvestre, although a fine teacher, was an outspoken agnostic and freethinker, but her attainments obviously overcame these liabilities. Following a farewell visit to the Bullochs in Liverpool, the family sailed for home in mid-May on the steamer *Russia*—almost a year to the day they had left the United States. There was no mistaking Teedie's feelings when the vessel passed Sandy Hook on May 25, 1870, and anchored within the shadow of Manhattan.

"New York!!!" he cheered. "Hip! Hurrah!"

Chapter 2

"I'll Make My Body"

Upon his return from Europe, Teedie Roosevelt slipped easily into a routine of tutors, lessons, and books. "I study English, French, German and Latin," he wrote his aunt Annie, and French was spoken at the family dinner table with the hope of improving the children's command of the language. Over the next few years he grew into a gawky, tousle-haired teenager who reminded acquaintances of a stork, a resemblance accentuated by a habit of reading at his desk while standing on one leg with the other drawn up. After the boy's voice broke, Mittie Roosevelt complained his laugh sounded like a "sharp, ungreased squeak" that "almost crushed the drum of my ear."

Teedie's health showed no signs of improving, and he suffered asthma attacks and stomach upsets with wearying regularity. He was shunted from place to place for a change of air—to his uncle in Philadelphia, to Saratoga, and to C.V.S. Roosevelt's summer home at Oyster Bay on the North Shore of Long Island—all to no avail. Nerves fraying, the boy's parents were losing patience with his persistent illnesses. "I am away with Teedie again much to my regret," the senior Roosevelt despondently wrote Bamie during the summer of 1870. "He had another attack, woke me up at night, and last night, to make a break I took him to Oyster Bay. It used me up entirely to have another attack come so soon after the last."

Following a particularly acute bout of asthma, Teedie was sent to Dr. A. D. Rockwell for a thorough checkup. The physician reported that he was "a bright, precocious boy . . . by no means robust" and prescribed "change, plenty of fresh air," and "more exercise." With this foundation to build upon, Teedie's father took him aside for a memorable man-to-man talk. "Theodore," said the elder Roosevelt, "you have the mind but not the body, and without the help of the body the mind cannot go as far as it should. You *must* make your body. It is hard drudgery to make one's body but I know you will do it."

"The little boy looked up," Corinne Roosevelt later recalled, and "with a flash . . . of white teeth" accepted the challenge. *"I'll make my body!"*

Roosevelt's adoring sister probably overdramatized this incident, but it was a turning point in the making of both his body and his character. In fact, in later years, some observers attributed his aggressiveness to overcompensation for his physical inferiority.* To help Teedie "make" his body, he was taken by his mother to a gymnasium operated by John Wood that was frequented by upper-class New Yorkers. Under Wood's direction, the boy's parents hoped that he would not only expand his chest and thus improve his breathing, but would make himself stronger and end the recurring ailments that afflicted him.

For the next several months, visitors to the gym were startled by the sight of Mrs. Theodore Roosevelt, elegantly turned out as always in white silks and muslins, sitting alone on a settee at the end of the room, seemingly oblivious to the pervasive smell of body sweat and liniment. With open approval, she looked on as

---

*"One can state as a fundamental law that children who come into the world with organ inferiorities become involved at an early age in a bitter struggle for existence which results only too often in a strangulation of their social feelings," observed Alfred Adler, the pioneer psychiatrist. "Instead of interesting themselves in an adjustment to their fellows, they are continually occupied with themselves and with the impression which they make on others. . . . As soon as the striving for recognition assumes the upper hand . . . the goal of power and superiority becomes increasingly obvious to the individual, and he pursues it with movements of great intensity and violence, and his life becomes the expectation of a great triumph." *Understanding Human Nature,* pp. 69, 191.

her son worked out nearby on a chest exercise machine. Relentlessly, he pulled the weights up, let them subside, and then pulled them up again.

Pleased with Teedie's progress, his father had Wood set up a private gym, complete with almost every type of athletic equipment, in the open-air porch adjoining the second-floor nursery at Twentieth Street. Over the next several years, the boy spent much of his spare time there, swinging on the bars, punching a bag, and working out with weights, all with the trancelike dedication of the confirmed bodybuilder. "One of my most vivid recollections," remembered Conie, "is seeing him between horizontal bars, expanding his chest by regular, monotonous motion—drudgery indeed."

Young Roosevelt was also pursued by his own demons. Years later, he told his friend Hermann Hagedorn that his compulsive effort to build himself up was spurred not only by the desire to please his father, but by the shock of recognition after chancing upon Robert Browning's poem *The Flight of the Duchess*. The poem described the effete son of a noble family as:

> . . . The pertest little ape
> That ever affronted human shape.
> . . . All legs and length,
> With blood for bone, all speed, no strength.

Suddenly, he saw himself as this "little ape," a puny, gangly lad with pipestem arms and legs. *All speed, no strength.* If he wanted to be like his heroes—Morgan's riflemen, the frontiersmen of his favorite stories, Washington's soldiers at Valley Forge—he must become strong. He began to re-create himself, and the weights on the chest machine became heavier, the hours on the bars longer. In the country, at Dobbs Ferry and at Oyster Bay, he accepted additional challenges: footraces, swimming, horseback riding, and tree-climbing contests with his brother and sister and many cousins.

Not long after Roosevelt became president, a dime novel entitled *From Ranch to White House: The Life of Theodore Roosevelt,* was published in the "Log Cabin to White House" series intended for young readers. Roosevelt was hardly born in a log cabin, but his boyhood, like that of Abraham Lincoln, became legend. Like Lincoln, he overcame adversity and built his character. But unlike Lincoln, he basked in the glow of this legend in his own lifetime and played a role in creating it. He made his struggle against poor health into the equivalent of Lincoln's rise from poverty. The metamorphosis from sickly, scrawny boy into masterful man became a lifelong model and standard of measurement of men, social groups, and nations.

Teedie's general health had so improved by August 1871 that he was allowed to go on a camping trip in the Adirondacks. He loved it. He slept on the ground, shot rapids, climbed mountains, and faced all the rigors of the summer woodsman. There is not a single mention in his diary of asthma or other illness during the entire month in the woods—the longest period of good health he had ever enjoyed—and nature and the outdoors became a passion.

The boy also pursued his zoological activities, and had grown knowledgeable enough to label the specimens in the Roosevelt Museum of Natural History with their proper scientific names. "I picked up a salamander (*Diemictylus irridescens*)," he recorded in his diary. "We saw a bald-headed eagle (*Halietus leucocephalus*)." He also studied taxidermy in a musty little shop on Broadway operated by John G. Bell, "a white-haired old gentleman, as straight as an Indian, who had been a companion of Audubon's." Now, he was even grubbier than most boys of his age: His hands were stained by arsenic and he usually reeked of formaldehyde and bird entrails. "When he does come into a room, you always hear the words 'bird' and 'skin,' " complained Conie. "It certainly is great fun for *him*." Upon one occasion, a well-meaning maid extracted an old toothbrush from his taxidermy kit that he had used to apply arsenic preservative soap to his animal skins, washed it, and put it back among his toilet articles. Luckily, he recognized the brush before using it.

In those days, there were still areas of wilderness on Long Island, and to help him expand his collection, the boy's father gave him his first gun, a 12-gauge shotgun of French manufacture. It was "an excellent gun for a clumsy and often absent-minded boy," Roosevelt recalled. "There was no spring to open it, and if the mechanism became rusty it could be opened with a brick without serious damage. . . ." But to his surprise, he found that he was unable to hit anything, while his companions shot birds or animals that were merely a blur to him.

"One day they read aloud an advertisement in huge letters on a distant billboard, and I then realized that something was the matter, for not only was I unable to read the sign, but I could not even read the letters," he recalled in his *Autobiography*. He told his father about this curious incident, and the elder Roosevelt sent him for an eye examination in which it was determimed that he was extremely nearsighted. "Soon afterwards, I got my first pair of spectacles, which literally opened an entirely new world to me," Roosevelt said. For the first time, he could distinguish things more than a few feet away from him and, marveling, declared: "I had no idea how beautiful the world was until I got those spectacles. . . ."

Nevertheless, he was chagrined to discover that his determined regimen of bodybuilding had not had much effect. Tutored at home, he had not mixed with other youngsters and had been spared their taunts about his spindly appearance and his thick eyeglasses. On a trip to a camp at Moosehead Lake in Maine after a severe asthma attack when he was fourteen, two boys of about his own age made fun of him. Back home, Elliott had been his protector but now he was on his own. The boys made life miserable for him, and Teedie finally decided to fight back. "I discovered either one singly could not only handle me with easy contempt but handle me so as not to hurt me much," he remembered.

Deeply wounded by the experience, Roosevelt persuaded his father to allow him to take boxing lessons. He began working out with John Long, an ex-prizefighter known to the elder Roosevelt, and although he always described himself as a "painfully slow and awkward pupil," developed the fighting instincts that were to dominate his character. Roosevelt continued to box even in the White House, and while sparring with a military aide, suffered a blow that cost him the sight of his left eye. Characteristically, he hid the injury so the young man would not be concerned about having injured the president.

Roosevelt's obsession with physical fitness gave him an aura of perpetual youthfulness. Throughout his life he was surrounded by the paraphernalia of bodybuilding: boxing gloves, weights, dumbbells, horizontal bars. And he exhibited the adolescent's uncertainties and nagging self-doubts. Later, he deliberately sought out dangerous situations—on the battlefield, in politics, in the depths of the Brazilian jungle—to prove himself. Roosevelt's universe was painted in primary colors and filled with dark conspiracies, fierce loyalties, and equally fierce hatreds. Some observers were charmed by his youthful joy in living; others thought him irritatingly immature.

"You must always remember that the President is about six," observed Cecil Spring Rice, an old friend. And on his forty-sixth birthday, he received greetings from his friend Secretary of War Elihu Root: "You have made a very good start in life, and your friends have great hopes for you when you grow up." A less benign view was taken by J. Laurence Laughlin, one of Roosevelt's former professors at Harvard. Upon being elected member of an Indian tribe, Laughlin noted disdainfully, the president joined, whooping and shouting, in a war dance "round and round" his office in the White House.

Toward the end of 1872, the Roosevelts embarked on a foreign adventure even more ambitious than their previous grand tour of Europe. In contrast to that trip, the children were excited by the prospect. The highlight was to be a two-month cruise up the Nile in the family's own dahabeah, or chartered houseboat, past Luxor and the splendors of Karnak, all the way to the First Cataract at Aswan, a round-trip of some 1,200 miles in all. The Holy Land, Beirut, and Damascus were next on the schedule, followed by stops in Constantinople, Athens, and finally Vienna.

Ostensibly, the trip resulted from the senior Roosevelt's appointment as an American commissioner to the Vienna Exposition of 1873, but he had numerous other reasons for taking his family abroad. Now independently wealthy—old C.V.S. Roosevelt had died two years before, leaving each of his sons more than $1 million—he no longer had to worry about business. Following the death of their father, James Alfred, the eldest of the Roosevelt brothers, established Roosevelt & Son in investment banking.

Under the able hand of James Alfred, who also headed the

family's real estate holding operation, the Broadway Improvement Company, the firm prospered by selling bonds of expanding American industries in the European market, including the financing of the transatlantic cable. James Alfred was a close friend and associate of James J. Hill, the railroad magnate, later the target of his nephew's most important effort at trust-busting. He was selective in his clientele, however, and is said to have personally thrown Jay Gould, the Mephistophelian Wall Street operator, bodily out of his office.*

Theodore was also planning a palatial new home far uptown at 6 West Fifty-seventh Street, just off Fifth Avenue, and next door to James Alfred, who lived at 4 West Fifty-seventh. The neighborhood around the Twentieth Street house was going commercial, and the old house was far too small for a man of his social obligations. Mittie and the children could remain abroad until the new house was ready for occupancy. There was another, more private reason for the move, however. Theodore was obviously embarrassed by the secret life being led by his brother Robert B. Roosevelt, and wished to put some distance between them. Uncle Barnwell, as he was called, lived next door with his wife, Lizzie, and their children, while he maintained his mistress, Minnie O'Shea, and a second family only a block or so away.†

Lusty and raffish, Robert Barnwell Roosevelt was an Elizabethan transplanted to the carefully ordered world of Victorian America. He enjoyed life and was at one time or another a sportsman, attorney, social reformer, political gadfly, newspaper publisher, and novelist. A leader in the fight against the Tweed Ring, which had a grip on the throat of New York City, he also served as a reform congressman and U.S. minister to Holland. He was a founder of the American conservation movement, author of the legislation creating the New York State Fisheries Commission, which he headed as well as the U.S. Fisheries Committee, and he launched the tradition of the hunter-naturalist best personified

---

*The ultimate commuter, James Alfred Roosevelt died in 1898 on the 4:32 P.M. Long Island train to Oyster Bay.

†Minnie O'Shea claimed to be the widow of an Irish soldier named Robert Fortescue, and her children by Robert B. Roosevelt bore the surname Fortescue. Following the death of Lizzie in 1889, RBR married Minnie and the "Fortescue" children were accepted by the Roosevelts, who steadfastly turned a blind eye toward their true origins. One served under TR in the Rough Riders.

by his nephew. In his love of the limelight, politics, and the outdoors, Theodore Roosevelt more resembled his uncle than his father.

The European phase of the Roosevelts' second trip abroad began with another visit to the Bullochs in Liverpool, where Teedie revealed a defiant streak of Yankee nationalism. The exhibits in the Liverpool Museum were "neither so well mounted or so rare" as those in the American Museum of Natural History, he wrote in his diary. Liverpool itself had "all the dirt of New York (and a good deal of its own beside)." The cab horses were poorer than those at home and "animals in general are treated much more badly." Oblivious of his eccentric appearance, he felt he was being pestered by the street urchins because they recognized him as an American. And he was irritated by a shopkeeper's refusal to sell him a pound of arsenic for use in his taxidermy work unless he would "bring a witness to prove that I was not going to commit murder, suicide or any such dreadful thing."

In Germany, Teedie's disdain for things European continued unabated. "Bonn is an old town, and like all old towns has both advantages and disadvantages," he observed. "The peasants, and old houses, and walls are very picturesque, but to offset this the system of drainage puts all the filth into the streets where it remains for some time before finding its way into some lower sewer. . . . Bologna is much less infected with dirt and beggars than Rome or Naples, but . . . as usual everybody combines to cheat you." But the boy was willing to concede that Europe had some attraction. The Rhone Valley was "wild and picturesque, the effect being still further enhanced by the beautiful cascades old Castles and especially the snug Swiss Hamlets which are to be seen nestling in among the hills."

Following a two-day passage across the Mediterranean from Brindisi, the Roosevelts gathered on deck before breakfast on the morning of November 27, 1872, for their first glimpse of Egypt. Teedie was completely caught up in the swirl of excitement that surrounded their arrival:

> Alexandria! How I gazed upon it! [he observed in his diary] It was Egypt, the land of my dreams; Egypt the most ancient of countries! A land that was old when Rome was bright, was old when Babylon

was in its glory. . . . As soon as our ship was in the harbour it was surrounded with arab felluccas with strange lateen sails and still stranger arab boatmen. . . . We got into one of these and rowed to shore, where, after getting through the Customs House (bribery and corruption) we took a carriage to the Hotel Abbott. I shall never forget that drive. On all sides were screaming Arabs, shouting Dragomen, shrieking donkey boys and braying donkeys. . . . There was a row between an Arab and a Negroe, in which the former was badly wounded and the latter was taken off to be bastinadoed.*

Following lunch, the Roosevelts, indefatigable tourists as always, drove out to Pompey's Pillar, where Teedie inspected the broken images of some old Egyptian gods that were strewn about. For the first time, words failed him. "On seeing this stately remain of former glory, I *felt* a great deal but I *said* nothing," he confided to his journal. "You cannot express yourself on such an occasion."

In Cairo, Mittie and the children visited mosques, gardens, bazaars, and antiquities while the elder Roosevelt searched for a comfortable dahabeah for the trip upriver. Teedie was fascinated by the Pyramids. He not only climbed to the top of the Great Pyramid of Cheops—"The view was perfectly magnificent"—but burrowed inside through a narrow, steeply sloping passage to the King's Chamber. "The atmosphere was perfectly stifling and now to add to the horrors [the passage] began to ascend. . . . I was continually falling from the floor being so smooth on so great an incline, and from being forced to bend double. . . . Numbers of Bats flitted above us with their sharp cries that seemed literally to pierce the ear."

The Roosevelts began their winter on the Nile on December 12, when they boarded the dahabeah *Aboo Erdan,* or *Ibis,* which the elder Roosevelt had chartered for two months at a cost of $2,000.† Teedie delightedly described the craft as "the nicest, coziest pleasantest little place you ever saw." There were small staterooms for every member of the family, a bath, and a main

---

*A form of punishment common in the Middle East in which the culprit is beaten on the soles of the feet with sticks

†A sum equal to about four times the yearly income of the average American family of the day

salon which doubled as a dining room. On deck, there were several divans that were shaded from the tropic sun by awnings. When the wind was right, the craft made about two or three miles an hour under sail. If there was no wind, the crew resorted to the age-old method of "tracking," in which sailors waded ashore, attached lines to the boat, and dragged it along against the current.

The pace of the dahabeah was hypnotically slow and the Roosevelts settled into a comfortable routine. Each morning began with two hours of lessons with Bamie in French and other subjects, leaving the rest of the day free for exploring historic sites along the riverbank. A visit to the temple of Karnak by moonlight produced a breathless reaction from Teedie similar to the one experienced at his first sight of Alexandria. "It was not beautiful only, it was grand, magnificent and awe-inspiring. It seemed to take me back thousands of years, to the time of the Pharohs." In the evening, they all gathered on deck, where Papa smoked a cigar and read aloud from a thick volume of Egyptian history or planned the next day's visit to monuments or a picnic.

Sometimes they joined with other Americans, who were also taking a Nile trip, for formal dinners and celebrations of Christmas and New Year's Day. Upon one occasion, the Roosevelts anchored near a dahabeah carrying Ralph Waldo Emerson and his daughter Ellen, and Theodore, Sr., took his children to visit the great man. Corinne remembered the poet-philosopher's "dreamy mystic face" and his "lovely smile—somewhat vacant, it is true, but very gentle."* Ellen Emerson was impressed by the Roosevelt brood. "Enchanting children," she wrote a friend. "Healthy, natural, well-brought-up and with beautiful manners."

For Teedie, with his fascination with birds and love of the outdoors, the voyage was a highlight of growing up. The Nile Valley in winter is a vast flyway and a shifting screen of birds passed before his astonished eyes: kestrels, doves, larks, plovers, herons, kingfishers, egrets, wild duck, cranes, kites, and hawks. In one day, he counted fifteen different species. Flocks of birds hovered about the *Aboo Erdan* and sometimes it looked as if someone had tossed a basket of papers from the boat and the pages, caught by a breeze, were speeding along ahead of it.

---

*Emerson apparently made no impression upon Teedie, however, for he fails to mention the encounter in his diary.

Back in Cairo, Teedie had picked up a guide to the birds of the Nile Valley and deliberately set to work to build a collection superior to that of most amateurs. Armed with a fine, new double-barreled shotgun, a Christmas present from his father, he pursued specimens for the Roosevelt Museum of Natural History with a dedicated single-mindedness that bordered on ruthlessness. Every day, he was out along the banks of the Nile, blasting away at any bird unfortunate enough to stray within range. His total bag that winter included as many as two hundred birds. At day's end, he skinned and stuffed the specimens, a ritual performed on deck before the curious eyes of the Egyptian crew.

Although the boy adopted the role of the aloof Young Naturalist, he often seemed more like Don Quixote. He had outgrown his clothes; skinny arms and legs stuck out of his sleeves and trousers, his boots were encrusted with Nile mud, his hair was uncut and worn "a-la-Mop," as he put it. Mounted on a fractious donkey, he rode about with his shotgun slung carelessly across his shoulder in a manner that could be "distinctly dangerous as he bumped absent-mindedly about," according to Corinne. Ellie, who had a talent for comic limericks, observed:

> There was an old fellow named Teedie,
> Whose clothes at best looked so seedy
> That his friends in dismay
> Hollered out, "Oh! I say!"
> At this dirty old fellow named Teedie.

Teedie's health improved dramatically under the hot desert sun and he did not suffer from asthma or "cholera morbus" during the entire two months on the Nile. "I think I have enjoyed myself more this winter than I ever did before," he wrote Edith Carow. Even his father was amazed at his energy as they tramped along the banks of the river together. "He is the most enthusiastic sportsman and infused some of his spirit into me," Theodore wrote his sister-in-law. "Yesterday I walked through the bogs with him at the risk of sinking hopelessly and helplessly . . . *but I felt I must keep up with Teedie.*"

Within a few weeks, the Roosevelts, in a rapid change of pace, had left the Nile and were leading a nomadic life in tent and on horseback in the hills of Judea. Late in January 1873, the *Aboo*

Erdan had reached the First Cataract at Aswan, and the elder Roosevelt was tempted to press on further into the heart of Africa. But time was growing short if he intended to show his family the Holy Land, Syria, Turkey, and Greece before arriving in Vienna by May 1 to take up his duties as American commissioner to the Exposition, so he reluctantly gave orders to turn back to Cairo.

The family's arrival in Palestine had almost biblical overtones. Theodore had made reservations for the first night at a monastery at Ramle, about halfway between the port of Jaffa and Jerusalem, but when they got there at sunset, the monks, finding women in the party, said they had "no room" for the travelers. "A long talk ensued," Teedie reported, before the monks relented enough to offer rooms for the men, but they absolutely refused to allow the women within the monastery's walls. "This difficulty was also overcome in time," according to Teedie—evidence of Papa's ability to take command in any situation.

Like many travelers, Teedie was disappointed to find most of the historical sites were far less grand than he had imagined. Viewing Jerusalem from the top of a nearby hill, he thought the city "remarkably small." He took in most of the places mentioned in the Old and New Testaments, but except for a sense of "awe" at being on Calvary where Christ was crucified, he adopted an attitude of scientific skepticism toward everything. "We saw . . . various other wonderful things which gave me the same impressions as when I saw the Bones of Saints (or Turkeys) in Italy." At the Wailing Wall, he thought "many of the women were in earnest but most of the men were shamming." He bathed in the Dead Sea and the Jordan—"what we should call a rather small creek in America"—and noted that bribes had to be paid to visit the birthplace of Jesus in Bethlehem, where the Greek and Latin clergy were contending for control of the holy places. On the other hand, the ruins of Baalbek, in Syria, were "with the exception of Karnak the grandest and most magnificent I have ever seen."

One of the pleasures of traveling with Father, recalled Corinne, was that wherever he went, he accumulated friends of all kinds, "and we his children, shared in the marvelous atmosphere he created." Arab sheikhs visited the caravan in which they rode, Gypsies danced "a wild sort of dance" before their tents, and several Arab women were persuaded to visit Mittie and the girls to tell them something of their way of life. When a sudden storm blew down the travelers' tents, the inhabitants of an Arab village offered them the hospitality of their homes.

With the return to a cooler, damper climate, Teedie's health began bothering him again. He battled asthma, stomach troubles, seasickness, and was often listless and morose. Athens lacked "the magnificent beauty of Baalbek and the gloomy granduer of Karnak." Constantinople made little impression on him. In Vienna, where his father was busy with the American display at the Exposition* and his mother fussed over the details of Bamie's European social debut, he was bored. "The last few weeks have been spent in the most dreary monotony," he observed. "If I stayed here much longer I would spend all my money on books and birds *'pour passer le temps.'* "

Theodore and Mittie had other plans for their offspring, however. Teedie, Elliott, and Corinne spoke French but did not know German, so their parents decided to send them to live for five months with a German family in Dresden, where they could pick up the language while the finishing touches were being put on the family's new home near Central Park. As soon as the elder Roosevelt's assignment in Vienna was completed, he returned to New York to oversee the project and to help steer the family business safely through the devastating financial Panic of 1873. Mittie and Bamie went off to Carlsbad to take the cure and then shopped for furnishings for their new house.

Filled with art treasures, theaters, and baroque buildings, Dresden, one of the world's loveliest cities, was called the "Florence of the Elbe." Teedie and Ellie—and later Conie—boarded with the family of a Dr. Minkwitz, a city counselor and member of the German Reichstag. Originally Corinne was sent to live with another family, but she was so unhappy at being separated from her brothers that Teedie, assuming a leadership role, prevailed upon his mother to allow his sister to join Elliot and himself at the Minkwitzes'.

It was a typically Teutonic household. Dr. Minkwitz was formal

---

*The American commission became involved in a scandal when the European press learned that its chairman, General T. B. Van Buren, an old friend of President Ulysses S. Grant, had been shaking down concessionaires. Van Buren was dismissed and the senior Roosevelt took over management of the U.S. exhibition.

but kindly; his wife was plump and the source of an unfailing supply of tea and cake; the three daughters were gay and well educated. The eldest, Fräulein Anna, taught the children German and arithmetic. Teedie, however, was captivated by the two sons, both students at the University of Leipzig who, as dashing members of the dueling corps, were scarred about the face. "One, a famous swordsman, was called *Der Rothe Herzog* (the Red Duke), and the other was nicknamed *Herr Nasehorn* (Sir Rhinoceros) because the tip of his nose had been cut off in a duel and sewn back on again."

Fräulein Anna insisted upon a rigorous work schedule. "The plan is this," Teedie wrote his father. "Halfpast six, up and breakfast which is through at half past seven, when we study till nine, repeat till half past twelve, have lunch, and study till three when we take coffee and have till tea (at seven) free. After tea we study till ten, when we go to bed. It is harder than I have ever studied before in my life, but I like it for I really feel that I am making considerable progress."

To encourage the children's facility in German, the Minkwitzes made a point of speaking to them only in that language, whether they understood or not. Teedie understood better than they knew. He picked up several observations that their hosts made about his parents and gleefully passed them on. "Father is considered 'very pretty' (*sehr hübsch*) and his German 'exceedingly beautiful,' " while his mother was "thought very pretty and beautiful" although she was criticized for speaking no German. The boy developed a fair fluency in the language and a lifelong fascination with the *Nibelungenlied* that matched his earlier interest in *The Saga of King Olaf,* and its romantic posturings stirred his blood. He also came away that summer with a favorable impression of the German people that lasted until World War I.

For their part, the Minkwitzes had some difficulty getting used to Teedie. He cultivated the look of a German student, wearing his hair long, and was even more untidy than before. Taking inventory of himself, he wrote Bamie: "Health; good. Lessons; good. Play hours; bad. Appetite; good. Clothes; greasy. Shoes; holey. Hair; more 'a-la-Mop' than ever. Nails; dirty." The obsessional nature of his scientific pursuits caused considerable consternation in the household, and he once alarmed the entire female contingent by producing a dead bat from his pocket. Not long after, he complained, "My arsenic was confiscated and my mice . . . thrown out the window."

In addition to his usual problems, Teedie came down with mumps and described his plight in a lighthearted note to his mother. "Picture to yourself an antiquated woodchuck with his cheeks filled with nuts, his face well oiled, his voice hoarse from gargling and a cloth resembling in texture and cleanliness a second-hand dust man's castoff stocking around his head; picture yourself that, I say, and you will have a good likeness of your hopeful offspring. . . ."

Headaches and bouts of asthma continued with increasing intensity, and Mittie, alarmed at the reports reaching her, raced to her son's side. She found him wheezing so badly he could not speak and was forced to sleep sitting up, as he had when a small child. Dresden had been struck by a cholera epidemic, so she decided to take all three children to the Swiss Alps and persuaded Fräulein Anna to accompany them. Teedie improved in the bracing, pine-scented air and to make up for lost time, asked the teacher to speed up his lessons. "Of course I could not be left behind," wrote Elliott, "so we are working harder than ever in our lives."

In Dresden, the Roosevelt children were members of the Dresden Literary American Club, an organization they formed along with a handful of other homesick expatriate youngsters. The club met on Sunday afternoons for the reading of stories and poetry that its members had composed. When this literary salon was over, the boys battered each other with boxing gloves that the elder Roosevelt had sent them. "If you offered rewards for bloody noses you would spend a fortune on me alone," Teedie reported.

The boy also began applying to others the code of behavior he learned from his father. A fellow must avoid profanity, excessive drinking, and maintain a high standard of personal ethics. Yet, at the same time, he must show proof of his virility in other ways to retain the respect of his fellows. "Did you hear that Percy Cushion was a failure?" Teedie wrote his father about one of his acquaintances. "He swore like a trooper and used disreputable language, so I gave him some pretty strong hints, which he at last took, and we do not see much more of him." Teedie added that he had met another boy, Edward Jacobs, who never swore and was "always ready to box or swim or do any other thing we propose." Disdain for the disreputable is obvious. Percy is a failure; Edward a success—and virile to boot. The future apostle of manly prowess and righteousness was already rendering judgments.

In October 1873, Mittie returned to Dresden to pick up her children for the return to America. The Minkwitzes, who had gotten over their earlier misgivings about Teedie, were sorry to see him leave. Mittie expressed concern about the boy's future to Fräulein Anna as they packed up his books and specimens. "You need not be anxious about him," she replied. "He will surely one day be a great professor, or who knows, he may become even President of the United States."

The return of the Roosevelts to New York was accompanied by numerous changes in their way of life. The Fifty-seventh Street house was far grander than anything the family had ever known. In place of the comparative simplicity of Twentieth Street, it was a showplace. Persian carpets covered the floors, the ballroom shimmered with mirrors and crystal, and ornate hand-carved woodwork was everywhere. The attic was set aside for the ever-expanding Roosevelt Museum of Natural History and there was a gymnasium on the top floor. The grandeur of the place can be gauged from the fact that the Roosevelts sent out five hundred invitations to Bamie's coming-out party in January 1874.*

Increasingly, the management of this sizable establishment fell upon Bamie's youthful shoulders, for her mother, who had been showing a tendency toward vagueness and was unable to handle money, now let things slip almost completely. For some time, she had been complaining of headaches and some form of intestinal trouble, which she referred to as "my horror." Mittie found it impossible to keep appointments. Invited to tea or other social affairs, she would spend so much time trying to make up her mind what to wear that when she finally decided, it was too late to go. Carriages would be summoned and then left standing in the street for hours until they had to be dismissed. More obsessed about cleanliness than ever, she had the furniture draped in sheets.

*The site is now occupied by the Bergdorf Goodman store.

When she said her prayers, a sheet was put down so she would not touch the floor. If her husband and children were angered by the eccentricities of their "dear little Motherling," they kept it to themselves. "To us, her children and her devoted husband, she seemed like an exquisite *objet d'art,* to be carefully and lovingly cherished," said Corinne.

In addition to moving to a new home in town, the elder Roosevelt joined the family colony at Oyster Bay and rented a white-pillared house on Cove Neck overlooking Long Island Sound. To the amusement of his friends, he called it "Tranquility" because it reminded him of the old Bulloch house at Roswell. "Anything less tranquil than that happy home . . . could hardly be imagined," recalled Corinne, who dubbed Oyster Bay "The Happy Land of Woods and Waters." Overrun by cousins and friends, the place bubbled with activity. The children swam and rowed, sailed, hunted, and explored the woods on foot and horseback. Sometimes Teedie went up to the crest of Sagamore Hill, named for the Indian chief Sagamore Mohannis, and declaimed poetry into the wind at the top of his voice. Some of the village boys teased him about his glasses—calling him "Four-eyes"—and he wanted to fight but was forbidden to do so by his parents.

Girls found Teedie clumsily appealing. While he condescendingly affected the role of experienced linguist, naturalist, and pugilist in their company, he had a sunny charm when he was not wrapped up in his books and experiments. In New York, his mother organized a Monday dancing class at Dodsworth's ballroom which sowed the seeds of many friendships. Edith Carow, who had grown into a thoughtful and attractive young girl, was still his favorite. They attended dancing class together, went rowing and for walks in the woods at Oyster Bay, and he painted her name on the stern of his rowboat. When Edie "dresses well," he observed, "and don't frizzle her hair, [she] is a very pretty girl."

But Edith was not his only choice of companion. He picnicked with the "very nice" Annie Murray, enjoyed a "lovely long row" in the moonlight with Nellie Smith, and went riding with Fanny Smith. Looking back many years later, Fanny said the characteristic she most remembered about Roosevelt "was the unquenchable gaiety which seemed to emanate from his whole personality. . . . As a young girl I remember dreading to sit next to him at any formal dinner lest I become so convulsed with laughter at his whispered sallies as to disgrace myself and be forced to leave the room."

The social graces did not always come easily. Fanny recalled that one winter, Elliott paid her a visit. While chatting with him on a windowseat, she suddenly noticed Teedie, "looking blue with cold," pacing up and down outside.

" 'Why Elliott, do you see Theodore out there? Why *doesn't* he come in!' I exclaimed.

"Elliott replied—and to this day the incident remains a mystery—that Theodore also had planned a visit but that suddenly he had been overcome by bashfulness and had decided to remain outside. We brought him in, where he became—as always—the 'life and soul of the party.' But the incident reminds me of the unexpected strain of self-depreciation which surprised one through the years."

When Teedie turned sixteen, his father decided that he should attend college and planned for him to enroll at Harvard in the autumn of 1876. First, however, the boy, whose education had been spotty, had to pass the entrance examinations. Although he was an omnivorous reader, a dedicated naturalist, spoke French and German, and possessed a knowledge of the world augmented by considerable travel abroad, his mind was a jackdaw's nest of unrelated facts. Also he was weak in Latin, Greek, and mathematics, subjects that loomed large in the Harvard entrance examinations. Because of his frail health, boarding school was regarded as out of the question, so the elder Roosevelt secured the services of a tutor, a young Harvard graduate named Arthur Cutler, to prepare his son for the examinations.

Teedie liked Cutler and the tutor was impressed with "the alert, vigorous character of young Roosevelt's mind." He placed special emphasis upon correcting the boy's deficiencies in mathematics, and putting in eight hours a day, he covered three years' work in two. Teedie also worked out in the new gymnasium, chinning himself and twirling on the parallel bars, boxed and wrestled and skated on the ice in nearby Central Park. Out skating one day, he fell upon his head and was knocked senseless "for several hours."*

The boy also spent long hours gathering, classifying, and mounting specimens for the growing Roosevelt Museum. "I worked with greater industry than either intelligence or success," he observed many years later, "and made very few additions to the sum of human knowledge, but to this day certain obscure ornithological publications may be found in which are recorded such items, as, for instance, that on one occasion, a fish-crow and on another an Ipswich sparrow, were obtained by one Theodore Roosevelt, Jr. at Oyster Bay, on the shore of Long Island Sound."

In the beginning, Elliott and West Roosevelt, a cousin, also joined in the lessons, but were quickly outdistanced by Teedie and dropped out. Not long afterward, Elliott's life began to come apart. He complained of excruciating head pains, was frightened of the dark, afraid to sleep alone. He was struck by frightening seizures during which he babbled incoherently or blacked out. The boy was repeatedly examined by doctors but they were baffled. Some descendants now suspect epilepsy, but there was no family history of such attacks. Others have suggested that the illness may have been psychosomatic because the seizures seemed to occur when Elliott was faced with challenges he believed he could not meet. The boy may also have been reacting to the loss of his leadership role to Teedie.

At his own request, Elliott was sent to boarding school at St. Paul's in Concord, New Hampshire, where one of his friends had enrolled. Out of the family orbit for the first time, he did well. "I think all my teachers are satisfied," he wrote his father in September 1875. Things soon began to go wrong, however. "Yesterday, during my Latin lesson, without the slightest warning, I had a bad rush of blood to my head," he wrote his father. "It hurt me so that I can't remember what happened. I believe I screamed out." Not long after, he collapsed and had to be brought home. Always convinced that physical problems could be overcome by an active outdoor life, the elder Roosevelt sent Elliott to "rough it" with one of his friends, an officer at Fort McKivett,

---

*Roosevelt was to have many similar accidents, causing some critics to claim that all these blows to his head had affected his mental stability.

in the Texas hill country. There he rode, shot, and hunted, living the kind of life envied by Teedie, and his head pains disappeared. But all plans for the boy to follow a systematic program of reading and study also vanished with them.

Teedie faced his examinations for Harvard with some trepidation, but he sailed through without difficulty. "Is it not splendid about my examinations?" he wrote Bamie. "I passed well on all the eight subjects I tried." The boy also appeared to be winning his battle for health. Asthma troubled him less frequently now, and although he was by no means robust, he had developed into a wiry youth, tanned to the waist from rowing in the sun and wearing long sideburns that emphasized a determined jaw. Proudly, he recorded his physical measurements in his diary: "Height: 5 feet, 8 inches; weight: 124 pounds; chest 34 inches; waist 26 1/2 inches."

That last summer before Teedie's departure for Cambridge was one of the most pleasant the Roosevelts had known. "My only regret is the rapidity with which it is going," the senior Roosevelt observed as he rocked on the porch of Tranquility and drew on his cigar. "Theodore's departure for Cambridge is one of the straws that show that the time is approaching when the birds will leave their nest and I am very glad for them all to have such pleasant memories connected with this summer. Our pleasures tie us all so much together. . . ."

Undoubtedly the father envied the son, for he had been denied the opportunity to attend college. In the weeks before Teedie's departure, they had long talks together about his future. The boy continued firm in his resolve to become a "scientific man of the Audubon, or Wilson, or Baird type." Yet the family hoped that his fascination with birds and animals was a passing fancy, that maturity would cause him to turn to some more practical interest such as business or the law. Eminently sensible as always, Roosevelt's father made no effort to persuade his son to seek another career against his will.

On September 27, 1876, he saw him off to Harvard from the little railroad station at Syosset on Long Island. With him, Teedie carried some parting advice. "Take care of your morals first," his father told him, "your health next and finally your studies."

Chapter 3

Harvardman

With a carpetbag bouncing at his side, Theodore Roosevelt, Jr., as he now styled himself, swung down from a horsecar as it rounded the sharp curve into Harvard Square. Passing through the low fence that surrounded Harvard Yard, he registered as a member of the Class of 1880. For the first time he examined what was to be the center of his world for the next four years. A handful of red-brick buildings were ranged haphazardly about the elm-shrouded Yard, while the newly completed, high-Victorian ramparts of Memorial Hall, which lay just outside its boundaries, brooded over the entire scene.

Roosevelt did not live in the Yard like most of his classmates. Fearing that the dampness of the ground-floor dormitory rooms allotted to freshmen might be detrimental to his health, his parents had dispatched the indefatigable Bamie to Cambridge to find quarters for her brother. She chose and furnished a second-floor room in a house owned by a Mrs. Richardson at 16 Winthrop Street, a narrow lane between Harvard Square and the Charles River. Upon his arrival, Roosevelt found his stuffed birds in place and his bowie knives crossed over the fireplace. Four large windows provided plenty of light and there was a fire burning in the grate to ward off the autumnal chill. Putting a photograph of his mother on the mantel, he settled into these "cosy and comfortable" quarters. He had a manservant to light the fire in his room in the morning and a woman to look after his laundry.

Harvard, during Roosevelt's years there, was basically a small provincial college with a student body of 821—including 246 freshmen—all members of a privileged caste. Year after year, the same names appeared upon the rolls: Blodgett and Cabot; Morison, Quincy, and Saltonstall. Most had grown up within a hundred-mile radius of Boston, attended the same select New England boarding schools, and associated only with each other. One outlander later recalled that he sat beside Josiah Quincy in class for four years, yet Quincy did not deign to speak to him once during the entire time. There were no foreigners among Roosevelt's classmates, no blacks, no Boston Irish, and no Jews.

"Roosevelt of New York" puzzled the languorous young men who set the tone of the college. Harvardmen prided themselves on indifference to the outside world. Undergraduates favored a "Harvard drawl" that was almost a yawn, walked in a "Harvard swing" that resembled a shamble, and professed a disdain for any form of enthusiasm. Overflowing with nervous energy, Roosevelt was constitutionally incapable of being indifferent about anything. He was considered too odd to gain instant acceptance, a feeling compounded by his thick eyeglasses and high-pitched, squeaky voice. Sometimes his thoughts outpaced his ability to express them and the result was an unintelligible explosion of words. "Eccentric," "half-crazy," and "different" were among the terms used to describe him.

Had Roosevelt arrived at Harvard without wealth and family connections, he might well have had a lonely time. Indeed, he made only a few close friends in college, and was not considered the outstanding member of his class. That accolade went to Robert Bacon, the strikingly handsome captain of the football team, heavyweight boxing champion, and varsity oarsman. In fact, the Class of '80 was known as "Bacon's class." When Roosevelt became president, he named Bacon, then a Morgan partner, to the post of assistant secretary of state and later secretary of state.

Henry Adams, who resigned as a history instructor the year after Roosevelt's arrival, bitingly described his students as "ig-

norant of all that men had ever thought and hoped." The spirit
of Harvard was expressed by George Pellew, the class poet, in a
song written for a Hasty Pudding show in 1880:

> . . . We ask but time to drift,
> To drift—and note the devious ways of man. . . .

One of the few places where this mask of indifference was
allowed to slip was in raucous behavior at the "girlie" shows at
Boston's Globe Theatre in Bowdoin Square. Primed on cheap
whiskey and beer, students hooted the performers and generally
made nuisances of themselves.* One Boston newspaper angrily
described the average Harvard undergraduate as "an immature
being, more gifted with brawn than brains." Arriving in Cincinnati after a particularly harrowing stand in Boston, the manager
of a troupe of female dancers is supposed to have inquired plaintively: "Is there a university here?"

"It was not considered good form to move at more than a walk,
[but] Roosevelt was always running," said one classmate. Foreshadowing the president who would regard the White House as
"a bully pulpit," he loved to argue and exhort, recalled another
Harvard contemporary. "He used to stop men in the Yard or call
them to him. Then he would block the narrow gravel paths and
soon make sparks from an argument fly. He was so enthusiastic
and had such a startling array of deeply rooted interests that we
all thought he would make a great journalist." In class, he interrupted professors with questions. One lecturer became so exasperated that he cried: "Now look here, Roosevelt, let me talk!
I'm running this course!"

William Roscoe Thayer, who was at Harvard with Roosevelt
and was later his biographer, recalled that he would drop into a
classmate's room looking for conversation, idly pick up a book,
and forgetting all about his host, quickly become immersed in it.
Thayer saw none of the "charm that developed later . . . he was
a good deal of a joke . . . active and enthusiastic and that was

---

*There was a lot of drinking at Harvard in those days, and it was expected that one
or two members of each class would die of alcoholism within a year or so of graduation.
James Giddes in the Pringle Papers.

all." Richard M. Saltonstall,* a downstairs neighbor at 16 Winthrop, remembered Roosevelt's ability to concentrate, seemingly oblivious to everything occurring about him. Once, he became so absorbed in a book as he sat before the fire that only the smell of the soles of his boots burning finally got his attention.

Richard Welling, who first saw Roosevelt when he was performing some rudimentary exercises in the gym, related one of the most telling anecdotes of the young man's college years. By official tests the strongest man at Harvard, Welling thought Roosevelt rather puny—"verily a youth in the kindergarten stage of development"—an opinion confirmed when he observed him actually skipping rope. But like others who took Roosevelt too quickly at face value, he was forced to change his mind when they went ice skating together later that winter.

Following a long ride into the country on a freezing horsecar, the skaters ventured out onto the pond, but it was too open to the fierce winds for good skating. The ice was rough and soon they were flailing their arms about like windmills in a gale to get warm. "Any sane man would have voted to go home, as the afternoon's sport was clearly a flop," Welling observed. But "Roosevelt was exclaiming 'Isn't this bully!'—and the harder it blew and the worse we skated the more often I had to hear 'Isn't this bully!' " For nearly three hours they remained out in the arctic cold, with the nearly frozen Welling praying that his companion would soon give in. Finally, Roosevelt said: " 'It's too dark to skate any more' (as though, if there had been a moon, we could have gone on till midnight). . . ." Welling intended this story to show Roosevelt's indomitability; others might have seen it as a sign of mental imbalance.

Roosevelt's first and best friend at Harvard was Henry Davis Minot, a sophomore and the son of a prominent Boston attorney. Having learned of Theodore's interest in birds through family friends and having a similar passion, Minot dropped by Roosevelt's room to introduce himself. Soon, they had their heads together talking about Swainson's thrush and the red-breasted nuthatch, and went birdwatching in the woods surrounding Cam-

---

*Saltonstall was the father of Leverett Saltonstall, later governor of and senator from Massachusetts.

bridge. When the college year was over, they agreed to make an expedition to the Adirondacks to look for rarer species. For the rest, Roosevelt snobbishly considered only a few of his classmates to be of the "gentleman sort." "On this very account," he wrote his sister Corinne, "I have avoided being very intimate with the New York fellows." And he carefully researched the background of acquaintances before extending the hand of friendship.

Bamie had often visited Boston and the summer colony at Bar Harbor in Maine and had excellent social connections in both places, so when it became known that the new man was the brother of the popular Miss Anna Roosevelt, he was invited to join a private eating club composed of a half dozen or so young Boston Brahmins. Rather than dining in the Commons in Memorial Hall, where America's gilded youth amused themselves by throwing rolls at each other, the club met at the home of a Mrs. Morgan, who had one eye and lived on Brattle Street. "Our quarters . . . are nice and sunny," he informed his parents. "For breakfast we have tea or coffee, hot biscuits, toast, chops or beef steak, and buckweat cakes." Unlike many of the socially elect, "Roosevelt was perfectly willing to talk to others when the occasion rose," recalled one classmate outside this charmed circle. The occasion arose only rarely, however.

Throughout his first year at Harvard, Theodore was in constant communication with his parents and sisters. A steady stream of letters passed between New York and Cambridge in which they admonished him to keep up with his studies while he replied to their queries about his health and schedule. These letters underscored the young man's love of his family. To Corinne, who had reported she was not feeling well, he wrote: "I want to pet you again awfully! My easy chair would just hold myself and Pussie." On his eighteenth birthday, he told his mother that because of his family, "I have never spent an unhappy day unless by my own fault!" And he wrote the senior Roosevelt, "I do not think there is a fellow in College who has a family who love him as much as you all do me. . . . I am *sure* that there is no one who has a Father who is also his best and most intimate friend, as you are mine. I have kept the first letter you wrote me and shall do my best to deserve your trust. I do not find it nearly so hard as I expected not to drink and smoke."

*   *   *

Harvard was dominated by the presence of its president, Charles W. Eliot, who had launched a flurry of reforms that were to transform it from a provincial college into a great university. "Our new President has turned over the whole University like a flapjack," observed Dr. Oliver Wendell Holmes, a member of the Board of Overseers. New buildings were going up and the teaching at the medical and law schools was being revamped. The faculty included William James, the pioneer psychologist and philosopher; Nathaniel Southgate Shaler, a distinguished geologist; James Russell Lowell, who taught romance languages; Charles F. Dunbar, America's first professor of political economy; and Charles Eliot Norton, a popular lecturer on the fine arts.

The free elective system was the most controversial of Eliot's reforms. Rather than being forced to study a rigid list of required subjects, upperclassmen were allowed to select from among a smorgasbord of courses set before them. Eliot believed that if given freedom of choice, they would choose the courses best for them. But he was not popular with his charges. An imperious figure, he strode across the Yard, head held high and looking neither to the left nor to the right as his black gown trailed out behind him. Even Roosevelt was intimidated. On one occasion, he called upon Eliot to petition for the use of a lecture room by a student group. "Mr. Eliot, this is President Roosevelt," he blurted out.

In later years, when the young man had indeed become "President Roosevelt," Eliot was critical of many of his policies. Even so, when Roosevelt returned to Cambridge in 1905 for his twenty-fifth reunion, Eliot felt duty-bound to invite him to stay at his home. Upon his arrival, perspiring and in need of a wash, Eliot recalled, the president dashed upstairs to his room, pulled off his coat, took a large pistol from his pocket, and slammed it down upon the dresser. Not long after, "he came running down the steps as if his life depended upon it.

"Now, are you taking breakfast with me?" Eliot asked.

"Oh, no," replied Roosevelt. "I promised Bishop [William] Lawrence I would take breakfast with him—and good gracious"—slapping his right hand to his side—"I've forgotten my gun!"

Rushing back up the stairs, the president of the United States retrieved his pistol and dashed off to breakfast with the Episcopal bishop of Massachusetts while the president of Harvard, appalled by this willful violation of the state law against carrying concealed weapons, shook his head in disbelief. "Very lawless," he muttered. "A very lawless mind."   ✦

Freshmen were not allowed electives, so Roosevelt took the required subjects: classical literature, Greek, Latin, German, mathematics, physics, and chemistry. "I think I am getting along all right in all my studies," he told his parents, except for mathematics and Greek, which he found "pretty hard" going. Iron discipline was already part of his life and he plotted his day with care. Typically, he rose at 7:15 A.M.; attended compulsory chapel (which was not abolished until 1886) at 7:45; had breakfast at 8:00; studied from 8:30 to 9:00; went to class from 9:00 to noon; studied from noon to 1:00 P.M., when he had lunch; studied again until 2:30, and went to class for an hour. The remainder of the afternoon until dinner at 6:00 P.M. was free. He studied from 7:00 to 8:30, and then visited friends or received visitors in his room. Usually, he spent a half hour or so writing letters or reading before the fire before going to bed at 11:00 P.M.

For recreation, he walked, sparred with "General" Lister, Harvard's boxing master, worked out in the college gymnasium, and rowed on the Charles River in a one-man shell. Asthma seemed less of a problem, although it remained a threat. In November 1876, he told his mother he was "a little under the weather," and in the following year made "a miserable failure" of a French examination because he had been "forced to sit up all the previous night with the asthma."

He kept a small menagerie in his room that included snakes, lobsters, and a large tortoise—it grew in size over the years with each telling of the story—which escaped from its pen and almost sent his landlady into hysterics. Robert Bacon was so repulsed by the thought of this collection that he refused to go near Roosevelt's room. Theodore organized a whist club, became a nonsinging member of the glee club by paying $5 for a season ticket to its concerts, and accompanied the football team to New Haven, where it was defeated by Yale. "The fellows . . . seem to be a much more scrubby set than ours," he noted. Besides Hal Minot, his friends included Charles G. Washburn, Dick Saltonstall, Bob Bacon, Harry Chapin, Henry Jackson, Harry Shaw, and Minot Weld. They called him Ted, or Ted-o, or Teddy.

To please his father, he taught a Sunday school class at Christ Episcopal Church in Cambridge although he was a member of the Dutch Reformed Church.* In Roosevelt's senior year, however, the church got a new rector, a Dr. Spaulding, who told him he would have to give up the class unless he became an Episcopalian. "As I have taught there 3 1/2 years I thought this rather hard," he said, and refused. The incident outraged some of the church membership. One Harvard professor angrily slammed the door in Spaulding's face and said he would no longer attend his church because of Roosevelt's dismissal. The young man was not unduly perturbed; he began teaching a mission class in East Cambridge.

With ready access to the best homes of Boston and suburban Milton and Chestnut Hill, Roosevelt was caught up in a social whirl of dinner parties, dances, and suppers. He took dancing lessons but "danced just as you'd expect him to dance if you knew him," recalled one debutante many years later. "He hopped." He enjoyed himself immensely. "I have been having a pretty gay time," he wrote Bamie shortly after the turn of the year. "Some of the girls are very sweet and bright, and a few of them are very pretty." The names of a Miss Fiske, a Miss Wheelright, a Miss Andrews, and a Miss Richardson—described as "the prettiest girl I have seen for a long while"—appear in his correspondence.

Roosevelt hadn't forgotten Edith Carow, however. Writing to Corinne about a sleighing party for "forty girls and fellows and two matrons" who were pulled in one huge sleigh along the snow-packed roads around Boston by a team of eight horses, he noted that "one of the girls looked quite like Edith—only not nearly as pretty as her Ladyship." Later, he invited Edith and several guests up to Cambridge. She flirted so successfully with Theodore and his friends that he told Corinne: "I don't think I ever saw Edith looking prettier; everyone . . . admired her little Ladyship immensely and she behaved as sweetly as she looked." And he asked

---

*Roosevelt's parents attended both Presbyterian and Dutch Reform churches, but their son, out of loyalty to his Dutch roots, attended the Reformed Church.

his sister to tell Edith "I enjoyed *her* visit *very* much indeed."

Roosevelt's interest in girls coincided with new attention to his appearance and wardrobe. He cultivated a set of smart side-whiskers and parted his hair in the middle, both considered marks of the effete snob. His suits were of a fashionable English cut, he wore a heavy watch fob, and carried a walking stick. He agonized over whether he should have a frock coat or a cutaway for afternoon wear and bought an elegant beaver hat.

The young man's active social life did not interfere with his studies. He received honor grades in all his courses except Greek. At a time when 50 was the passing grade at Harvard, he averaged 75 and ranked in the upper half of his class. It was a highly creditable performance for a young man without previous exposure to classroom instruction and discipline.

In later years, Roosevelt was critical of the education he received at Harvard. "There was very little in my actual studies which helped me in after-life," he declared. "There was almost no teaching of the need for collective action, and of the fact that in addition to, not as a substitute for individual responsibility, there is a collective responsibility." Instead, his education had exalted the individual who had "no duty to join with others in trying to make things better for the many by curbing the abnormal and excessive development of individualism in a few." The result was "acquiescence in a riot of lawless business individualism which would be as destructive to real civilization as the lawless military individualism of the Dark Ages." These liberal views were, however, hardly reflective of his thinking when he was a student at Cambridge.

Once classes were over for the year, Roosevelt and Hal Minot set out at once on their long-planned expedition to the Adirondacks. For a week they skirted the shores of Lakes St. Regis and Spitfire, and found the birds in fine plumage and excellent voice. Minot had to leave, but Theodore stayed on for another week and was rewarded by hearing for the first time the song of the hermit thrush, which he described as "the purest natural melody to be heard in this or, perhaps, any land." In full Keatsian flight, he wrote that he had been out with a small party hunting deer on a lake in the heart of the wilderness:

. . . The night was dark, for the moon had not yet risen, but there were no clouds, and as we moved over the surface of the water with the perfect silence so strange and almost oppressive to the novice in this sport, I could distinguish dimly the outlines of the gloomy and impenetrable pine forests by which we were surrounded. We had been out for two or three hours but had seen nothing; once we heard a tree fall with a dull, heavy crash, and two or three times the harsh hooting of an owl had been answered by the unholy laughter of a loon from the bosom of the lake, but otherwise nothing had occurred to break the death-like stillness of the night; not even a breath of air stirred among the tops of the tall pine trees. Wearied by our unsuccess we at last turned homeward when suddenly the quiet was broken by the song of the hermit thrush; louder and clearer it sang from the depths of the grim and rugged woods, until the sweet, sad music seemed to fill the very air and to conquer for the moment the gloom of the night; then it died away and ceased as suddenly as it had begun. Perhaps the song would have seemed less sweet in the daytime, but uttered as it was, with such surroundings, sounding so strange and so beautiful amid these grand but desolate wilds, I shall never forget it.

Back in Oyster Bay, Roosevelt plunged into work on what was to be his first publication—*The Summer Birds of the Adirondacks in Franklin County, N.Y.*—which also bore Minot's name. Printed at the expense of the young naturalists, this four-page pamphlet listed ninety-seven different species and was reviewed favorably in a leading ornithological journal. This was followed in 1879 by *Notes on Some of the Birds of Oyster Bay, Long Island,* his second publication. It also focused family attention on Theodore's choice of career, for in his sophomore year at Harvard he would be able to take natural history courses as electives, which would point him toward the scientific life. That summer there was considerable discussion with his father about the choice of a career. In his *Autobiography,* he recalled:

My father had from the earliest days instilled into me the knowledge that I was to work and make my own way in the world, and I had always supposed that this meant that I must enter business. But . . . he told me that if I wished to become a scientific man I could do so. He explained that I must be sure that I really intensely desired to do scientific work, because if I went into it I must make it a serious career; that he had made enough money to enable me to take up such a career and do non-remunerative work of value *if I intended to do the very best work there was in me;* but that I must not dream of taking it up as a dilettante. . . . After this conversation I fully intended to make science my life-work.

In keeping with this decision, Roosevelt chose two natural history courses—elementary botany and comparative anatomy—upon his return to Cambridge in September 1877.* William James was his instructor for the latter course, and he found it "extremely interesting." His two other electives were French and German, while rhetoric, constitutional history, and themes, in which students were required to write several papers during the year, were his required courses. Despite this heavy load, Roosevelt participated in practically every aspect of life at Harvard. He joined the staff of the *Advocate,* the college magazine; he boxed, wrestled, and moved up to the first rung of Harvard's hierarchy of exclusive clubs by being one of the forty members of his class invited to join the Institute of 1770. The social elite of this group were automatically chosen for Delta Kappa Epsilon, or "Dickey," and the men who would be chosen in their junior year by the clubs were culled from this group. Porcellian was at the apex of the pyramid, and it was the target of Roosevelt's aspirations.

Experience had made him more tolerant of his classmates. "My respect for the mental quality of my classmates had much increased lately . . . as they no longer seem to think it necessary to confine their conversation exclusively to athletic subjects," he wrote Corinne the day after his nineteenth birthday. "I was especially struck by this the other night, when after a couple of hours spent in boxing and wrestling with Arthur Hooper and Ralph Ellis, it was proposed to finish the evening by reading aloud from Tennyson, and we became so interested in *In Memoriam* that it was past one o'clock when we separated."

Politics was also about to shoulder its way into his life. Except for his participation in a torchlight parade for Rutherford B. Hayes, the Republican candidate, during the presidential campaign in 1876—won by Hayes in what is now generally agreed to

---

*"Natural history" embraced a variety of subjects in those days, including botany, zoology, comparative anatomy, physiology, and embryology as well as geology, geography, and meteorology.

have been the first stolen presidential election in American history—he had shown no interest in national affairs. But during the young man's sophomore year, his father agreed to take an active role in government. Like most members of his class, Theodore Roosevelt, Sr., regarded politics as "a dirty business" and the haven of grafters and boodlers, but he put aside his misgivings to help further the cause of civil service reform by accepting appointment as collector of customs in New York.

Today, when governmental bureaucracy at all levels is under attack for gross inefficiency and a lack of responsiveness, it is difficult to recall the incandescent passions stirred only little more than a century ago by the fight for civil service and the simple faith of liberals in its all-encompassing benefits. With an evangelical fervor worthy of the medieval crusaders, they believed that politics could be purified of the prevailing corruption and the temptations of office removed by merely ending the spoils system. To men like the elder Roosevelt, the cause of civil service reform took on the zeal of the abolitionist movement in the previous generation.

Well educated, affluent, and often patronizingly virtuous, these reformers were naïve about the true nature of political power in industrial America. In reality, they were trying to preserve the values of an older, simpler, and less urban nation that had already all but vanished. Until the Civil War, politics had for the most part been the domain of men like themselves—men whose personal and business affairs allowed them time to deal with the affairs of the communities. While carefully protecting their own interests and those of their class, they also had a sense of noblesse oblige.

Rapid industrialization and the astounding growth of the urban lower class due to repeated waves of immigration produced a revolt against this essentially upper-middle-class and aristocratic control of government. The vacuum was filled by professional politicians who served as middlemen between the contending forces and built organizations to get out the vote that were fueled by patronage and graft. The reformers saw only the corruption of the system and blamed it on the moral depravity of the lower class and the professional politicians. Their remedy was a return to government by the "best men"—which left no place for the ordinary citizen.

The Republican party had its roots in the reformism of the abolition movement, but following the Civil War it was controlled by professional spoilsmen known as "Stalwarts.". Patronage was the mother's milk of politics as far as they were concerned, and their leader was Senator Roscoe Conkling of New York, one of the most flamboyant figures ever to stride across the American political stage. Six feet three inches tall and with a mane of blond curls, he wore florid vests and white flannel trousers and gave the appearance of strutting even while sitting down. "Lord Roscoe" was a master of political invective and charged that the reformers, with their unrelenting demands for "snivel service," utterly failed to comprehend that "parties are not built up by deportment, or by ladies' magazines or by gush. . . ."

Conkling made a strategic error during the 1876 presidential campaign. Having lost the Republican nomination to Hayes at the Cincinnati convention—where Theodore Roosevelt, Sr., had made a well-received anti-Conkling speech—he had petulantly sat on his hands during the campaign. When the votes were counted, it appeared Hayes had been soundly defeated by Samuel Tilden, the Democratic candidate, but party leaders managed to turn defeat into victory. Returns from several southern states were disputed and in exchange for promises to remove federal troops from South Carolina and Louisiana, where they propped up Reconstruction governments, southern politicos swung the disputed election to Hayes. Immediately after taking office, the new president, labeled "Rutherfraud" and "His Fraudulency" by disgruntled opponents, ordered the troops withdrawn. Without the support of U.S. Army bayonets, carpetbag rule collapsed and the freed slaves were left to the mercy of their former masters.

To the amazement of the reformers, Hayes soon declared himself in favor of civil service reform. Defying the party bosses, he named Carl Schurz, a German-born reformer, as secretary of the interior, forbade federal workers to take part in politics, and moved to take control of the New York Customs House, long a cornucopia of patronage and graft. The president had two reasons for zeroing in on the Customs House: Businessmen steadily complained about the scandalous conditions prevailing there and it was a direct blow at Conkling, for its 2,500 jobs were the source of much of his power.* Conkling threatened revenge, but an un-daunted Hayes demanded the resignation of Chester Alan Arthur, a Conkling lieutenant who filled the lucrative post of Collector, and nominated the elder Roosevelt in his place.

"I will take the office not to administer it for the benefit of a party," Roosevelt told the press, "but for the benefit of the whole people." Despite this bold front, he was doubtful about his chances of being confirmed by the Senate, as required by law, given Conkling's hostility. In fact, he secretly hoped to be rejected because of the abuse certain to be heaped upon anyone trying to "purify" the Customs House. Besides, he was not feeling well and was complaining of persistent abdominal pains.

Conkling accepted the challenge and over the next two months, the nation was treated to the spectacle of a Republican senator fighting an appointment made by a president from his own party. A pawn in this clash of titans, the elder Roosevelt could do little but grimly hold his counsel. From Cambridge, young Theodore wrote Bamie: "Tell Father I am watching the 'Con-trollership' movement with the greatest interest," and asked that his subscription to the anti-Conkling *New York Tribune* be re-newed.

Through senatorial courtesy, which gives a senator a veto over federal appointments in his state, Conkling delayed action on the nomination with the hope of preventing the Senate from voting before it went out of session. If he succeeded, Roosevelt's name would have to be resubmitted in the following session. "It looks as if Father will not get the collectorship," young Theodore noted in early November. "I am glad on his account, but sorry for New York." Hayes refused to withdraw the nomination, however, and on December 12, the Senate rejected Roosevelt by a margin of 31 to 25.

"A great weight was taken off my shoulders when Elliott read the other morning that the Senate had decided not to confirm me," he wrote his son. "No one can imagine the relief." But he felt "sorry for the country . . . as it shows the power of the par-tisan politicians who think of nothing but their own interests. I

---

*These jobholders included Herman Melville, who was drearily finishing out his life as a customs inspector on the Hudson River waterfront at $4 per day.

fear for your future. We cannot stand so corrupt a government for any great length of time."*

Word that the elder Roosevelt was not well had reached Theodore and that very same day he wrote Bamie he was "very uneasy about Father. Does the Doctor think it very serious? I think a travelling trip would be the best thing for him; he always has too much work on hand. Thank fortune my own health is excellent, and so, when I get home, I can with a clear conscience give him a rowing up for not taking better care of himself. . . . The dear old fellow never thinks of himself in anything."

Two days later, on December 18, the senior Roosevelt collapsed. Several doctors were hastily summoned and diagnosed his illness as acute peritonitis. Family members believed he had strained himself in some way and were convinced the injury had been aggravated by the political rebuff he had suffered. In reality, he had cancer of the stomach. For several days, Roosevelt was in severe pain, but as Christmas approached and his son came home from college, his spirits improved. "Xmas. Father seems much better," the young man wrote in his diary. On January 2, 1878, he returned to Harvard with his father's parting assurances that "I have never caused him a moment's pain . . . that . . . I was the dearest of his children to him."

Not long after Theodore's departure, his father's condition worsened and the pain was excruciating. Although he was only forty-six, his hair and beard turned gray almost overnight. Every member of the family took turns watching over him—Mittie, Corinne, Bamie, Elliott, his brother James Alfred, who came in from next door—to read to him, to hold his hand, and administer sedatives to dull the pain. "I have just sat with him some seven hours," Corinne wrote Edith Carow. "He slept most of it, but at times was in fearful agony. Oh Edith, it is a most frightful thing to see the person you love best in the world in terrible pain, and not be able to do a single thing to alleviate it."

With midterm examinations in the offing, the family agreed that Theodore should not be informed of the severity of his fa-

---

*Conkling's triumph was only temporary. The following summer, President Hayes fired Chester Arthur by executive order and persuaded enough Republicans to ignore senatorial courtesy and join the Democrats in confirming the president's nominee for the collectorship. This crushing defeat marked the beginning of the end of Lord Roscoe's power.

ther's illness. One morning the senior Roosevelt was relatively free from pain, and Mittie read him a letter from their son that had just arrived. "You would have felt more than repaid for the exertion of writing such a cheery, long letter to him, if you could have seen the expression on his face," Aunt Annie wrote Theodore. "His whole face lighted up with a beautiful smile when she read out of the two examinations you have passed. . . . I have not seen him look so pleased and like himself for a long time."

Much of the burden of caring for the dying man fell upon Elliott and for days he hardly left his father's room. He ate and slept little and was close to collapse. "Elliott gave unstintingly a devotion which was so tender that it was more like that of a woman and his young strength poured out to help his father's condition," Corinne remembered. Throughout the rest of his life, he was haunted by his father's pitiful "cries for ether" and "the mercy of a chloroform sleep."

On the morning of February 9, the end appeared near. An urgent telegram was dispatched to Theodore telling him to come home immediately and he caught the night boat to New York. One can only imagine the agonies suffered by the young man as he faced the loss of his father, and the trip seemed an eternity. As the news that Theodore Roosevelt, Sr., was dying spread through the city, the poor—men, women, and children—took up a silent vigil in front of the family townhouse. "Newsboys from the West Side Lodging House, little Italian girls from his Sunday School class, sat for hours on the stone steps," Corinne recalled.

"The agony in his face was awful," Elliott wrote of his father's last hours. "Ether and sedatives were of no avail. . . . He would be quiet a minute then with face fearful with pain would clasp me tight in his arms. . . . The power with which he would hug me was terrific and then in a second he would be lying, white, panting, and weak as a baby in my arms with sweat in huge drops rolling down his face and neck. . . . Oh my God my Father what agonies you suffered." When Theodore arrived early the next morning, he found flags flying at half-staff all over the city. His father had died shortly before midnight and all he could do, as the new head of the family, was try to comfort his distraught mother, brother, and sisters.

Overcome with grief, Roosevelt tortured himself for not having done more for his beloved father in his illness, for not even having arrived in time to bid farewell to him. Alone in his room, he drew a slash of black ink down the lefthand side of the page of his diary for February 9, 1878, and wrote: "My dear Father. Born Sept. 23, 1831." For the next two days, the pages are blank. When he took up his pen again on February 12, his anguish flowed out upon the page:

> He has just been buried.* I shall never forget these terrible three days; the hideous suspense of the ride on; the dull inert sorrow, during which I felt as if I had been stunned, or as if part of my life had been taken away, and the two moments of sharp, bitter agony, when I kissed the dear, dead face and realized that he would never again on this earth speak to me or greet me with his loving smile, and then when I heard the sound of the first clod dropping on the coffin holding the one I loved dearest on earth. . . . I feel that if it were not for the certainty, that as he himself has so often said, "he is not dead but gone before," I should almost perish. With the help of my God I will try to lead such a life as he would have wished.

Upon his return to Harvard, young Roosevelt tried to present a cheerful face to his fellows, but for weeks and months he privately mourned his father. "Every now and then it seems to me like a hideous dream," he told Hal Minot. Night after night, he sat alone in his room on Winthrop Street, pouring into his diary his feelings of loss, pain, and despair. There was a dark undercurrent of frenzy about it all, a hint of madness and terror. "Sometimes when I fully realize my loss, I feel as if I should go wild. . . ." "He was everything to me; father, companion, friend. . . ." "If I had very much time to think I believe I should go crazy. . . ." "Oh Father, Father, how bitterly I miss you, mourn you and long for you. . . ." "How little I am worthy I am of such a Father. . . . How I wish I could ever do something to keep up his name."†
This pattern would be repeated throughout Roosevelt's life.

---

*In the family plot in Greenwood Cemetery in Brooklyn

†*Harper's Weekly* (March 2, 1878) lauded the senior Roosevelt as an "American citizen of the best type—cheerful, hearty, sagacious, honest, hopeful; not to be swerved by abuse, by hostility, by derision. . . ."

Any change in his emotional and personal situation resulted in a mixture of anxiety and melancholia. Sunday, with all its memories, was the worst day for him. "Have been thinking over the many, many lovely memories of him," he wrote one Sunday. "Had another square breakdown." Once, he attended services at the small Presbyterian church at Oyster Bay where the family worshipped and was convinced that in the dim light he had seen his father. "I could see him sitting in the corner of the pew as distinctly as if he were alive, in the same dear old attitude. . . . Oh, I feel so sad when I think of the word 'never.' "

What would he do with himself? He was having doubts about becoming a natural scientist, and the inheritance he received from his father—about $125,000—left him comfortably fixed. For help he turned to Bamie. He loved his mother, but he had never sought counsel from her; Elliott had been shattered by his father's illness; his uncles were not close enough to him. Bamie must take Father's place; must stand by him. "My own, sweet sister, you will have to give me a great deal of advice and assistance now that our dear Father is gone, for in many ways you are more like him than any of the rest of the family." But his sister was only three years older than himself and had little experience of the world. Working out his own fate in the end, he resolved to "study well," to finish Harvard, and to carry on "like a brave Christian gentleman."

Theodore's father had always impressed upon his son the virtue of hard work and, as if to cauterize his grief, the young man threw himself into his studies. Invitations poured in from sympathetic friends, but he rejected them all. Grades did not mean much to him—"I most valued them for *his* pride in them"—but he kept up a furious pace, "grinding like a Trojan." When his sophomore year ended, he had his reward. In spite of a heavy course load, he won honors in six of his eight subjects and, exceeding his freshman record, had an overall average of 89. "I am well satisfied," he wrote. "My dearest father would have been pleased." He was also invited to join Alpha Delta Phi, or A.D., one of the most exclusive Harvard clubs.

That summer at Oyster Bay, Roosevelt tried to bury his sorrow in a frenzy of riding, rowing, and swimming. He rowed all the way across Long Island Sound to Rye Beach and back again in a single day, a distance of over twenty-five miles. He took long rides on his horse, Lightfoot, once cantering more than twenty miles over the back roads. He went sailing in fair weather and foul. As the sail luffed in a quickening breeze, he pulled the sheet to his stomach until the canvas bellied out and waited for the tension in his arms and legs to relax. The boat, swinging on its tack, skimmed away, out onto the waters of the Sound.

But wherever he turned there were reminders of his father. In the house, "every nook and corner . . . every piece of furniture . . . is in some manner connected with him." Leafing through his boyhood diary of that winter on the Nile, he found that "every incident is connected with *him*."

Later that summer, Edith Carow visited Corinne and she and Theodore seemed drawn to each other. On August 19, he took her sailing; on the twentieth, they went rowing; on the twenty-first, they drove to Cold Spring and spent "a lively morning" together picking water lilies. Next day, they went sailing again and in the evening had tea with the rest of the family. "Afterwards," Theodore recorded in his diary, "Edith and I went up to the summerhouse." And then there is silence. Edith Carow is not mentioned again in his diary for several weeks.

Over the next few days, the young man appeared to be consumed by a fierce rage. He rode Lightfoot "so long and hard" that he feared he had injured the animal. Angered by a neighbor's dog as it ran after him, barking and nipping at his horse while he was out on a morning ride, he pulled out a pistol and shot the animal. He justified this act by claiming the dog's owner had been warned once before. Out sailing with some cousins, he blazed away with a new rifle at anything that bobbed on the water.

What had happened in the summerhouse?

From comments Roosevelt made years later, it is clear that he and Edith had had a romance that summer and quarreled. "We both of us had . . . tempers that were far from being of the best," he told Bamie. Edith remained silent about the incident except to volunteer upon one occasion that Theodore "had not been nice." Family members say he had asked her to marry him and she had refused. Some theorize that the Carows objected to the match because Edith was only seventeen; others claim that shortly before his death, Theodore's father had discouraged an engagement because Edith's father, Charles Carow, was a confirmed alcoholic who was drinking himself to death.

Time had clearly come for a change of scenery and Roosevelt

decided to spend the rest of his vacation in the woods of northern Maine. Arthur Cutler, his former tutor, had told him about Bill Sewall, who served as a guide for hunters at Island Falls, on Lake Mattawamkeag near the Canadian border, and would be an excellent companion in the wilderness. Island Falls was so thoroughly isolated that Roosevelt and the rest of the party, which included his cousins Emlen and James West Roosevelt and a Dr. Will Thompson, required two days to reach the place, covering the last thirty-six miles in a buckboard.

Sewall was certainly an impressive figure. Thirty-three years old, well over six feet tall, and with a full, reddish-brown beard and flashing eyes, he projected an aura of robust simplicity. He neither smoked nor drank, studied his Bible every morning, read Whittier, Scott, and Longfellow, and recited poetry as he tramped through the forest or paddled his canoe along the lake. Cutler had asked Sewall to keep an eye on Roosevelt to make certain that he didn't overdo things, and upon first sight, the woodsman thought him a "thin, pale youngster with bad eyes and weak heart." Before long, however, he realized that Roosevelt was "different" from every guest that he had had before.

For the next three weeks, they were out in the woods together in all kinds of weather. Despite what Sewall described as Roosevelt's asthmatic "guffle-ing," he noted that the visitor was always in high spirits and eager for action. One day they covered twenty-five miles on foot, which, Sewall observed, "is a good fair walk for any common man." "We hitched up well, somehow or other, from the start," he recalled many years later. "He was fair-minded, Theodore was. And then he took pains to learn everything. . . . The reason he knew so much about everything, I found, was that wherever he went he got right in with the people . . . he was quick to find the real man in very simple men. He didn't look for a brilliant man when he found me; he valued me for what I was worth."

Roosevelt himself was captivated by Bill Sewall and the woodsman became his friend, companion, and counselor. One need not indulge in Freudian speculation to surmise that he became a father figure to the youth, who was still mourning the loss of his own father. Part of the bond between them was a mutual passion for the old Norse sagas, and to Roosevelt's eye, Sewall seemed a

Viking sprung from the pages of *King Olaf*. Sewall told him it
might well be true. Long ago, he explained, a baby had been
found on a sea-wall near a town on the English coast after the
wreck of a Viking ship. "That's where the name comes from,
they say. I'm not so sure but what there's something in it."

Over the next several years, Roosevelt made repeated trips to
Island Falls, making the acquaintance of woodsmen, hunters, and
lumbermen, and formed his first contacts with ordinary folk. Forty
years later, he wrote that Sewall and his family represented "the
kind of Americanism—self-respecting, duty-performing, life-en-
joying—which is the most valuable possession that one generation
can hand on to the next."

Theodore returned to Harvard directly from Maine in a much
more cheerful frame of mind. There were still twinges of melancholy when he thought of his father—"Oh, Father I sometimes
feel as though I would give half my life to see you but a moment!"
he confided to his diary on his twentieth birthday. But with un-
intended humor, this gloomy entry is followed by the notation
that he had shot "2 gray squirrels."

Once again, he undertook an ambitious plan of study that
included zoology, geology, German, Italian, and political economy and required twenty hours of recitations and laboratory
work. He also joined the Rifle Club, the Natural History Club,
and helped organize the Finance Club. And he found himself in
the enviable position of being able to choose between Porcellian
and Alpha Delta Phi, the two most exclusive clubs at Harvard.
Reluctantly, he turned down the "Porc," which he really wanted,
because he had already accepted the membership extended by
A.D. at the end of the previous semester.

Much to his surprise and delight, he eventually got the opportunity to become a member of Porcellian. During a drunken quarrel in the Yard, a "Porc" man taunted an A.D. man by saying
that Teddy Roosevelt, if he had been given the chance, would
have chosen his club first. When the incident became public, A.D.
coolly announced that inasmuch as Mr. Roosevelt had not signed
its constitution, he was free to reconsider his choice. "Of course
by this arrangement, I have to hurt somebody's feelings, and I
have rarely felt as badly as I have during the last 24 hours," he

noted. "It is terribly hard to know what is the honorable thing to do." In the end, he selected the most prestigious club.

Not long afterward, Roosevelt celebrated his initiation as a "Porc" man by getting " 'higher' with wine than I have ever been before." The throbbing hangover that followed caused him to resolve never again to repeat the experience. "Wine makes me awfully flighty," he noted. Nevertheless, as he told Bamie, "I am delighted to be in, and have great fun up there. There is a billiard table, a magnificent library, punch-room &c, and my best friends are in it." They included Dick Saltonstall, who was closest to him now that Hal Minot had left college,* and he was a frequent visitor to Saltonstall's home in Chestnut Hill.

And there he met the girl of his dreams.

---

*Minot's father had insisted that his son study law in his office in Boston.

Chapter 4

"Teddykins" and "Sunshine"

"See that girl? I am going to marry her. She won't have me, but I am going to have *her!*"

So said Theodore Roosevelt to a friend, his teeth flashing a determined grin as he pointed across the room to a fair-haired young lady with gray-blue eyes and a turned-up nose. She was named Alice Hathaway Lee.

Theodore met Alice, whose cheerful brightness had won her the nickname "Sunshine," on October 18, 1878, while weekending with the Saltonstalls in Chestnut Hill. It was love at first sight. The Lees and Saltonstalls were cousins and lived in adjoining mansions on Essex Street, and Roosevelt first saw Alice as she was coming down the garden path between the two houses. The image of her radiant beauty was etched forever upon his memory. "I loved her as soon as I saw her sweet, fair, young face," he wrote later.

The next morning, Theodore, Dick Saltonstall and his sister Rose, and "the very sweet, pretty" Miss Lee walked in the woods together. In the afternoon, they drove to a friend's home in Milton, where "we spent the evening dancing and singing, driving back about 11 o'clock," Roosevelt noted in his diary. On Sunday, he managed to be alone with Alice after church and they went "chestnutting" together. Three weeks later, he returned, ostensibly to visit the Saltonstalls but in reality to see Alice. She was "as sweet and pretty as ever, and I enjoyed myself most heartily."

Until now, Theodore had been undecided as to what he wanted in a woman. "I do so wonder who my wife will be!" he had mused in his diary. " 'A rare and radiant maiden,' I hope; one who will be as sweet, pure and innocent as she is wise." By Thanksgiving, he knew he wanted Alice. "Win her I would, if it were possible," he vowed.

But "pretty Alice" had other ideas. The daughter of George Cabot Lee, the most proper of Boston bankers, she had numerous admirers, and at seventeen was in no hurry to choose one above the others. She was the second oldest of five girls and one son of Lee and his wife, Catharine Russell Hathaway, also a member of a prominent Boston family. Lee was the principal partner in Lee, Higginson and Company, an old-line Boston banking and investment firm, and keeper of the company's vault, which was buried beneath its State Street offices. In the Boston of that day, the symbol of a secure investment was to have it "as safe as in Lee's vaults."

"Enchanting," "flowerlike," and "radiant" were words commonly used to describe Alice. Rather tall and slender for a girl of that era—she was five feet seven inches, which made her nearly as tall as her suitor—she wore her honey-colored hair in fashionable "water-curls" about her temples, carried herself gracefully, and like Theodore's mother, often dressed in white. In every way, she was different from Edith Carow. Alice was fair, flirtatious, and bright without being intellectual; Edith was darker, serious, literary, and, as Theodore well knew, possessed an erratic temper. In both looks and temperament, the young women were natural rivals.

With his usual tenacity, Theodore laid siege to the heart of the bewitching Alice, but she was taken aback by the ardor of his pursuit. Besides, his kinetic personality was altogether different from what she was used to. One girl who knew Theodore at the time recalled that he was "studious, ambitious, eccentric—not the sort to appeal to at first." A faint smell of formaldehyde clung to him, and his thick glasses and the sudden explosion of words when he got excited were disconcerting. Sometimes he would produce natural history specimens from his pockets to prove a point. Although they were soon on friendly terms, Alice kept him at arm's length. One friend later said she "had no intention of marrying him—but she did."

Alice seemed both fascinated and disturbed by this unusual suitor, and their first exchange of correspondence reflects her wariness. Wishing to have a photograph of his beloved, the young man had, at Thanksgiving, invited Alice to join him for a "tintype spree" in Boston on a coming Saturday. Inasmuch as she could not come to the photographer's studio alone, he asked Rose Saltonstall to accompany them. Having anxiously awaited a reply for several days, he reminded her of the invitation. "You *must* not forget our tintype spree," he wrote. "I have been dexterously avoiding forming any engagements for Saturday."

Alice's reply was prim and proper.* "We have by no means forgotten our little spree, but as neither of our Mothers like us to go into town on Saturdays if we can possibly help it, we think it had better be put off until Spring. . . . Rose sends her regards to the *genteel* young man of Cambridge." And she reminded him "that you said you would not show this note" to anyone.

With this slight encouragement, Theodore stepped up the tempo of his calls at Chestnut Hill, and the Lees soon began calling him "Thee," the name he had taken over from his father. The courtship was carried on under the rigid conventions of the day. A girl might accept a book or flowers, but to accept a gift of jewelry was the sign of a fast woman. Proper young ladies did not even think of allowing a man to kiss them before they were engaged. Both Theodore and Alice were undoubtedly virgins. Taking note of the carousing of his classmates in the seamier precincts of Boston, he wrote in his diary, "Thank Heaven, I am

---

*Until recently, it was believed Roosevelt had destroyed most of the letters of his courtship of Alice, but after the death of their daughter Alice Roosevelt Longworth in 1980, a cache of letters was discovered. The present account is the first to make extensive use of them. Michael Teague, a friend of Mrs. Longworth, believes she withheld the letters because she was embarrassed by their sentimentality. They were "the sort of language," she once said, "which aroused every bit of the New England peasant in me and made me want to dry my hands." The letters are now in the Theodore Roosevelt Collection in the Houghton Library at Harvard. See Teague, "Theodore Roosevelt and Alice Hathaway Lee: A New Perspective," *Harvard Library Bulletin,* Summer 1985.

at least perfectly pure." He asked for God's help in staving off temptation "and to do nothing I would have been ashamed to confess" to his father. And when he learned that his black sheep cousin Cornelius Roosevelt had "distinguished himself by marrying a French actress!" he was outraged. "He is a disgrace to the family—the vulgar brute. P.S. She turns out to be a *mere courtezan! a harlot!*"*

That winter, Theodore and Alice went skating and sleighing together and read to each other before the fire. Alice may have had mixed feelings about him, but her adoring nine-year-old brother looked forward to his visits and fund of wild tales about bears and wolves. "Snowed heavily all day long," reads an entry in Roosevelt's diary. "But in spite of the weather I took a long walk with pretty Alice . . . spent most of the remainder of the day teaching the girls the five step and a new dance, the Knickerbocker. In the evening we played whist and read ghost stories."

When spring came, Theodore had his horse, Lightfoot, sent up to Cambridge and rode over to Chestnut Hill almost daily. The young couple played tennis, a game at which the willowy Alice excelled, and went riding and sailing together. Theodore, Alice, and Rose finally undertook their long-postponed "tintype spree" in May, and the resulting pictures show him elegantly turned out with bowler hat, walking stick, and luxuriant sidewhiskers. Alice wears a lace-fronted dress and a feathered hat and looks dreamily into the camera.

Under the watchful eye of Mrs. Saltonstall, he gave a luncheon in his rooms on Winthrop Street attended by Alice and several friends, and afterward took his guests over to the Porcellian. "What a royally good time I am having," he noted. "I can't conceive of a fellow possibly enjoying himself more." But Alice remained temptingly elusive.

*　　　*　　　*

---

*Cornelius Roosevelt was the classic "remittance man." The son of TR's uncle S. Weir Roosevelt, he led a life of wine, women, and song before taking flight to Paris, where he was founder of a branch of the family known as the "Paris penal colony." The family took legal action to prevent him from sharing in the estate of his grandfather, C.V.S. Roosevelt, but supplied him with enough money to keep him from returning to New York City to plague them. He lived on until 1902, having happily survived the frigid blasts of family disapproval, on borrowed money which he never repaid. To amuse his friends, he liked to open his mouth wide enough to hold a billiard ball. Once, one got stuck.

For the time being, Theodore was anxious to mask his obsession with Alice and kept up an active social life that included a number of other Boston girls. When Alice was mentioned in his letters home, he always carefully coupled her name with a reference to the "dear honest"—and very homely—Rose Saltonstall. And sometimes his thoughts turned to his old flame, Edith Carow. Upon one occasion, he warily asked Corinne to give Edie his love "if she's in a good humor; otherwise my respectful regards. If she seems *particularly* good-tempered, tell her that I hope that when I see her at Christmas it will not be on what you might call one of her off days."

By rising early and studying for several hours before and after breakfast, Theodore kept up with his courses and had afternoons and evenings free to pursue Alice. His grades in his junior year were among his best at Harvard—averaging 87—and he was first in the class in zoology and political economy. He found time for ribald sessions at "Dickey" and dinners of partridges and burgundy at the "Porc"—with his intake of wine carefully limited— and just after the turn of the year, he was elected to the Hasty Pudding Club. He presented learned papers to the Nutall Ornithological Club and the Natural History Club, and delivered a joint paper with Bob Bacon on "Municipal Taxation" before the first meeting of the Finance Club. "We little suspected that we were being addressed by a future President of the United States and his Secretary of State," Professor J. Laurence Laughlin observed many years later.*

Theodore also went out for the lightweight boxing championship and, having advanced to the finals, was matched on March 22, 1879, against the defending champion, C. S. Hanks. This fight inspired one of the most vivid legends that grew up around Roosevelt. He was skillfully carrying the fight to his opponent

---

*Laughlin also remembered that in 1913, he called upon his former student, now out of the White House and editor of *The Outlook* magazine, to lobby for his support of a bill establishing the Federal Reserve system. He sent in his card and Roosevelt immediately emerged from his office and boomed out to everyone: "That's the man that taught me my Economics!"

Upon being informed of Laughlin's mission, Roosevelt said he didn't want to make a speech on banking reform. "When it comes to finance or compound differentials, I'm all up in the air." Laughlin observed that in view of the statement "you just made that I taught you Economics, that is hardly a great tribute to my course or my teaching." Laughlin to Henry Pringle.

when the referee signaled the end of a round. Roosevelt dropped his guard, but Hanks landed a strong blow on his nose, which spouted blood. The spectators began hissing, and according to this story, Roosevelt raised his arm to signal for silence. "It's all right," he declared. "He didn't hear the referee." And he walked over to his opponent and shook hands. Although he lost the fight, this gesture of good sportsmanship was supposedly the most remembered event of his college years. Some biographers have embroidered the tale by having Alice Lee and several of her friends huddling in their furs in the freezing balcony of the gym.

This is a good story but it never occurred. Many years later, George F. Spalding, one of the spectators, told the *Harvard Alumni Bulletin* that rather than being bloodied by a blow after the end of a round, Roosevelt slipped and fell to his knee during the fight and his opponent inadvertently struck him lightly on the chest. The other man immediately "ducked his head in apology," but not before some spectators had begun to hiss. " 'Damn it, fellows, he didn't mean it,' " said Roosevelt. Spalding also pointed out that Alice could not have been present in the balcony because not only did the gym not have one, but ladies were not permitted to attend boxing matches at the time.*

With the semester rapidly drawing to a close, Theodore anxiously noted that soon he would be far away while Alice would be fair game for all the eligible young men of Boston. To prevent his love from being stolen from him, he decided the time had come to propose marriage and apparently chose Class Day, June 20, 1879, for this momentous event. The night before, he was so tense that at the "Dickey" strawberry festivities, he "got into a row with a mucker and knocked him down, cutting my knuckles pretty badly against his teeth." Brushing the "mucker" off like lint from a sleeve, he made no further mention of the incident.

Next day, Theodore, now outwardly calm, attended parties in the rooms of several classmates, lunched at the "Porc," went to a tea dance in the new gym, and dined at the Pudding. For two hours that evening, from eight to ten o'clock, he and Alice oc-

---

*Later, Roosevelt erroneously claimed to have won the lightweight boxing championship at Harvard. See *Letters of Theodore Roosevelt,* Vol. I, p. 67.

cupied a windowseat in Hollis Hall overlooking the Yard as the Glee Club performed. Colored lanterns swung gently in the trees and Alice's face was hauntingly beautiful in the shadows. The refrain of one of the songs, "Gentle Maiden, Dance Ever With Me," must have struck a plaintive chord in his heart: "Can'st thou refuse me? Can'st thou but choose me?" But Alice did refuse him—or at least answered evasively—and he resolved to resume his suit in the autumn.

Theodore set two tasks for himself that summer. Upon his return to Oyster Bay, he began training Lightfoot to work in harness so he would be able to pull a sporty little two-wheel carriage, known as a tilbury or "dog cart," he had just purchased. He would cut a bold figure when he went calling again at Chestnut Hill. Conveniently, the narrow seat would hold only two people— and they would have to sit very close together. Throughout the summer, he worked hard trying to emulate his father's skill with whip and reins, but never achieved the senior Roosevelt's elegance and grace of movement.

Elliott was at home that summer and the two brothers spent much time together. The younger Roosevelt was working for his uncle James Gracie at a bank downtown and was taking his place among New York's gay society of paddock and ballroom. Handsome and charming, he easily made friends of both sexes, and one acquaintance, John S. Wise, described him as "the most loveable Roosevelt." Theodore was always challenging him to physical contests and noted in his diary: "As athletes we are about equal; he rows best; I run best; he can beat me sailing or swimming; I can beat him wrestling or boxing; I am best with the rifle, he with the shotgun, &c. &c."

Theodore's other goal that summer was to climb Mount Katahdin, at 5,267 feet the highest peak in Maine. From the window of Bill Sewall's cabin at Island Falls, he had seen the mountain, which had been a challenge to him ever since his first visit the previous year. Arthur Cutler and his cousin Emlen Roosevelt were also vacationing at Island Falls when Theodore arrived late in August, and agreed to join Theodore, Sewall, and the latter's partner, Will Dow, in the climb. Before reaching the base of the mountain, the climbers had to cover some of the most difficult terrain in the Northeast. They slashed their way through virgin wilderness, sliding and slipping over rain-slicked rocks and slogging through swamps.

Although weighted down with a forty-five-pound pack and wearing moccasins because he had lost one of his shoes while fording a stream, Theodore reached the summit along with Sewall and Dow little the worse for wear. Cutler and Emlen had collapsed and given up far below. That night, heavy rains turned the party's tents and bedding into a sodden mess, but Theodore proudly noted in his diary: "I can endure fatigue and hardship pretty nearly as well as these lumbermen."

Still eager for challenges, Roosevelt suggested to Sewall that he join him in an expedition to the remote Munsungan Lake. This ten-day adventure made the Katahdin climb seem like child's play. They set out in a pirogue, or dugout canoe, but a good part of the time, they were up to their hips in icy water as "again and again we had to wade up rapids, dragging the heavy dug-out after us." As he later wrote his mother:

> The current more than once carried us off our feet, and swept us into the pools of black water; now we would lift the boat over ledges of rock, now unload and carry everything around waterfalls, and then, straining every muscle, would by main strength drag her over shoal places. . . . So we plodded wearily on till nightfall, when we encamped, and after dinner [hardtack, salt pork, and tea because game was scarce] crept under our blankets drenched through, but too tired to mind either cold or wet. Next day was a repetition of this; hour after hour we waded on in perfect silence. . . .

For most sportsmen, this ordeal would have been enough outdoor living for a long time, but not for Roosevelt. As soon as he had rested, he and Sewall were off again. Over the next three days, they covered another 110 miles by wagon and on foot. "As usual it rained," he noted, "but I am enjoying myself exceedingly, am in superb health and as tough as a pine knot." By the time Sewall saw him off on the train to Boston, the woodsman would have been pardoned if his attitude resembled that of Richard Welling on that cold day on the ice at Cambridge as he gritted his teeth and wondered how long he must listen to Roosevelt's maddening cries of "Isn't it bully! Isn't it bully!"

* * *

Planning for the final assault on the coquettish Alice, Theodore arranged a light schedule for his senior year at Harvard—only five subjects compared to nine as a junior. He also lost no time in showing off his new dogcart at Chestnut Hill. "The horse, harness, cart and robes all looked beautifully and I was exceedingly proud of the whole turn-out and especially of my pretty companion, who looked too perfectly bewitching for anything," he told his mother. But if he had expected this stylish rig to influence Alice, he was disappointed. She was not tempted into a commitment.

Word of Roosevelt's romantic quest soon leaked out and he became a figure of fun. A "Dickey" musical burlesque written by Owen Wister* singled him out as:

> The cove who drove
> His Tilbury cart . . .
> Awful tart
> And awful smart,
> With waxed mustache and hair in curls:
> Brand-new hat.
> Likewise cravat.
> To call upon the dear little girls!

With Thanksgiving and the first anniversary of his vow to win Alice approaching, Theodore enlisted the Roosevelt women in his campaign to persuade her to be his. Early in November, his mother invited Mr. and Mrs. Lee and Alice and Rose to visit the family in New York. A few weeks later, Bamie and Corinne came to Chestnut Hill, where both the Lees and the Saltonstalls gave dinners in their honor. The major event of the weekend was an elaborate lunch arranged by Theodore at the Porcellian for thirty-four guests, including Alice and her mother. "Everything went off to perfection," he noted, "the dinner was capital, the wine was good, and the fellows all gentlemen."

But at Alice's coming-out party the week after Thanksgiving, Theodore was dismayed to be treated like just another of the

---

*Then a Harvard sophomore and, later, author of *The Virginian* and Roosevelt's close friend young men who circled about her "like moths around a candle," according to one of her cousins. Obviously, his suit was not progressing well, for Alice was hardly mentioned in his diary for weeks. "Oh the changeableness of the female mind!" he burst out in a letter home.

Several pages from this period are ripped out of the diary, and some entries are obliterated by ink blotches. Terrified that someone else might run off with Alice's affections, he ordered a pair of dueling pistols from France. "I did not think I could win her," he confessed later, "and I went nearly crazy at the mere thought of loosing her." Night after night, he wandered despondently through the woods near Cambridge, sometimes not going to bed at all. It was almost a repetition of the frenzy that had followed his father's death. A worried classmate notified the family, and his cousin West Roosevelt was dispatched to talk with him.

Returning to New York for the Christmas holidays, Theodore launched into a dizzying whirlwind of social activities apparently with the hope of putting Alice at least temporarily out of his mind. He attended several dinner parties, and, on Christmas Eve, called on ten different girls. On Christmas Day, assuming one of his father's duties, he oversaw the festivities at the newsboys' shelter. On the day after Christmas, he lunched with Edith Carow and noted, "She is the most cultivated, best read girl I know." But if Edith were entertaining second thoughts about having turned down his proposal, it was already too late.

That evening, to Roosevelt's delight, Alice, her sister Rose, and Rose and Dick Saltonstall arrived in New York City to spend a week with the Roosevelts. For Theodore, life suddenly took on a dazzling brilliance. "It is perfectly lovely having the dear, sweet Chestnut Hillers with us—and so natural," he exulted. There were dinner parties, dances, and evenings at the theater, and as he squired Alice about town, he had "an uproariously jolly time." On New Year's Day, 1880, Elliott gave a luncheon at Jerome Park, and everyone drove out to Long Island over the snow-packed roads. Holding Alice in his arms as they waltzed away the afternoon, he sensed that he had won the girl of his dreams.

On Sunday, January 25, 1880, an exuberant Theodore, back at Harvard, wrote in his diary:

At last everything is settled; but it seems impossible to realize it.
I am so happy that I dare not trust in my own happiness. I drove
over to the Lees determined to make an end of things at last; it
was nearly eight months since I had first proposed to her, and I
had been nearly crazy during the past year; and after much pleading my own sweet, pretty darling consented to be my wife. Oh,
how bewitchingly pretty she looked! . . . Oh, how I shall cherish
my sweet queen! How she, so pure and sweet and beautiful can
think of marrying me I cannot understand, but I praise and thank
God it is so.

Hurrying down to New York, Theodore broke the news to the
family, and remained there only long enough to order an engagement ring. "My sweet little queen, how I long to be with
you," he wrote Alice. "I am so happy, that I hardly dare trust in
my own happiness; last Sunday evening seems almost like a
dream. . . ." He also told Edith of his engagement, which was to
be formally announced on Valentine's Day in Chestnut Hill. The
effect upon her can only be imagined. She had built her life about
him and now he was out of reach, seemingly forever. Edith tried
to conceal her shock and disappointment, but it was clearly evident to Corinne Roosevelt, her best friend.

The Roosevelts warmly welcomed Alice into the family. For
her part, Alice said, in replying to a note from Mittie Roosevelt,
"it seems like a dream" to have "such a noble man's love. But I
do love Theodore deeply and it will be my aim both to endear
myself to those so dear to him and retain his love." Both Corinne
and Bamie sent notes of welcome. "There is hardly another girl
in the world that I would consign him to, but I gladly do so to
you for I know and see that your influence over him is so sweet
and lovely," said the younger sister.

On February 5, the engagement ring arrived from New York
and, as Theodore told Alice, he could hardly wait to get to Chestnut Hill to "put it on your sweet little hand myself." The question
of the wedding date was still to be settled, with the happy couple
wishing to be married as early as the following autumn. Theodore
expected "a battle royal" with Alice's parents, who felt their
daughter was too young to marry and would have welcomed a
long engagement. But this roadblock was removed when Bamie
suggested that during the first winter in New York, the newlyweds

could live with her and Mittie on Fifty-seventh Street. The Lees agreed to this arrangement, and the wedding was set for October 27, 1880—Theodore's twenty-second birthday.

There was also the question of Theodore's future career. Money was not a problem—his inheritance produced an annual income of about $8,000, at a time when a streetcar conductor earned $2.75 for a fifteen-hour day—and Alice had her own money. Even before he had first met her, his enthusiasm for a career in science had been waning and the romance ended it. The prospect of three years of foreign study, a necessary academic requirement, with its consequent separation from Alice, made him "perfectly blue." And she was unlikely to be satisfied with being the wife of a scientist. In the summer between his junior and senior years, his thoughts turned toward studying law, "preparatory to going into public life." He sought the advice of Professor Laughlin, who told him that young men like himself were needed more in politics than in science. William Roscoe Thayer later recalled Theodore told him, "I am going to try to help the cause of better government in New York City; I don't know exactly how. . . ."

Roosevelt's engagement removed the last vestiges of doubt about his future career. "I shall study law . . . and must do my best, and work hard for my little wife," he noted in his diary on March 25, 1880, and made plans to enroll in Columbia Law School following graduation from Harvard. In later years, he insisted that he gave up his boyhood dream of being a scientist because Harvard had dulled his interest by emphasizing the laboratory and the minutiae of biology over field work. "I had no more desire or ability to be a microscopist and section-cutter than to be a mathematician," he declared. But he had told Hal Minot that, as a result of the engagement, "you can perhaps understand a change in my ideas regards science."

The spring of 1880 was one of the happiest in Roosevelt's life and contrasted sharply with his emotional desperation of only a few months before. He was almost a daily visitor to Chestnut Hill and could barely restrain himself. "When we are alone, I can hardly stay a moment without holding [Alice] in my arms or kissing her," he confided to his diary. "She is such a laughing, pretty little witch; and yet with it all she is so true and so tender."

He called her his "darling little sunshine," his "sweet, pretty darling," "my own sunnyfaced queen," "my sweet mistress." She called him Thee, and then Teddy, and sometimes "Teddykins."*

But the prospective bridegroom still had to finish college. Even though he spent most of his final year at Harvard concentrating on winning Alice's hand, Theodore completed his courses with flying colors. He also assumed an even more active role in extracurricular activities. He became president of A.D.Q, vice president of the Natural History Society, president of O.K., secretary of the Pudding, editor of the *Advocate,* and librarian of Porcellian. He graduated *magna cum laude,* on June 30, 1880, was awarded a Phi Beta Kappa key, and ranked twenty-first in a graduating class of 177. "Only one gentleman stands ahead of me," he noted.

One day, while browsing through the Procellian library, he chanced upon a copy of William James's *Naval History of Great Britain,* and his nationalistic sentiments were aroused by the bombastic anti-Yankee tone of its account of the War of 1812. Finding the American counterpart, J. Fenimore Cooper's *History of the Navy of the United States,* no more unbiased, he decided to write an accurate and objective history of his own. Despite all the distractions of his senior year, he had completed a draft of the first two chapters by graduation, but later acknowledged they were "so dry that they would have made a dictionary seem light reading by comparison."

The most curious aspect of his final months at Harvard was his choice of topic for his senior essay: "The Practicability of Equalizing Men and Women Before the Law." One looks in vain for the source of his support for what was considered a radical cause. He had shown no previous interest in it and there was no history of feminism in the Roosevelt family. Was he thinking of the strong-willed, efficient Bamie? Perhaps it was a reflection of his father's reformist views, or the result of discussions with outspoken feminist bluestockings in the drawing rooms of Boston and Cambridge. Maybe Alice, despite her sheltered life, was aware of the growing demand for women's rights and influenced her adoring

---

*In later years, he disliked being called Teddy—possibly because it was a link to Alice—although he tolerated it as a popular political tag. "No one in my family has ever used it, and if it is used by anyone it is a sure sign that he does not know me," he declared.

future husband. No matter what the inspiration, he plunged in, and the very first sentence struck a tone that would be the keynote of his political career: "In advocating any measure we must consider not only its justice but its practicality."

> Viewed purely in the abstract, I think there can be no question that women should have the equal rights with men [he contended]. Even as the world now is, it is not only feasable [sic] but advisable to make women equal to men before the law. . . . A cripple or a consumptive in the eye of the law is equal to the strongest athlete or the deepest thinker; and the same justice should be shown to a woman whether she is, or is not, the equal of man. A son should have no more right to any inheritance than a daughter should have. Property is but little more likely to be mismanaged in the hands of a woman than in those of a man. Especially as regards the laws relating to marraige [sic] there should be the most absolute equality preserved between the two sexes. *I do not think the woman should assume the man's name.*

With the days winding down toward graduation, Theodore examined his life at Harvard and pronounced himself satisfied. "I have certainly lived like a prince for my last two years in college," he wrote. "I have had just as much money as I could spend; belonged to the Porcellian Club; have had some capital hunting trips; my life has been varied; I have kept a good horse and cart; I have had a dozen good and true friends in college, and several very pleasant families outside; a lovely home; I have had but little work, only enough to give me an occupation, and to crown all infinity above everything else put together, I have won the heart of the sweetest of girls for my wife. No man ever had so pleasant a college course."

There was a cloud hanging over Theodore's happy existence, however, and he told no one about it—not even Alice—until almost the end of his life. Shortly before graduation, he had a complete physical checkup by the college physician, Dr. Dudley A. Sargeant, and the doctor had alarming news. Sargeant detected an irregular heartbeat and told the young man to choose a sedentary occupation, even advising him to avoid running up and down stairs if he wished to enjoy a long life. In building up his body, he had increased his endurance and helped control his asthma, but had also strained his heart. Roosevelt quickly made his choice. "Doctor," he said, "I am going to do all the things you tell me not to do. If I've got to live the sort of life you have described, I don't care how short it is."

To introduce Alice to New York society, Mittie Roosevelt arranged a series of teas, parties, and "at homes" that spring, among them a small family dinner attended by her daughters and a few friends, including James Roosevelt, a member of the Hudson River branch of the family. Recently widowed, "Squire James" was considered a possible suitor for Bamie's hand, but that independently minded young lady had quickly rejected the idea of an arranged marriage with a man old enough to be her father. Besides, James was attracted to her friend, the beautiful Sara Delano. "He talked to her the whole time." Mittie laughed after all her guests had gone. "He never took his eyes off her!"

And Sara Delano was well worth a second look. Tall and stately, she was twenty-six, had dark eyes, an abundance of auburn hair, and classic features marred ever so slightly by a strong chin— indicative of stubbornness. A few years later, she might have served as a model for Charles Dana Gibson's coolly elegant idealization of the American girl. Plainly smitten, James Roosevelt invited her and the Roosevelt women to visit Springwood, his estate on the Hudson at Hyde Park. Sara later regarded this visit as one of the most important events in her life. Springwood was in full floral glory, and as they wandered about the grounds, she and James reached an understanding. "Had I not come then," she told her son Franklin, "I should now be 'Old Miss Delano' after a rather sad life!"*

For ten days in July, Alice visited Oyster Bay, and as Theodore showed her his boyhood haunts, he vowed that "she shall always be mistress over all that I have." He resolved to build a large house on the summit of Sagamore Hill, where as a boy he had recited poetry into the wind, that would be suitable for the sizable family that he hoped he and Alice would enjoy. The house would be called Leeholm after his beloved.

Theodore accompanied Alice to Chestnut Hill and then to Bar

*Sara and James were wed at Algonac, the Delano family estate near Newburgh, New York, on October 7, 1880, three weeks before the marriage of Theodore and Alice.

Harbor in Maine. Once again, he exercised strenuously while Alice won the Mount Desert Ladies' Tennis Tournament. The excitement proved too much and, following a dance celebrating Alice's nineteenth birthday on July 29, he came down with an attack of the dreaded "cholera morbus." "Very embarrassing for a lover, isn't it?" he confided to Corinne. "So unromantic, you know; suggestive of too much unripe fruit." In Alice, he found such a willing nurse that he claimed it was a delight to be sick.

In mid-August, Theodore joined Elliott for a rugged, six-week western hunting trip they had been planning for some time. Perhaps he rationalized this extended absence from Alice by looking upon the expedition as a last bachelor fling with his brother, a farewell to boyish irresponsibility. Maybe it was a defiant gesture in the face of Dr. Sargeant's warning about the perilous state of his health. Island Falls was no longer enough of a challenge; he required new and more arduous adventures. On the eve of his departure for the West, he told Alice:

> I hope we have good sport, or at any rate, that I get into good health. I am feeling pretty well now. . . . Sweet blue-eyed queen, I prize your letters so! Do write me often. Get plenty of sleep and as much exercise and lawn tennis as you want; and remember that the more good times you have—dancing, visiting or doing anything else you like—the happier I am. The more attention you have the better pleased I'll be. . . . I do love you so, and I have such complete trust in your love for me; I know you love me so that you will *like* to get married to me—for you will always be your own mistress and mine too.

Using Chicago as a base, the Roosevelt brothers traveled into Illinois, Iowa, and up into the Red River country of Minnesota. Out West for the first time, Theodore was fascinated by everything he saw. The prairie extended to the horizon like an ocean, so vast it seemed infinite. Smoke from a distant settlement looked as if it came from a ship, hull down below the horizon. Flocks of wild geese and ducks filled the skies, rising and falling and changing direction with the wind. "The farm people are pretty rough but I like them very much," he wrote Bamie. "Like all rural Americans they are intensely independent."

Grouse, quail, snipe, doves, ducks, and hawks fell to the guns of the hunters as they tramped through the windswept waves of yellow grass. "[We] are travelling on our muscle and don't give a hang for any man," Theodore reported. But the hunting was not as good as he had hoped, and his asthma returned. Both his guns broke, and he was bitten by a snake and thrown out of a wagon on his head. Also he noticed something about Elliott he had not seen before and discreetly tried to alert the family. The tone was jocular, but there was an edge to his message: "As soon as we got here he took some ale to get the dust out of his throat; then a milkpunch because he was thirsty; a mint julip because it was hot; a brandy smash to 'keep the cold out his stomach'; and then sherry and bitters to give him an appetite. . . ."

Letters from Alice awaited him in Chicago. "Don't you think I am pretty good to write you every day?" said one. "I suppose you laugh and say, these funny letters, they sound just like Alice." But she had received no letters from him for some time and longed "so for some nice quiet little evenings with you alone, it makes me so homesick to think that I shall not see you for so long, for I love to be with you so much."

The last few weeks before the wedding passed in a kaleidoscope of activity. Upon his return east at the end of September, Theodore hastened to Chestnut Hill to be with "my own heart's darling." The separation had increased Alice's attraction. "I cannot take my eyes off her; she is so pure and holy that it seems almost profanation to touch her, no matter how gently and tenderly; and yet when we are alone I cannot bear her to be a minute out of my arms. . . ." After four days of "living in a dream of delight," he returned to New York to enroll at Columbia Law School and to lavish $2,500 on jewelry for his bride. "I have been spending money like water for these last two years," he noted, "but shall economize after I am married."

There were also additional signs of a shift in their emotional roles, for it was Alice who was now expressing longing for him. "How I wish it was three weeks from to-day, our wedding day," she wrote him on October 6. "I just long to be with you all the time and never separate from you, even for three weeks, Teddykins. . . . I should die without you now Teddy and there is not another man I ever could have loved in the whole world." The next day, worried about a report that he did not look well, she urged him to "*please go* to the doctor and see if he can't give you some thing to cure you, every time I think of you I get more and more blue, do take good care of yourself. . . ."

Alice wrote again on October 10. "I just long for it [the wedding] and without the slightest fear, as I shall be the happies[t] girl that you ever saw. . . ." Thanking him for his wedding present of a sapphire ring, she scolded: "But what a very extravagant boy you are, Teddy." And on October 17, he sent his last letter to his "dearest love" before their marriage:

> You are too good to write me so often, when you have so much to do; I hope you are not all tired out with the work. But at any rate you will have two weeks complete rest at Oyster Bay, and then you shall do just as you please in *everything*. Oh my darling, I do hope and pray I can make you happy. I shall try very hard to be as unselfish and sunny tempered as you are, and I shall save you from every care I can. My own true love, you have made my happiness almost too great; and I feel I can do so little for you in return. I worship you so that it seems almost desecration to touch you; and yet when I am with you I can hardly let you a moment out of my arms. My purest queen, no man was worthy of your love; but I shall try very hard to deserve it, at least in part.

The wedding took place on a perfect Indian summer day. Guests crowded the Unitarian Church in Brookline, among them a large New York contingent that included Edith Carow and Fanny Smith. Elliott was best man and Corinne a bridesmaid. "It was the dearest little wedding," Fanny wrote. "Alice looked lovely and Theodore *so* happy and [he] responded in the most determined and Theodorelike tones." Everyone returned to the Lee house to toast the happy couple in champagne. Edith kept her thoughts to herself, but defiantly "danced the soles off her shoes." Late in the afternoon, the newlyweds left for a hotel in Springfield, where the groom had reserved a suite of rooms for the night. The next day they were to go on to Oyster Bay for a two-week honeymoon.

Always the proper Victorian, Theodore drew a discreet curtain over the wedding night. "Our intense happiness is too sacred to be written about," he noted tersely in his diary.

Chapter 5

"I Have Become a Political Hack"

For a moment, Theodore Roosevelt paused in the doorway of the family home on Fifty-seventh Street, filled his lungs with winter air, and set off for Columbia Law School, in the shabby old Schermerhorn mansion on Great Jones Street. The routine never varied. Punctually at 7:30 A.M., he left home no matter what the weather, and strode the fifty-four blocks down Fifth Avenue with a copy of Blackstone's *Commentaries* tucked under his arm. Elbows pumping and heels striking the pavement like a hammer, he arrived in time for class at 8:30.

Alice and Theodore had returned to New York from their honeymoon at Oyster Bay in mid-November 1880. For two weeks, they had the entire place to themselves except for a pair of maids, a manservant, and a big black-and-white collie named Dare. The larder had been well stocked by Bamie and their every need was met by the servants. Life was "a perfect dream of delight," Theodore wrote his mother. "How I wish it could last forever." Like travelers after a stormy voyage, the newlyweds were now snug in a safe harbor. Breakfast was at ten, dinner at two, and tea at seven. Riding, tennis, and long walks in the woods filled the days. In the evenings, they curled up before the fire and he read aloud from *The Pickwick Papers, Quentin Durward,* and the poems of Keats. On November 2, Roosevelt voted for the first time in a presidential election, riding over to the village of East Norwich to cast his ballot for the Republican ticket of James A. Garfield and Chester A. Arthur, whom his father had been named to replace as collector of customs in 1877.

Eleven days later, they were enthusiastically welcomed to the Roosevelt home by his mother and sisters and took up residence in the apartment set aside for them on the third floor. Theodore immediately assumed the role of head of the family and presided over the dinner table. Were the couple, she finishing her teens and he just out of them, happy with this arrangement? Very— according to Theodore's diary. "I can never express how I love her," he wrote. If there were any strains, he mentioned them only obliquely. Alice was a "teasing, laughing, pretty witch" who jogged his arm while he was writing, making him feel "rather bad tempered." But he forgave all. There was much in their relationship to suggest that he regarded Alice as the classic Victorian "child bride" who was never allowed to grow up—like Dora in *David Copperfield.* In fact, his favorite photograph of Alice was taken when she was fourteen.*

Roosevelt took his father's place at the Newsboys' Lodging House, and was elected a trustee of such family charities as the Orthopedic Hospital and the New York Infant Asylum. He joined the National Prison Association and the Free Trade Club. But he felt himself ill-suited for a life of good works. "I tried faithfully to do what Father had done but I did it poorly," he told a friend many years later. "In the end I found out that we each have to work in his own way to our best."

Although Roosevelt plunged into his law studies, he was less than inspired by them. The heroes of the bar were the attorneys for big corporations, and the teaching of law emphasized the protection of property rights. Lectures based upon Blackstone and similar texts—the case method was almost unknown—were the foundation of teaching at Columbia. The lectures of Timothy W. Dwight, the reigning deity at the school, were the keystone

---

*Following Alice's death, he entrusted the picture to his aunt Anna Gracie for safekeeping. She later gave it to Alice's daughter, Alice Roosevelt Longworth, and it was found among her papers.

of the first-year course of study. Having expounded at great length on the subject of the day, Dwight put specific cases to his students, and they were called upon to deduce answers from the material presented during the lecture. It was a situation made to order for Roosevelt's taste for arguing with his teachers.

As at Harvard, some classmates were annoyed. "The pertinacity in which he interrupted the kindly Dwight" still rankled one of them decades later. But Professor John W. Burgess,* of the School of Political Science where Theodore took a course, said he "seemed to grasp everything instantly [and] made notes rapidly and incessantly." Another classmate recalled that far from angering Professor Dwight, he was one of the old man's "favorites." Roosevelt "never seemed to have the air of an attorney or of a professional student," he added, but "still it was evident that he was one of the best men there, considered as a man."

On his long walks back and forth to Great Jones Street, Theodore had plenty of time to ponder the law and its meaning and decided that it was not for him. He compared the moral law and the common law and was uncomfortable with the discrepancy between them. In his *Autobiography,* he observed:

> Some of the teaching of the law books and of the classroom seemed to me to be against justice. The *caveat emptor* side of the law, like the *caveat emptor* side of business, seemed to me repellent; it did not make for social fair dealing. The "let the buyer beware" maxim, when translated into actual practice, whether in law or business, tends to translate itself further into the seller making his profit at the expense of the buyer, instead of by a bargain which shall be to the profit of both. It did not seem to me that the law was framed to discourage as it should sharp practice. . . .

Had he not had an income, Roosevelt acknowledged that he might well have adopted a different attitude. "I had enough to get bread," he said. "What I had to do, if I wanted butter and jam, was to provide the butter and jam." Inasmuch as money-making was not his primary objective, he could afford to choose

---

*Burgess was one of the most prominent Social Darwinists who applied Charles Darwin's theory of evolution to other fields. This new gospel held that just as the higher animals dominate the lower, the Anglo-Saxon had risen above lesser peoples. In his book *Political Science and Comparative Constitutional Law,* Burgess celebrated American nationalism and extolled the American Anglo-Saxon political system as the best the world had produced.

how he would earn his "butter and jam." Consequently, he took a fresh interest in his book *The Naval War of 1812*. When the last class was done for the day, he crossed Lafayette Square to the Astor Library to dig out such works as *Niles' Register,* the *London Naval Chronicle,* and the published memoirs of various naval officers, which provided him with details for the book. The work progressed slowly because even as he carried out his research, he had to teach himself the fundamentals of naval terminology, tactics, and gunnery in the age of fighting sail.

Roosevelt did not allow law school or his historical research to get in the way of an active social life. New York was a far cry from the staid Boston society Alice had known and everyone was dazzled by her beauty and vivacity. "Alice is universally and greatly admired and she seems to grow more beautiful day by day," her husband noted. "Oh, how happy she has made me." Soon, the Roosevelts were part of what Edith Wharton called "the little inner group of people who, during the long New York season, disported themselves together daily and nightly with apparently undiminished zest."

Society was still dominated by the aboriginal New York families, surrounded by friends and cousins, and wrapped in self-satisfied brownstone security. The new rich, the Vanderbilts, the Oelrichs, and the Millses, were just being accepted by the Knickerbocker families, and a merger of money and social position was in process. After all, didn't Vanderbilt sound very much like Van Rensselaer? To Alice, New York seemed a breathless round of receptions, dinners, society balls, and theater parties followed by "jolly little suppers" at Delmonico's. Wearing a white gown and with flowers in her hair, Alice scored a triumph at the glittering Patriarchs Ball and the Family Circle Dancing Class—productions of Ward McAllister, chamberlain to Mrs. William B. Astor—*the* Mrs. Astor. In fact, Alice and Theodore were now related, if somewhat tenuously, to the doyenne of Manhattan society; James Roosevelt Roosevelt, the son of "Squire James" Roosevelt by his first wife, had married Mrs. Astor's daughter Helen.*

Alice fitted in well with the Roosevelts. She joined a sewing circle, held Tuesday afternoon teas, and became a member of the

---

*According to some sources, James disliked the appendage "Junior" and had given his first son his distinctive name to avoid it. Friends called him "Rosy" Roosevelt.

Fifth Avenue Presbyterian Church, which was attended by the family. "Now we are one in *everything*," said her husband. Alice sensed how much the ties between Theodore and his adoring mother and sister meant to them and showed a willingness to share him. She was soon on intimate terms with her mother-in-law. Corinne recalled that she was "gentle and lovely" with Mittie, and they often went out on drives with her. One afternoon, she saw Alice waiting in the drawing room when she was supposed to have gone out an hour or so before. Asked why she was still there, Alice replied she was waiting for "Little Motherling," who was always late "but not generally quite as late."

Alice and Corinne became close companions, shopping and making social calls together. With Bamie, Alice was always reserved, showing the deference to her that was exhibited by everyone. "The first impression of Alice was that she was young and attractive but without great depth," Corinne observed many years later. "This feeling on the part of the family changed swiftly and it [was] . . . universally conceded that her abilities lay below the surface and had she lived she might have risen to any occasion." Theodore's own adoration of her was unabated. When Alice went to visit her parents in Chestnut Hill, he wrote her one Sunday: "I have to read the Bible all to myself, without my pretty Queenie standing beside me in front of the looking glass combing out her hair. There is no pretty, sleeping little rosebud face to kiss and love when I wake in the morning. . . ."

Owen Wister, an old friend from Harvard, presented a graphic picture of life in the Roosevelt household. Stopping by one evening, he found Theodore working over a tactical diagram for his book at a stand-up desk in the library with one leg raised in his familiar stork position. Suddenly, Alice dashed in and exclaimed: "We're dining out in twenty minutes and Teedy's drawing little ships!" There was a bustle of activity as Roosevelt raced upstairs to shave, followed by a cry of exasperation as he cut himself. When the wound wouldn't stop bleeding, he was surrounded at once by all the Roosevelt women, who took "measures to save his collar from getting stained."

To the law and literature, Theodore added politics. He decided the time had come to redeem his promise "to try to do something to help the cause of better government in New York." Perhaps the most extraordinary thing about his multifaceted life was this choice of a career. Well-bred, wealthy young men simply did not take an active role in politics. For most of his contemporaries, "better government" meant philanthropic good works or membership on a reform committee. A gentleman might contribute to a campaign fund—without inquiring too closely how the money was spent. Since Boss Tweed's day, he did not become directly involved in the messy business of politics. The proper course for Roosevelt would have been to become active at the bar or in business, cultivate influential men, and then in the fullness of time he might be handpicked by the railroads and banks as their candidate for the United States Senate.*

Logically, Roosevelt might have made a career as an author, he might have entered publishing or a law office, he might have disciplined himself and become an academic. But undoubtedly inspired by his father's humiliation at the hands of the politicians, he was determined to become part of what he called "the governing class." Yet this is only part of the story. In the final analysis, Roosevelt made a career of politics, and studied and mastered politics because he loved power and liked politics. Americans worship ambition and success but only rarely confess that they are motivated by the will to power. Consequently, he had to mask his ambition and desire for power behind altruistic motives. In the end, he succeeded even in convincing himself.

Roosevelt became a Republican without hesitation. In 1880, "a young man of my upbringing could join only the Republican party, and join it I accordingly did," he later declared. In his boyhood, the Republican party had saved the Union, and his father and his father's friends had all been Republicans. But political parties were closed corporations in those days, and as a friend put it, he "had to break into the organization with a jimmy."

The men I knew best were the men in the clubs and of social pretension and the men of cultivated taste and easy life [Roosevelt

---

*U.S. senators were elected by the state legislatures rather than by direct election until the ratification of the Seventeenth Amendment to the Constitution in 1913, and they were usually dominated by big business or the railroads, who made certain that those chosen were sympathetic to their interests.

recalled]. When I began to make inquiries as to the whereabouts of the local Republican Association and the means of joining it, these men—and the big business men and the lawyers also—laughed at me, and told me that politics were "low"; that the organizations were not controlled by gentlemen; that I would find them run by saloon-keepers, horse-car conductors, and the like, and not by men with any of whom I would come into contact, and moreover, they assured me that the men I met would be rough and brutal and unpleasant to deal with.

Roosevelt persisted, however, and found his way to Morton Hall, clubhouse of the Twenty-first District Republican Association.* Morton Hall was a rather grand name for a dingy, barnlike room over a saloon on East Fifty-ninth Street. Reeking of cigar smoke and splattered with tobacco juice, it was furnished with rough wooden benches, a few sticks of furniture, and the inevitable brass spittoons. Gloomy portraits of former President Grant and Congressman Levi P. Morton, whose name graced the hall, were the only adornments. Most of the members were wary of the young aristocrat with his Harvard drawl, fashionable clothes, and eyeglasses worn on a ribbon, but Jake Hess, the stout German Jew who was the district leader, tolerated Roosevelt with a "distant affability." Hess was not only a Republican ward boss, but also city commissioner of charities and corrections and, well aware of the philanthropic work of Roosevelt's father, appreciated the value of his name to the organization.

Theodore worked hard at trying to bridge the social gap between himself and the other members by going around to the clubhouse often enough that they would get used to his presence and he would get used to them. Sometimes he dropped by in evening clothes before going to a dinner or ball. "I insisted on taking part in all the discussions," he later told Beatrice Webb, the English social reformer. "Some of them sneered at my black coat and tall hat. But I made them understand that I should come dressed as I chose." Following the business of the evening, "I

---

*The Twenty-first District was bounded on the south by Fortieth Street, on the north by Eightieth Street, on the east by Lexington Avenue, and on the west by Seventh Avenue. Columbia College was located within its boundaries. The district was popularly known as the "Diamond Back" district because so many of its residents were supposedly so wealthy that their diet consisted chiefly of diamondback terrapin.

used to play poker and smoke with them." In the beginning, his cousin Emlen accompanied him but dropped by the wayside, remarking: "I did not like the personnel of the organization." In spite of the roughness of his companions and their off-color stories, Roosevelt was fascinated by the men he met. As Bill Sewall had remarked, "Wherever he went he got right in with the people."

Theodore struck up a particularly warm friendship with Joe Murray, one of Hess's chief lieutenants and the man who was to give him his start in politics.* A burly, red-faced Irishman who had been a drummer boy in the Union Army, Murray was originally a Democrat. He began as a street brawler on the Tammany payroll in an era in which both parties stole ballot boxes and roughed up opposition poll watchers. Both men were eye-openers for each other. Prudish to the point of self-righteousness and with the prejudices of his class, Roosevelt regarded politicians as little more than sticky-fingered grafters and was surprised to find that Murray was "by nature a straight man, as fearless and as stanchly loyal as any one whom I have ever met." On the other hand, Murray, like Sewall, could see below the surface peculiarities and liked Roosevelt for his energy and unreserved friendliness.

This, however, did not prevent Murray from squelching Theodore's first political initiative. Early in 1881, he made an effort to put the organization on record in favor of a bill to reform the city's street-cleaning operations, introduced in the State Assembly at Albany by a nonpartisan citizens committee. Roosevelt made a speech in behalf of the bill and spoke with such "ginger" that he was applauded several times. But when the votes were counted, Murray and all his supporters "stood stiffly with the machine" and he was left with only about a half dozen votes out of several hundred cast. The defeat was not unexpected and Roosevelt took it good-humoredly. Besides, as soon as the law school term ended, he and Alice embarked on their long-delayed honeymoon, a four-month tour of Europe. "Hurrah! for a summer abroad with the darling little wife," he proclaimed in his diary.

---

*In later years, a grateful TR made certain that Murray occupied a comfortable place on the public payroll—first as superintendent of the New York Customs House, and then as a member of the New York City Excise Board, deputy superintendent of public buildings in Albany, and finally assistant commissioner of immigration at Ellis Island.

*   *   *

This exuberance was short-lived. Hardly had the steamer *Celtic* cleared New York harbor than Alice, who had never been on an ocean voyage before, became violently seasick. "We had a beautiful passage, very nearly as gay as a funeral," Theodore wrote Rose Lee with more than a touch of sarcasm. Alice was in bed almost the whole voyage, and "I fed her at every blessed meal she ate; and held her head when, about 20 minutes later, the meal came galloping up into the outer world again. I only rebelled once; that was when she requested me to wear a mustard plaster *first* to see if it would hurt. . . . [Periodically] Alice would conclude that she was going to die, and we would have a mental circus for a few minutes. . . ."

Alice improved immediately after they landed in Ireland, their first stop. Theodore found it "a beautiful country," but there was an underlying wretchedness. On the way to Cork, the couple discovered "a man lying on the road insensible from sheer hunger!" Under Roosevelt's direction, some peasants "revived him after a while; and I had him fed and sent to Cork and gave him ten shillings." In England, they spent a few days visiting the Bullochs in Liverpool, followed by eleven days of sightseeing in London.

This was Alice's first visit and Theodore had the pleasure of showing her all the sights. They trotted from museum to monument to the theater, all at a dizzying pace. "Baby . . . is the most intensely interested and indefatigable sightseer and the sweetest, daintiest little travelling companion I have ever known," he noted.* On a visit to the Zoological Gardens, Alice, mystified by the lions' manes and obviously thinking of poodles, astounded her husband by asking who shaved them. It became a family joke and he often told it over his wife's protestations.

A visit to the National Gallery caused Roosevelt to adopt the role of art critic, although as he noted, Alice "has a far keener appreciation of most of the pictures than I have. . . . I like Doré— though he's very apt to paint by the square mile." There were "some very fine Rembrandts which ('half darkness and half light')

---

*In a note probably added to this letter after Alice's death, he wrote: "She was so young and innocent that I used often to call her my 'baby wife'; but she added to her pretty innocence the sweetness and strength of a true woman."

have always been favorites of mine. Rubens I care less for; I do not believe that you can get a 'grand Greek Aphrodite' by merely exhibiting a scantily attired Dutch housewife, however well painted." And some of the Turners "were simply evidences of an unsound mind; one in particular, a 'fire at sea,' ought to have been styled, 'chaos, personified and with the colic.' "

Four days were spent in Paris sampling the restaurants—"How delicious the food is"—and the shops. Upon visiting the tomb of Napoleon, Theodore called it the most "impressive sepulchre on earth . . . it certainly gave me a solemn feeling to look at the plain, red stone bier which contained what had been once the mightiest conqueror the world ever saw." In Venice, they stayed in an old palazzo where their room looked out on the Church of San Giorgio Maggiore, and they shared breakfast each morning on the balcony with the pigeons. In the evening, they were out on the canals, "and it seems like fairyland to float lightly and noiselessly along in the gondola, the many colored lights gleaming on the water, and the sounds of guitars and of singing," he wrote his mother. "How beautiful it is! And what a lovely summer we are having!"

Troubling news from home interrupted this idyl. President Garfield was wounded by an assassin on July 2, 1881, while entering a Washington railroad station. As the stricken Chief Executive crumpled to the ground, the assassin cried out: "I am a Stalwart and Arthur is President!" At first, it appeared that he would recover, but his condition worsened and he was not expected to survive. "Frightful calamity for America," Theodore noted in his diary. ". . . This means work in the future for those who wish their country well." "My God! Chet Arthur!" declared reformers in dismay when Conkling's old lieutenant took the oath of office after Garfield died on September 19.

In Switzerland, Alice took over the task of writing to Mittie. "Teddy enjoys Switzerland more than any other place," she wrote. "He did not feel well while he was in Italy but now he is all right again. He takes a great deal of exercise, which has done him good." This exercise was in preparation for the ascent of the nearly 15,000-foot-high Matterhorn. The mountain had first been conquered only sixteen years before by a team that had lost four of its members. Although the route to the top had been eased by the installation of chains and ropes at critical points, it was still a difficult climb, particularly for an amateur who had been warned by his doctor not even to run upstairs.

With only two guides assisting him, Roosevelt began his climb of the Matterhorn on August 3, 1881. "The mountain is so steep that snow will not remain on the crumbling jagged rocks," he reported. Foot by foot, he struggled up the escarpment and, at the end of the first day, rested at a hut about halfway up. Early the next morning, Roosevelt reached the summit, "after seeing a most glorious sunrise that crowned the countless snow peaks and the billowy white clouds with a crimson iridescence." By late afternoon, he was having tea with Alice in Zermatt, and claiming that he was "not nearly so tired" after all this exertion.

Throughout his trip to Europe, Roosevelt carried the manuscript of his naval history about with him—calling it his "favorite *chateau-en-espagne*"—and worked on it sporadically. "I have plenty of information now, but I can't seem to get it into words," he lamented to Bamie. A chance meeting with Simeon E. Baldwin, a Connecticut jurist and member of the International Law Association, brought about an offer to help him obtain original records from the Navy Department. And a farewell visit to his uncle James D. Bulloch in Liverpool proved a godsend. Uncle Jimmie, an experienced naval officer, sharpened his understanding of the realities of war at sea and cleared up some of the technical difficulties that had stymied him.* Now, the book proceeded at a rapid pace, and by the time of his return home on October 2, 1881, it was nearing its end. Summarizing the trip for Bill Sewall, Roosevelt wrote: "I have enjoyed it greatly, yet the more I see the better satisfied I am that I am an American. . . ."

Refreshed by his trip to Europe, Theodore established his future priorities. "Am working fairly hard at my law, hard at politics, hardest of all at my book," he noted in his diary. G. P. Putnam's Sons had agreed to publish it and the completed manuscript was

---

*TR also persuaded his uncle to write an account of his adventures as head of the Confederate Secret Service in Europe and helped him arrange his papers for the task. The book, *The Confederate Secret Service in Europe,* was published in two volumes in 1883 and is still the key work on the subject. For a brief account of Bulloch's operations, see Miller, *Spying for America: The Hidden History of U.S. Intelligence,* Chapter 7.

due in their hands by Christmas. Simeon Baldwin had been true to his word and the library was now crowded with crates of log-books, shipbuilding contracts, and official correspondence. To Alice, who was visiting her parents, he wrote that he missed her but was glad she was away "for I am so busy that I could not be any company at all for you. I must get this Naval History through and off my mind; it worries me more than I can tell now, and I wish it were finished."

For a young man of twenty-three, *The Naval War of 1812* was a remarkable achievement. It was the first operational history of the conflict based upon extensive research in original sources, and has been the most lasting of all of Roosevelt's thirty-eight books. A century after its appearance, it is still a standard work on the subject. While there is never any doubt as to which side the author favored, he labored to create a book whose authority would be accepted on both sides of the Atlantic.

Roosevelt's monumental research shows in the intense detail he put into the book. Examining the number of guns, weight of broadside, and size of crews, he shows which side had the advantage going into an engagement. The sound of gunfire and the smoke of battle filters through almost every page. In describing the battered condition of the *Niagara,* Oliver Hazard Perry's flag-ship at the Battle of Lake Erie, Roosevelt wrote:

> Every brace and bowline was shot away, and the brig almost completely dismantled; her hull was shattered to pieces, many shots going through it, and the guns on the engaged side were by degrees all dismounted. Perry kept up the fight with splendid courage. As the crew fell one by one, the commodore called down through the skylight for one of the surgeon's assistants; and this call was repeated and obeyed till none were left; then he asked: "Can any of the wounded pull a rope?" and three or four of them crawled up on deck to lend a feeble hand in placing the last guns. . . .

Roosevelt had more on his mind than to recapture the glories of a bygone age of fighting sail, however. While he was completing his book, events were unfolding that were to give it an added dimension. In the years immediately following the Civil War, industry and agriculture had undergone tremendous growth and American entrepreneurs were looking for new frontiers to conquer. Foreign trade was increasing and the merchant marine, devastated by the *Alabama* and her consorts, was beginning to revive. Commercial expansion inspired fears that rivalries over access to foreign markets might precipitate a conflict with one or more European powers. There was talk of rebuilding the navy, which consisted of little more than a handful of decrepit sailing vessels and rusting monitors left over from the war. Admiral David D. Porter compared them to "ancient Chinese forts on which dragons have been painted to frighten away the enemy."

The Naval War was an argument for a strong navy. In 1812, naval unpreparedness had invited catastrophe, Roosevelt pointed out. Had the United States possessed a well-manned and strong fleet, the British would not have provoked her into war. Such a navy would have cost less, in both human and material terms, than the actual cost of the war. He blamed the "criminal folly" of Thomas Jefferson and James Madison for allowing the ships built by their Federalist predecessors to rot in the navy yards. Rather than prepare for war, they turned the other cheek to British insults and, when finally forced to fight, the nation had to depend upon an untrained militia. To Roosevelt, Jefferson was "perhaps the most incapable Executive that ever filled the presidential chair . . . it would be difficult to imagine a man less fit to guide the state with honor and safety through the stormy times that marked the opening of the present century."

Yet even as Theodore worked on the book, his priorities were being dramatically reordered. The Twenty-first District Republican organization was engaged in choosing a nominee for state assemblyman and a bitter fight divided the membership. Because the Republicans dominated the district, winning the nomination was tantamount to election. The incumbent, William Trimble, was Jake Hess's man and a henchman of Senator Conkling, so it appeared that his renomination would be almost automatic. These qualifications alone were grounds enough for Roosevelt to oppose him, but Trimble had also voted against the street-cleaning bill. Plunging into a fight against him, Roosevelt vowed "to kill our last year's legislator."

Unknown to Theodore, Joe Murray had also turned against Trimble. Taking his own private polls, he had found that the incumbent would be a weak candidate in the silk-stocking Twenty-first District because of his open ties to the machine and the general outcry against the Stalwarts following the assassination of President Garfield. But when he warned Hess that Trimble would lose the seat to the Democrats, he was contemptuously brushed aside. "He'll be nominated anyway," the leader declared. "You don't amount to anything."

Following the old political adage, Murray decided to get even rather than get angry. Without Hess's knowledge, he quietly rounded up enough votes to nominate his own man. Impressed with Roosevelt's zest for the rough-and-tumble of politics, well-regarded name, and potential as a vote-getter—he would bring out the Columbia College and Fifth Avenue crowd to augment the machine's own loyalists—Murray let him know that he, too, was opposed to Billy Trimble. Would Mr. Roosevelt care to run?

"No, I would not dream of such a thing," the young man replied. "It would look as though I had had selfish motives in coming around to oppose this man."

"Well," said Murray, "get me a desirable candidate."

"Oh, you won't have any trouble," Roosevelt answered and promised to help find a suitable person.

When the two men met again the following evening, Roosevelt said he had not found a candidate.. With the nominating convention only three days off, Murray impatiently pressed him to run.

"In case we can't get a suitable candidate, will you take the nomination?" he asked. "Yes, but I don't want it," Roosevelt told him. "In the meantime I want you to promise me that if you can find a suitable man, have no hesitancy about nominating him."

When they met again the next night, Murray reached out to shake Roosevelt's hand. He was surprised that instead of taking it, the young man eagerly seized both his hands. "Mr. Murray, I have done you a great injustice," he declared. Roosevelt explained he thought Murray was toying with him and had sought the advice of a friend, former Assemblyman Edward Mitchell. " 'Joe is not in the habit of making statements that he cannot make good,' " Mitchell had told him. " 'There is one thing I can tell you—you have fallen into very good hands.' "

The blue haze of cigar smoke already hung thickly over Morton Hall when Jake Hess gaveled the Twenty-first District caucus to order on the evening of October 28, 1881. Looking over the delegates, Joe Murray was certain he had fifteen votes for Roosevelt to ten for his opponent. A lawyer named Hays led off with a forty-five-minute speech, placing Trimble's name in nomination. When it was over, Murray shuffled to his feet and simply declared, "Mr. Chairman, I nominate Theodore Roosevelt." The final tally surprised even Murray—16 to 9 for Roosevelt. Not long afterward, the nominee outlined his platform in his diary. ". . . Strong Republican on State matters, but Independent on municipal matters." Thus, at the very beginning of his political career, Roosevelt established the tension between party loyalty and independence that was to dominate his life in politics.

With the election set for November 9, only eleven days away, Theodore immediately launched his campaign. Accompanied by Murray and Hess, who now cheerfully backed him, the candidate set out to call upon the Sixth Avenue saloonkeepers who had considerable political influence in the neighborhood. Much to the alarm of his backers, Roosevelt showed a disturbing tendency to speak his mind. Let Joe Murray relate what happened:

> We started in a German lager-beer saloon. . . . The saloonkeeper's name was Carl Fischer. . . . We had a small beer and Hess introduced [Roosevelt] to Fischer and Fischer says, "By the way Mr. Roosevelt, I hope you will do something for us [saloonkeepers] when you get to Albany. We are taxed much out of proportion to grocers, etc., and we have to pay $200 for the privilege [of operating]."
> "How much do you say you pay?"
> "Two hundred dollars," replied Fischer.
> "Why that's not enough!"

Murray and Hess quickly hustled the candidate out onto the sidewalk and assessed the damage. A few more encounters of this sort could well cost them the election. "We came to the conclusion that we had better stop the canvass right then and there," said Murray. "I says, 'Mr. Roosevelt you go see your personal friends. . . . Hess and I will look after this end. You look after the other end.' "

Roosevelt's backing on "the other end"—among the so-called

"better element"—was solid. Every member of the family, except for his uncle James Alfred, overcame a reluctance to have anything to do with politics and supported his candidacy. If Alice was asked her opinion on whether her husband should seek office, there is no record of it. Business and social leaders eagerly supported him out of the novelty of seeing one of their own kind running for office, respect for his father's memory, and their own distaste for the Democratic candidate, Dr. W. W. Strew, who had recently been fired as superintendent of the lunatic asylum on Blackwell's Island.

Joseph Choate, one of the most prominent attorneys of the day, informed Roosevelt that "some of your father's friends" had volunteered to help underwrite his campaign expenses.* Professor Dwight and several members of the Columbia faculty and Elihu Root, another leading light of the bar, were among the signers of a manifesto supporting his candidacy. And *The New York Times* weighed in with an editorial that declared: "Every good citizen has cause for rejoicing that the Republicans of the Twenty-First Assembly District have united upon so admirable a candidate for the Assembly as Mr. Theodore Roosevelt."

Reflecting his assurance of victory, the candidate told the press he would go to Albany "untrammelled and unpledged" and that he "would obey no boss and serve no clique." Writing to Alice, still at Chestnut Hill, he seemed remarkably off-handed about the campaign. "The canvass is getting on superbly; there seems to be a good chance of my election," he wrote on November 5, "but I don't care anyway." This letter, in fact, was a strange mixture of yearning for "my prettiest and sweetest little wife" and anxiety over financial difficulties he had gotten himself into through a lavish life-style:

> I so longed for you when I received your darling letter that I could hardly contain my desire to see you. Oh, my sweetest true-love pray for nothing but that I may be worthy of you; you are the light and sunshine of my life, and I can never cease thanking the Good God who gave you to me. I could not live without you, my sweet-mouthed, fair-haired darling. . . . You are all in all, my heart's

---

*Following the election he wrote Choate that "I feel that I owe both my nomination and election more to you than any other man." *Letters,* Vol. I, p. 55.

darling, and I care for nothing else; and you have given me more than I can ever repay. . . . My book is all entirely finished except the remodelling of the first chapter. So everything is getting on well except financially. I confess I am in by no means a good condition from a monetary point of view, and in awfully bad odor with Uncle Jim.*

This letter may have been a subtle hint to Alice to ask her father for financial assistance. Apart from the large marriage settlement she had received, Mr. Lee also provided his daughter with a generous allowance that was a substantial contribution to the couple's financial well-being. It is unknown whether such assistance was sought or given, but not long after, when Theodore invested $20,000 in G. P. Putnam's publishing company, his first check bounced; he had only half that amount in his account.

Roosevelt voted early on the morning of November 9, and then worked in the library, absorbed in his book. Out on the street, Hess and Murray mobilized their voters and protected the polls against any Democratic attempts to finagle the election. A group of brawny Columbia football players volunteered their services as poll watchers, and Murray dispatched them to the roughest areas of the district. When the ballots were counted, Roosevelt scored a personal triumph. As expected, he had beaten his opponent, but the margin of victory, 3,502 votes to 1,528, was nearly twice the usual majority won by Republican candidates in the district. At the age of twenty-three, he now stood at the brink of a political career that was to last the rest of his life.

Luck and family position had much to do with the young man's successful first venture into politics, but he owed his victory to more than sheer good fortune. Roosevelt had paved the way with careful groundwork, and when an opportunity presented itself, he had seized it. "I have become a political hack," Roosevelt informed one of his Harvard classmates.†

---

*James Alfred Roosevelt, the family's financial adviser

†Election Day had hardly passed before Roosevelt was subjected to the demands of his district. The woman suffrage party informed him that it hoped "to find you as able and forceful a promoter as Mr. Trimble of the rights and interest of the women of your constituency." TR Scrapbook.

And James E. Seaman, writing in "behalf of the Colored People of the Twenty-First Assembly District" advised him that he "and two faithful lieutenants [had] worked all day" in favor of his candidacy." TR Scrapbook.

Chapter 6

The Gentleman from New York

Theodore Roosevelt burst upon Albany with the same effect he had at Harvard and within a few months became one of the most prominent figures in New York politics. Ordinarily, a freshman legislator ranked somewhere between junior committee clerk and a janitor, but Roosevelt immediately attracted the attention of the New York City newspaper correspondents. On the evening of January 2, 1882, they had taken up their usual place near the cigar stand in the lobby of the Delevan House, the capital's unofficial political headquarters, to observe the comings and goings of the legislators, when they noticed him. Although it was bitter cold outside, he was not wearing an overcoat.

Roosevelt's "step across the hotel corridor was quick and vigorous, his whole manner alert," noted George Spinney of *The New York Times*. "He was a good hearted man to shake hands with and he had a good, honest laugh. You could hear him for miles, and it was not an affected laugh." He had "little Dundreary* whiskers" and his glasses dangled on a silk cord that ran

---

*A reference to Lord Dundreary, the archetypically dim English aristocrat in the play *Our American Cousin*

over his ear. "His teeth seemed to be all over his face. He was genial, emphatic, earnest but green as grass." There was a general agreement among the reporters that he was "an uncommon fellow, distinctly different." But the unresolved question was whether or not he would be what Tammany sachem George Washington Plunkett called "a morning glory"—a reformer who "looked lovely in the mornin' and withered up in a short time."

At twenty-three the youngest member of the Assembly, Roosevelt was greeted with a mixture of disdain, curiosity, and amusement. Isaac Hunt, who represented Jefferson County in upstate New York, painted a vivid picture of his first appearance at a Republican caucus in the ornate state capitol that same evening:

> All of a sudden, the door opened and in rushed Mr. Roosevelt. He made his way up and sat right down in front of the chairman of the conference. He had on an enormous overcoat and a silk hat in his hand. As soon as opportunity was given, he addressed the chairman and pulled off his overcoat and he was in full dress. He had been to a dinner. He had on his eyeglasses and his gold fob. His hair was parted in the middle and he addressed the chairman in the vernacular of the FFVs of New York. We almost shouted with laughter to think that the most veritable representative of the New York dude had come to the Chamber. After a little our attention was drawn upon what he had to say, because there was force in his remarks. . . .

Hunt became Roosevelt's closest friend and ally in the Assembly, but the majority, both Democrats and some of his fellow Republicans, regarded him as an arrogant, blue-blooded dilettante. Some called him "Oscar Wilde," after the foppish Englishman then making a tour of America. The Democratic newspapers called him "his Lordship" and "the exquisite Mr. Roosevelt," with the *New York World* noting that his trousers were cut so tightly that when he was making his "gyrations" before an audience "he only bent the joints above the belt." "He is just a damn fool," growled Tom Alvord, a Republican from Onondaga County who had been Speaker the year Roosevelt was born. There are "sixty and one-half [Republican] members in the Assembly," he declared. "Sixty plus that damned dude."

In the privacy of a diary that he kept intermittently over the five months of the session, Roosevelt furiously gave as well as he got. "A number of republicans, including most of their leaders, are bad enough but over half the democrats, including almost all the City Irish, are vicious, stupid looking scoundrels with apparently not a redeeming trait," he wrote. "There are some twenty five Irish Democrats in the house, all either immigrants or the sons of immigrants, and coming almost entirely from the great cities—New York, Brooklyn, Albany, Buffaloe. They are a stupid, sodden vicious lot, most of them being equally deficient in brains and virtue. . . . The average catholic Irishman of the first generation as represented in this Assembly, is a low, venal corrupt and unintelligent brute. . . ."

Some of Roosevelt's attitude can be traced to the prejudices of his class, his anger at being singled out for harassment like a new "boy at a strange school," and sheer frustration at the vagaries of the political process. Having no need to defend a newly acquired fortune or excuse a suspicious past, he brought to Albany the stern moral code of his father, a code that rejected amorality in business and public activities as well as in personal life. Only slowly did he become aware of the important social services rendered by the political machines and politicians to their constituents in return for votes.

In his calmer moments, Roosevelt observed the denizens of the strange new world into which he had entered with the same clinical thoroughness with which he had taken field notes on the birds of the Nile Valley. Analyzing the background of his fellow legislators "to find out whether they were their own masters or were working under the direction of someone else," he calculated that at least a third were completely corrupt and sold their votes to the highest bidder. There was also a brisk trade in so-called "strike" bills, which the more venal members introduced in order to blackmail corporations into buying them off. Cash and other favors were openly exchanged in the lobbies.

The first order of business when the Assembly was gaveled to order on January 3, 1882, was the election of a Speaker. With the Democrats having a clear majority of 67 of the 128 members, this should have been a perfunctory matter. But the eight Tammany Democrats from New York City were at odds with the Democratic majority, who followed Samuel Tilden. Under the instructions of the Tammany leader, "Honest John" Kelly, they backed their own candidate, John J. Costello, while the Tilden Democrats supported Charles E. Patterson of Troy. The Republicans chose Tom Alvord as their standard-bearer. Roosevelt opposed Alvord, whom he called "a bad old fellow," and supported General George H. Sharpe, another former Speaker.* But once Alvord was named by the caucus, he was bound to support him. Although the Tammany holdouts knew they could not elect their own man, they hoped to cut a deal with one side or another in exchange for a variety of political plums. "A long deadlock is promised us," Roosevelt predicted.

For the next five weeks, the members did nothing but trudge up and down Capitol Hill from the Delevan House and back again to cast futile votes for Speaker. The Assembly could not be called into session until the deadlock was broken; no committees could be appointed, no bills submitted. The entire process was "both stupid and monotonous," Roosevelt noted, seething with anger. His temper was not helped by separation from Alice. The young man's impatience was clearly evident to a newsman who watched him go through a stack of morning newspapers at breakfast. "He threw each paper as he finished it, on the floor unfolded," recalled William Hudson of the *Brooklyn Eagle,* "until at the end there was, on either side of him, a pile of loose papers as high as the table."

Roosevelt made his maiden speech on January 24. Upon learning that several Republicans were discussing the possibility of joining with the Democratic majority to break the Tammany deadlock, he leaped to his feet to object on grounds that it was good politics to let the Democrats tear themselves apart. "As things are today in New York, there are two branches of Jeffersonian Democrats; the Tilden Democrats and the Kellyites," he declared. "Neither of these alone can carry the State against the Republicans. . . . We have no interest in helping one section against the other; combined they have the majority and let them make all they can out of it!

---

*During the Civil War, Sharpe had headed the Bureau of Military Information under General Grant, the most effective intelligence organization of the war. Later, he was sent abroad as a diplomatic agent to gather information aimed at linking Judah P. Benjamin, the former Confederate secretary of state and then living in England, to the assassination of Abraham Lincoln. For a full account of his activities, see Miller, *Spying for America,* Chapter 8.

"While in New York I talked with several gentlemen who have large commercial interests at stake," Roosevelt continued, "and they do not seem to care whether the deadlock is broken or not. In fact, they seem rather relieved! And if we do no business till February 15th, I think the voters of the State will worry along through without it."

Having said his piece, he sat down amid the warm applause of the Tammany members, who, much to his amusement, thought he was on their side. Like most inexperienced speakers, he later noted, he had been nervous, but profited from the advice of an old country lawmaker: " 'Don't speak until you are sure you have something to say, know just what it is; then say it, and sit down.' " Isaac Hunt thought his friend "spoke as if he had an impediment in his speech, sort of as if he were tongue-tied. He would open his mouth and run out his tongue. . . ."* Roosevelt himself was pleased with his performance, a feeling reinforced by the *New York Evening Post,* which said he had made "a very favorable impression." On the other hand, the Democratic *New York Sun* ridiculed him as "a blonde young man with eyeglasses, English side-whiskers and Dundreary drawl" and emphasized his "quaint" pronunciation of "r-a-w-t-h-e-r r-e-l-i-e-v-e-d."

The deadlock was finally broken on February 14, 1882, and Charles Patterson was elected Speaker. When the committee assignments were made, Roosevelt found that, for a freshman legislator, he had been given a plum, a highly sought-after seat on the Committee on Cities. "Just where I wished to be," he exulted. Even so, he made little effort to conceal his disdain of his fellow committee members. "The Chairman is an Irishman named Murphy, a Colonel in the Civil War, a Fenian; he is a tall stout man, with a swollen red face, a black mustache, and a ludicrously dignified manner; always wears a frock coat (very shiney) and has had long experience in politics—so that, to undoubted pluck and a certain knowledge of parliamentary forms, he adds a great deal of stupidity and a decided looseness of ideas as regards the 8th commandment. . . .† Altogether the Committee is about as bad

---

*"I always think of a man biting ten-penny nails when I think of Roosevelt making a speech," George Spinney, of *The New York Times,* recalled.
†Thou shalt not steal.

as it could possibly be; most of the members are corrupt, and the others are really singularly incompetent."

Alice's arrival in Albany brightened his mood and with their "books and everything" they settled in at the Kenmore House, a residential hotel just across the square from the capitol. Theodore invited several of his colleagues to meet her. "All the men were perfectly enchanted," he noted with pride. "They praised my sweet little wife." Hunt described Alice as "a very charming woman, tall, willowy-looking. I was very much taken with her." The only shadow over the couple's happiness was Alice's failure to become pregnant.

The Assembly had not even completely organized itself when Roosevelt submitted several bills aimed at increasing the efficiency of the New York City government. Only one—designed to reform the method of electing aldermen—was passed, but the proposals created the impression in the public mind that he was a dedicated reformer fighting the "black horse cavalry," the more venal of the machine politicians. Always in need of colorful copy, the legislative correspondents gravitated toward the outspoken young man from the Twenty-first District and his name began to appear prominently in their dispatches from Albany. It was the beginning of a mutual love affair between Roosevelt and the press that lasted for the next forty years.

To the majority of the members, however, Roosevelt was an irritating gadfly. He never doubted the moral certainty of his positions. Every issue was a clash between the forces of light and darkness. "There is increasing suspicion that Mr. Roosevelt keeps a pulpit concealed on his person," observed one newspaper. He always seemed to be trying to get the floor to protest some outrage. Standing at his place, he stretched far forward over his desk, with his face reddening, calling out, over and over again, "Mister Spee-kar! Mister Spee-kar!" Sometimes he would shout for forty minutes, his voice shifting from tenor to screeching falsetto, until Patterson recognized him.

Upon one occasion, "Big John" McManus, a beefy onetime prizefighter and Tammany stalwart, proposed that Roosevelt be tossed in a blanket. Getting word of the prank, the intended victim marched up to McManus, who towered over him, and declared: "By God! McManus, I hear you are going to toss me in a blanket.

By God! If you try anything like that, I'll kick you, I'll bite you, I'll kick you in the balls.* I'll do anything to you—you'd better leave me alone." McManus dropped the plan.

Roosevelt's reputation as a reformer had grown to the point where supporters of legislation targeted for shakedowns by the "black horse cavalry" requested him to sponsor the measures. One such bill was a proposal by the Manhattan Elevated Railway, which was planning additional terminal facilities in New York City. Because of the large amount of money involved, company lobbyists feared the bill might become hostage to blackmail and asked him to shepherd it through the Cities Committee. Roosevelt examined the proposal carefully and concluded that even though the company was controlled by Jay Gould, the notorious financial buccaneer, the terminal was "an absolute necessity" not only for the railway but the city. He consented to sponsor it on condition that the company agree to do "nothing improper" to secure passage.

The bill had no sooner come before the Cities Committee, of which he was then acting chairman, when Roosevelt observed that some of the most corrupt members had already combined to block action upon it. Using one pretext or another, they refused to report the bill out, either favorably or unfavorably, for a vote by the entire Assembly. Exasperated, he decided to force the bill out of committee, but inasmuch as there were "one or two members . . . who were pretty rough characters," he knew ordinary parliamentary methods might be inadequate:

> There was a broken chair in the room, and I got a leg of it loose and put it down beside me where it was not visible, but where I might get at it in a hurry if necessary. I moved that the bill be reported favorably. This was voted down without debate by the "combine," some of whom kept a wooden stolidity of look, while others leered at me with sneering insolence. I then moved that it be reported unfavorably, and again the motion was voted down by the same majority and in the same fashion. I then put the bill in my pocket and announced that I would report it anyhow. This almost precipitated a riot, especially when I explained, in answer

---

*George Spinney's account. Someone erased "balls" from the transcript and substituted five dashes but the original word can be discerned.

to statements that my conduct would be exposed on the floor of
the Legislature, that . . . I should give the Legislature the reasons
why I suspected that the men holding up all report of the bill were
holding it up for purposes of blackmail. The riot did not come off;
partly, I think, because the opportune production of the chair-leg
had a sedative effect.

The chair leg was of no avail in dealing with the entire legislature, however, and the bill was stonewalled on the floor of the
Assembly. "All the hungry legislators were clamoring for their
share of the pie," reported one newspaper. Finally, a railway
lobbyist told Roosevelt that the bill would be taken out of his
hands and turned over to some "older and experienced" member—meaning that payoffs would be made. As soon as the necessary arrangements were made, Roosevelt angrily noted, the bill
sailed through the legislature without delay.

In the meantime, *The Naval War of 1812* was published and
hailed by critics on both sides of the Atlantic. *Harper's* said:
"Professional readers will be inclined to accept Mr. Roosevelt's
monograph as the most accurate, as it certainly is the most cool
and impartial, and in some respects the most intrepid, account
that has yet appeared of the naval actions of the war of 1812."
The New York Times described the book as "an excellent one in
every respect, and shows in so young an author the best promises
for a good historian—fearlessness of statement, caution, endeavor
to be impartial, and a brisk and interesting way of telling events."
The Navy Department quickly recognized the book's value. It
decreed that at least one copy was to be placed on board every
ship in commission and it became a textbook at the fledgling Naval
War College. The book also became a weapon in the debate over
rebuilding the navy; less than a year after its publication, in March
1883, Congress appropriated $1.3 million for the construction of
its first state-of-the-art warships—the protected cruisers *Atlanta,
Boston,* and *Chicago.* Roosevelt's book went through several edi-
tions and received the ultimate compliment by being accepted in
Britain as the authoritative work on the subject. William L.
Clowes, a British naval historian who was preparing a multivolume
official history of the Royal Navy, was so impressed that he asked

Theodore to write the volume on the War of 1812—a singular honor for an American.*

In searching about for a career, Roosevelt had deplored the idle lives of some of his Harvard classmates, "fellows of excellent family and faultless breeding, with a fine old country place, four-in-hands, tandems, a yacht, and so on; but oh the decorous hopelessness of their lives." Just such a life seemed to be in store for his brother Elliott. Restless and unable to concentrate very long on any project, Elliott drank heavily. Although supposedly employed by one of the family firms, he spent most of his time at his club, riding to hounds, playing polo, and having affairs. "He drank like a fish and ran after the ladies," Edith Carow observed many years later.

Elliott was a close friend of Sara Delano Roosevelt and she and her husband, James Roosevelt, invited him to stand as godfather for their son, who had been born on January 30, 1882, and would be named for Sara's favorite uncle, Franklin Hughes Delano. Elliott protested that he was unworthy to serve as the child's godfather and agreed only after my "dear mother . . . persuaded me that I should accept the high honor you offer me." And so, on March 20, 1882, Sara wrote in her diary: "At 11 we took darling Baby to the chapel in his prettiest clothes and best behavior. Dr. Cady christened him 'Franklin Delano.' Baby was quite good and lovely so we are proud of him."

Elliott was a frequent visitor to the Delano and Roosevelt homes on the Hudson River, and at a house party at Algonac, the Delano family estate near Newburgh, he met a beautiful debutante named Anna Rebecca Hall. Tall and slender, Anna was a stunning blonde of nineteen with large, haunting blue eyes. In London a few years later, Robert Browning is said to have become so infatuated with her that he begged to be permitted to sit and gaze at her while she was having her portrait painted. The eldest of the four handsome Hall sisters, she was linked to two of New York's most prominent families, the Livingstons and the Ludlows.

There was a streak of instability in the family, however, which

---

*In this version, TR toned down his criticism of the British and tried so hard to be neutral that when he discussed impressment, he sometimes seemed to side with them.

in Anna's father, Valentine G. Hall, Jr., manifested itself in the form of religious fanaticism. Anna's childhood at the family estate at Tivoli, on the Hudson near Rhinebeck, was out of a gothic novel. Old Mr. Hall was a despot who ruled the lives of his wife and children with an iron hand. They were not even permitted to shop for their own clothes; instead, dresses were brought to the house and the women were allowed to choose from among them. Gaiety was considered sinful and piety was rigidly enforced. Hall also insisted that his daughters walk with a stick across their backs held in place by the crooks of their elbows, which gave them a distinctive, regal bearing.

To Anna, the handsome and fun-loving Elliott must have appeared like a dashing cavalier, laughing and tossing a sword in the air. Although Anna's family background caused some misgivings, the Roosevelts encouraged the courtship with the hope that marriage and a family would help provide Elliott with some stability. The young couple were married in December 1883, in a ceremony described by the *New York Herald* as "one of the most brilliant social events of the season." Theodore Roosevelt was his brother's best man.

Late in March 1882, Isaac Hunt told Roosevelt that during an investigation of about a dozen insolvent insurance companies, he had uncovered irregularities indicating that the receivers were milking the companies of hundreds of thousands of dollars in trumped-up fees and expenses. Further inquiry revealed that every one of these corrupt receivers had been appointed by Justice Theodore R. Westbrook of the State Supreme Court. In fact, reported Hunt, Westbrook's son and a cousin were among the attorneys involved and had already taken a rake-off of upwards of $15,000 from one company. Inasmuch as he was already deeply involved in the insurance investigation, Hunt suggested that his colleague launch a campaign to impeach Westbrook.

But the Manhattan Elevated affair had made Roosevelt wary of taking up the crusader's lance. The impeachment of a supreme court justice was "a very serious matter," he told Hunt. Besides, he was worried about Alice, who had just returned to New York City for gynecological surgery, which her doctor said would help her conceive a child. Nevertheless, Judge Westbrook's name rang

a bell. Three months before, *The New York Times** had published
a story linking Westbrook to a plot to depress the shares of the
Manhattan Elevated company to facilitate its purchase by Jay
Gould at a rock-bottom price, a $15 million fraud that defrauded
the ordinary stockholders. Westbrook was so closely tied to
Gould, the *Times* continued, that he had even held court in
Gould's office. Curiously enough, Roosevelt also recalled, West-
brook had never made any effort to refute the charges.

Suspicions aroused, Roosevelt visited the newspaper's offices
on Park Row and asked to see the evidence on which the exposé
was based. Henry Lowenthal, the city editor, readily made the
documents available, and the young legislator pored over the record until the early morning hours. Most damning in Roosevelt's
eyes was an unpublished letter from Westbrook to Gould in which
he stated, "I am willing to go to the very verge of judicial discretion to protect your vast interests." Having armed himself with
his research, the young man returned to Albany primed to hunt
big game. On March 29, he rose in the Assembly and offered a
resolution empowering the Judiciary Committee to launch an
investigation of Westbrook's conduct in the Manhattan Elevated
Railway case.

Roosevelt's words reverberated about the chamber "like a
bursting bombshell," Hunt later recalled. But the machine, although taken unawares, quickly rallied to squelch Roosevelt's call
for an inquiry. Assemblyman John Brodsky, a Republican lawyer
from the East Side, leaped to his feet to declare that he wanted
to debate the resolution. Under the rules, this meant it would be
temporarily laid aside—giving the opposition time to marshal its
forces—and considerable pressure was also brought to bear upon
Roosevelt.

Most of Roosevelt's colleagues regarded him as a man holding
a stick of dynamite with the fuse burning. Gould was the most
powerful man in the United States. He owned 10 percent of the
nation's railroads, controlled the Western Union Telegraph Company and the *New York World,* which could make or break political reputations. Over lunch, a member of a prominent law firm
cautioned Roosevelt that reform might be all very well, but if he

---

*Then owned by George Jones, Edith Wharton's father

really wanted to advance in life it was time he left politics and identified himself with "the right kind of people," those who "controlled others." Appalled, Theodore asked whether this meant he should give in to the "ring"? Newspaper talk of a political ring was naïve, the attorney told him. In reality, the "inner circle" was one of businessmen, judges, lawyers, and the leaders of both political parties. In later years, Roosevelt said this was the first glimpse he had of this "combination . . . which I was in after years so often to oppose."

Angrily ignoring the advice, he pushed ahead. He waited a few days to make his move, possibly because he was awaiting word of Alice's successful recovery from the surgery. On the afternoon of April 5, he took the floor only twenty-five minutes before the Assembly was to adjourn for the day—and most of the "black horse cavalry" had departed for their favorite watering holes— and moved that his resolution be taken up. Old Tom Alvord vehemently shouted, "No! No!" but the Assembly voted 48 to 22 in favor of the proposal. As Roosevelt rose to speak, reported George Spinney of the *Times,* "the House for almost the only time during the session grew silent, and prepared to listen to every word uttered. . . ."

"Mr. Speaker, I have introduced these resolutions fully aware that it was an exceedingly important and serious task," Roosevelt began, but he contended there was sufficient evidence to warrant an investigation. Following a detailed account of the Manhattan Elevated case, he stated that "the men who were mainly concerned in this fraud are known throughout New York as men whose financial dishonesty is a matter of common notoriety." To make certain that everyone knew to whom he was referring, he boldly named Gould and his associates and called them "sharks" and "swindlers." Turning to Westbrook, he said the justice had placed "the whole road in the hands of the swindlers. . . .

"We have a right to demand that our judiciary should be kept beyond reproach," Roosevelt continued, "and we have a right to demand that if we find men against whom there is not only suspicion, but almost a certainty that they have been in collusion with men whose interests were in conflict with the interests of the public, they shall, at least, be required to bring positive facts with which to prove there has not been collusion; and they ought themselves to have been the first to demand such an investigation."

For ten minutes, Roosevelt held the floor and the only sound was his voice, broken by the resounding blow of his right fist periodically striking the palm of his left hand for emphasis. There was no applause when he finished, only a muffled hum, for no one in recent years had been courageous enough—or so foolish— as to attack the fearsome Gould directly. Had a vote been taken immediately after Roosevelt's speech, Spinney thought the stunned Assembly would have overwhelmingly approved the resolution calling for an investigation of Westbrook's conduct.

But Alvord quickly grabbed the floor, and with his flinty eye on the clock, which showed only fifteen minutes left of the session, talked away the time until adjournment. Taking the pained attitude of a father admonishing a wayward son, he gravely urged "the young man from New York" to consider "the terrible wrongs" that he claimed he had seen committed during his many years in Albany by inquiries such as the one Roosevelt requested. On and on he chattered, brushing off Roosevelt's request that he give way for a motion to prolong the session. Finally, at two o'clock, the gavel fell and the old man sank back into his chair, having at least temporarily outmaneuvered the youthful reformer. "That damn dude," snorted Alvord. "The damn fool, he would tread on his own balls just as quick as he would on his neighbor's."*

That night, in the bars and behind closed doors in the Albany hotels, the opposition plotted to bury Roosevelt's resolution. Tammany leader John Kelly dispatched a messenger from New York on the night train to tell the faithful to stand fast, and Gould mobilized his forces. On the other hand, with the exception of the *World,* nearly all the comments from the press were laudatory. "There is a splendid career open to a young man of position, character and independence like young Mr. Roosevelt who can denounce the legalized trickery of Gould and his allies . . . without being restrained by the cowardly caution of the politician," declared the *Times.*

---

*Hunt's statement. In the original typescript, the word "balls" was erased and "chickens" substituted.

"By Godfrey, I'll get them on the record yet!" an ebullient Roosevelt declared as the members trooped back into the chamber the following morning. Realizing that few of them would vote against the resolution if their votes were on record, he wanted such a vote. But first, he had to win a two-thirds margin on a motion to lay aside the current business. In his excitement, he forgot to ask for the ayes and nays and the chair called for a standing vote instead. The members bobbed up and down so rapidly that nobody could keep count. The final tally by a deputy clerk, 54 to 50 against the motion, was patently faked.

Later that afternoon as the session droned on, Roosevelt was suddenly on his feet again. Stridently calling, "Mister Spee-kar! Mister Spee-kar!" he demanded another vote—and remembered this time to call for the ayes and nays. The tally was 55 to 49 in favor of the motion but still short of a two-thirds majority. Nevertheless, Roosevelt had clearly won a victory, for the vote showed that a majority was in favor of an investigation. Writing to Alice that evening, Theodore mixed the exultation of political battle with concern over her condition:

> I have drawn blood by my speech against the Elevated Railway judges, and have come in for any amount both of praise and abuse from the newspapers. It is rather the hit of the season so far, and I think I have made a success of it. Letters and telegrams of congratulation come pouring in on me from all quarters. But the fight is severe still, and today I got a repulse in endeavoring to call up the debate from the table. How it will turn out in the end no one can now tell. I hope you are getting well by this time, my poor pretty patient darling. I wish I could be with you while you have your nervous fits, to cheer you up and soothe you: I know just how my little pink wife looks, and I just long to hold her in my arms. . . .

The final vote was delayed until the end of the Easter recess, and a relentless drumfire of newspaper editorials mobilized civic leaders and public opinion behind Roosevelt. When the issue was finally decided on April 12, the Assembly voted by the overwhelming margin of 104 to 6 to order an inquiry by the Judiciary Committee into Westbrook's actions. Roosevelt and his associates realized, however, that they had won only a battle, not the war, because the committee was stacked in favor of Westbrook. Roosevelt even had to insist that the investigators hire outside counsel to conduct the proceedings.

Later on, Isaac Hunt was asked if Roosevelt had remained cool during this crisis. "No, he was just like a jack coming out of the box," he replied. "There wasn't anything cool about him. He yelled and pounded his desk and when they attacked him, he would fire back with all the venom imaginable. He was the most indiscreet guy I ever met." While the committee was conducting its inquiry, Gould's lobbyists handed out large amounts of cash to influence legislators, and Hunt reported that the entire Democratic majority lined up behind Westbrook. A fellow lawyer from Hunt's hometown reminded him that the jurist had once decided a case in his favor. "You don't want to go to work and destroy a good judge like Judge Westbrook," he was told. Roosevelt, Hunt estimated, could have gotten $1 million to halt his attacks on Westbrook.

Unable to rid themselves of Roosevelt by conventional means, Gould's henchmen tried to blackmail him. One night while he was out walking in New York City, a woman apparently tripped and fell to the pavement near him. Helping her to her feet, he called a cab, but became suspicious when she insisted that he escort her home. Theodore made a note of the address, paid the cab driver to take her there, and then sent a detective to investigate. The man reported having found a number of men waiting to pounce on Roosevelt.

On May 31, the Judiciary Committee brought in a majority report that whitewashed Westbrook while a minority recommended that he be impeached. Gould's men had done their work well. Up until the very night before, according to Hunt, a majority of the committee had been in favor of impeachment. Three pivotal members changed their votes at the last moment after pocketing payoffs of $2,500 each, he said. Roosevelt squirmed in his seat as the clerk read the majority report, which conceded Westbrook had occasionally been "indiscreet and unwise" but was guilty of merely using "excessive zeal" in trying to save Manhattan Elevated from destruction. No sooner had the clerk finished, than Roosevelt was on his feet to move adoption of the minority report.

"I cannot believe that the Judge had any but corrupt motives in acting as he did in this case," he heatedly told the Assembly.

"He was in corrupt collusion with Jay Gould. . . . There cannot
be the slightest question that Judge Westbrook ought to be impeached." Speaking clearly and slowly—if in a slightly trembling
voice—he warned that no one could clear Westbrook. "He stands
condemned by his own acts in the eyes of all honest people. All
you can do is to shame yourselves and give him a brief extension
of his dishonored career. You cannot cleanse the leper. Beware
lest you taint yourselves with his leprosy."

The speech was "powerful, wonderful," said Hunt. "But you
could see you could not change one of them. They were adamant.
They had been lined up and they knew exactly how many votes
they had." Following a lengthy debate, the Assembly rejected
Roosevelt's motion and voted, 77 to 35, in favor of the majority
report that cleared Westbrook. Looking back nearly four decades
later, Theodore summed up the result: "Big business of the kind
that is allied with politics thoroughly appreciated the usefulness
of such a Judge, and every effort was strained to protect him. We
fought hard—by 'we' I mean some thirty or forty legislators, both
Republicans and Democrats—but the 'black horse cavalry' and
the timid good men, and the dull conservative men, were all
against us. . . ." Two days later, the legislature adjourned and he
joined Alice at Oyster Bay.

By Rooseveltian standards, the summer of 1882 was less hectic
than usual. There was no hunting expedition to Maine or the West
or a tour of Europe. Instead, Roosevelt remained close to home.
He joined the New York National Guard and was commissioned
a second lieutenant and caught up on his law studies in the office
of his cousin John Ellis Roosevelt. Over the Fourth of July, he
and Alice visited her family in Chestnut Hill, where in one day
he played ninety games of tennis and took the part of a bear while
playing with the neighborhood children. "My acting became too
realistic, and the smaller ones began to have a horrible suspicion
that perhaps I really *was* a bear, and a stampede . . . ensued."

In September, Roosevelt and Alice, Mittie and Corinne, who
had married a wealthy young Scotsman named Douglas Robinson,
rented a house in the Catskills near Richfield Springs. Alice was
so enchanted with the countryside that Roosevelt considered pur-
chasing a farm in the area. The only drawback was the distance

from New York, but as he told Bamie, if he continued in politics and pursued his writing, that would be no handicap. However, in spite of the success of his first tempestuous year in Albany and the good reviews received by *The Naval War,* he was not at all certain that he would continue to follow those paths. If he were to turn to a career in law or business, it would be better to remain in the city. In the end, the young couple moved into a brownstone at 55 West Forty-fifth Street, next to the Robinsons.

Although Roosevelt professed to be uncertain about his future in politics, his first year in the Assembly had been a triumph. True, Westbrook had escaped justice,* but Roosevelt could console himself with the fact that the affair had made him the best-known political figure in the state though only two years out of Harvard and still short of his twenty-fourth birthday. Republican newspapers and magazines were unanimous in praise of his courage, and reform groups honored him with testimonials. Carl Schurz, a leading reformer, declared that he had, almost alone in the Assembly of 1882, "stemmed the tide of corruption in that fearful legislative gathering." And he had the psychic compensation of knowing that he had succeeded in the one area where his father had failed—politics.

The Westbrook affair also provided Roosevelt with an invaluable short course in practical politics. In this rough-and-tumble school, he had learned to intrigue and maneuver, to hold his temper in debate, to balance conflicting ambitions and intricate relationships, and to turn the insatiable need of journalists for colorful copy to his own advantage. Isaac Hunt observed that the onetime dude—derided only a few months before as a half-member—was now "considered a full-fledged man worthy of any man's esteem."

Roosevelt's record was good enough to guarantee his reelection in November 1882 by a better than two-to-one margin. This showing was even better than it appeared because it occurred in the face of a Democratic sweep of the state. Grover Cleveland, the Democratic gubernatorial candidate, even carried the usually

---

*Not long afterward, Westbrook was visiting Troy, in upstate New York, on legal business when he was found dead in his hotel room. There were rumors that he had committed suicide, but they were never confirmed.

rock-ribbed Republican Twenty-first District. "All Hail, fellow survivor of the late Democratic Deluge!" Roosevelt wrote his friend Billy O'Neil after the Democratic sweep. "As far as I can judge the next House will contain a rare set of scoundrels, and we Republicans will be in such a hopeless minority that I do not see very clearly what we can accomplish, even in checking bad legislation. But at least we will do our best."

Chapter 7

"There Is a Curse on This House!"

"I rose like a rocket," Theodore Roosevelt later said of his second term in Albany. This meteoric rise began on New Year's Day 1883, when he was chosen by the Republican caucus as its candidate for Speaker. The Stalwarts, or party regulars, went all out to block the "Young Reformer" but as Roosevelt noted, he had the support of the "country Republicans, all of them native Americans, and for the most part farmers or storekeepers or small lawyers." The nomination was an empty honor, however, because the Democrats had an overwhelming majority in the Assembly. Nevertheless, he emerged from the struggle as minority leader and chief of a reformist faction known as the "Roosevelt Republicans."

Roosevelt owed his ascendancy to the vacuum created by the fall of Roscoe Conkling, the Roosevelt family's longtime political foe, and to the presence of Grover Cleveland in the governor's chair. The accompanying disarray of the Republican leadership enabled Roosevelt to work together with the Democratic governor in the cause of reform without fear of retaliation.

Conkling's eclipse had begun soon after James Garfield became president in 1880 and named James G. Blaine, a Conkling rival who headed his own faction of the Republican party, the "Half Breeds," as his secretary of state and closest political adviser. Lord Roscoe angrily claimed that Garfield had shortchanged him on some promised political appointments, which had gone to Blaine instead, and broke with the president. Once again, the job of collector of the New York Customs House became the centerpiece of a struggle between Conkling and a Republican president. Under prodding from Blaine, Garfield dismissed Conkling's man and history seemed to repeat itself as Lord Roscoe again urged the Senate to turn down the presidential nominee. But the assassination of Garfield by a self-identified spoilsman and Stalwart ended Conkling's reign as a political boss.

The Republicans were badly split among Stalwarts, Half Breeds, and reformers in the wake of this struggle, and Grover Cleveland won the governorship the following year. Cleveland's rise to prominence was even more rapid than that of Theodore Roosevelt. Corpulent and amiable, he was a prosperous Buffalo lawyer who had served as sheriff and in a few other minor offices. When the Republicans chose a flagrant boodler as their candidate for mayor of Buffalo, the Democrats offered the nomination to Cleveland. As mayor, he dealt harshly with grafters, winning a reputation for rocklike integrity at the very moment the reformers were seeking a gubernatorial candidate. Cleveland united all the Democratic factions behind him and, helped by a general revolt among the electorate against the Republican machine, was swept into the governorship.

The alliance between Roosevelt and Cleveland began early in the session when the governor asked him and several other Republican reformers, including Isaac Hunt and Billy O'Neil, to his office to seek support for a civil service reform bill tied up in the Judiciary Committee by a coalition of Democratic and Republican spoilsmen. Cleveland told Roosevelt and the others that if they could get the measure to the floor, he would mobilize the Cleveland Democrats behind it. Roosevelt eagerly accepted the challenge.

One day, when Hunt was temporarily in charge of the committee and some of the opponents were absent, he managed to spring the bill loose. Once it reached the floor, Roosevelt made the main speech in support of the legislation. "My object in pushing this measure," he declared, was "to take out of politics the vast band of hired mercenaries whose very existence depends on their success, and who can almost always in the end overcome the efforts of them whose only care is to secure pure and honest government." Cries of "Nonsense!" and "No! No!" greeted his remarks, but the combined votes of the reformers on both sides of the aisle were enough to win passage of the bill.

This episode showed that Roosevelt had learned enough about practical politics to understand that to win his objectives, he must sometimes find common cause with the Democrats. It also made it clear that he had developed the insight and ability to clarify and define policy that made others look to him for leadership. Moreover, an impression was created in the public mind of an alliance between Roosevelt and Cleveland—a theme reinforced by a Thomas Nast cartoon in *Harper's Weekly* showing the slim young legislator and the stolid-looking governor working over a sheaf of reform measures. In later years, Hunt noted that passage of civil service reform helped win the presidency for Grover Cleveland in 1884, but added: "Mr. Roosevelt was as much responsible for that law as any human being."

In Albany, Roosevelt was also brought face to face for the first time with the poverty and social injustice that existed on the dark underside of America. Like most of his class, Roosevelt's social and economic views were conventionally conservative and shaped by the laissez-faire doctrines he had learned at Harvard. In his early years in politics, he favored low wages, low taxes, and a low level of services. When a measure was introduced to reduce the working time of streetcar conductors from fifteen to twelve hours a day, he opposed it on grounds that the measure was socialistic and interfered with the workings of the free market. And in opposing a proposal by the Hatters' Union to abolish competitive production in the prisons, he showed more solicitude for the tax-payers than for free labor.

Roosevelt lacked understanding of the effect of the Industrial Revolution upon the ordinary workingman. Sheltered from the economic realities and consequences of industrial capitalism, he had no knowledge of the growing impersonality of relations between employee and employer, the intense competition for subsistence-level jobs, and the grinding boredom of most unskilled work. In 1882, the prevailing wretchedness was worsening as the nation slipped into an economic depression that lasted for several years. In New York City alone, one third to one half of the work force was jobless and wages were reduced to as little as $2 a day.

To Roosevelt and his class, the degradation of the workingman was a result of natural law—of character—rather than economic or social injustice. If there was to be any improvement in the condition of the poor, it was not to come through ameliorative action by the state, but as a result of the individual's own efforts or through private charities such as those his father supported. Yet when the workers turned to unions and strikes to help themselves, these efforts roused a chorus of protest from the upper classes. "Professional agitators," Roosevelt would write in 1885, were "always promising to procure by legislation the advantages which can only come to workingmen . . . by their individual or united energy, intelligence, and forethought."

Years later, Roosevelt regretted his youthful hostility to unions and most labor legislation. "One partial reason—not an excuse or a justification, but a partial reason—for my slowness in grasping the importance of action in these matters," he said in his *Autobiography,* "was the corrupt and unattractive nature of so many of the men who championed popular reforms, their insincerity, and the folly of so many of the actions which they advocated." To Roosevelt's credit, he was, unlike many of his contemporaries, capable of growth and understanding.

Samuel Gompers, the English-born Jew who headed the Cigar Makers' Union and later founded the American Federation of Labor, provided young Roosevelt with his first lesson in the need for social change. For years, the union had denounced the manufacture of cigars in the tenements of the Lower East Side as detrimental both to the health of the workers and to those who smoked their products. Cigar manufacturers, on the other hand, preferred that the work be done at home to save themselves the cost of renting factory space.

In 1882, Gompers had succeeded in having a bill introduced into the Assembly banning tenement manufacture of cigars, but opponents shunted the measure to a special committee with the expectation that it would be killed there. In fact, Roosevelt was appointed a member primarily because of his antilabor record. "The businessmen who spoke to me about [the bill] shook their heads and said it was designed to prevent a man doing as he wished and as he had a right to do with what was his own property," he noted.

In the beginning, Gompers remembered, Roosevelt refused to believe in the existence of the conditions described to him. But when the other two members of the committee took no interest in the measure, the union leader persuaded Roosevelt to accompany him on an inspection tour so he could see the situation for himself. Roosevelt reacted with astonishment and revulsion. He had known in the abstract about such conditions as a result of his father's philanthropic work, but was appalled by what he saw.

"I have always remembered one room in which two families were living," he said. ". . . There were several children, three men and two women in this room. The tobacco was stowed about everywhere, alongside foul bedding, and in a corner where there were scraps of food. The men, women and children in this room worked by day and far into the evening, and they slept and ate there. They were Bohemians, unable to speak English, except that one of the children knew enough to act as interpreter."

Instead of opposing the bill, Roosevelt became its most vociferous champion. "My first visits to the tenement-house districts in question made me feel that, whatever the theories might be, as a matter of practical common sense I could not conscientiously vote for the continuance of the conditions which I saw," he recalled. "These conditions rendered it impossible for the families of the tenement-house workers to live so that the children might grow up fitted for the exacting duties of American citizenship."

With strong backing from Roosevelt, the bill passed the Assembly only to die in the Senate when a lobbyist stole the only copy from the files in the closing hours of the session. The following year, it was passed by both houses, but Governor Cleveland was reluctant to sign it. Gompers asked Roosevelt to appear before the governor in support of the bill, and he made the main argument for it—"indeed, almost the only argument"—and successfully allayed Cleveland's misgivings. The governor told Roosevelt that "although he felt very doubtful, yet that, in view of the state of the facts as I had set forth, he would sign the bill."

When the measure was declared unconstitutional by the courts, in 1884, Roosevelt was again instrumental in obtaining passage of a modified bill designed to meet the judicial objections. And once again, it was overturned by the State Court of Appeals, which declared that it could not perceive "how the cigar-maker is to be improved in his health or his morals by forcing him from his home and its hallowed associations and beneficient influences to try his trade elsewhere."

Having seen the actual conditions in the tenements with his own eyes, Roosevelt was so shaken by this ruling that it jarred his faith in the judiciary. "It was this case which first waked me to a dim and partial understanding of the fact that the courts were not necessarily the best judges of what should be done to better social and industrial conditions," he later declared. "The judges who rendered this decision . . . knew nothing whatever of the needs, or of the life and labor, of three-fourths of their fellow citizens in great cities. They knew legalism, but not life."

Throughout the remainder of his legislative career, Roosevelt maintained a faith in the unfairness of draining the strong to support the willfully indolent, but he supported measures to regulate the working conditions of women and children, to require safety measures in factories and various trades, and to establish a state labor bureau. But on the question of wage increases for policemen, firemen, and city workers, which he regarded as politically inspired, he hewed to the conservative line.

Unlike during the previous session, Alice did not accompany her husband to Albany, and he roomed alone at the Kenmore. "I felt as if my heart would break when I left my own little pink darling, with a sad look in her sweet blue eyes, and I have just longed for her here in this beastly Hotel," Roosevelt wrote his wife shortly after his arrival in Albany. Every weekend, he promised, if his work as minority leader and four committee assignments allowed, he would return to New York City to be with his "blessed little wife."

For her part, Alice now had her own house to furnish, organize, and run—the only home the young couple were ever to have. Fanny Smith, a frequent visitor, remembered the Forty-fifth Street house as "a small pleasant place" where the Roosevelts welcomed friends "with the kind of generous warmth that characterized them both." Lengthy instructions establishing the duties of the servants had been written out by Aunt Anna Gracie, and the master of the house read them with fascination. Upon reaching the injunction that "every morning the cook should meet the ashman with a pail of boiling water," he was mystified. "What," he demanded, "has the poor ashman done to deserve a daily scalding?"

The weekends that Alice and Theodore spent together in their new home were undoubtedly among his most precious memories. "Back again in my own lovely little home, with the sweetest and prettiest of little wives—my own sunny darling," he confided to his diary that winter. "I can imagine nothing more happy in life than an evening spent in my cosy little sitting room, before a bright fire of soft coal, my books all around me, and playing backgammon with my own dainty mistress."

Yet, for all his delight in the comforts of home, Roosevelt was not reluctant to return to the legislative fray in Albany. Not everything went his way at that session, however. He lashed out at the New York City Board of Aldermen as "miserable and servile tools"; went after Jay Gould again by calling on the attorney general to dissolve the Manhattan Elevated Railway Company; introduced bills to end public support of institutions run by both the Protestant and Catholic churches, to control the liquor traffic by raising license fees, and to bring back the whipping post for wife and child beaters—only to see all his proposals suffer defeat in the Assembly.*

In March 1883, Roosevelt exhibited the most striking example of the independent turn of mind that always marked his political career. Aroused by Gould's financial manipulations, he joined most members of the Assembly in voting for a bill to roll back an attempt to raise the fare on the Manhattan Elevated from five

---

*"It is much to be hoped," the *New York Morning Journal* observed sarcastically, "now that he [Roosevelt] has opened the way for a discussion of much-needed reforms, that he will not stop at the whipping post. . . . The public can be brought around to view with complacency the re-establishment of the stocks, the ducking stool and the pillory; and . . . there will be little trouble in paving the way for the thumbscrew and the rack." February 19, 1883.

cents to ten cents. Despite the popularity of the measure, Governor Cleveland courageously vetoed it as a breach of the contract contained in the railway's franchise, which guaranteed a fixed return upon investment. "The State should not only be strictly fair," he solemnly declared, "but scrupulously fair."

The Assembly was thrown into an uproar by Cleveland's message. Everyone clamored to be heard, but the minority leader's insistent cries of "Mister Spee-kar! Mister Spee-kar!" won him the floor. The governor's brave decision to veto such a popular measure, Roosevelt declared, had made him realize he had blundered in agreeing with the majority to soak the Manhattan Elevated. With unusual humility, he confessed that "I weakly yielded, partly in a vindictive spirit towards the infernal thieves and conscienceless swindlers who have had the elevated railroad in charge and partly to the popular voice of New York.

"For the managers of the elevated railroad I have as little feeling as any man here," he continued. "If it were possible, I would willingly pass a bill of attainder on Jay Gould and all of Jay Gould's associates. . . . I regard these men as furnishing part of the most dangerous of all dangerous classes, the wealthy criminal class. Nevertheless, it is not a question of doing justice to them, it is a question of doing justice to ourselves. . . .

"We have heard a great deal about the people demanding passage of this bill. Now, anything the people demand that is right is almost clearly and most emphatically the duty of this Legislature to do; but we should never yield to what they demand if it is wrong. . . . If the people disapprove our conduct let us make up our minds to retire to private life with the consciousness that we have acted as our better sense dictated. . . . I would rather go out of politics having the feeling that I had done what was right than stay in with the approval of all men, knowing in my heart that I had acted as I ought not to."

This speech was the most memorable of Roosevelt's legislative career. While the acid words "the wealthy criminal class" were etched on the public imagination, the voice of a future president was heard in these remarks. Plainly, he said, the duty of a public official is not to be a rubber stamp, blindly supporting or voting against measures based on whether or not they had the support of his constituents. Instead, it was his duty to inform himself on public questions and make decisions based upon a measure's merit rather than its popularity.

In openly reversing himself on the Five-Cent Fare Bill, he showed as much courage as Cleveland had in vetoing it. In fact, his two closest colleagues, Hunt and O'Neil, had urged him to remain silent. "What do you want to do that for, you damn fool!" they declared, and parted company with him on the issue. Press reaction to Roosevelt's position was almost entirely negative. While the *New York Tribune* said he had acted with "characteristic manliness," the *Sun,* in a more representative editorial, sneered that "the popular voice of New York will probably leave this weakling at home hereafter."

Undoubtedly as a result of the drubbing he took from the press on the Five-Cent Fare Bill, Roosevelt's self-control snapped. He was chagrined a few days later, when the Assembly overrode on a straight party vote a Committee on Privileges and Elections decision to decide a contested election in favor of a Republican, Henry L. Sprague. Furious, he took the floor to tender his resignation from the committee and to vent his rage upon the "Sodom and Gomorrah of the Democracy. . . . The difference between our party and yours," he shouted across the aisle, "is that your bad men throw out your good ones, while with us the good throw out the bad!" On and on the tirade continued, for a full fifteen minutes. When Roosevelt finally took his seat, his resignation was ignored and the session continued as if nothing had happened. "Young Mr. Roosevelt," noted the *New York Observer* in terms that were widely approved, "has been very silly and sullen and naughty. . . ."

Success had obviously gone to the young man's head. "I came an awful cropper," he conceded years later, "and had to pick myself up after learning by bitter experience the lesson that I was not all-important and that I had to take account of many different elements in life. It took me fully a year before I got back the position I had lost." Perhaps Grover Cleveland made the soundest assessment of the Roosevelt of those years. "There is great sense in a lot of what he says, but there is such a cocksureness about him that he stirs up doubt in me all the time. . . . Then he seems to be so very young."

*   *   *

That summer, the Roosevelts received the news for which they had long been waiting—Alice was pregnant. Some of her breathless excitement is clearly evident in a note she dashed off to Bamie on July 22, 1883:

> Dr. Wynkoop has just gone. He says I have every symptom of having a child and that I ought to engage my nurse from the 5th of February as it is very likely to come between that time and the 15th or the 20th. I am so delighted. . . .

Roosevelt's own reaction must have been equally ecstatic, for he desperately wanted children, too. Once their initial excitement had subsided, the young couple turned their attention to Leeholm, the house he planned to build on a grassy hilltop overlooking Long Island Sound at Oyster Bay. Roosevelt had already purchased ninety-five acres of land and was steadily adding to his holdings. Now, the architects were instructed to provide a spacious home with a multitude of gables, dormers, windows, chimneys, and a wraparound porch where he and Alice could sit on rocking chairs and contemplate the sunset. Inside, they wanted ten bedrooms, not counting those for the servants, a large, comfortable library, and big fireplaces for logs. Everything was to have an aura of permanence, of family. There, Roosevelt planned to live an expansive life with his beloved Alice, surrounded by many children.

All this excitement, coming on top of the strenuous legislative session, proved too much for the prospective father. He suffered a recurrence of asthma and "cholera morbus," complete with vomiting and stomach cramps. Later, Roosevelt described the whole period as "a nightmare." He felt a desperate need to throw off the shackles of domesticity, to escape "a restless caged feeling" of being hemmed in by sickness and domestic affairs.

The opportunity presented itself in the person of a retired naval officer, Commander H. H. Gorringe, who had just returned from the Dakota Bad Lands, where he had opened a camp for wealthy hunters on the Little Missouri River just east of the Montana line. The vast buffalo herds that had once populated the plains were decimated, but there were still opportunities to bag one of the great beasts, he told Roosevelt. Would he care to accompany him on his next trip? He certainly would, and began laying plans for the expedition.

But a few days before they were to leave for the West, Gorringe announced that he couldn't go. Roosevelt faced a dilemma. Alice was four months pregnant and already showing signs of a difficult pregnancy. Mittie remarked on how "very large" she looked; there was a puffiness around her eyes and her ankles were swollen. But Roosevelt, imagination inflamed by the prospect of a buffalo hunt, decided to go west anyway. The reaction of Alice, lonely and feeling neglected, can only be surmised.

On September 2, immediately after his departure, Roosevelt, in an attempt to mollify his wife, assured her of his devotion. "I have been miserably home-sick for you all the last forty-eight hours; so home-sick that I think, if it were not that I had made all the preparations, I should have given up the journey entirely," he declared. "I think all the time of my little laughing, teazing beauty, and how pretty she is, and how she goes to sleep in my arms, and I could almost cry I love you so. You sweetest of all little wives! But I think the hunting will do me good; and I am very anxious to kill some large game—though I have not much hopes of being able to do so." With his mother, he was considerably more frank. "I feel like a fighting cock," he exuberantly wrote during a layover in Chicago.

A few nights later, Theodore Roosevelt swung down from a Northern Pacific train, quickly unloaded a duffle bag and gun case, and watched as the train chugged away into the darkness. It was three o'clock on a cool September morning, and he had just arrived at the town of Little Missouri in the Bad Lands of Dakota. No one awaited him at the depot. In fact, there was no depot, only a few clumps of dusty sagebrush underfoot. In the distance, he made out a handful of darkened shacks, and shouldering his gun case and dragging the bag, he headed for the largest of these, which bore a sign identifying it as the Pyramid Park Hotel. Loud hammering on the door finally aroused the alcohol-befuddled innkeeper. With considerable grumbling, he showed the newcomer up to the second floor "bull pen," a large loft containing fourteen canvas cots. Thirteen were filled with huddled forms, and for twenty-five cents, the young man was given possession of the empty one.

Eager to bag a buffalo, he was up early the next morning to look over the land and to engage a guide. The town—"Little Misery" to its inhabitants—sprawled along the Little Missouri River. It was surrounded on three sides by scarred and precipitous clay buttes etched by the wind into the weird and barbaric shapes that gave the Bad Lands its name. Roosevelt sensed a somber, almost haunting beauty in the desolate landscape, especially at sunset when the last light of day exploded into a rainbow of colors. Herds of Texas longhorns were taking the place of buffalo on the sprawling grasslands, and there were a few lonely ranches scattered about. Bands of restive Indians still roamed free and the eerie howl of wolves was heard in the night.

Roosevelt had some difficulty in securing a guide because most Bad Landers were reluctant to accompany a tenderfoot with thick glasses out into this wilderness. He persisted, however, and eventually found a Canadian named Joe Ferris, who agreed to act as his guide. For several days, they tracked a small buffalo herd over the rough country. They ran out of food, except for a few biscuits, were soaked by interminable rain, and awoke in the cold morning air with their teeth chattering. "Isn't this bully!" Roosevelt shouted to his companion. "By Godfrey, but this is fun!" Nevertheless, he got his buffalo—climaxing the hunt with an ecstatic Indian war dance around the carcass and a bonus of $100 for the astounded Ferris.

To Alice, he wrote a graphic account of his adventures that is worth quoting at length because it shows Roosevelt at his best:

Darling Wifie,
Hurrah! The luck has turned at last. I will bring you home the head of a great buffalo bull, and the antlers of two superb stags. For eight days my bad luck was steady. On the ninth it culminated. We started out quite early in the morning, and before we had gone a quarter of a mile a rattlesnake struck at one horse and barely missed him; we shot him and took his rattles. This was the fourth we had seen on the trip; two we saw while on horseback, and shot the other I met while crawling on all fours after an antelope, having almost crawled on to him when I was warned by the angry, threatening sound of the rattle. A little while afterwards my horse suddenly went down through the apparently solid bottom of a dry creek we were crossing as if it had been a trap door; and it took twenty minutes of as hard work as I ever had before we managed

to haul it by main strength out of the quick sand. A mile farther on we got into a country very much broken up by deep, narrow ravines; the country being very barren and desolate. While cautiously making our way through this country we found the very fresh tracks of a large buffalo bull; I dismounted at once and followed them cautiously some little distance, when I caught sight of him feeding at the bottom of a steep gulley; I crawled up to the edge, not thirty yards from the great, grim looking beast, and sent a shot from the heavy rifle into him just behind his shoulder, the ball going clean through his body. He dropped dead before going a hundred yards. . . . I am in superb health, having plenty of game to eat, and living all day long in the open air. With a thousand kisses for you, my own heart's darling I am ever your loving,

Thee

Deeply impressed by what he had seen in the Bad Lands, Roosevelt decided to invest "very cautiously at first" in the cattle business. If ranching were successful, he told Alice, it "will go a long way towards solving the problem that has puzzled us both a good deal at times—how I am to make more money as our needs increase, and yet try to keep in a position from which I may be able at some future time to again go into public life, or literary life. But, my own darling, everything will be made secondary to *your* happiness, you may be sure."

What he did not tell her was that before leaving the Bad Lands, he put $14,000 into acquiring his own cattle herd. He reached an agreement with Sylvane Ferris, the guide's brother, and Bill Merrifield, his partner, that called for them to purchase four hundred head of cattle on his account and to run them on the government land adjacent to their ranch, the Maltese Cross ranch at Chimney Butte on the Little Missouri.* Nobody bought land in those days; the usual practice was to take squatter's rights to the public domain. The sum of $14,000 was in addition to the $20,000 he had already put into Leeholm and represented a significant portion of his capital. When his uncle James Alfred Roosevelt learned of the cattle-raising venture, he heartily disapproved of it, but financial caution was not one of Theodore Roosevelt's trademarks.

---

*Sylvane Ferris was impressed by the way Roosevelt did business; he simply handed over a check. "All the security he had for his money was our honesty."

*   *   *

Roosevelt handily won reelection to a third term in the Assembly that November, and with the Republicans having taken a majority of the seats, he launched a campaign for the speakership. "I am a Republican pure and simple," he told potential supporters, "neither a 'half breed' nor a 'stalwart,' " and no political clique would do his thinking for him. He met with all the newly elected assemblymen, and one wearily emerged from a session to proclaim: "He's a brilliant madman born a century too soon." Polls taken on the eve of the Republican caucus indicated a Roosevelt victory, but the bosses waged an all-out campaign against him, and he was defeated when some supporters defected at the last moment.

When the session opened, Roosevelt, who had been named chairman of the Cities Committee, introduced a series of measures designed to weaken Tammany control of New York City. They included a far-reaching proposal that became known as the "Roosevelt bill," to increase the power of the mayor at the expense of the aldermen, who, in his opinion, were "merely the creatures of the local ward boss or of the municipal bosses." Not content with that, he also persuaded the Assembly to appoint a five-man investigating committee with himself as chairman, to probe corruption in the city government. The committee held hearings in New York every Friday, Saturday, and Monday, which allowed him to spend more time with Alice.

Looking into the conduct of the sheriff's office in New York City, the committee found corruption was rampant. Not only were taxpayers being robbed but prisoners in the infamous Ludlow Street jail were being subjected to petty blackmail. One inmate told of having spent five years in a cell under a toilet that leaked effluvia on him because he couldn't pay the $250 demanded for a better cell. Soup was carried in the same pails used for scrubbing; bedbugs and lice were found in the food. Roosevelt angrily charged that the situation was "revolting" and introduced a bill to prevent overcrowding and to improve conditions in the prison.

With the baby due toward the middle of February, Alice left the Forty-fifth Street house and moved back in with her in-laws, occupying the couple's old apartment on the third floor. Corinne had also returned to the house to keep her company. Although Alice was feeling "crampy" and unwell, Roosevelt's letters to her reflected a husband's normal concern unshadowed by anxiety. "I love you and long for you all the time," he wrote her on February 6, "and oh *so* tenderly, doubly tenderly now. . . ."

Alice was always waiting downstairs to greet him when he came home, and then she would call upstairs: "Corinne! Teddy's here. Come share him." On Monday, February 11, he returned to Albany despite his wife's discomfort and the fact that Mittie was in bed with what appeared to be a severe cold. Corinne and her husband, Douglas Robinson, were away on a visit to Baltimore,* but Bamie and Aunt Anna Gracie were on hand, so there was little reason to worry. The baby was not expected until three days later, on February 14, and he would return well in time for the birth. If he felt any uneasiness, he pushed it aside. Some observers see in this another manifestation of the avoidance syndrome that appeared in critical moments of his life. In 1878, when his father was dying, he remained at Cambridge; and again, in 1898, he left his second wife barely recovering from a near-fatal illness to go off to war in Cuba.

Shortly after her husband's departure, Alice wrote what was to be her last letter to him. It is scrawled in pencil, and her usually elegant handwriting is difficult to read:

> Darling Thee,
> I hated so to leave you this afternoon. I don't think you need feel worried about my being sick as the Dr told me this afternoon that I would not need my nurse before Thursday—I am feeling well tonight but am very much worried over the baby your mother, her fever is still very high and the Dr is rather afraid of typhoid, it is not the least catching. I will write again to-morrow and let you know just how she is—don't say any thing about it till then. I do love my dear Thee *so* much, I wish I could have my little new baby soon
> ever Your loving wife
> Alice

The next day, February 12, Alice went into labor and at about 8:30 P.M. was delivered of an eight-and-three-quarter-pound baby

---

*Before leaving, Corinne remembered having told Alice: " 'Don't you dare have your baby until I get back.'

" 'I promise,' answered Alice, dimpling and smiling gaily." Letter to Pringle, September 18, 1930.

girl. "You ought to have been a little boy," said Aunt Anna, who assisted at the birth, as she took the baby from the doctor and wrapped it in flannel. "I *love* a little girl," murmured Alice, exhausted by the agonies of a first delivery. She briefly held the baby in her arms, and kissed it before falling asleep.

In Albany the next morning, fellow assemblymen clustered about the jubilant Roosevelt, shaking his hand and congratulating him after a telegram announcing the birth of the child had reached the capitol. While the baby was in good health, the mother was "only fairly well"—but that was not unexpected under the circumstances. The new father requested a leave of absence to begin later in the day and before departing reported out fourteen bills from his Cities Committee. A few hours later, there was another telegram, ominously bidding him to come down to New York immediately. "I shall never forget when the news came and we congratulated him on the birth of his daughter," Ike Hunt recalled. "He was full of life and happiness—and the news came of a sudden turn as he took his departure."

Thick fog enshrouded Roosevelt's train as it crept down the Hudson toward New York, its bell seeming to toll a dirge. From the river came the mournful sound of foghorns. "Suicidal weather," *The New York Times* labeled it. "There is something comfortless and unhappy . . . something suggestive of death and decay in the dampness that fills the world." Roosevelt could do nothing but read the two telegrams over and over again and stifle the urge to panic. Only six years before, he had raced home in response to a similar telegram only to find his father dead. What horrors awaited him now? The train finally reached the city shortly before midnight and he groped his way home through the fog. He was relieved to see a light burning in Alice's third-floor bedroom. At the door he was met by a distraught Elliott with words similar to those with which he had greeted Corinne an hour before: "There is a curse on this house. Mother is dying and Alice is dying, too!"

Alice had been diagnosed as suffering from Bright's disease, a kidney ailment, which had gone undetected throughout her pregnancy,* while Mittie was in the last stages of typhoid. The

---

*The American public had first become aware of Bright's disease two years before when the Associated Press reported that President Arthur was suffering from the illness,

baby was alive and well, however. Racing up the steps to Alice's room two at a time, Roosevelt discovered that she was barely conscious and hardly recognized him. He held her in his arms for the next two hours as if struggling to prevent her from slipping away from him. Someone murmured that if he wished to see his mother before she died, he should hurry down to her room on the second floor. In the stillness of the morning, he stood with his sisters and brother at her bedside in the same room in which their father had died, and he echoed Elliott's words. "There *is* a curse on this house!" Mittie died at 3:00 A.M., and Roosevelt returned to Alice's room in a daze to take her in his arms again. The vigil continued through the long night and into the following afternoon, when Alice died. It was St. Valentine's Day—the fourth anniversary of their engagement. She was twenty-two years old.

Tense and white-faced, Theodore Roosevelt sat with Bamie, Corinne, Elliott, and Alice's father in a front pew of the Fifth Avenue Presbyterian Church for the double funeral on February 16. Two rosewood coffins covered with roses and lilies of the valley rested before the altar, and the huge church was filled with New York's social and political elite. Sobs were heard as Reverend Dr. John Hall, himself near the breaking point, eulogized the dead women and prayed for the bereaved and the four-day-old baby. Burial was in Greenwood Cemetery alongside Mittie's mother, Martha Bulloch, and Theodore Roosevelt, Sr. No one knows what went through Roosevelt's mind as the clods fell upon the coffins, but Arthur Cutler, his tutor, told Bill Sewall he was "in a dazed, stunned state. He does not know what he says or does."

The shock was traumatic. Roosevelt drew a large cross in his diary for February 14, 1884, and wrote beneath it: "The light has gone out of my life." For four years, the yellow-haired girl from

---

which was almost always fatal to adults, but the report was denied. The symptoms of the malady, which may often precede diagnosis by years, include spasmodic nausea, depression, and indolence. Arthur died of complications resulting from Bright's disease in 1886. See Reeves, *Gentleman Boss,* pp. 317–318. Today, Bright's disease is easily diagnosed and cured.

Some of Alice's family and friends were convinced her physician had been criminally negligent in failing to detect her illness and deal with it. See statement of Mrs. Robert Bacon in the Pringle Papers.

Chestnut Hill had bathed his life in sunlight; now he was alone. Perhaps his grief was compounded by a look of silent reproach he may have fancied seeing in Alice's face.

But life had to go on. The following day the baby, held by her father and wearing a locket containing a lock of her mother's hair, was christened Alice Lee.* Within a week, Roosevelt had regained enough control of himself to return to Albany, where his colleagues stood in silent tribute when he took his seat. The baby was left in Bamie's capable care. To Carl Schurz, who had expressed condolences, he wrote: "Your words of kind sympathy were very welcome to me, and you can see I have taken up my work again; indeed I think I should go mad if I were not employed. . . ."

For the remainder of the session, he threw himself into his work, holding hearings, shuttling back and forth to New York, bringing bills to the floor, writing reports. Exhibiting a heretofore unsuspected willingness to compromise, he succeeded in obtaining enactment of most of the reform legislation he had espoused. Outlook broadened by experiences, Roosevelt learned that in the imperfect world of politics, the choice is often not between right and wrong but a trade-off between evils. He also came to realize that not all his opponents were cynical grafters.

"I looked the ground over and made up my mind that there were several other excellent people there, with honest opinions of the right, even though they differed from me," he later told the journalist Jacob Riis. "I turned in to help them, and they turned to and gave me a hand. And so we were able to get things done. . . . That was my first lesson in real politics. It is just this: if you are cast on a desert island with only a screwdriver, a hatchet, and a chisel to make a boat with, why, go make the best one you can. It would be better if you had a saw, but you haven't. So with men."

Of his personal tragedy, he would say nothing. "You could not talk to him about it," recalled Hunt. ". . . He did not want anybody to sympathize with him." In later years, Roosevelt closed

---

*The lock of hair was preserved by Mrs. Gracie and was discovered in the cache of papers found following Alice Longworth's death in 1980. Although it has darkened with age, it still retains a curl and light tones can be seen.

the door on his life with Alice. He did not mention her in his
Autobiography or talk about her to anyone, including their daughter. Unwilling even to utter Alice's name, he called the child Baby
Lee. Some letters from their courtship were destroyed, and photographs and souvenirs of their honeymoon and marriage were
ripped from his scrapbooks. But before shutting Alice away in
his memory, he wrote a final tribute in a privately published
memorial volume:

> . . . She was beautiful in face and form, and lovelier still in spirit;
> as a flower she grew, and as a fair young flower she died. Her life
> had been always in the sunshine; there had never come to her a
> single great sorrow; and none ever knew her who did not love and
> revere her for the bright, sunny temper and her saintly unselfish-
> ness. Fair, pure, and joyous as a maiden; loving, tender, and happy
> as a young wife; when she had just become a mother, when her
> life seemed to be just begun, and when the years seemed so bright
> before her—then, by a strange and terrible fate, death came to
> her. And when my heart's dearest died, the light went from my
> life for ever.*

The presidential campaign of 1884 came as a relief for Roosevelt, for it allowed him to submerge his grief and energize him
for the fight to prevent James G. Blaine from winning the Republican nomination. If any politician embodied the values—or
lack of them—of the Gilded Age, it was Jim Blaine. Speaker of
the House, senator from Maine, secretary of state, and perennial
presidential hopeful, the "Plumed Knight" was a dominant force
in the Republican party and in the affairs of the nation for thirty
years. A magnetic leader with eyes that flashed sparks and a voice
that hypnotized the galleries, he pursued the voters up and down
the byways of the Republic. Year after year, election after election, his name echoed over the land like a drumbeat: "Blaine!
Blaine! James G. Blaine!" But he could never disabuse enough
voters of their conviction that he was a crook.

---

*Roosevelt's future silence concerning Alice may be explained by a letter he wrote
Corinne regarding a friend whose life had taken a tragic turn:
"I hate to think of her suffering, but the only thing for her to do now is to treat the
past as the past; the event as finished and out of her life; to dwell on it, and above all to
keep talking of it with anyone, would be both weak and morbid. She should try not to
think of it. . . ." Robinson, *My Brother, Theodore Roosevelt,* pp. 240–241.

Regarding Blaine as "decidedly mottled," and opposed to another term for President Arthur, who had profited from the elder Roosevelt's political martyrdom, Roosevelt supported the dark horse candidacy of the reform-minded Senator George F. Edmunds of Vermont. At the state convention at Utica, he outmaneuvered the Blaine and Arthur factions and was successful in helping elect four pro-Edmunds men, including himself, as delegates-at-large to the national convention to be held in Chicago in June.

Roosevelt was in his element amid the controlled madness of a political convention. Working around the clock with Henry Cabot Lodge, a delegate-at-large from Massachusetts, he tried to organize the hundred or so scattered Edmunds delegates into a force that would hold the balance of power between Blaine and Arthur. He was everywhere on the floor of the cavernous Exposition Hall beside Lake Michigan, and made the acquaintance of leading Republicans from all over the country. In later years, they would remember the energetic young New Yorker in a "nobby" straw hat and his forcible seconding speech for John R. Lynch, a Mississippi black, who was nominated as temporary chairman by the Edmunds forces.

"It is now, Mr. Chairman, less than a quarter of a century since, in this city, the great Republican Party for the first time organized for victory and nominated Abraham Lincoln, of Illinois, who broke the fetters of the slave and rent them asunder forever," Roosevelt declared, his slight frame shaking with the effort to be heard above the din. "It is a fitting thing for us to choose to preside over this convention one of that race whose right to sit within these walls is due to the blood and treasure so lavishly spent by the Republican Party. . . ."

In the end, however, the Blaine steamroller easily crushed the opposition and the Plumed Knight was nominated on the fourth ballot. Pushing his way through the throng, an Ohio congressman named William McKinley reached Roosevelt's side to urge him to make a unity speech in favor of Blaine. Roosevelt flatly refused. In the New York and Massachusetts delegations, knots of grim-faced men sat silently with their arms folded throughout the demonstration in support of the nominee. Later, reformers such as Carl Schurz, Henry Ward Beecher, and Charles Francis Adams, unable to stomach their own party's choice, bolted to the Democratic candidate—Grover Cleveland—and expected Roosevelt to join them.

The Chicago convention had thrust Roosevelt into the national spotlight, but it also created the most agonizing dilemma of his political career. As a delegate to the convention, he was pledged to support the party's nominee. Yet, if he did so, he would compromise his principles and be ostracized by his fellow independents, or "mugwumps." On the other hand, if he didn't support Blaine, the party regulars, who had long memories, would make certain he had no future in politics.

Initially, some thought Roosevelt had indicated he would join the bolt, but after much soul-searching, he and Lodge supported Blaine. "I am by inheritance and by education a Republican; whatever good I have been able to accomplish in public life has been accomplished through the Republican party," he declared. "I have acted with it in the past, and wish to act with it in the future; I went as a regular delegate to the Chicago convention, and I intend to abide by the outcome of that convention." Having made this declaration of party loyalty, Roosevelt could have sat out the campaign, but he insisted upon stumping for Blaine. Even Roscoe Conkling had refused to do that. When asked to speak in behalf of Blaine, the imperious Lord Roscoe snapped: "No thanks. I don't engage in criminal practice."

The mugwumps were shocked and dismayed by Roosevelt's decision. Men who had praised him as the conscience of the Republican party and its hope for the future now considered him a turncoat. William Roscoe Thayer, a Harvard friend, was "dumbfounded." Some attributed Roosevelt's decision to ambition, claiming he was seeking party backing for a run for mayor of New York City; others, including his Lee relations in Boston, blamed it on the "evil genius" of the ambitious Cabot Lodge, who was running for Congress. Whatever the reason, Roosevelt's decision to support Blaine meant a break with his father's gentlemanly view of politics. He also made it abundantly clear that he had decided to be a political professional. Had he bolted to Cleveland, he would have been devoid of influence within the Republican party, a mere "morning glory."

In the end, Roosevelt suffered defeat on all fronts. Not only had his decision to opt for party regularity angered the reformers, but Blaine was narrowly defeated by Cleveland, returning the Democrats to the White House for the first time since before the Civil War. The Plumed Knight had failed to quickly disavow a supporter's charge that the Democrats were the party of "rum, Romanism and rebellion," and it cost him New York's vital Catholic vote. Gloomily surveying his prospects following the debacle, Roosevelt wrote Lodge: "Blaine's nomination meant to me pretty sure political death if I supported him; this I realized entirely, and went in with my eyes wide open. I have won again and again; finally chance placed me where I was sure to lose whatever I did. . . . I do not believe that I shall ever be likely to come back into political life."

Having lost the support of the reformers and without the solace of national office to make up for it, he resolved to leave New York behind and make a career in the Dakota Bad Lands. The Fifty-seventh Street house was sold, and Baby Lee was left with Bamie. Now, Roosevelt was in the wilderness both literally and figuratively.

Chapter 8

"The Wine of Life"

Theodore Roosevelt's abrupt decision to go west had a certain logic about it. The adventurous ranch life of the Bad Lands not only provided relief from his haunting memories of Alice and time to lick his political wounds, it seemed like a good business move. Beef was in demand in the cities of the East, and the undulating grasslands of the northern plains, reaching down from Canada like an enormous thumb, were viewed as a land of opportunity. One of the best-selling books of the day was *The Beef Bonanza: or How to Get Rich on the Plains,* and Roosevelt decided to put its teachings into practice. Over the next several years, he steadily increased his investment in ranching until it totaled about $85,000. He nearly lost his buckskin shirt, but the West put its brand upon him for life.*

"I heartily enjoy this life, with its perfect freedom, for I am very fond of hunting and there are few sensations I prefer to that of galloping over these rolling, limitless prairies, rifle in hand, or

---

*In addition to the $125,000 he had inherited from his father, Roosevelt received an additional $62,500 after the death of his mother, so the Bad Lands investments took half his capital.

winding my way among the barren, fantastic and grimly picturesque deserts of the so-called Bad Lands," he wrote Cabot Lodge.
Putting his roots in even deeper, Roosevelt purchased another
thousand head of cattle and established his headquarters at the
Elkhorn Ranch, overlooking a bend of the Little Missouri. Cot-
tonwoods grew along the riverbanks, softening the forbidding
landscape. Bill Sewall and Will Dow were persuaded to come out
to Dakota from Maine to manage the place under an arrangement
that gave them salaries, part of any profits, and no liability for
losses.

Roosevelt saw grandeur and mystery in the Bad Lands. "The
country is growing on me more and more," he wrote Bamie. "It
has a curious, fantastic beauty of its own." Somehow it *"looks*
just exactly as Poe's tales and poems *sound."* Other visitors were
less impressed; one described the area as "Hell with the fires gone
out." In later years, Roosevelt was convinced that had he not gone
to the Dakota Territory, he would never have become president.
The accuracy of this statement is impossible to gauge, but his
time in the West had a tremendous influence both upon his character and the way the public perceived him.

For Roosevelt, the West was a new and uncharted world. Background, family, education, position—all counted for little, and
he had to make his way in a rough democracy of equals. He was
thrown together with cowboys, gamblers, gunmen, peace officers,
soldiers, and a few European aristocrats who had come west to
make fortunes. It was a world right out of the books of Captain
Mayne Reid that had captivated him as a small boy. Long days
in the saddle in good weather and foul also helped complete his
struggle for health. Before he went West, observers still spoke of
him as slender and frail-looking, but on the range he developed
the bull neck, massive shoulders, and barrel chest that transformed him into "one hundred and fifty pounds [of] clear bone,
muscle and grit." And one newsman observed that the voice that
had piped, "Mister Spee-kar! Mister Spee-kar!" in Albany "is
now hearty and strong enough to drive oxen."

Roosevelt had three major liabilities in politics: he was an aristocrat, he was an intellectual, and he was an easterner. Altogether,
he spent only about three years in the Bad Lands, a period interrupted by sometimes lengthy stays in the East. Yet he so suc-

cessfully identified himself with the West that for the remainder of his life, the public thought of him as a rough-riding cowboy rather than a New York dude. This western experience removed the stigma of effeminacy, ineffectuality, and intellectualism that clung to most reformers.

The idea for the Rough Riders was rooted in Roosevelt's stay in the West, and the regiment was crucial to the success of his political career. But Roosevelt was not a *cowboy;* he was a *ranchman* who employed cowboys. While he worked with cowboys, there was a sharp social distinction. Roosevelt's hunting knife came from Tiffany's and his Winchester rifle was handsomely engraved. And he insisted upon being addressed as "Mr. Roosevelt"—not Teddy—a name he discouraged after Alice's death. Thirty years later, he wrote with nostalgic eloquence of a departed era:

> We led a free and hardy life with horse and rifle. We worked under the scorching midsummer sun, when the wide plains shimmered and wavered in the heat, and we knew the freezing misery of riding night guard around the cattle in the late fall roundup. In the soft springtime the stars were glorious in our eyes, each night before we fell asleep; and in the winter we rode through blinding blizzards, when the driven snowdust burned our faces. There were monotonous days, as we guided the trail cattle, or the beef herds, hour after hour at the slowest of walks; and minutes or hours teeming with excitement as we stopped stampedes or swam the herds across rivers treacherous with quicksand, or brimmed with running ice. We knew toil and hardship and hunger and thirst; and we saw men die, violent deaths as they worked among the horses and cattle, or fought in evil feuds with one another, but we felt the beat of hardy life in our veins, and ours was the glory of work and the joy of living.

Roosevelt was only a fair shot, an average horseman, and never learned to handle a rope very well, but repeated examples of courage hastened his acceptance by hard-bitten Bad Landers. One night he rode into the town of Mingusville, across the Montana line, and went into a hotel barroom to find a drunk striding up and down before the cowed patrons with a cocked revolver in each hand. He had already put several shots into the face of the clock over the bar. As soon as he spotted the bespectacled Roosevelt, he called him "Four-eyes" and bellowed that "Four-eyes is going to treat!" Regarding it as a joke, Roosevelt took a seat behind the stove, where he joined in the laughter at his own expense. But the drunk repeated the command at pistol point. Rising from his chair as if to obey, Roosevelt remarked resignedly, "Well if I've got to, I've got to." Suddenly, he lashed out with a right to the jaw, a left, and another right. The drunk sank to the floor and his guns went off harmlessly.*

On another occasion, Roosevelt was gossiping with a few friends in the office of the *Bad Lands Cowboy,* the local newspaper, and was angered by the steady flow of off-color stories told by "Hell-Roaring" Bill Jones, a fellow with a short fuse and a well-known readiness to use his pistol. Roosevelt himself resorted to an infrequent "damn!" but preferred "by Godfrey!" in times of stress and had no taste for obscenity. Finally, he could take no more. "I can't tell why in the world I like you," he snapped at Jones, looking him straight in the eye. "You're the nastiest-talking man I ever heard." Hell-Roaring Bill's hand fell on his pistol and the others began looking for the exits. But Jones had developed a rough admiration for Roosevelt. Sheepishly, he let his hand drop and replied apologetically: "Mr. Roosevelt . . . I don't mind saying that maybe I've been a little too free with my mouth."

Nevertheless, Roosevelt sometimes amused his neighbors. During a roundup, he called one of his men to "hasten forward quickly there!" Everyone was thrown into paroxysms of laughter, and for some time afterward, Dakotans were summoning each other with "Hasten forward quickly there!" Even so, he won his spurs, eating his share of trail dust, working as hard as any of his cowboys, serving as a deputy sheriff who captured three thieves at gunpoint, and helping the ranchers to organize the Little Missouri Stockmen's Association to deal with rustlers, marauders, and other problems. And when he went away, they never forgot him. They

---

*This story sounds too good to be anything but pure fiction, but is confirmed by witnesses. Although Roosevelt occasionally told "stretchers," such as having been light-weight boxing champion at Harvard, he had, compared to most public men, an extremely high level of veracity. "Roosevelt prepared records of many of his deeds for the attention of posterity," says Edward Wagenknecht in *The Seven Worlds of Theodore Roosevelt.* "I believe that in general he came as close to telling the truth as a man can come in talking about himself" (p. 99). In his *Autobiography,* Roosevelt omitted things he didn't want to talk about rather than lie about them.

flocked to him when he called upon them to follow him to Cuba; they supported him when he became president. And when he raised the banner of political insurgency, they followed him again.

"Black care rarely sits behind a rider whose pace is fast enough," Roosevelt once wrote and he kept up a fast pace in the Bad Lands in a not always successful attempt to outdistance melancholia. As usual, his mercurial temperament alternated between elation and depression. In passing moments, he considered remaining in the Dakota Territory for the rest of his life and making a full-time career out of ranching, hunting, and writing. Bill Sewall observed that he was "very melancholy at times" and claimed his life had no meaning. "You have your child to live for," Sewall protested. "Her aunt can take care of her a good deal better than I can," he replied. And Bill Merrifield, his partner in the Maltese Cross, also recently widowed, tried to console him, but he would not be persuaded. "Don't talk to me about time will make a difference—time can never change me in that respect."

There was an ever-present reminder of Baby Lee at the Elkhorn Ranch in the person of Sewall's daughter, who was the same age as his own child. "The poor little mite of a Sewall girl . . . has neither playmates nor play toys," he wrote Bamie. ". . . I wish the poor forlorn morsel had some play toys. If you go in town ever . . . get and send out to me a box with the following toys, all stout and cheap; a big colored ball, some picture blocks, some letter blocks, a little horse and wagon and a rag doll."

Long days on the trail and on hunting expeditions were followed by equally long hours with pen and paper. Roosevelt turned his western adventures into a trilogy of well-received books: *Hunting Trips of a Ranchman* (1885), *Ranch Life and the Hunting Trail* (1888), and *The Wilderness Hunter* (1893). His interest in natural history served him well, and the books set a new style for the genre. Unlike most hunting books of the time, they were far more than mere narratives of trophies bagged. Roosevelt provided vivid portraits of the land and the people and animals that inhabited it. "No writer until then combined quite so well the offices of the hunter and those of the naturalist," says Paul Cutright, a leading environmentalist. The account of the private life of the grizzly bear, in *The Wilderness Hunter,* was the most detailed that had yet appeared, and the material on the bighorn sheep in *Ranch Life* had no rival.

Reviewers called the books "tinglingly alive, masculine and vascular" and said they "could claim an honourable place on the same shelf with . . . Walton's *Compleat Angler.*" Favorable comparisons were also made with Francis Parkman's classic, *The Oregon Trail.* These books, along with Frederic Remington's art and Owen Wister's novels—*The Virginian* is dedicated to Roosevelt— gave Americans the "wild West" of popular lore. Nevertheless, none of his books measured up to Roosevelt's own ambitions. Writing to Wister, he observed: "I wish I could make my writings touch a higher plane, but I don't well see how I can, and I am not sure that I could do much by devoting more time to them. I go over them a good deal and recast, supply or omit, sentences and paragraphs, but I don't make the reconstruction complete the way that you do."

The full flavor of Roosevelt's wilderness tales shows best in a grizzly bear hunt that appears in *Hunting Trips of a Ranchman.* The grizzly was the most dangerous animal in North America, and he knew in his heart that he must bag one to make his stay in the West complete. Hunting grizzlies on foot in timber was not for the faint-hearted, however. Two hunters had recently been killed by a single bear, and wounded bears were known to charge a hundred yards despite several bullets in the heart. While Roosevelt said he had never hunted tigers like his brother, he observed that a grizzly bear would make "short work of either a lion or a tiger, for the grizzly is greatly superior in bulk and muscular power to either of the great cats, and its teeth are as large as theirs, while its claws, though blunter, are much longer."

Roosevelt and Merrifield were in the Bighorn Mountains when they discovered the tracks of a huge bear. "It gave me rather an eerie feeling in the silent, lonely woods, to see for the first time the unmistakable proofs that I was in the home of the mighty lord of the wilderness," Roosevelt wrote. Keeping a sharp lookout ahead and to the side, the hunters ventured deeper into the gloomy half-light of the pine forest following a trail of broken limbs and twigs. Suddenly, they spotted their quarry not ten feet away. The bear was monstrous, nine feet tall and weighing over twelve hundred pounds.

"He had heard us, but apparently hardly knew exactly where

or what we were, for he reared up on his haunches sideways to us," Roosevelt continued. "Then he saw us, and dropped down again on all fours, the shaggy hair on his neck and shoulders seeming to bristle as he turned toward us. As he sank down on his forefeet I had raised my rifle; his head was bent slightly down and when I saw the top of the white bead fairly between his small, glittering, evil eyes, I pulled the trigger. Half-raising up, the huge beast fell over on his side in the death throes, the ball having gone into his brain, striking fairly between the eyes as if the distance had been measured by a carpenter's rule."

Some writers have been critical of Roosevelt's enthusiasm for hunting and charge that he indulged in excessive killing merely for the sake of killing. Edmund Morris quotes extensively from an 1884 diary, noting that he killed "170 items in just forty-seven days."* But Paul Schullery, who made a detailed examination of the diary, points out that virtually all the animals were used for camp food, which could not be hauled in. Roosevelt rarely hunted alone, he adds, and his bag fed the entire party. Fifty of the "items" were trout and many more were grouse, ducks, and rabbits, which were used for food. "Two hungry, hard-working hunters can consume an enormous quantity of fresh meat in forty-seven days. . . ."

Hunting was more than blood and slaughter to Roosevelt. He enjoyed the totality of the wilderness experience, and in the preface to *The Wilderness Hunter,* wrote:

> For . . . [the hunter] is the joy of the horse well ridden and the rifle well held; for him the long days of toil and hardship, resolutely endured, and well crowned at the end with triumph. In after-years there shall come forever to his mind the memory of endless prairies shimmering in the bright sun; of vast snow-clad wastes lying desolate under gray skies; of the melancholy marshes, of the rush of mighty rivers; of the breath of the evergreen forest in summer; of the crooning of ice-armored pines at the touch of winter. . . .

In the year between Roosevelt's first visit to the Bad Lands and his return, the camp at Little Missouri had been abandoned and

---

*The Rise of Theodore Roosevelt, p. 286. Since Alice's death his diaries had become "a monotonous record of things slain," according to Morris.

everything moved across the river to the newly created town of Medora. The place was the brainchild of one of the most bizarre figures attracted to the American West—Antoine-Amédée-Marie-Vincent Manca de Vallombrosa, the Marquis de Morès, or the "Emperor of the Bad Lands," as he was dubbed by the popular press. The marquis was French but his title, awarded his forebears in the fourteenth century, was Spanish. Twenty-five years old, tall, dashing, and with a flaring black moustache that gave him the air of a stage villain, he was a graduate of the French military academy at Saint-Cyr, had served in the army as a cavalry officer, and had killed at least two men in duels. He went about armed with a pair of large pistols and a rifle, and had a bowie knife tucked in his boot. A royalist with aspirations to the throne of France, he told his new neighbors he would use the fortune he planned to make in the Bad Lands to achieve this goal.

De Morès did not lack imagination. Visiting America following his marriage to Medora von Hoffman, the daughter of a Wall Street financier, he conceived a bold scheme to revolutionize the beef industry. Why not slaughter cattle on the range and ship dressed beef east by the new refrigerator cars? That would be cheaper than sending live cattle to Chicago, do away with the middleman, and reduce the cost to the consumer, which in turn would increase the demand for beef. Backed by the millions of his father-in-law, the marquis organized the Northern Pacific Refrigerator Car Company and brought in hundreds of construction workers to build a new town, which he called Medora after his wife, a slaughterhouse, and a packing plant. Topping it off was a twenty-five-room frame house on a bluff overlooking the town, the "chateau," where he lived like a feudal lord. It had a grand piano upon which Madame de Morès played Liszt and Verdi, a well-stocked wine cellar from which iced champagne was served in silver goblets, and it was staffed by twenty servants.

In the beginning, the marquis fared brilliantly in his venture. He took on the Armours and Swifts and other members of the Chicago beef trust, and they were forced to take notice of him. Throwing off ideas like a human pinwheel—some good, others wildly impractical—he planned to make pottery from the Bad Lands clay, open a stagecoach line between Medora and Deadwood in the Black Hills, and ship salmon from the Columbia

River to the tables of New York. Not everyone was caught up by the spirit of adventure that swirled about him like a brilliantly colored cape, however.

Behind his back some Dakotans called de Morès that "crazy Frenchman" and others were impatient with his arrogance. "It takes me only a few seconds to understand a situation that other men have to puzzle over for hours," he told one Bad Lander. "I seem to see every side of a question at once." And he brushed aside western traditions when they got in his way. Westerners prided themselves on independence from rules and regulations, but there were certain unwritten customs that had to be honored—and he ignored them. He brought in fifteen thousand sheep despite the protests of ranchers that the animals nibbled the grass so closely that the roots were killed and the pasture ruined for cattle and game. He began buying up land and fencing it, when traditionally in the Bad Lands no one owned land and cattle grazed on the open range. When these fences were cut, he ordered them restrung.

Before long, some of the wilder cowboys were threatening to kill the marquis. He went to Mandan, the nearest town with a magistrate, and asked for advice. "Why, shoot," was the judicial reply. Not long afterward, de Morès and his men caught three of the cowboys outside Medora after they had shot up the town— not an unusual occurrence—and one was killed. The Frenchman's enemies claimed the dead man had been shot down in a cold-blooded ambush, and he was charged with murder. The marquis pleaded self-defense and the charges were dismissed before he came to trial. Nevertheless, bad blood remained and the killing was to haunt him for the rest of his time in the Dakota Territory.

As two of the area's leading citizens, Roosevelt and de Morès maintained a cordial if rather formal relationship. They attended the meetings of the local stockmen's association and traveled together to Montana to volunteer for a vigilante campaign against rustlers and horse thieves.* Roosevelt dined several times at the "chateau" and enjoyed the excellent food and wine and the com-

---

*The vigilantes were members of a secret society known as the "Stranglers," because they hanged suspected malefactors on the nearest tree. Roosevelt and de Morès were told they were too well known to become members of a secret society.

pany of Madame de Morès. The two men had a common interest
in politics, horses, books, and hunting, but there was tension
between them. Roosevelt was irritated by the marquis's preten-
sions, his unrestrained arrogance and several attempts to encroach
on land occupied by Roosevelt's cattle.

Once, when the marquis, his wife, and Roosevelt were all in
New York City at the same time, he asked Bamie to invite them
to dinner. "By all means bring them," she replied. "But please
let me know beforehand whether you and the Marquis are on
friendly terms at the moment or are likely to spring at each other's
throat." In later years, she observed that "Theodore did not care
for the Marquis, but he was sorry for his wife. . . ."

Relations between these two strong-willed men steadily deteriorated, and in 1885, Roosevelt received what he perceived as a
challenge to a duel from the Frenchman. The episode was triggered by another indictment against de Morès on a charge of
killing the cowboy three years before. Convinced that Roosevelt
was among those in league against him, de Morès wrote him from
his cell while awaiting trial:

My Dear Roosevelt
My principle is to take the bull by the horns. Joe Ferris is very
active against me and has been instrumental in getting me indicted
by furnishing money to witnesses and hunting them up. The papers
also publish very stupid accounts of our quarrelling—I sent you
the paper to N.Y. Is this done by your orders? I thought you were
my friend. If you are my enemy I want to know it. I am always on
hand as you know, and between gentlemen it is easy to settle
matters of that sort directly.
                                                    Yours very truly
                                                    Morès

Regarding the note as a threat and preliminary to a challenge,
Roosevelt asked Sewall to be his second if there was a duel. The
marquis was an experienced duelist and far more experienced
with a pistol while Roosevelt had never handled a sword. Even
so, Roosevelt told Sewall that for honor's sake he could not back
off. Noting that as the challenged party he would have choice of
weapons, he said he would choose Winchester rifles at twelve
paces. Neither man would probably have emerged alive from such
a suicidal encounter. Sitting down on a log outside his cabin at

the Elkhorn Ranch, Roosevelt drafted a reply to the marquis's note. It was conciliatory but unflinching:

> Most emphatically I am not your enemy; if I were you would know it, for I would be an open one, and would not have asked you to my house or gone to yours. As your final words, however, seem to imply a threat it is due to my self to say that the statement is not made through any fear of possible consequences to me: I too as you know, am always on hand, and ever ready to hold myself accountable in any way for anything I have said or done.
>
> Yours very truly
Theodore Roosevelt

No duel ensued, even though the marquis was never known to have backed down in any situation in which he believed he was right. Thus, it is to be presumed that he realized that Roosevelt was not his enemy. He was certainly mistaken about Joe Ferris furnishing money to prosecution witnesses. There were no bankers in Medora, and Ferris, who Roosevelt had helped open a general store, was merely acting as one.

The marquis was found innocent of the charges against him, but time was already running out on his enterprises. The stagecoach line between Medora and Deadwood lost money and the pottery never materialized. The sheep turned out to be the wrong breed for the climate and half of them died. The railroads, undoubtedly working hand in glove with the beef trust, refused to grant him the same rebates on freight rates they gave his competitors, adding to his costs. And range-fed beef turned out to be less popular with consumers than beef that had been fattened in the stockyards of Chicago. The marquis's father-in-law withdrew his financial backing and soon the packing plant closed. Not long after, just as winter was settling in on the Bad Lands in 1886, de Morès and his wife left Medora for good. The short-lived reign of the Emperor of the Bad Lands was over.*

*   *   *

---

*Back in France, the marquis claimed the Chicago beef trust was dominated by Jews and announced himself the victim of "a Jewish plot." Turning to politics, he organized a movement that mixed socialism with a rabid anti-Semitism that fed the mania which led to the Dreyfus affair. In 1896, he was killed by North African tribesmen while carrying out a wild scheme to unite the Muslims in a holy war against the British and the Jews.

Throughout his years in the Dakota Territory, Roosevelt kept up a running correspondence with Henry Cabot Lodge, who became his best friend. In many respects, Lodge served as the elder brother he never had. "I can't help writing you, for . . . there are only one or two people in the world, outside my own family, whom I deem friends or for whom I really care," Roosevelt once told him. Eight years older than Roosevelt, Lodge had a background and interests similar to those of his friend. He was a member of an old and wealthy Boston family, a fellow Porcellian, was one of the first three men awarded the doctor of philosophy degree by Harvard, had already published five books, and was politically ambitious.

But there were numerous differences between the two men. Lodge's drooping eyelids and superior manner repelled people, and he had a voice that rasped like the tearing of a bed sheet. Nannie Lodge, his beautiful wife, spent much of her time placating people her husband had offended. More calculating and cautious than Roosevelt, the austerely elegant Boston Brahmin supplied a check to his friend's impetuosity. He recognized the high qualities in the young Roosevelt, fed his ambitions, sustained him when others—and Roosevelt himself—doubted his political future, and played a vital role in advancing his career.

In September 1885, Roosevelt came in from the Dakota Territory to attend the New York State Republican convention at Saratoga, where a gubernatorial candidate was to be chosen. His influence was peripheral—the mugwump press was still denouncing him for defecting to Blaine the previous year—but he added a new dimension to his attacks on the Democrats during the ensuing campaign on behalf of the unsuccessful Republican candidates. It was an article of faith among Republicans that had southern blacks been permitted to cast ballots in 1884, Blaine rather than Cleveland would have been elected president. In preventing blacks from voting, Roosevelt declared, southerners usually argued they were unfitted to exercise their political rights; they did not want them, and if they wanted them, they should not have them.

"Can these be called fair answers?" he asked. "The second proposition is untrue, the third we ourselves should do our best to render untrue; and the first cannot be advanced in good faith as a reason for fraudulently disfranchising a class of our fellow citizens by any man who honestly believes in our American theory of government." Today, Roosevelt's views on civil rights would be considered paternalistic, but they were far more advanced than those of most of his contemporaries—and continued to be so.

Roosevelt's denunciations of some southerners led him into a nasty exchange of letters with former Confederate President Jefferson Davis that reflected no credit upon him. In an article in the *North American Review,* he compared Davis unfavorably with Benedict Arnold, and Davis protested. Roosevelt's reply,* written in the third person, was perfunctory and ill became a youth of twenty-six in dealing with a man a half-century his senior. In later years, he regretted his rudeness. "I answered with an acerbity which being a young man, struck me as clever," he wrote in 1905. "It does not strike me as in the least so now." Nevertheless, it emphasized a streak of ruthlessness that friends and enemies sensed at other points of his career.

The house at Oyster Bay was completed and he had reached an understanding with Bamie, guardian of Baby Lee, giving her the use of the place while he made his city headquarters at her home, 422 Madison Avenue. Sometimes he visited his aunt Anna at her Long Island estate, where he playfully cuddled Elliott's baby daughter Eleanor. Bamie seemed headed for spinsterhood, and the consuming passion of her life was the welfare of her widowed brother. She often gave teas and "evenings" at home, summer dinner dances and hunt breakfasts at Oyster Bay that were designed to ease him back into the life he had formerly led.

Roosevelt adopted the life-style of a Long Island country squire, played polo, and joined the Meadowbrook Hunt. The challenge and excitement of riding to hounds appealed to him, and he was as proud of his pink coat as he was of the beaded buckskin shirt he had worn in the Bad Lands. Hunting with the

---

*"Mr. Theodore Roosevelt is in receipt of a letter purporting to come from Mr. Jefferson Davis, and denying that the character of Mr. Davis compares unfavorably with Benedict Arnold. Assuming the letter to be genuine Mr. Roosevelt has only to say that he would indeed be surprised to find that his views of the character of Mr. Davis did not differ radically from that apparently entertained in relation thereto by Mr. Davis himself. Mr. Roosevelt begs leave to add that he does not deem it necessary that there should be any further communication whatever between himself and Mr. Davis." *Letters,* Vol. I, p. 93.

Meadowbrook on the day before his twenty-seventh birthday, he was leading the pack when his horse missed a five-foot fence, fell, and rolled over on him. Roosevelt's left arm was broken and his face badly cut up, but he managed to remount and was in at the kill. When he returned to his stables, looking "pretty gay, with one arm dangling, and my face and clothes like the walls of a slaughterhouse," he found Baby Lee and her nurse waiting for him. He hopped down from his horse and playfully ran toward the child, but she screamed in terror at the bloody apparition and fled.

That evening, face patched and arm in splints, Roosevelt presided over a hunt ball at his home. He twirled one of the guests— Edith Carow—about the floor with cheerful abandon. "I don't grudge the broken arm a bit," he wrote Lodge a few days later. "I am always willing to pay the piper when I have a good dance; and every now and then I like to drink the wine of life with brandy in it." Not long afterward, he and Edith became secretly engaged.

Following his marriage to Alice, Roosevelt had retained a casual friendship with Edith, but Alice complained that Edith was cool and distant. Perhaps Edith's reserve was based upon the fact that as the Roosevelts' star was ascendant, her own was sinking. The death of her father, who had drunk up most of the family money, left Edith, her mother, and sister Emily teetering on the brink of genteel poverty. In fact, Mrs. Carow decided to remove herself and her daughters to Europe, where living costs were cheaper. After Alice's death, Edith seems to have made up her mind to land her old beau—and she was a very determined woman. "She wanted him very much," a relative was later quoted as saying. "She was passionately in love with him." Perhaps Roosevelt, very much the Victorian and still faithful to the memory of Alice, understood this, for he told Bamie that he wanted to be warned when Edith visited the Madison Avenue house so he could avoid her.

This warning system broke down, and either by chance or design, Roosevelt and Edith confronted each other one day on the stairs in the front hallway. What passed between them is unknown, but undoubtedly this meeting of former lovers triggered conflicting memories and emotions. They had known each other since the nursery; now they were estranged. Yet, as awkward words of greeting passed between them during these moments on the staircase, long-suppressed feelings apparently came to the surface. Soon, they were seeing each other again, first privately and then in public. For days, his diary contained no entry but the letter *E*. Roosevelt discovered that Edith had blossomed physically since their last encounter, and she was more tolerant and sympathetic than the angular and moody young girl of the past. She found him completely different from the youth she had known, a masterful man of the world.

As his loyalty to Alice's memory weakened, Roosevelt tormented himself with recriminations. "I have no constancy—no constancy," he angrily told a friend after pacing his room half the night. A man loved a woman, a woman loved a man until death—and beyond. To love another when the loved one had died only two years before was, in Roosevelt's mind, a form of infidelity. But he was young, in love, and very much wanted a family and a home. On November 17, 1885, he asked Edith to marry him, and she accepted.

To allow for a proper period of mourning for Alice, they agreed to keep their engagement a secret even from Roosevelt's sisters and to wait a year before being married. Victorian propriety would be maintained and it provided time to sort out several problems: what to do about Baby Lee; allow Edith to help get her mother settled, and decide where they would live, for Edith had no intention of becoming a rancher's wife in the wilds of Dakota. In the meantime, Roosevelt returned to the Bad Lands to look after his ranch while Edith joined her mother in London.

One morning early in April 1886, Dr. Victor H. Stickney, the only physician in the cowtown of Dickinson, about forty miles east of Medora, was on his way home for lunch when he spotted "the most bedraggled figure I'd ever seen come limping down the street." Covered with mud, his clothes in shreds, and "all teeth and eyes . . . he stopped me with a gesture, asking me whether I could direct him to a doctor's office." Stickney identified himself as a physician and the stranger declared, " 'By George, then you're exactly the man I want to see . . . my feet are blistered so badly that I can hardly walk. I want you to fix me up. . . .' "

The battered stranger was Theodore Roosevelt and the story he told Stickney—which became one of his favorite after-dinner tales—concerned his last and most exciting adventure in the Bad Lands. Not long after he had returned to the Elkhorn Ranch from New York, Bill Sewall reported that the boat they used to carry supplies across the Little Missouri had been stolen. Roosevelt immediately suspected Mike Finnegan, who lived nearby with two equally bad characters, a half-breed named Burnsted and an old German named Pfaffenbach. The trio were known horse thieves and only one jump ahead of a posse of vigilantes intent on a lynching, so they had probably taken the boat and disappeared downstream while the river was at flood.

The boat itself was worth only about $30, but Roosevelt was angered by the theft and as a Billings County deputy sheriff, he resolved to catch the thieves. "In any wild country where the power of the law is little felt or heeded," he declared, "and where everyone has to rely upon himself for protection, men soon get to feel that it is in the highest degree unwise to submit to any wrong without making an immediate and resolute effort to avenge it upon the wrongdoers at whatever cost or risk of trouble." The decision to follow the desperadoes into the wildest part of the Bad Lands during the worst weather of the year was a challenge not to be undertaken lightly, however.

Sewall and Dow, both experienced woodsmen, immediately began building another boat, and the craft slid out into the Little Missouri at the end of March. Roosevelt, who thought the chase might make a good magazine story, brought along a camera as well as several books to read. For three days and nights, they were battered by a howling wind and subzero temperatures. At times they drifted between walls of ice that rose ten feet on both sides of the river. Finally, on the third day, about a hundred miles downstream from the Elkhorn Ranch, they spotted the missing boat moored against the riverbank and saw smoke curling upward from a nearby campfire.

Pfaffenbach, in camp alone, was taken by surprise and said the others were out hunting. When Burnsted and Finnegan returned, they were immediately covered and told to throw down their weapons. "The half-breed obeyed at once, his knees trembling," wrote Roosevelt. "Finnegan hesitated for a second, his eyes fairly wolfish; then as I walked up within a few paces, covering the center of his chest [with a cocked rifle] so as to avoid overshooting, and repeating my command, he saw that he had no show, and, with an oath, let his rifle drop and held his hands up beside his head."

Most westerners would have immediately hanged the thieves on the nearest tree, or shot them on the spot and then returned to the ranch by foot. But Roosevelt was determined to bring his prisoners in to face justice, and the only way to do that was to follow the course of the river and hope for a thaw. Over the next six days, the captors poled their way toward Mandan, the first sizable town downriver, about 150 miles away. Little headway was made because of the freezing temperatures and the ice floes. Every night, they had to take turns guarding the prisoners, because if they were tied up their arms and feet would freeze. "There is very little amusement in combining the functions of a sheriff with those of an Arctic explorer," Roosevelt grimly observed.

Once, they were almost sucked under the ice jam and huge blocks of ice fell into the river, threatening to capsize the craft. Food ran low, game was scarce, and they were reduced to eating flour mixed with muddy river water. A rather strange camaraderie grew up between captors and captives, and all the men talked and joked together. Roosevelt even borrowed from the prisoners a copy of a dime novel called *The History of the James Brothers*. But his main reading during these six days was a French translation of *Anna Karenina,* which he read from cover to cover. He pronounced the book of "very more interest than I have found any other novel for I do not know how long."*

---

*Writing to Corinne after his arrival in Dickinson, he said:

"I hardly know whether to call it a very bad book or not. There are two entirely distinct stories in it; the connection between Levines story and Annas is of the slightest, and need not have existed at all. Levines and Kitty's history is not only very powerfully and naturally told, but is also perfectly healthy. Annas most certainly is not, though of great and sad interest; she is portrayed as being prey to the most violent passion, and subject to melancholia, and her reasoning power is so unbalanced that she could not possibly be described otherwise than as in a certain sense insane. Her character is curiously contradictory; bad as she was however she was not to me nearly as repulsive as her brother Stiva; Vronsky had some excellent points. I like poor Dolly—but she should have been less of a patient Griselda with her husband. You know how I abominate the Griselda type. Tolstoi is a great writer. Do you notice how he never comments on the actions of his personages? He relates what they thought or did without any remark whatever as to whether it was good or bad, as Thucydides wrote history—a fact which tends to give his work an unmoral rather than an immoral tone." *Letters,* Vol. I, p. 98.

Later that spring, he read *War and Peace* while on a roundup, and regarded it as a masterpiece, although he was critical of Tolstoi's indictment of war. Like almost everyone

Finally, they reached a cow camp, where Roosevelt told Sewall and Dow to continue on to Mandan with the boats while he took the prisoners overland to Dickinson. He hired a wagon and team from a nearby rancher—who was surprised that he hadn't already strung up his captives—and piled them in. Not completely trusting the ranchman, he walked behind the wagon with his Winchester at the ready. The long-awaited thaw had set in, and the prairie was now a vast sea of mud. "I trudged steadily the whole time behind the wagon through the ankle-deep mud," Roosevelt related. " . . . Hour after hour went by always the same . . . hunger, cold and fatigue struggling with a sense of dogged, weary resolution. At night, when we put up at the squalid hut of a frontier granger, I did not dare to go to sleep. . . . I sat up with my back against the cabin door and kept watch over them all night long. So after thirty-six hours sleeplessness, I was most heartily glad when we at last jolted into the long, straggling main street of Dickinson, and I was able to give my unwilling companions into the hands of the sheriff."*

Following his usual pattern, Roosevelt mixed intellectual work with his strenuous outdoor life. In between the capture of the three thieves and the annual cattle roundup, he wrote a biography of Senator Thomas Hart Benton, the western expansionist. The contract for the book, part of the American Statesmen series, was procured for him by Cabot Lodge, who was doing George Washington for the series. Roosevelt's task was complicated by the fact that he had almost nothing in the way of research materials. Writing to Lodge at one point, he wryly implored him to hire a research assistant in Boston to look up some of the basic data he needed. ". . . Being by nature both a timid and, on occasions, by choice a truthful man, I would prefer to have some

---

who has read the novel, he fell in love with Natasha. She "is a bundle of contradictions, and her fickleness is portrayed as truly marvelous; how Pierre could ever have ventured to leave her alone for six weeks after he was married I cannot imagine." *Letters,* Vol. I, pp. 103–104.

*Under Dakota law, Roosevelt received a fee as deputy sheriff for making the arrests and mileage for the three hundred miles covered—a total of about $50. Finnegan and Burnsted were each sentenced to three years in the territorial penitentiary at Bismarck, while Roosevelt dropped the charges against Pfaffenbach, saying, "He did not have enough sense to do anything good or bad." Following his dismissal, the old man expressed fervent gratitude, which moved TR to observe it was the first time anyone had thanked him for calling them a fool.

foundation of fact, no matter how slender, on which to build the airy and arabesque superstructure of my fancy—especially as I am writing a history."

The book was written at high speed. On April 29, he told Bamie that he had completed only one chapter; early in June, he announced that the entire manuscript, totaling some 85,000 words, was nearly finished. "Some days he would write all day long," recalled Sewall. "Some days only part of the day . . . Sometimes he would get so he could not write. Then he would take his gun and saunter off." Much of the book was written in a room over Joe Ferris's store in Medora, where he often worked past midnight by the light of a kerosene lamp.

Thomas Hart Benton is a journeyman effort, but every page bears the imprint of Roosevelt's experiences in the West. The first three chapters are as good a brief statement of our early western development as has been written. But the book is more notable for what it reveals about the author than the subject. Roosevelt always wrote best making a point, correcting a misconception, demolishing an error—in short, when struggling if not against an opponent of flesh and blood, then against a defective idea. In *Benton,* he vigorously praised manifest destiny, championed the Union, and lashed out against the abolitionists— and by indirection at the mugwumps—for extremism.

In supporting Benton's anti-British stand in the Oregon boundary dispute of the 1840s, he revealed his own views on territorial expansion: "By right we should have given ourselves the benefit of every doubt in all territorial questions and have shown ourselves ready to make prompt appeal to the sword whenever it became necessary as a last resort." Unsurprisingly, some critics, like *The Nation,* found "too much muscular Christianity, minus the Christian part" in Roosevelt's philosophy. Sales were brisk, however, and the book remained the standard work for two decades.*

Throughout that summer, he and Edith exchanged frequent letters, most of which she burned after his death. The fragments that survived are full of tenderness and youthful ardor: "You know

---

*American Statesmen authors were offered the choice of a $500 flat fee for their book or a percentage of the profits. I could not readily determine TR's choice, but I hope he selected the latter in view of *Benton*'s record of solid sales.

all about me, darling," she wrote. "I have never loved anyone else. I love you with all the passion of a girl who has never loved before." She was enchanted by his picture of life out west and did not think him "sentimental in the least to love nature; please love me too and believe I think of you all the time and want so much to see you. . . ."

While Roosevelt professed to have little interest in reentering politics, he was inclined to accept when sounded out about assuming the presidency of the New York City Board of Health. And when relations with Mexico were strained by a border incident, he wrote the War Department, offering to raise several companies of mounted riflemen from among the "harum-scarum rough riders" of the Dakotas. "I haven't the least idea there will be any trouble," he told Lodge, "but as my chance of doing anything in the future worth doing seems to grow continually smaller I intend to grasp at every opportunity that turns up." The New York job failed to materialize and there was no war with Mexico.

Not long after he had finished *Benton,* Roosevelt was "savagely irritated" to learn that toward the end of August 1886, *The New York Times* had published a small social item announcing the engagement of former Assemblyman Theodore Roosevelt and Miss Edith Carow. A week later, it was followed by a retraction, probably inserted at the insistence of his elder sister. When Roosevelt learned of the exchange, he wrote "Darling Bamie" a letter full of guilt and embarrassment—a letter so remarkable it was suppressed by the Roosevelt family for nearly a century.

"The statement itself is true," he wrote. "I am engaged to Edith and before Christmas I shall cross the ocean to marry her." Yet, still troubled by the memory of the beautiful, dead Alice Lee, he added: "I utterly disbelieve in and disapprove of second marriages; I have always considered that they argued weakness in a man's character. You could not reproach me one half as bitterly for my inconstancy and unfaithfulness, as I reproach myself. Were I sure there were a heaven my one prayer would be I might never go there, lest I should meet those I love on earth who are dead. . . ." And, uncertain of Edith's attitude toward the child of her rival, he regarded Baby Lee as some sort of peace offering to regain his sister's goodwill, telling Bamie that she could

retain custody of the child to whom she had grown greatly attached.

Both Bamie and Corinne were shocked by this letter. Although
they had known Edith Carow since childhood—she and Corinne
had been best friends—they felt threatened by her. Even though
Roosevelt had told Bamie that "you will be giving me the greatest
happiness in your power if you will continue to pass your summers
with me," they realized that Edith, unlike Alice, was possessive
and feared she would come between them and their adored
brother. "They didn't want it at all," Alice Longworth later recalled, "because they knew her all too well, and they knew they
were going to have a difficult time with her, that she would come
between them and Father."

The situation in the Bad Lands was also growing ominous.
Overgrazing and drought were undermining the prospects for
making a success of ranching. "Overstocking may cause little or
no harm for two or three years, but sooner or later there comes
a winter which means ruin to the ranches that have too many
cattle on them," Roosevelt warned. Little rain had fallen that
summer and the grass was parched, adding to the impending
threat of disaster. With large numbers of cattle being sent to
market, prices were falling. Will Dow, who had gone to Chicago
with the fall shipment, reported that the price offered was less
than it cost to raise and transport them. Sewall and Dow decided
that "we were throwing away Roosevelt's money," and the two
woodsmen, who had never been comfortable in the Bad Lands
anyway, told him they had decided to return to Maine.

Roosevelt made no effort to persuade them to reconsider. Anxious to appease his sisters over his coming marriage, he, too,
wished to hurry east. He placed his herd under the management
of Ferris and Merrifield, and left for New York at the end of
September. The day before his departure, he and Sewall took a
long walk together on the prairie. Roosevelt asked if he should
go into politics or the law, and Sewall advised him to go into
politics, "because such men as he didn't go into politics and they
were needed in politics. If you go into politics and live," he added,
"your chance to be President is good." Roosevelt threw back his
head and laughed. "Bill, you have a good deal more faith in me
than I have myself. That looks a long ways ahead to me."

*   *   *

Hardly had Roosevelt unpacked his bags in New York City when he was visited by a group of influential Republicans who urged him to become the party's candidate for mayor. The honor was a dubious one, since they were looking for a sacrificial lamb. No one gave the Republican nominee much of a chance in the three-way race for City Hall in which the other candidates were Henry George, originator of the "single tax" and candidate of the United Labor Party, and Abram S. Hewitt, the Democratic nominee. Under no illusions about the reason for his nomination or prospects for victory, Roosevelt accepted anyway, mainly to reestablish himself in politics.

Henry George is one of those curiosities periodically churned up by the bubbling cauldron of American politics.* Born in Philadelphia, he had dropped out of school before he was fifteen, worked as an itinerant printer, and in 1879 self-published his masterwork, *Progress and Poverty*. In this immensely popular book, he blamed depressions and the poverty of ordinary Americans on wild land speculation, and espoused the single tax—a massive levy on the profits of increased real estate values—which would obviate the need for all other taxes.

George's slashing indictment of the injustice of the American economic system drew huge crowds and alarmed the so-called "better element," already badly frightened by a wave of labor unrest that was sweeping the country. Only a few months before, on May 4, 1886, eight policemen had been killed by a bomb while trying to break up a demonstration in Chicago's Haymarket Square protesting police brutality against strikers at the Mc-Cormick Harvester plant. Although the actual perpetrators of the outrage were never caught, an inflamed public blamed foreign-born anarchists who were active in the turbulent Chicago labor movement.†

---

*Henry George's granddaughter and biographer is Agnes de Mille, the noted choreographer.

†Like many Americans, Roosevelt was outraged by the bombing. "My men are hardworking, labouring men, who work longer hours for no greater wages than many of the strikers; but they are Americans through and through," he wrote Bamie. "I believe nothing would give them greater pleasure than a chance with their rifles at one of the mobs. When we got the papers, especially in relation to the dynamite business they became more furiously angry and excited than I did. I wish I had them with me, and a fair show at ten

Richard Croker, the wily boss of Tammany Hall, seeing an opportunity to attract the votes of frightened businessmen and other respectable citizens, had given the Democratic nomination to Hewitt, a wealthy reform-minded industrialist, rather than to the usual Tammany wheelhorse, and he was expected to win. Roosevelt did not fear defeat, which was a foregone conclusion, but dreaded not making a good showing. "If I make a good run, it will not hurt me," he wrote Lodge, who was running for Congress in Boston, "but it will if I make a bad one as is very likely."

Roosevelt never entered any fight without making a maximum effort, and he campaigned all over the city with his usual energy. For the next three weeks, the "Cowboy Candidate" worked eighteen-hour days, appeared at four and five rallies a night, cajoled voters, exuded confidence in the presence of newspaper reporters, and shook outstretched hands wherever he went. Most newspapers supported Hewitt, but the *Times* backed Roosevelt, stating that "he excites more confidence and enthusiasm than has been inspired by any candidate in a mayoralty contest within the memory of this generation of voters." "He is very bright and well considering the terrific strain," Bamie wrote Edith. "It is such happiness to see him at his very best once more."

But in spite of all the enthusiasm, Roosevelt knew he did not have a chance. Writing to Lodge on the eve of the election, he acknowledged: "I have made a rattling canvass" but the "timid good" were slipping away to Hewitt. The mugwumps, he contemptuously noted, were acting "with unscrupulous meanness and a low, partisan dishonesty and untruthfulness which would disgrace the veriest machine heeler. May Providence in due season give me a chance to get even with some of them!" So certain was he of defeat that he and Bamie secretly booked passage under the names of "Mr. and Miss Merrifield" on the *Etruria*, which was to sail for England on November 6, four days after the election.

---

times our number of rioters; my men shoot well and fear very little." *Letters,* Vol. I, pp. 101–102.

Eight anarchists were found guilty of murder, although there was no evidence linking them to the bombing. Seven were sentenced to hang and one was given life imprisonment. Of the seven, one committed suicide, four were hanged, and two were given life terms. Six years later, the three surviving anarchists were pardoned by Governor John P. Altgeld of Illinois on grounds that they had not received a fair trial. Altgeld's political career was destroyed by this act of clemency.

Roosevelt was defeated as expected—"worse even then I had feared," he telegraphed Lodge, who had finally won his own seat in Congress. He trailed both Hewitt and George, winning about 15,000 fewer votes than Republican candidates usually received in New York City. Fearing both the single-tax messiah and the youthful Republican standard-bearer, many of the party faithful deserted to the Democratic candidate. Roosevelt took consolation in the fact that "at least I have a better party standing than ever before." He had paid his dues and the Republicans now owed him.

Roosevelt and Bamie slipped on board their ship a few days later. They spent their last night in New York addressing the formal announcements of Theodore's engagement and forthcoming marriage. Not long after sailing, the "Merrifields' " disguise was penetrated by an engaging young British diplomat named Cecil Spring Rice, and they enjoyed his company on the crossing. When Theodore Roosevelt and Edith Carow were married at St. George's Church, Hanover Square, on December 2, 1886, "Springy" was best man.

Chapter 9

"I'm a Literary Feller"

Brisk winds blew in off Long Island Sound, chopping white-capped V's into the gray water as a carriage carrying Theodore and Edith Roosevelt drew up before the house at Oyster Bay near the end of March 1887. In deference to his new wife, Roosevelt had changed the name of the estate from Leeholm to Sagamore Hill.* No servants greeted them, for the house was closed for the winter. Edith had been a guest there, but she looked about with fresh interest as she moved along the gloomy paneled hall lined with the mounted heads of big game shot by her husband and through the rooms filled with white-shrouded furniture. Now it was her task to transform this sprawling house, built for another, into a comfortable home for her own family, for Edith was already pregnant with their first child.

The newlyweds had enjoyed a four-month European honeymoon, carefully plotted by Roosevelt to avoid places associated with his travels with Alice Lee. They spent considerable time in

---

*"Sagamore Hill takes its name from the old Sagamore Mohannis, who as chief of his little tribe, signed away his rights in the land two centuries and a half ago," TR wrote in his *Autobiography*, p. 328. Sagamore means chief.

England, where, thanks to introductions arranged by Cecil Spring Rice, they were taken up by society. Roosevelt was put down for the Athenaeum and St. James clubs, and dined with distinguished political and literary figures, including Sir James Bryce,* George Otto Trevelyan, and Robert Browning. They were invited to weekends at great country houses. Winter was passed on the Riviera, and then the Roosevelts went on to Rome, Florence, Venice, and Milan. Looking back, Edith recalled these weeks as the most romantic time of her life. "I love you all the time in my thoughts and think of our honeymoon days, and remember them all one by one, and hour by hour," she wrote her husband years later.

In Florence, they were shocked to learn that the worst blizzards recorded in frontier history had struck the Bad Lands that winter and the range had been transformed into a frozen hell. Thousands of cattle, already weakened by the summer drought, were wiped out, including most of Roosevelt's own herds. Faced with the prospect of financial disaster after several years of lavish spending, Roosevelt nearly panicked. He considered closing, or even selling, Sagamore Hill "and going to the ranch for a year or two" to recoup his fortunes. "If I stay east I must cut down tremendously along the whole line," he wrote Bamie. Should he sell his hunter, reduce the gardening staff, or try to grow his own hay? He decided not to lay in a stock of French claret because he felt he couldn't afford it. Edith, unhappily familiar with making do on a limited budget, worked on plans to run Sagamore with about half the money Bamie had had at her disposal.

The honeymooners were also troubled by the question of Baby Lee's custody. When Edith learned of her husband's offer to turn the care of the three-year-old child over to Bamie, she immediately objected. "I hardly know what to say about Baby Lee," Roosevelt wrote his sister with considerable embarrassment. "Edith feels more strongly about her than I could have imagined possible. However, we can decide it all when we meet."

Immediately after their return to New York, the Roosevelts hastened to Bamie's home at 689 Madison Avenue, where they would live until Sagamore Hill was reopened. Little Alice, all

---

*Bryce mentioned TR favorably in his classic *American Commonwealth,* calling him "one of the ablest and most vivacious of the younger generation of American politicians."

blond curls and blue eyes and wearing her best dress and a pink sash, greeted her new stepmother with a bunch of pink roses nearly as big as herself. From the beginning, Edith insisted that the child call her "Mother." Final details of her custody were worked out at a family conference. To soften the blow of leaving Bamie—her "Aunt Bye"—she was left with her until May while the family prepared for the move out to Long Island. "It almost broke my heart to give her up," Bamie recalled years later. "Remember, darling, if you are very unhappy you can always come back to me," she told the sad-eyed little girl as she prepared to send her to her new home.

Roosevelt went west to assess the damage from the blizzard as soon as he could get away. Even though he had expected the worst, he was shocked by the widespread devastation that greeted his eyes. The melting snows slowly revealed a grisly scene of death and destruction. The rotting carcasses of steer covered the plains, and they packed the gullies and creek bottoms near the ranch like cordwood. Men had ridden out into the storm to look after their cattle and never returned. Children had lost their way between ranch house and stable and were frozen to death only a hundred yards from their homes. Only the wolves and coyotes had thrived that winter. Most of the ranchers were bankrupt, and those who were left were soon to express their frustration in radical Populism. Only a handful of starving cattle remained of Roosevelt's herd, and he faced the loss of almost his entire investment.

"I am bluer than indigo about the cattle; it is even worse than I feared," he wrote Bamie. "I wish I was sure I would lose no more than half the money . . . I invested out here." The more he looked, the more he was depressed. "For the first time I have been utterly unable to enjoy a visit to my ranch," he told Cabot Lodge. "I shall be glad to get home." Before returning to New York, Roosevelt began liquidating his investment in the Bad Lands—and his direct association with the West. Although he came back to the Elkhorn Ranch several times, it was only for short hunting trips, never again to stay very long.*

---

*By the time the Bad Lands operation was finally liquidated in 1899, TR had suffered a loss of about $23,500. If one includes the lost interest of 5 percent on his total investment, Hermann Hagedorn estimates his total loss at about $50,000. *Roosevelt in the Bad Lands*, Appendix III.

*   *   *

For the moment, Roosevelt's career was "in irons"—to use a nautical term that would have been familiar to the author of *The Naval War of 1812*. Like a square-rigger losing headway before the wind, he seemed to be drifting backward onto a dangerous shore. Douglas Robinson, who was now managing his financial affairs, warned that he was digging into his capital at the alarming rate of $2,500 a year—and there was no more heinous sin in Knickerbocker society than to touch capital. Faced with the need to make money, he cast about for a way to increase his income, but the prospects were bleak. The Democrats were in firm control in New York City, Albany, and Washington, so there was no possibility of a political appointment to relieve his financial distress. "I intend to divide my time between literature and ranching," Roosevelt told a reporter who inquired about his political plans upon his return from Europe.

In fact, he appeared to go out of his way to complete his banishment from political life. Some of his supporters from the mayoralty election tendered him a dinner at Delmonico's on the evening of May 31, 1887. Expecting a few light remarks, they had settled in with their after-dinner brandy and cigars only to be shocked by an angry tirade against everything and everyone. President Cleveland, the mugwump press, the Immigration Department, the Anti-Poverty Society—nothing was safe from the torrent of invective. Roosevelt seemed intent on discharging a cargo of bile that he had carried since the election, but this scorched-earth policy only convinced critics of his basic instability. "It was a mistake ever to take him seriously as a politician," observed E. L. Godkin in the *New York Post*. And the satirical magazine *Puck* told young Mr. Roosevelt, "You are not the timber of which Presidents are made."

The speech revealed that traits that would be exhibited by Roosevelt as president were already in place. The vigorous expression of his opinions, his directness, the moral righteousness of his own views had crystallized. Over the years, he gained experience but essentially he remained unchanged. Still muttering imprecations against the mugwumps—"If I get an opportunity I am going to sail into . . . [them] with a sword dipped in vitriol," he told Lodge—Roosevelt went into exile at Sagamore Hill.

Working "like a couple of dusty, not to say grimy beavers," the Roosevelts rearranged furniture, hung animal heads, and shelved hundreds of books in the sprawling house. Edith resigned herself to the heavy masculinity of the place, but appropriated one of the front rooms as her refuge from the cares of running a large household. Roosevelt's own private sanctum, the Gun Room, which contained his collection of weapons and western memorabilia, was on the third floor and faced west over the treetops to the blue waters of the Sound.

Life at Sagamore quickly settled into a pattern. Because of financial constraints, Edith's pregnancy, and her preference for solitude, the Roosevelts had few visitors that summer except for Cecil Spring Rice, who had been assigned to the British legation at Washington. Bamie had a standing invitation to visit, but due to the tension between herself and her sister-in-law, she rarely made an appearance. Every morning, Edith organized the housework, sewed, or answered letters; in the afternoon, she went walking in the woods with her husband or he took her out rowing. While he pulled on the oars, she read aloud from Browning or Thackeray or another of their favorite authors. Less frequently, they rode over to the estate leased by Elliott Roosevelt and his wife, Anna, at Hempstead. Elliott now showed every sign of being a confirmed alcoholic and there were several embarrassing incidents.

Roosevelt's enforced leisure provided him with his first opportunity to get to know his little daughter and he found her "too good and cunning for anything." Alice watched him shave, supervised his tennis, and demanded that he carry her down to breakfast piggyback every morning with the cry "Now, pig! Now, pig!" On rainy days, they sat on the floor together and built forts from her piles of blocks. In June 1887, Roosevelt began work on his fourth book, a biography of Gouverneur Morris, a Revolutionary era political and diplomatic figure. Once again, Lodge had come to his rescue by persuading John T. Morse, the editor of the American Statesmen series, to commission Roosevelt to write a companion volume to his *Benton*. "Theodore . . . *needs the money*," he told the reluctant Morse.

Intelligent, sardonic, and a *bon vivant,* the peg-legged Morris appealed to Roosevelt as an "entertaining scamp," but he quickly ran into a problem. "The Morrises won't let me see the old gentleman's papers at any price," he told Lodge. "I am rather in a quandary." Fortunately, there was enough material in the public domain, including Morris's dramatic diary of his service as American minister in Paris during the French Revolution, to enable the ever-resourceful author to flesh out the work.

Written at Roosevelt's usual fever pitch, the book was finished by early September. While much of it is heavy going, the pages that cover Morris's career in France and his involvement in a plot to save King Louis XVI and Marie Antoinette from the guillotine are bright and sprightly. Like his earlier works, it reveals much of his own thinking and character. There is the usual contempt for Jefferson and his works, sideswipes at secondary figures such as "the filthy little atheist" Thomas Paine, and the account of the French Revolution fairly crackles with his hatred of political disorder and mob violence. The Federalists, the creators of his own political philosophy, are lauded for their nationalism and sound economic policies, but sharply criticized for their opposition to western expansionism, their disloyalty during the War of 1812, and their aristocratic disdain of democracy.

In words that foreshadowed Roosevelt's future brief against the blind conservatism of some Republican leaders, he wrote: "In a government such as ours, it was a foregone conclusion that a party which did not believe in the people would sooner or later be thrown out of power. . . . This distrust was felt, and of course excited corresponding and intense hostility. Had the Federalists . . . freely trusted in the people the latter would have shown their trust was well founded; but there was no hope for leaders who . . . feared their own followers."

Roosevelt argued that a nationalist should also be a democrat. This marriage of nationalism and democracy was the cornerstone of his political philosophy. Previously, nationalism had been linked to conservatism, as in the case of the Federalists, while democrats, such as Jefferson, tended toward states' rights and laissez-faire. Although *Gouverneur Morris* was regarded by viewers as the weakest of Roosevelt's books, it served as the standard life until the family finally released Morris's papers four decades later.

Undoubtedly recalling the tragic fate of Alice Lee, Roosevelt awaited the birth of his second child with a mixture of anticipation and fear. The resulting tension brought about a surprise recurrence of his asthma, but it vanished just as quickly as it had come amid the excitement surrounding the arrival of the baby. Edith was not expected to give birth until the latter half of September, but on the evening of September 12, she suddenly went into labor. The nurse who was to help with the delivery had not arrived, but fortunately Dr. James West Roosevelt lived nearby and was in attendance when a baby boy "came with a rush" at 2:15 A.M. "Edie is getting along well," her husband reported to Bamie later that day. "The boy is a fine little fellow about 8½ pounds." He was christened Theodore Roosevelt, Jr.—and there was much happiness in the family, for unlike the birth of little Alice there was no funeral four days later.

The baby was the center of the Roosevelts' universe—to Edith because he was her first child; to her husband because he was his heir and namesake. The new mother's recovery was slow. But to the distress of Bamie, always on call for family emergencies, Edith turned for help to her old Irish nurse, Mary Ledwith, known to everyone as Mame, rather than to her sister-in-law, obviously to prevent Bamie from putting down roots at Sagamore. The boy was "a very merry lovable little fellow" who was soon crawling about the floor "just like one of Barnum's little seals," the proud father observed. "He plays more vigorously than anyone I ever saw." Alice was also fascinated by her new brother. She refused to allow her tiny rocking chair to be moved from beside his crib and boasted: "*My* little brother's a howling polly parrot." She watched intently as he "eats Mama" and refused to go to Chestnut Hill for a periodic visit to her Lee grandparents unless she could take him with her.

Over the next decade, Edith was either getting over a baby or having one. Ted was followed by four more children: Kermit, born in 1889, Ethel in 1891, Archibald in 1894, and Quentin in 1897. As the number of children increased, there was considerable rivalry among them for their parents' attention. Looking back over her long life on her ninetieth birthday, Alice observed her childhood was hardly a period of unalloyed joy. She said she had really been the "ugly duckling" of the family. "My brothers used to tease me about not having the same mother. They were very cruel about it and I was terribly sensitive."

For several years, she suffered from what was later suspected to be a mild case of polio that left one leg shorter than the other. Every night before Alice went to bed, her stepmother would "stretch each foot in a steel contraption that rather resembled a medieval instrument of torture, and for several years I wore braces on each leg from ankle to knee," Alice wrote in her memoirs. "Even when walking on level pavement they used to catch and throw me." Eventually, she was able to discard the braces and there was no hint in later years that she had ever been lame.

Roosevelt was the center of his children's world. He marked their toy horses and cattle with the brand of his western ranch and told them stories of his own boyhood heroes. When it rained, they played hide-and-seek and he hid in the Gun Room. "When we approached in our search he would moan or growl, whereat we would scuttle away and run downstairs again in a state of delighted terror," Alice recalled. Once when the children were down with the measles, their father made each a miniature ram and monitor from pasteboard and, together with some toy ships from Ted's collection, re-created the Civil War battle of Mobile Bay.

"They were, of course, absorbed spectators," he wrote. "In the battle Ted's monitor was sunk; and as soon as I left to dress," Ted began the battle over again, Alice looking on from the bed. This time Ted intended that Alice's monitor should sink while Alice was alert to see that no such variation took place.

" 'And now bang! goes a torpedo, Sisser's monitor sinks!' declared Ted.

" 'No, it didn't sink at all!' insisted Alice. 'My monitor always goes to bed at seven, and now it's three minutes past!' "

As she grew older, Alice became something of a rebel, doubtlessly compensating for her problems. Although she was required to pray for "my mother who is in heaven," her father never spoke to her about Alice Lee. That door to his heart was locked and barred and Alice could never find the key. Only Bamie ever discussed her mother with the child, telling her how pretty she had been and how much everyone had loved her. Edith's own view was caustic. Had Alice lived, she contended, she would eventually have bored her husband to death.

Little Alice's relations with her stepmother became increasingly strained over the years. When Edith was displeased, her glare was enough to send arctic chills up and down the spine of the target, but discipline only made Alice more rebellious. At nine, she decided she didn't want to be a girl any longer, and repeatedly expressed the intention—usually before visitors—of cutting her hair short, wearing trousers instead of skirts, and of giving birth to a monkey.

Edith blamed this fractious spirit on Alice's visits to her Lee grandparents. They doted on the child, and in Chestnut Hill she was always the center of attention, not one of numerous brawling children. Grandmother Lee would merely smile indulgently while Alice jumped up and down on the living room sofa—"something I should never have dreamed of doing at home"—and spoiled her with expensive presents. While Edith was struggling to make do with hand-me-downs for her other children and worrying about ever-mounting butcher's and grocer's bills, Alice had a wardrobe of new clothes and a pony and cart. "We'd better be nice to Alice," Roosevelt once wryly advised his wife. "We might have to ask her for money."*

Summers at Sagamore were "the seventh heaven of delight," according to Roosevelt, and he found his growing brood of children increasingly entertaining. Wearing plus fours and golf stockings, he led ragtag bands of his own offspring—the "bunnies"—and their myriad cousins and friends on long cross-country "scrambles" that were more obstacle courses than nature walks. "Under or over but never around" was his motto. These outings usually ended with a picnic and an overnight stay in the woods. He entertained the group by telling ghost stories while they gathered about the campfire—and would pounce on an unwary member of his audience with a bloodcurdling yell as fitting climax to a terrifying tale. Sometimes the children accompanied him up to the Gun Room, where he read aloud. Roosevelt's method of teaching the children to swim was typically direct. Upon learning that his niece Eleanor, Elliott's shy and retiring daughter, could not swim, he told her to jump off the dock. Fear of displeasing him overcoming her fear of the water, she leaped in and came up gasping and sputtering.

---

*George Cabot Lee left his grandchild $50,000 in trust and Alice's grandmother provided her with a trust that yielded $10,000 annually.

What was Edith Roosevelt's reaction to her sometimes outrageous husband and unruly brood? One detects a certain ambivalence. Her preference for solitude and his for company created strains that required adjustments on both sides. Most of the time she was kind and considerate, but she could also be demanding and possessive. Some family friends thought her ruthless in the way she took over her husband's life, pushing his adoring sisters out of it. "We've seen enough of Corinne and Douglas," she told her husband upon one occasion, "and I don't think we'll ask them down for a while."

The strain was also increased by Edith's constant insecurity about money, undoubtedly the result of being forced to live in straitened circumstances following the death of her father. When her chronic neuralgia—in reality migraine headaches—was acting up, neither her spouse nor the children knew what to expect, and her affection for them was tempered by a sharp tongue. During one "scramble," Roosevelt permitted the youngsters to go swimming with their clothes on. When the company arrived back home, Edith immediately sent them to bed with a dose of ginger syrup. "There's nothing I can do," the president of the United States told the protesting juveniles. "I'm lucky that she didn't give me a dose of ginger, too." Still, they complemented each other, as both noted admiringly over the years, and their marriage was a happy one. On the thirty-second anniversary of his engagement to Edith, Roosevelt told his youngest son, Quentin: "I really think I am just as much in love with her as I was then—she is so wise and good and pretty and charming."

Not long after the birth of Theodore, Jr., Roosevelt again felt the urge for open country and to enjoy a holiday away from home, packed his buckskin shirt, and hunting rifles for a return visit to the Bad Lands. For several weeks he hunted antelope in the broken country west of the Little Missouri, killed two black-tailed deer with a single fluke shot, and fought a raging prairie fire with the split and bleeding carcass of a steer. But he was shocked to find the once teeming Bad Lands virtually empty of big game, and some of the migratory birds had not returned to the valley. Elk and grizzlies were now following the buffalo into oblivion. The land had been denuded of its rich carpet of grass, the wild-plum bushes poisoned by heaps of cow dung, and the creekbeds cracked and dry. What had once been a paradise was becoming a wasteland.

Vaguely at first, and then more clearly, Roosevelt started to sense the dangers threatening the large game animals that inhabited the western part of the nation. Upon his return to New York, he began discussing the problem with George Bird Grinnell, editor of the magazine *Forest and Stream*. They were convinced that quick action should be taken to protect these animals from indiscriminate slaughter and formed a club of wealthy and likeminded sportsmen. It was named the Boone & Crockett Club after Daniel Boone and Davy Crockett, two of Roosevelt's heroes, and he was elected president.

Initially, the Boone & Crockett, the first club of its kind in the nation, was primarily concerned with promoting hunting in "the wild and unknown" areas of the country. But the settlement of the West made it difficult to carry out this objective, and the club turned its energies toward preservation rather than the hunting of large game. This, in turn, led it into forest and land conservation, a transformation signified by the formation of a Committee on Parks. The membership included some of the nation's leading lawmakers and eminent scientists, which gave it considerable influence upon public opinion and on Congress.

The Boone & Crockett's attention was drawn almost immediately to conditions in Yellowstone National Park, which had been created by Congress in 1872. The law had failed to include an enforcement provision, and promoters were building resorts and proposing that a railroad be run through the most scenic areas. They were cutting timber and "improving" the attractions by defacing rock formations and killing off buffalo, elk, and other game animals. Under Roosevelt's leadership, the Boone & Crockett was in the forefront of a wide-ranging campaign to prevent Yellowstone from becoming an ecological disaster area. In 1894, these efforts were crowned with success when Congress approved the Park Protection Act that saved Yellowstone from further despoliation.

From his father, Roosevelt had inherited a veneration for trees—the elder Roosevelt would not allow one to be cut down unless diseased—that was almost religious in intensity. Writing to Bamie from Milan during his second honeymoon, he had remarked that the cathedral "gives me a feeling I have never had elsewhere except . . . in the vast pine forests where the trees are very tall and not too close together." When the American Forestry Association began to struggle to save vast stands of timber that faced the same threat as Yellowstone, he mobilized the Boone & Crockett in the fight. Together, they succeeded in lobbying Congress to approve, in 1891, a law upon which the national forest reserve system is based. Little more than a decade later, as president, Roosevelt became the champion of conservation—an idea that had begun with a lonely hunt in the Bad Lands.

With the apparent collapse of his political career, Roosevelt drove himself to make a success as a writer and a historian. "I shall probably never be in politics again," he told a former legislative colleague. "I should like to write some book that will really take rank as in the very first class." For some time, he had been thinking about writing a multivolume history of the conquest of the North American continent—a sweeping drama of heroism, hardship, warfare, and struggle—and early in 1888, he began work upon *The Winning of the West* with the intention of making his mark among serious historians of the American past.

Every morning Roosevelt climbed the steep stairs to the Gun Room, took his seat at his desk—which he had turned to the wall so he would not be distracted by the view—and began to fill the blank sheets before him. Originally, he planned six or eight volumes that would track the westward movement—from Daniel Boone, "a tall, spare, sinewy man with eyes like an eagle's and muscles that never tired," to the peopling of Arizona and New Mexico in his own time. Only the first part of this grand design was actually completed—the first two volumes in 1889, the third in 1894, and the fourth in 1895—and he only carried the story as far as the Louisiana Purchase. Significantly, when he became successful in politics, he dropped the project.

In contrast to the *Benton* and *Morris* biographies, where Roosevelt's research was shallow, he dug deeply into original sources. "I realize perfectly that my chance of making a permanent literary reputation depends on how I do this big work," he wrote his publisher, George H. Putnam. He scoured the major eastern libraries, sought information in various state and government archives, and begged material from collectors. He burrowed into seldom-used manuscript collections, blew the dust off old diaries, and turned up documents that had never been consulted by scholars before.

The book took over Roosevelt's life, but progress was slow—often painfully slow. "Writing is horribly hard work to me," he complained. Wrestling with words and phrases and paragraphs, trimming awkward passages, struggling to paint word pictures with his pen like an artist working with brush and canvas, Roosevelt often exhausted himself. "It seems impossible to write more than a page or two a day," he told Bamie. He wrote out his thoughts, scratched them out, revised them, and then struck them out again. The strain was multiplied by the need to turn out a continuous stream of magazine articles to deal with Edith's gnawing sense of financial insecurity.

Unlike the professional historians who came to dominate historical writing in the United States, Roosevelt rejected the German-inspired "scientific" method exalted by university-trained scholars.* He conceived of history as a branch of literature rather than a pseudoscience. History, he argued, should be a work of art and imagination rather than a dry compendium of facts. Always a moralist—Thomas B. Reed, the Speaker of the House, once said that Roosevelt thought he had discovered the Ten Commandments—he argued that the historian should make moral and value judgments and not remain aloof from the heat and dust of the arena. Events do not occur in a vacuum, and he believed historians should relate them to the human experience.

Roosevelt wrote in the mode of the "gentleman historian" typified by Francis Parkman and William H. Prescott. In fact, he asked Parkman, whose seven-volume *History of France and England in North America* was the most impressive achievement yet made by an American historian and his model, if he could dedicate *The Winning of the West* to him. "Of course I know that you would not wish your name to be connected in even the most

---

*Roosevelt vociferously expressed these views in great detail in his address upon becoming president of the American Historical Association in 1912. See "History as Literature," *Works*, Vol. XII, pp. 3–24.

indirect way with any but good work," he assured Parkman, "and I can only say that I will do my best to make the work creditable."

Fascinated by the way in which nations became great, Roosevelt sought in *The Winning of the West* to discover the sources of American nationality. Inspired by Darwinian theory, he saw the westward movement as a distinct series of evolutionary episodes. The West was not won in one sudden pell-mell rush, but over and over again. The process was ongoing, and the stages were recorded like the strata in a geological specimen. First came the fur trappers, Indian traders, and explorers. Then the hunters, next the surveyors with their compasses and chains, and finally permanent settlers. Having cleared the forests with their axes, they plowed the fields and established common principles of English law, justice, and government. English, German, Irish, Scotch, and Huguenots all participated in the western movement, he noted, but in a single generation they had been welded into a new species of man—the American. In short, a free people, settling on free land, had founded a nation built on free institutions.

Roosevelt's combination of scholarly research with the narrative power of the novelist paid off. The writing throughout *The Winning of the West* is clean and straightforward; some passages crackle with excitement. For him, the book was confirmation of the validity of the strenuous life. Readers hear the whizzing of Indian arrows, the steady creaking of Conestoga wagons rolling westward, and the distant thunder of war drums; they smell the acrid smoke of campfires and the black powder flashing from long rifles. Roosevelt excelled in such set pieces as the rallying of the frontiersmen before the battle of Kings Mountain during the Revolution:

On the 26th [of September 1780] they began the march, over a thousand strong, most of them mounted on swift, wiry horses. They were led by leaders they trusted, they were wonted to Indian warfare; they were skilled as horsemen and marksmen, they knew how to face every kind of danger, hardship and privation. Their fringed and tasselled hunting shirts were girded in by bead-worked belts, and the trappings of their horses were stained red and yellow. On their heads they wore caps of coonskin or minkskin, with the tails hanging down. . . . Every man carried a small-bore rifle, a tomahawk, and a scalping-knife. . . . Before leaving their camping ground . . . they gathered in an open grove to hear a stern old Presbyterian preacher invoke on the enterprise the blessings of Jehovah. Leaning on their long rifles, they stood in rings around the black-frocked minister, a grim and wild congregation, who listened in silence to his words of burning zeal as he called on them to stand stoutly in the battle and to smite their foes with the sword of the Lord and of Gideon. . . .

The Winning of the West provided Roosevelt with the literary and financial success that he was so earnestly seeking. The two-volume first printing quickly sold out, and the reviews were overwhelmingly favorable. Frederick Jackson Turner, whose own hypothesis that America defined itself by pushing out to the frontier was still four years in the future, declared that the effect of Roosevelt's experiences in the Bad Lands was stamped on every page, and he had depicted America's western expansion "as probably no other man of his time could have done." In particular, Roosevelt was applauded for pointing out that "a new composite nationality . . . a distinct American people" had emerged beyond the Alleghenies.

The book was not without shortcomings, however. Roosevelt tended to generalize from his own time in the Bad Lands about other, earlier Wests where his experiences were of questionable validity. Obsessed with the struggle and human drama of westward expansion, he acknowledged having neglected economic and institutional history. Roosevelt's suggestion—and Turner's—that there was great equality on the frontier because land was readily available was also flawed. Nevertheless, *The Winning of the West* was a major contribution to the narrative history of the West, and a generation of Americans eagerly devoured it.

While Roosevelt was pushing ahead with *The Winning of the West,* the pages going almost directly from his hand to the printer, national affairs took an unexpected turn that was to propel him back into politics. In December 1887, Grover Cleveland ignored the advice of Democratic leaders and unexpectedly called for a reduction of the prohibitively high tariffs enacted during the Civil War to protect American industry against foreign competition. Cleveland argued that this Chinese wall of tariffs was not only costly to consumers, but also contributed to the growth of trusts.

"Our present tariff laws, this vicious, inequitable and illogical source of unnecessary taxation, ought to be at once revised and amended," he declared with characteristic directness. The tariff was "a burden upon those with moderate means and the poor, the employed and the unemployed, the sick and well, the young and old."

Cleveland's proposal for tariff reform was a windfall for the Republicans. Hungering for the spoils of office, they grasped eagerly at an issue that might bring victory in 1888. The president was branded as a free trader who would do away with the protective tariff—called the "American tariff"—which the Republicans claimed had made the United States prosperous. Cleveland, it was charged, was favoring British manufacturers over American capitalists and laborers alike. "There's one more President for us in Protection," exulted James Blaine—and there was no doubt in anyone's mind whom the "Plumed Knight" had in mind.

Roosevelt paid little attention to these developments. "I'm a literary feller, not a politician these days," he told Professor Brander Matthews of Columbia. He expected Blaine to win the Republican nomination handily and saw no benefit to himself in it. "It is unfortunate but it is true" that "Blaine is the choice of the bulk of the Republicans," he and Lodge agreed. But there was considerable unease among party leaders about Blaine's public reputation and they denied him the nomination. Unable to win it for himself, he played kingmaker for a compromise candidate, Benjamin Harrison of Indiana. Short, stout, and reserved, Harrison was a Civil War political general and an ex-senator. He had a handshake like "a wilted petunia," in the words of one recipient, and was mainly known as the grandson of President William Henry Harrison. For their part, the Democrats renominated Cleveland with little enthusiasm.

Roosevelt thought the tariff issue was too complex for the average voter to comprehend but applauded the choice of Harrison over Blaine. "I am myself more and more encouraged over the political prospects," he told Lodge. Now anxious to get back into politics, he volunteered his services to the Republican National Committee, and early in October was dispatched for twelve days of stump speeches in Michigan and Minnesota, where he attacked Cleveland and praised the protective tariff. "I . . . act as target and marksman alternately with immense zest," he wrote Cecil Spring Rice, who was trying to puzzle out the mysteries of his first American presidential campaign.

Cleveland's assault on the protective tariff represented a challenge to large corporations that were organized after the Civil War and cemented the alliance of big business and the Republican party. John "One Price" Wanamaker, the Philadelphia merchant prince, was put to work "frying the fat" out of businessmen as campaign finance chairman. An Ohio industrialist named Marcus Alonzo Hanna raised $100,000 on his own. Tom Platt, New York's unchallenged Republican boss, came away with a neat pile of $10,000 checks from a meeting with J. P. Morgan and other Wall Street moneymen. So much money was raised to ensure Harrison's election—close to $4 million, a phenomenal sum at the time—that the election of 1888 was known as the "Boodle Campaign."

A sizable portion of these funds was used to buy votes in the critical states of New York and Indiana. In New York, Platt purchased the secret support of Tammany Hall for Harrison. Matthew Quay, the Republican boss of Pennsylvania, dispatched his loyal legions to Indiana to stuff ballot boxes, and another twenty thousand votes were bought on the open market at the going rate of $15 in gold or $20 in greenbacks. Although Cleveland carried the country by over a hundred thousand popular votes, Harrison narrowly captured New York and Indiana, and their electoral votes gave him the presidency. "Providence has given us the victory," intoned the president-elect. Matt Quay knew better. "Think of the man!" he sputtered. "He ought to know that Providence hadn't a damned thing to do with it!"

Following the election, Roosevelt pressed ahead with the final chapters of *The Winning of the West,* while keeping a weather eye on Washington. He had just turned thirty and time was slipping away. The Republican victory had opened the opportunity for a government appointment, but he feared he was persona non grata with the new administration. "I would like above all things to go into politics," he confessed to Lodge, "but . . . that seems impossible." Roosevelt's fears were well warranted. Lodge, by now a growing force within the Republican party, had suggested to Jim Blaine, the new secretary of state and gray eminence of the

Harrison administration, that his deserving friend be named assistant secretary, but the proposal was rejected. Blaine knew his man only too well.

> My real trouble in regard to Mr. Roosevelt is that I fear he lacks the repose and patient endurance required in an Assistant Secretary [he told Lodge]. Mr. Roosevelt is amazingly quick in apprehension. Is there not danger that he might be too quick in execution? I do somehow fear that my sleep at Augusta or Bar Harbor would not be quite as easy and refreshing if so brilliant and aggressive a man had hold of the helm. Matters are constantly occurring which require the most thoughtful concentration and the most stubborn inaction. Do *you* think that Mr. T.R.'s temperament would give guaranty of that course?

Lodge kept Blaine's comments to himself and appealed directly to Harrison in Roosevelt's behalf. "I had a little talk with the President about you and he spoke very pleasantly but he is a reserved person," he informed his friend. Harrison, also concerned about Roosevelt's willful, impulsive, and somewhat erratic independence, was reluctant to give him a place in his administration. Lodge persisted, however, and Tom Reed and Blaine's son Walker also put in a word for him. Harrison finally gave in and offered to name Roosevelt one of the three United States Civil Service commissioners. Lodge was not at all certain that Roosevelt should take it. The job paid only $3,500 a year, enjoyed little prestige, and was guaranteed to make enemies among the politicians, while any laxity would arouse the ire of vigilant reformers.

Roosevelt accepted with alacrity, and was soon on his way to Washington.

Chapter 10

"We Stirred Up Things Well"

Matthew F. Halloran, executive secretary of the Civil Service Commission, never forgot his first encounter with Theodore Roosevelt. Hearing an unexpected commotion in his outer office on the morning of May 13, 1889, he looked up as an energetic figure burst in upon him in a high-speed blur. "I am the new Civil Service Commissioner, Theodore Roosevelt of New York," he declared briskly. "Have you a telephone? Call the Ebbitt House. I have an appointment with Bishop Ireland. Say I will be there at ten o'clock."

"I jumped up with alacrity," Halloran continued. "There stood an energetic, athletic appearing man . . . slightly above medium height, broad-shouldered, full-chested, and wearing a becoming brown mustache. Behind large-rimmed eye-glasses flashed piercing blue-gray eyes. . . . The dazzling smile with its strong white teeth . . . is still a most vivid recollection."

Within minutes of being sworn in, Roosevelt appropriated the most imposing of the offices reserved for the three commissioners and was ready to do battle with the horde of politicians eager to reward their followers with jobs on the public payroll. The new member confidently accepted this challenge. In fact, he had thrown down the gauntlet to the spoilsmen even before his arrival in Washington. "You can guarantee that I intend to hew to the line and let the chips fall where they will," he told a New York City newsman. And the chips were soon falling thick and fast.

Over the next six years, Roosevelt kept the Civil Service Commission, which had dozed comfortably in a wing of the old City Hall on Judiciary Square, in a constant state of turmoil. Swooping down without warning upon far-flung outposts of his empire, he investigated rumors of fraud, held hearings, issued reports, made speeches, and wrote magazine articles to dramatize the cause of reform. When politicians attacked the commission, he returned the attack "blow for blow." If they cut off appropriations, he retaliated by refusing to conduct civil service examinations in their districts. Roosevelt's feuds and quarrels catapulted the worthy but dull cause of civil service reform onto the front pages of the newspapers—and propelled Roosevelt into the national limelight. This battle restored him to the good graces of the mugwumps, although he remained disdainful of them. For his part, President Harrison, who was less than enthusiastic about his performance, tartly observed that Mr. Roosevelt "wanted to put an end to all the evil in the world between sunrise and sunset."

Roosevelt's fellow commissioners—Charles Lyman, a Republican, and Hugh S. Thompson, the sole Democrat—were considerably older than he and of equal rank, but they tacitly accepted the leadership of their energetic young colleague. One member of the commission recalled that "every day I went to the office . . . it was as to an entertainment. I knew something was sure to turn up to make our work worthwhile, with him there." Even Roosevelt's method of giving dictation signified action. Matthew Halloran recalled that he strode up and down his office, glasses askew, rapidly unburdening himself of a stream of words. Telling points would be emphasized "by driving his right fist vigorously into his left hand. . . . 'By Godfrey!' was the expression which gave explosion to his excitement, while 'By Jove' satisfied his milder surprise."

The Civil Service Commission had the impossible task of enforcing an imperfect law with inadequate means. Established in 1883 to administer the Pendleton Act, which required open competitive examinations in filling some government jobs, it was a facade behind which politics and patronage operated as usual. While the law covered civil service appointments, its control of promotions and removals was not as clearly stated. The commission's staff was also too small to handle the hundreds of examinations, inquiries, and complaints flowing into Washington. To make up for it, Roosevelt developed a network of correspondents to bring cases of fraud to his attention. "Do let me know if there is any crookedness going on at Baltimore," he wrote Charles Joseph Bonaparte,* one of the city's leading reformers.

If a congressman stopped by the office to try to hurry up a constituent's appointment, Roosevelt would ask Halloran for the list of eligibles for that job. "Running his eye quickly down the page until he reached the name of the eligible in question, he would count rapidly the number of names ahead and would declare in characteristic fashion, 'Now, Mr. Congressman, [there are] —— names ahead of your man. He determined his standing by the rating he obtained in the examination and nothing can advance him out of his turn. You may be assured he will be certified when reached. . . .' " While going over the list, Roosevelt would try to find the names of other eligibles from the congressman's district who ranked higher and invite his visitor's attention to them—neatly diverting him from his original mission.

Most of Roosevelt's predecessors had been content merely to supervise the grading of examination papers and maintain registers of those eligible for appointment to government jobs, and were careful not to arouse the anger of congressmen or department heads by trying to enforce the law rigorously. Their duties were not onerous, they enjoyed leisurely lunches at an excellent oyster house on Louisiana Avenue near the office, and were invited to the less exclusive White House receptions.

Civil service reform had been a cardinal principle of Roosevelt's political creed since his service in the New York State Assembly, and he was convinced the abuse of patronage for selfish and partisan purposes was the curse of American politics. His political

---

*Charles Joseph Bonaparte was the grandson of Jerome Bonaparte, Napoleon's youngest brother, who had fallen in love with and married Elizabeth Patterson, the beautiful daughter of a Baltimore merchant, while visiting America. Angry over the marriage, Napoleon had had it annulled by the Catholic Church. Charles Bonaparte later served as secretary of the navy and then attorney general in TR's Cabinet.

creed was a mix of conservatism and liberalism, and waging war against the ungodly in the name of the merit system ideally suited him, for it neatly combined the conservative gospel of efficiency in government with the liberal dream of an open society. "No republic can permanently endure when its politics are corrupt and base," he declared. "The spoils system, the application in politics of the degrading doctrine that to the victors belong the spoils, produces corruption and degradation. . . . The spoils-monger and the spoils-seeker invariably breed the bribe-taker and the bribe-giver, the embezzler of public funds, and the corrupter of voters."

Roosevelt not only propagandized for civil service reform, but he improved and revised the examinations. "Plain, commonsense questions" ought to be asked, he declared, and he thought experience was more important than theoretical knowledge. For example, he felt that marksmanship and the ability to ride and read cattle brands were of more importance than spelling to border patrolmen in the western states. Upon learning that at Roosevelt's direction the appointees for such jobs had been chosen at a target shoot, Charles Bonaparte wryly observed that Roosevelt had been "remiss." He did not get the best men. He should have had the men shoot at each other and given the job to the survivors.

The Pendleton Act had originally placed some fourteen thousand federal jobs under the merit system, and shaking down government workers for political contributions was barred, although the practice was hardly stamped out. Grover Cleveland had extended civil service coverage to another seven thousand workers, while packing the rest of the bureaucracy with Democratic faithful. Following the election of Harrison, Republican stalwarts demanded their place at the public trough, and although the new president declared his intention in his inaugural address of enforcing the Pendleton Act "fully and without evasion," open season was declared upon Democratic jobholders. Thousands of Republicans swarmed into Washington or beseeched their congressmen by mail for jobs. "Hundreds of Offices," "Places to Suit All Classes," "Take Your Choice," trumpeted one Republican paper to the faithful.

The task of clearing away Democratic deadwood was given to John Wanamaker, who had been named postmaster general in reward for his skill in "frying the fat" out of the fat cats when he was Harrison's campaign finance chairman. In the first six weeks of the new administration, Wanamaker did his job with such zeal that cartoonists soon were portraying him with a tape measure over his arm, calling out "Cash!" as if he were holding a bargain sale of post office jobs in his store. In all, he removed some thirty thousand fourth-class postmasters—one every five minutes—and replaced them with deserving Republicans. "There was never in our history a grosser violation of distinct promises and pledges than the partisan devastation of the post office under this administration," *Harper's Weekly* angrily charged.

Most of these removals took place before Roosevelt arrived in Washington, but a collision with Wanamaker was inevitable. Within a week after his arrival in Washington, Roosevelt was looking about for a target that would focus attention on the commission and zeroed in on that familiar fountainhead of corruption, the New York Customs House. Following up rumors reaching Washington, he found that some of the questions for a recent examination had been leaked to candidates who had paid $50 each for the information. He issued a scathing report accusing the local examiners of fraud and called for the dismissal of three officials and the prosecution of at least one of them. Such on-the-spot investigations became a regular feature of Roosevelt's methods, and spoilsmen and reformers both were now on clear notice that he did not "intend to have the Commission remain a mere board of head clerks."

Next, with the two other commissioners in tow, he set out to investigate breaches of the civil service law in the Middle West. Newsmen quickly noted that the tour would begin in Indianapolis—President Harrison's hometown—where his good friend William Wallace was the city postmaster. The commissioners, acting upon information supplied by local civil service reformers, forced Wallace to fire three crooked employees, one of them reputed to be a professional gambler, whom he had rehired after they had been dismissed by his Democratic predecessor. In Milwaukee, where corruption was rampant, Roosevelt called Postmaster George H. Paul "about as thorough-faced a scoundrel as

I ever saw." Similar attention-getting tactics were applied in Chicago and Grand Rapids.

"We stirred up things well," a triumphant Roosevelt told Lodge upon his return to Washington from his "slam among the post offices." The outcry from the spoilsmen was "furious," however, and there were demands for his ouster. From the White House there was only silence, but it was intimated that the energetic young commissioner ought to tread more carefully—to enforce the law "rigidly" where people supported it and handle it "gingerly" elsewhere. Roosevelt was not at all certain that the president was really upset, however. "It is to Harrison's credit, all that we are doing to enforce the law," he said. "I am part of the Administration; if I do good work it redounds to the credit of the Administration." Soon, however, he was not so sanguine.

In summer, Washington became a ghost town as the wives and children of the upper reaches of the Washington bureaucracy abandoned the torpid and muggy capital. Edith, pregnant with her second child, decided to remain at Sagamore until after the baby came in November. Meantime, her husband, who disliked hotels and found it necessary to economize, accepted an invitation from the Lodges to stay at their home on Connecticut Avenue. Leading a bachelor life, Roosevelt renewed his acquaintance with Cecil Spring Rice and made friends he would keep for life—among them Tom Reed, one of the most powerful men in Congress; Henry Adams, historian and sardonic observer of the Washington scene; his father's old friend, John Hay, and the German diplomat Baron Hermann Speck von Sternberg,* among others. To Adams, Roosevelt, with his talk of wiping out the evils of patronage, was an object of pity. Over dinner one night, he shook his head as the young commissioner animatedly discussed his work, and thought, "The poor wretch."

The pressures of his reform campaign kept Roosevelt in Washington most of the summer, and there were only a few flying visits to Oyster Bay to see Edith and the "bunnies." "I am rapidly sinking into a fat and lazy middle age," he chaffed, although he

---

*In 1903, TR was instrumental in having Speck von Sternberg, who had an American wife, appointed German ambassador to the United States.

was not yet thirty-one. An anguished Edith, pregnant and saddled with the task of managing the household, the farm, and the children, wrote: "It is four weeks tomorrow evening since you left. . . . I miss you every minute, awake or asleep, and long for you beyond all expression. After this we will be together if all goes well. I am looking forward so much to the time when the baby comes and you are with me."

On October 10, Roosevelt received a telegram announcing that Edith had prematurely given birth to another boy. Immediately leaving for home, the anxious father found that he had arrived at the Thirty-fourth Street ferry too late to catch the last train to Oyster Bay. He crossed to Long Island City and, ignoring the cost, chartered a special train—arriving at Edith's bedside at four o'clock in the morning. In a reversal of form, Bamie had been invited in, and both mother and child were doing well under her care. "Edith can't say enough of what a comfort you were to her," Roosevelt wrote his sister after she had returned to New York. "Again and again she says, 'Oh that *dear* Bysie.' "

The baby was christened Kermit, after a branch of Edith's family. Little Ted proved to be more vocal than the baby at the ceremony held in the parlor at Sagamore. Who was the man wearing "Mame's clothes?" he loudly asked after seeing the Episcopal priest's voluminous surplice. And when the baby cried, he announced, "Baby bruvver Kermit miaous." He was quickly removed across the hall to the library before his penetrating comments broke up the solemnity of the occasion.

In late December, Edith came down to Washington to inspect the tiny house her husband had rented at 1820 Jefferson Place N.W.,* a narrow little street just around the corner from the Lodges' home. All that Roosevelt felt he could afford, it was only about one tenth the size of Sagamore Hill, and a dressing room would have to double as a guest room if they had overnight visitors. Extra furniture was shipped down from Sagamore, and Edith soon had everything "homelike and comfortable." Like all newcomers to the city, she found Washington expensive. To rent even a small house would cost $2,400 a year, the food bill for the growing family might amount to $1,000, to which must be added

---

*The house is still standing a century later, but has been converted into offices.

the cost of coal and gaslight and servants. Roosevelt's entire salary of $3,500 would not even cover the bare necessities.

Nevertheless, as 1889 drew to an end, Roosevelt thought it had been a good year. Edith had given birth to another son, he was back in politics with both feet, and *The Winning of the West* appeared to be a critical and financial success. To Lodge, he wrote:

> What funnily varied lives we do lead, Cabot! We touch two or three little worlds, each profoundly ignorant of the others. Our literary friends have but a vague knowledge of our actual political work; and a goodly number of our sporting and social acquaintances know us only as men of good family, one of whom rides to hounds, while the other hunts big-game in the Rockies.

Washington's social season officially began with the president's New Year's Day reception on January 1, 1890. Rain was falling as the Roosevelts drove past the long lines of ordinary citizens waiting on Pennsylvania Avenue to enter the White House and shake Harrison's limp hand. Little had changed, Edith noted, since she had visited the Executive Mansion thirteen years before as a youthful sightseer. Palms, red poinsettias, and pink azaleas had been brought in to brighten things up, but the place remained rather drab. Caroline Harrison, the president's wife, had spent much of her first year in the White House trying to deal with a frightening rat problem. Hordes of rats had taken over the place and all but ate with the family. Exterminators were called in, and the floor and walls of the kitchens and pantries had to be torn out and replaced.*

"Hail to the Chief" heralded the president's arrival in the Blue Room at eleven o'clock. First to be received were the Cabinet, then the diplomatic corps, justices of the Supreme Court, and members of Congress. Ranking officers of the army and navy, resplendent in gold braid, were followed by the departmental heads, including the civil service commissioners. "Nobody got a heartier reception than the three Commissioners," reported *The*

---

*The rats remained a problem, however, and during Cleveland's second administration a servant reported that while on his rounds he discovered that "a great rat had forced his way into [a canary's cage and] had just killed the poor little canary. . . ." Billings, "Social and Economic Life in Washington in the 1890's," *Records of the Columbia Historical Society of Washington, D.C.,* 1966–1968.

Washington Post, "Mr. Roosevelt being very demonstrative in his protestations of good wishes to the President for the coming year."

Over the succeeding months, Edith explored the capital, a pleasant southern town of trees, marble buildings, and well-paved streets with a population of about 200,000. One English visitor spoke of an air "of comfort, of leisure, of space to spare, of stateliness you hardly expected in America. It looks a sort of place where nobody has to work for his living, or, at any rate, not hard." Government officials breakfasted at eight or nine o'clock, lunched at noon, and left at four o'clock for a hearty dinner at home or dined at one of the many restaurants. The choice places to dine included Willard's Hotel, the Ebbitt House, Harvey's, and Hall's on the Potomac waterfront, where the bar was graced by a huge painting of a nude Venus. On Sundays, the well-to-do promenaded along Connecticut Avenue with the diplomatic colony, and compliments were exchanged in a dozen languages. But behind the mansions that graced the patrician squares and circles, the alleys were filled with shanties, home to a large share of the city's black population of some sixty thousand.

Politics and society were inseparably entwined, and the Roosevelts were immediately caught up in a swirl of teas and receptions, large and small dinners, theater parties, and champagne suppers. Summed up by one wit as the four G's—"giggle, gabble, gobble and go"—society was dominated by women. While a man might have standing because of his job, his actual position in society was dependent upon the social skills of his wife. For Edith, with her preference for privacy, such a life was a chore, but she took it up for the sake of her husband's career.

Particularly wearying was the "great cardboard exchange" in which the ladies of official Washington left home four days a week, cardcases in hand, for their round of calls. In the late afternoons, the streets were filled with conveyances of ladies making calls. A call consisted of a brief exchange of pleasantries and possibly, if there were enough time, a cup of tea before a woman was off to her next stop. If she did her duty, she might make as many as 1,800 calls in a single season.

Wealth meant little, so members of Congress and the bureaucrats with modest means could be popular if they offered good conversation and a varied list of guests. The scale of entertaining at Jefferson Place was certainly not grand—mostly tea parties and Sunday-evening suppers for no more than eight—but the most interesting people in Washington turned up there. Although Roosevelt lamented that he could not afford champagne, he was always a good storyteller and kept guests enthralled with tales of his cowboy days in the Bad Lands. "There was a vital radiance about the man," recalled Mrs. Winthrop Chanler, a frequent guest, "a glowing unfeigned cordiality toward those he liked that was irresistible."

Tom Reed, now the all-powerful Speaker of the House, came regularly to Jefferson Place, often arriving arm in arm with Cabot Lodge. Well over six feet tall and weighing about three hundred pounds, Reed had an enormous, clean-shaven baby face that gave him the air of a New England Buddha. Roosevelt had exercised his limited influence with western congressmen to round up votes for Reed during the intraparty contest for the speakership with William McKinley of Ohio, and Reed was an invaluable ally in Roosevelt's battle to keep his job amid demands for his ouster.* Before the session was over, he would seize such absolute control of the House that he was dubbed "Czar" Reed.

Reed was the ablest debater in American politics, and noted for his cool sarcasm. When one member renowned for windiness proclaimed a passionate interest in being right rather than president, Reed quietly intoned: "The gentleman need not be disturbed; he will be neither." And when another member haltingly proclaimed, "I was thinking, Mr. Speaker, I was thinking . . ." Reed expressed the hope "that no one will interrupt this commendable innovation."

The Roosevelts were among the handful of Washingtonians welcome at the elegant Romanesque manor houses across Lafayette Square from the White House† where Henry Adams and John Hay half-wistfully, half-contemptuously observed democ-

---

*Roosevelt lobbied for Reed while telling McKinley that one day in the future he would vote for him as president of the United States. McKinley "was as pleasant as possible—probably because he considered my support worthless," he reported to Lodge. *Roosevelt-Lodge Correspondence,* Vol. I, p. 99.

Reed was no reformer but supported Roosevelt's efforts. "Well, I didn't know you were in love with Civil Service Reform," one colleague told him. "I don't like it straight," the Speaker replied, "but mixed with a little Theodore Roosevelt, I like it well."

†The site is now occupied by the Hay-Adams Hotel.

racy in action. Adams, grandson and great-grandson of presidents, made a career of cynical detachment while men he regarded as inferior to himself occupied the white-pillared house of his ancestors. Hay, on the other hand, was a handsome lad from Illinois and onetime secretary to Abraham Lincoln who had tossed aside a bright career in journalism—he had traced the Chicago Fire to Mrs. O'Leary's cow—to marry the plump daughter of a wealthy Cleveland family. Without seeming to try, he fell into powerful positions presidency after presidency, finally becoming secretary of state.

Adams regarded Roosevelt with alternate fascination and repulsion as he followed the brash young man's ascent to the White House. "What is man that he should have tusks and grin!" he asked and attributed to Roosevelt "that singular primitive quality that belongs to ultimate matter—the quality that medieval theology assigned to God—he was pure act." On the other hand, he was captivated by Roosevelt's "sympathetic little wife" and noted with amusement the subtle manner in which she controlled her husband. "He stands in such abject terror of Edith," Adams observed.

The Roosevelts were among the regulars who gathered for late breakfast and conversation about the newest in art, literature, and politics in Adams's book-lined living room. Besides the widowed Adams and Hay and his wife, the group included sculptor Augustus Saint-Gaudens;* John LaFarge, an artist and Adams's old friend; Clarence King, a brilliant geologist; Nannie and Cabot Lodge; and Senator J. Donald Cameron of Pennsylvania and his petite, Titian-haired wife, Elizabeth, the most beautiful woman in Washington. Rich and repulsive, Cameron was tolerated in this charmed circle only because Adams was secretly and hopelessly in love with his wife. Sexual passions seemed to rage beneath the surface of this sedate group: John Hay and Nannie Lodge, who was almost as striking as Lizzie Cameron, are said to have also

---

*Saint-Gaudens was then at work on the majestic hooded bronze figure commissioned by Adams to brood over the grave of his late wife, Clover, and eventually his own, in Washington's Rock Creek Cemetery. Unstable and depressed, Clover Adams had committed suicide in 1885. Although the figure is commonly called "Grief," Adams despised the name and told the sculptor that he wanted a work that expressed "acceptance of the inevitable." It was Eleanor Roosevelt's favorite sculpture and she often spent hours alone contemplating its mystery.

been clandestine lovers. And Clarence King had a secret black common-law wife who bore him five children.

These frequent visits to Lafayette Square took Roosevelt past the White House, and a friend coming upon him one day observed that he "looks precisely as if he had examined the building and finding it to his liking, has made up his mind to inhabit it." Roosevelt later admitted that it was during this period that he began to cast appraising eyes at the White House. "I used to walk past the White House, and my heart would beat a little faster as the thought came to me that possibly—possibly—I would some day occupy it as President."

Roosevelt's thoughts of future glory ran headlong into a period of family crisis, financial difficulty, and a fight with John Wanamaker that threatened to end his political career. While his elder brother was making his way in the world, Elliott Roosevelt's heavy drinking and frivolous life had become a scandal. Theodore thought Elliott should check himself into a sanitarium for a cure under medical supervision. "Half measures simply make the case more hopeless," he despairingly wrote Bamie, "and render the chance of a public scandal greater."

Instead, Elliott gathered up his wife, Anna, and two children, Eleanor and Elliott, Jr., and went off to Europe. Soon, Anna wrote home that she was pregnant again, and was afraid her husband might injure her in one of his drunken rages. As usual in times of family crisis, Bamie was dispatched to deal with the problem. She persuaded Elliott to seek treatment with a Viennese specialist, and within a few months he had recovered sufficiently to rejoin his wife in Paris. There, his second son, Gracie Hall Roosevelt, was born.*

In New York, the "nightmare of horror" took a new turn. Katy Mann, a former maid in Elliott's household, claimed that he was the father of her illegitimate child. She demanded $10,000 in child support, a large sum at the time, and threatened to file a paternity suit if it were not paid. Elliott flatly denied the charge, then began to equivocate. The family hired "an expert in likenesses" to look

---

*Hall Roosevelt also was an alcoholic, much to the distress of his sister Eleanor, and drank himself to death in 1941.

over Katy Mann's baby, and he came away convinced her story was true. Torn between affection for his "dear old beloved brother" and anger at his shameless conduct, Theodore now found himself in the unhappy position of go-between in a sordid paternity case.

Elliott collapsed under these fresh pressures and started drinking again. He was violent and suicidal, and Bamie, following the advice of her brother, had him placed in an asylum outside Paris, while Anna returned to America with the three children. Not long afterward, the Katy Mann affair faded from view, indicating that hush money was paid, although by whom and how much is an unresolved question.

Roosevelt's anxieties about his brother coincided with a final break with John Wanamaker, a break that blighted the remaining years of the Harrison administration. In the spring of 1891, Roosevelt learned that workers in the post office in Baltimore were being shaken down $5 to $10 each to finance pro-Harrison candidates in an upcoming Republican primary. Wary of a direct challenge to Wanamaker, Roosevelt directed complaints to the postmaster general. But when Wanamaker ignored them, he launched an investigation of his own. With his usual flair for the dramatic, Roosevelt descended upon Baltimore, where politics were notoriously raucous, on Election Day to observe enough illegality on the part of federal employees to keep the courts busy for months.

Old Guard Republicans immediately bombarded Harrison with demands for Roosevelt's dismissal, and the "little gray man in the White House" replied that he would wait for a full report on the incident before taking action. This was a none-too-subtle warning to Roosevelt to soft-pedal his findings if he wished to remain in office. Nevertheless, the report, written at his usual breakneck pace, bristled with allegations of flagrant fraud. Some of the testimony was cheerfully Rabelaisian.

"How do you do your cheating?" Roosevelt asked one ward heeler.

"Well, we do our cheating honorably," was the reply.

Roosevelt recommended to Wanamaker and the president that twenty-five employees of the Baltimore post office, all of them Harrison supporters, be immediately dismissed from their jobs.

Later he began to have second thoughts about crossing swords with the postmaster general, and allowed his colleagues to persuade him to withhold the report until the summer vacation period, when public impact would be less.

For the moment, the crusader put down his lance and followed Edith to Sagamore, where she was awaiting the birth of another child. The very thought of it angered some of her friends. "When I think of their very moderate income and the recklessness with which she brings children into the world without the means either to educate or provide for them, I am quite worked up," Lizzie Cameron told Henry Adams. "She will have a round dozen I am sure. . . . It is a shame." The baby, a sturdy girl who was named Ethel, was born on August 13, 1891. She was "a jolly naughty whacky baby" who reminded her father of his stocky Dutch ancestors and he called her "Elephant Johnny."

At about the same time, he forwarded the Baltimore report to the White House and Post Office Department. One look and stricken officials put it under lock and key. Wanamaker claimed that the evidence against the twenty-five malefactors was "inconclusive," and ordered postal inspectors to make an independent investigation of the commissioner's charges. Roosevelt practically exploded when he received the news. "You may tell the Postmaster-General for me that I don't like him for two reasons," he thundered at the messenger. "In the first place, he has a very sloppy mind, and in the next place he does not tell the truth!"

While the politicians fought over the spoils of office, signs of popular unrest were flickering in the southern and western skies like heat lightning on a distant horizon. From the sun-baked cotton lands of the South to the hot, dusty prairies of the West, restive farmers and ranchers were muttering in discontent. Falling prices, drought, blizzards, overproduction, the protective tariff, low crop prices, and high interest rates were combining to ruin them. Business, banking, even labor, were organizing—and now the farmers followed their example. Taking the advice of Mary Lease, one of the most outspoken critics of the status quo, "to raise less corn and more hell," the disaffected farmers organized the Populist party.

The Populist revolt swept agrarian America like a religious

revival. "It was a crusade, a Pentecost of politics, in which a tongue of flame sat upon every man," declared one witness. Wall Street was accused of imposing upon America "a government of Wall Street, by Wall Street and for Wall Street." There was strength in union, and the Populists went beyond the southern and western farmers to embrace other groups: advocates of woman suffrage, Socialists, single-taxers, "free" silverites, and reformers of every stripe. Nothing like it had been seen before in American politics. There were even efforts to reach across racial lines to blacks, which sent shivers up and down the spines of southern Bourbons. In Nebraska, a young Democrat named William Jennings Bryan urged his party to fuse with the Populists. In 1890, the new party swept a dozen western and southern states, and sent a score of homespun lawmakers to Washington to startle staid easterners. Upon seeing Senator William Peffer of Kansas, whose long, flowing beard reminded some observers of a Hebrew prophet, Theodore Roosevelt called him "a well-meaning, pin-headed, anarchistic crank."

The attention of the Roosevelts was fixed, however, on Elliott's disintegration. Theodore and Bamie, with Anna's consent, moved to have him declared legally insane and his fortune of $170,000 placed in trust for the benefit of his family. The Roosevelts hoped to avoid a public scandal, but the news leaked to the newspapers, where it made front-page headlines, much to the chagrin of family elders, who blamed Theodore. "It is all horrible beyond belief," he told Bamie. "The only thing to do is to go resolutely forward." Elliott, however, decided to fight, and published denials of insanity in both the Paris and New York editions of the *Herald*.

With the hope of preventing the scandal from mushrooming, Roosevelt went to Paris to persuade his brother voluntarily to sign his property over to his wife. Several anxious weeks followed until he reported on January 21, 1892: "Won! Thank heaven I came over. . . ." Elliott, "utterly broken, submissive and repentant," had agreed to the trust and to return to the United States for treatment of his alcoholism and a period of probation and separation from his wife, during which he would establish himself in a steady occupation. Douglas Robinson, his brother-in-law, found a place for him managing his family's coal and timber interests in Virginia, and Elliott voluntarily went into exile.

The separation was especially poignant for eight-year-old Eleanor Roosevelt, who idealized her father. To the little girl, her mother was "one of the most beautiful women I have ever seen," but she saw in Anna's eyes only disappointment in the child's plainness. Because of Eleanor's solemnity, Anna called her "Granny" and told visitors she was "old fashioned." Hungering for praise and affection, she longed for her beloved father. Missing him desperately and understanding little of the reason for his absence, Eleanor blamed her mother for the separation. She lived for Elliott's letters to his "Little Nell," and carried them about as a talisman.

This crisis unfolded against a background of increasing financial pressures upon Edith and Theodore. When the family returned to Washington after Ethel's birth, it was to a larger and more comfortable house at 1215 Nineteenth Street, N.W.,* just around the corner from their previous residence. But the need to support four children placed additional strains on an already precariously balanced family budget. Edith, whose business sense was better than her free-spending husband's, only managed to keep up with their bills by merciless scrimping. "The repairs of the carriages . . . hangs over me like a nightmare," she declared at one point. To save money, she made tooth powder from ground-up fishbones, burned alum, and other ingredients. Referring to his Harvard Club bill, she asked, "Are your dues ten or twenty dollars? Could you resign?" She placed him on an allowance, but to no avail. "Every morning Edie puts twenty dollars in my pocket, and to save my life I never can tell her afterward what I did with it," he told a friend as he fumbled for money to pay for a book and came up with only twenty-five cents.

Black moments of doubt and uncertainty about the future dominated Roosevelt's days. Harrison and Wanamaker were ominously silent about the Baltimore report, the spoilsmen were clamoring for his head, and the burden of paying for two households and the twice-yearly migration between them had brought him to the edge of bankruptcy. Had he done wisely in opting for a career in public service? How would he pay for the education of his children? "The trouble is that my career has been a very

---

*This house, too, is still standing and also has been converted into offices.

pleasant, honorable and useful career for a man of means," he declared, "but not the right career for a man without the means."

Unable to bear further delay in resolving the Baltimore case, Roosevelt unburdened himself of his complaints against Wanamaker and the president in a fiery speech before the executive board of the New York City Civil Service Reform Association on March 8, 1892. "Damn John Wanamaker!" Roosevelt raged beyond the hearing of reporters. Nearly a year had passed since he had demanded the ouster of the twenty-five miscreants, and nothing had been done except for Wanamaker's diversionary investigation of his charges. Although the postal inspectors' report had been pigeonholed, Roosevelt was convinced it would support his findings. But how could he pry it loose?

Carl Schurz, the veteran reformer, suggested that he demand an investigation of the matter by the House Civil Service Committee. The committee—controlled by the Democrats and delighted by the intraparty squabble among the Republicans in a presidential election year—was quick to comply. Over the next several weeks, the air was filled with charges and countercharges. Wanamaker contended that Roosevelt had browbeaten witnesses into making confessions, and that the funds collected in the Baltimore post office were not for election expenses but merely for the purchase of a pool table for the workers. For his part, Roosevelt accused Wanamaker of "slanderous falsehoods" and stood by his report "not only in its entirety, but paragraph by paragraph."

To President Harrison, who watched grimly as his administration was torn asunder, Roosevelt wrote: "I have used every effort to avoid a conflict with the Post Office Department. It has now become merely a question of maintaining my own self-respect and upholding the civil service law." Harrison must have finally had his fill of the righteous Roosevelt, but refrained from dismissing him out of fear of alienating the mugwumps. In the end, a majority of the investigating committee found that the evidence produced by Wanamaker's own inspectors corroborated Roosevelt's findings, and accused Wanamaker of willful negligence in failing to enforce the law.

While the victory was sweet for Roosevelt, it was a severe blow to Harrison's chances for reelection. Roosevelt tried to make amends for the damage he had done Harrison's prospects by publishing an article praising the president's foreign policy and campaigning strenuously for him in the West, but it was to no avail. The rise of Populism;* the McKinley Tariff, which resulted in higher prices for consumers; labor unrest; and the defection of the mugwumps to Grover Cleveland, again the Democratic candidate, resulted in an overwhelming defeat for Harrison. Although Cabot Lodge managed to win a Senate seat, the Republicans lost control of both houses of Congress.

Six months of uncertainty followed for Roosevelt. He desperately wanted to remain in Washington, but had little hope of reappointment because he had assailed Cleveland as an enemy of civil service reform. Nevertheless, he urged Carl Schurz to intercede with Cleveland. Schurz duly advised the president-elect that if he wished to deal a blow to the spoils system, he could "hardly find a more faithful, courageous and effective aide than Mr. Roosevelt." Much to Roosevelt's surprise, Cleveland did not bear a grudge and invited him to stay on "for a year or two."

For the most part, the remainder of Roosevelt's time in Washington was a repetition of the previous service on the commission. There was the same battle against the spoilsmen—this time they were Democrats rather than Republicans—and the same effort to attract the attention and support of the president. Cleveland, however, was more supportive of civil service reform than Harrison, telling department heads to cooperate with the commission and even going so far as to drop a Democratic commissioner to whom Roosevelt objected.†

He worked fitfully on the third volume of *The Winning of the West,* rode in Rock Creek Park with Lodge, led his children on "scrambles" among the rocks, and worried about getting fat. He struck up an acquaintance with the visiting Rudyard Kipling, al-

---

*The Populist presidential candidate, James B. Weaver, polled over a million votes, but trailed Cleveland and Harrison.

†The commissioner, George D. Johnston, an irascible onetime Confederate officer, carried a pistol. After Johnston objected vociferously to Roosevelt's office being recarpeted before his, TR had a private chat with the president. Cleveland offered Johnston a consular post as far from Washington as possible.

ready famous as the author of *Barrack-Room Ballads*. At first, the two men crossed swords, but soon entered what became a lifelong friendship. "I curled up on the seat opposite" Roosevelt at the Cosmos Club on Madison Place, Kipling later recalled, "and listened and wondered until the universe seemed to be spinning around and Theodore was the spinner."

On those evenings when he and Edith had no engagements, they read. If Roosevelt disliked a book or article, he let out a yell that reverberated around the house. After reading a story by Henry James, he indignantly informed Bamie: "I think it represents the last stage of degradation. What a miserable little snob Henry James is. His polished, pointless, uninteresting stories about the upper classes of England make one blush to think he was once an American. . . ."

Sometimes Roosevelt would bring his children to the office. Alice, with flaxen braids down her back, would sit on one side of his desk and Ted, Jr., small and bespectacled, on the other side, while the younger children cuddled on his lap. It was a touching family scene, Martin Halloran recalled. When one of the children dived into a wastebasket in search of canceled postage stamps and threw its contents all over the floor, their father wasted no time in making the chastened offender pick up every last scrap.

Edith claimed, however, that he was of little use in the practical matters of rearing the children. Once when she was unwell, her husband volunteered to shepherd the family on their semiannual move from Oyster Bay to Washington. He got the five children and two nurses safely on the train to Long Island City and on the ferry across to East Thirty-fourth Street. "Then," he later told his wife, "everything became confused." He was walking to the horsecar, holding Ted and Kermit by the hand, when Ted slipped and fell head first into a mud puddle. He was covered from head to foot with mud, too much to be wiped off with a handkerchief. What should he do? A friendly policeman came to the rescue and helped him scrape the worst of the mud off the boy. By the time this operation was completed, the nurses and the other three children had disappeared.

Roosevelt soon had more pressing problems. Within months of Cleveland's inauguration, the nation was plunged into the worst economic depression since the Civil War. Factories and businesses closed, railroads went bankrupt, farm prices and industrial wages plummeted, banks failed, and an army of the unemployed, led by Jacob Coxey, an Ohio quarry owner, marched on Washington to demand in vain that Congress put the jobless to work building roads.* Strikes and violence flared across the country.

In the face of this disaster, Cleveland, a firm believer in laissez-faire, followed the traditional policy of noninterference in economic affairs. Every appeal for remedial legislation was rejected; the storm would have to blow itself out. Federal troops were ordered to Chicago to break a strike by Pullman workers, and Eugene V. Debs, the head of the American Railway Union, was jailed. From the hinterland, where the flames of Populism still burned, there was a demand for inflationary measures to revive the economy, including for free coinage of silver. This would increase the amount of money in circulation, thereby reducing the value of the dollar and facilitating the repayment of mortgages and other debts in cheaper money.

Today, it is impossible to understand the heated passions stirred a century ago by the struggle over "free silver." Gold and silver were not merely precious metals, but symbols of the battle between the agrarian West and South and eastern financial interests. The amount of money in circulation was limited by the gold in government reserves, but gold production had remained static since the Civil War, while the population had doubled. Farmers, workers, and debtors claimed this created great hardship because they had to work harder and longer to repay their obligations. A farmer who took out a $5,000 mortgage in 1870—then the equivalent of 2,500 bushels of wheat—had to produce 5,000 bushels to repay the loan twenty years later.

Free silverites argued that the logical solution was to increase the supply of money by allowing free, unlimited coinage of silver. They were supported by silver mine operations, who were pro-

---

*Coxey was arrested for trespassing on the Capitol lawn before he could finish his speech. His proposal was not as radical as it sounded at the time, and during the Great Depression of the 1930s, Franklin Roosevelt adopted the policy of putting the unemployed to work on public works projects. Fifty years after the march, in 1944, Coxey returned to Capitol Hill to complete his speech. It was heard by about two hundred curious and uncomprehending spectators.

ducing an increasing supply of the metal. "Sound" money advocates rejected this remedy as dangerous radicalism that would lead to ruinous inflation. In 1890, the silverites had prevailed and under the Sherman Silver Purchase Act, the government began buying silver. But following the Panic of 1893, Cleveland, convinced that the law was one of the major causes of the upheaval, persuaded Congress to repeal it. This proved to be a Pyrrhic victory, for it had no effect on the economy and splintered the Democratic party.

Roosevelt suffered fresh financial distress during this turmoil. He learned that Douglas Robinson, who was managing his finances, had overstated his income the previous year and understated his expenditures. Once again, he talked of selling Sagamore Hill or of sending the family home while he lived alone in a rented room in Washington. Financial disaster was averted, however, by an unexpected legacy from one of Edith's uncles.

Over the years, Roosevelt had grown bored with the Civil Service Commission. It was, he noted, "a little like starting to go through Harvard again after graduating." When a group of reformers approached him in August 1894 with the suggestion that he run as the fusion candidate for mayor of New York, he was inclined to accept but sought Edith's approval before giving them an answer. She had recently given birth to another child, Archibald Bulloch Roosevelt, and was flatly opposed to leaving Washington. Besides, she thought they could ill afford the cost of a campaign that he might not win. Although his salary as commissioner was paltry, it was better than no salary at all, particularly in a period of economic upheaval. With considerable regret, Roosevelt decided not to run.

The problem of his brother Elliott also continued to plague him. Not long after Elliott began his period of probation in Virginia, his life had begun to unravel. Anna Roosevelt died of diphtheria at the end of 1892, and a year later Elliott, Jr., succumbed to the same illness. Little Eleanor fantasized that now her father would return and carry her away, but Anna, mistrusting her husband, had designated her own mother as the children's guardian. Whatever efforts Elliott had made at reform vanished with the death of his wife, and moving back to New York, he took an apartment with his mistress on West 102nd Street. Reports reached the family of one drunken escapade after another. "Poor fellow," sighed his brother. "If only he could have died instead of Anna."

In mid-August, about the same time Roosevelt was mulling over whether or not to run for mayor, Elliott went berserk with delirium tremens, tried to leap from a window of his apartment, and died during a convulsive attack. "Theodore was more overcome than I have ever seen him—cried like a child for a long time," noted Edith. The bitterness he had felt toward his brother had faded, and he told Bamie: "He would have been in a straight jacket had he lived forty-eight hours longer. But when dead the poor fellow looked very peaceful, and so like his old generous, gallant self of fifteen years ago. The horror and terrible mixture of sadness and grotesque, grim evil continued to the very end; and the dreadful flashes of his old sweetness, which made it even more hopeless. I suppose he has been doomed from the beginning. . . ."

In the wake of these disasters, Roosevelt fled west in one of his black introspective moods, certain that he had made a mistake in refusing to run for mayor. "The prize was very great; the expense would have been trivial and the chances of success very good," he wrote Bamie.* "But it is hard to decide when one has the interests of a wife and children to consider first." With Lodge he was more candid. "The last four weeks, ever since I decided not to run, have been pretty bitter ones for me. I would literally have given my right arm to make the race, win or lose. It was the one golden chance, which never returns."

Eventually, Edith learned from Bamie just how intensely her husband had wanted to make the race for mayor, and she was overcome by guilt and remorse. "He should never have married me and then would have been free to make his own course," she

---

*Bamie was now temporarily living in London at the request of her fifth cousin James Roosevelt Roosevelt. "Rosy" Roosevelt had been rewarded with the post of first secretary of the American legation after making a $10,000 donation to Cleveland's campaign. His wife, Helen, had just died, and Bamie arrived to look after his children, Helen and James Roosevelt Roosevelt, Jr., heirs to the Astor fortune. She made so many friends that Thomas F. Bayard, the American minister, asked her to serve as his official hostess. Bamie, it was said, wanted to marry Rosy, but he was involved with his English mistress, whom he later married.

told her sister-in-law. "I never realized for a minute how he felt over this, or that the mayoralty stood for so much to him. . . . I am too thankful he is away now for I am utterly unnerved as a prey to the deepest despair. . . . If I knew what I do now I should have thrown all my influence in the scale . . . and helped instead of hindering him." For the first time in nearly eight years of marriage, the Roosevelts' relationship was in trouble. Yet, through mutual love and understanding, they managed to surmount the crisis.

In place of Roosevelt, the Republicans nominated William L. Strong, a successful dry-goods merchant without political experience. New York was undergoing one of its periodic crusades against graft and corruption, and he was overwhelmingly elected by a coalition of reform Republicans, Democrats, and "good" people generally. Strong offered Roosevelt the post of street cleaning commissioner, but he turned it down.* To Lodge, he intimated that it was not the sort of office he could "afford to be identified with." Instead, his eye was on one of the seats on the Board of Police Commissioners. He was eager to get back into New York politics; the challenge of helping clean up the corruption-ridden police department appealed to his combative instincts, and the job paid the respectable sum of $5,000 annually. In mid-April 1895, Strong offered him the post and he accepted.

The effectiveness of Roosevelt's six years on the Civil Service Commission is a matter of debate. Some critics charge that his noisemaking accomplished little of substance. On the other hand, his boundless energy gave the commission influence it had never had before. A total of 26,000 jobs were removed from the category of political plum and placed under the merit system, and new tests for applicants were devised. For the first time, women were put on the same competitive level with men, which increased their numbers in government jobs. On a personal level, Roosevelt had restored his relations with the reform wing of the Republican

---

*TR persuaded Strong to appoint Joe Murray, who had given him his start in politics, to the post of commissioner of excise. *Letters,* Vol. I, pp. 418–419. Following Murray's appointment, he told him not to try to evade the civil service law in filling jobs. "Don't buck up against it, whether you believe in it or not," Roosevelt advised. *Ibid.,* p. 431.

party and won the admiration of ordinary Americans for taking on the spoilsmen. Yet, one problem defied even his strenuous efforts, and haunts the bureaucracy a century later: how to promote efficiency among government workers and make them responsive to the needs of the people they are supposed to serve.

Chapter 11

"It Is a Man's Work"

Theodore Roosevelt turned the corner into Mulberry Street on the morning of May 8, 1895, and hurried toward New York City's dingy old Police Headquarters with three perspiring men trailing along behind him. Several reporters assigned to the police beat watched from the front steps of the building across the street where they had their offices. Roosevelt recognized an old friend, Jacob Riis of the *Evening Sun,* among them and shouted: "Hello, Jake!" Waving to the newsmen to follow the new Board of Police Commissioners, he bounded up the front steps of Police Headquarters.

"Where are our offices?" demanded Roosevelt of the startled police officers. "Where is the Board Room? What do we do first?"

Making his way to the second floor, he found the outgoing board waiting, stiff, formal, and dignified, for their successors. Quickly dispensing with ceremony, Roosevelt shook hands, called a meeting of the new Police Board, and had himself elected its president. Only one member, Major Avery D. Andrews, a West Pointer and nominal Democrat, supported him for very long, however. The others, Frederick D. Grant, a Republican and son of the former president, and Andrew D. Parker, an attorney and anti-Tammany Democrat, were soon jealous of his increasing fame. "Thinks he's the whole board," Parker sniffed to Lincoln Steffens of the *Evening Post*.

Over the next two years, Roosevelt was the delight of newspaper cartoonists and reporters, the pride of reformers, and the despair of Tammany Hall, crooked policemen, and saloonkeepers. Often, he was at odds with his own party and his colleagues on the Police Board. Under the guidance of Riis, author of *How the Other Half Lives,* a vivid account of appalling conditions in New York's slums, he poked about parts of the city he would never have seen and became even more aware of the horrible circumstances in which many of its citizens existed. "My experience in the Police Department taught me that not a few of the worst tenement-houses were owned by wealthy individuals who hired the best and most expensive lawyers to persuade the courts that it was 'unconstitutional' to insist on the betterment of conditions," he later recalled.

"We have a real Police Commissioner," wrote Arthur Brisbane of the *World* not long after Roosevelt's arrival. "His teeth are big and white, his eyes are small and piercing, his voice is rasping. . . . His heart is full of reform [and] he looks like a man of strength . . . a determined man, an honest, conscientious man, and like the man to reform the force."

Roosevelt assumed his new position as the Police Department was emerging from another periodic investigation of the graft and corruption that blot its history. The inquiry had been touched off by Reverend Dr. Charles H. Parkhurst, who charged from his pulpit in Madison Square Presbyterian Church that New York was "a very hotbed of knavery, debauchery and bestiality," from which the politicians and the police enriched themselves. Tammany Hall, he thundered, was nothing less than "a commercial corporation, organized in the interest of making the most out of its official opportunities."

To prove these charges, Parkhurst disguised himself as a visiting sport and toured the city's Tenderloin district. He dropped in on a thieves' den on the East River, a fan-tan parlor and opium den in Chinatown, and a sampling of brothels—all countenanced by the police. The shocked divine revealed that in one house of ill repute located only a few hundred yards from his own church, he

had seen five naked women perform "a circus" or "sort of gym-
nastic performance." Parkhurst's fulminations led to the creation
of the Lexow Committee, which confirmed his charges of whole-
sale corruption in the Police Department and contributed to the
election of Mayor Strong.

Roosevelt wasted no time in grabbing the Police Department
by the scruff of its neck and giving it a good shaking up. The first
to go was Superintendent Thomas F. Byrnes, the city's top cop,
who had amassed a fortune of $350,000, which he claimed derived
from stock market tips from Jay Gould, among others. His ouster
was followed by that of his deputy, Inspector Alexander "Club-
ber" Williams, who said his yacht and Connecticut estate were
the result of shrewd speculation in Japanese building lots. A raft
of other policemen of all ranks found to be on the "take" from
gamblers, saloonkeepers, brothel owners, confidence men, and
assorted criminals were also rooted out. On the other hand,
Roosevelt never vilified the average policeman, whom he characterized as a bluff and sometimes heavy-handed but basically
good-natured Irishman or German victimized by a system over
which he had no control. Good men were given generous praise
and promotions.

Under no illusions about the difficulty of cleaning up the department, Roosevelt nevertheless threw himself into the job with
animal vigor. "I have the most important and corrupt department
in New York on my hands," he told Bamie. "I shall speedily assail
some of the ablest, shrewdest men in this city, who will be fighting
for their lives, and I know how hard the task ahead of me is. Yet,
in spite of all the nervous strain and worry, I am glad I undertook
it; for it is a man's work."

Roosevelt was a member of the Police Board for less than two
years, but in this brief period he established a sense of profes-
sionalism heretofore unknown to the department and made a
sweeping series of changes, reforms, and innovations that out-
lasted his tenure. He centralized executive control, which reduced
political influence on police matters; broadened the use of special
squads for more effective crime fighting; brought civil service and
new professional standards to police recruitment; extended the
employment opportunities of women; opened a police academy
for trainees; improved discipline; brought in such new technology
as the Bertillion system of identification, the telephone, and

horse-drawn patrol wagons, and standardized police weapons. As Jay Stuart Berman, who has made the most thorough study of Roosevelt's police activities has observed, the concepts and practices that he fostered in 1895 still serve as guidelines for modern efforts to upgrade police administration in the United States.

Every morning, Roosevelt bicycled to his office from Bamie's house on Madison Avenue, which he and Edith had rented. Late at night, he wandered the city streets with a black cloak over his evening clothes and a hat pulled down over his eyes in search of crime or a policeman asleep on duty. Jake Riis and Lincoln Steffens often accompanied him on his midnight rounds. "A good many [policemen] were not doing their duty," Roosevelt noted, "and I had a line of huge frightened guardians of the peace down for reprimand or fine." Then, as the first light glimmered in the east, he would slip into Mike Lyons's restaurant on the Bowery for ham and eggs before returning to headquarters for a few hours' sleep on his office couch in preparation for another strenuous day.

A *World* reporter produced a graphic portrait of Roosevelt as he presided at a departmental trial of policemen found derelict in their duty:

> When he asks a question, Mr. Roosevelt shoots it at the poor trembling policeman, as he would shoot a bullet at a coyote . . . he shows a set of teeth calculated to unnerve the bravest of the Finest. His teeth are very white, they form a perfectly straight line. The lower teeth look like a row of dominoes. They do not lap over or under each other, as most teeth do, but come together evenly. . . . They seem to say: Tell the truth to your Commissioner, or he'll bite your head off. . . . Generally speaking this interesting Commissioner's face is red. He has lived a great deal out of doors, and that accounts for it. . . . Under his right ear he has a long scar. It is the opinion of all the policemen . . . that he got that scar fighting an Indian out West. It is also their opinion that the Indian is dead.

Policemen were soon watching nervously over their shoulders at night for a dark-cloaked figure with blinding white teeth. Street peddlers capitalized on the public's fascination with Roosevelt and sold whistles shaped like "Teddy's Teeth," which pranksters blew at red-faced policemen. The well-publicized tales of Roosevelt's nightly prowls earned him the nickname "Haroun-al-Roosevelt" after the legendary caliph of Baghdad.

Roosevelt's handling of the press was masterly. Newsmen came to him because he had a ready stock of inside gossip and was a source of highly quotable comments on almost any issue, foreign or domestic. Some newsmen, particularly the ambitious Steffens, later one of the leading muckrakers, were encouraged to believe that they were helping him run the Police Department. He would lean out the window of his office and yell across Mulberry Street to summon Steffens for a private tirade against the latest Tammany Hall outrage, or allow himself to be persuaded from taking some action that the reporter thought dangerous. Steffens was convinced that Roosevelt was more calculating than he appeared, however. Most people believed "he never thinks, that every act is born of the impulse of the moment." They were wrong, said Steffens. "He thinks before he acts."

Having always pursued high adventure and hand-to-hand combat with his enemies, Roosevelt enjoyed himself immensely as police commissioner. Upon one occasion, a notorious anti-Semitic preacher arrived from Germany and demanded police protection while he made speeches denouncing the Jews. "A great many of the Jews became alarmed and incensed about his speaking here, and called upon me to prevent it," Roosevelt recalled in his *Autobiography*. "Of course I told them I could not—that the right of free speech must be maintained unless he incited to riot. . . . On thinking it over, however, it occurred to me that there was one way in which I could undo most of the mischief he was trying to do." Gleefully, Roosevelt assigned a Jewish sergeant and a detail of forty Jewish policemen to guard him—and the fanatic was reduced to a figure of ridicule.*

Out-of-town newspapers were unanimous in their praise for New York's vigorous new police commissioner. Residents of other cities troubled with the same problems were impressed by his

---

*Unlike many members of his class, such as Henry Adams and his brother Brooks, who ranted about the coming "enslavement of everyone to the Jews," Roosevelt showed no signs of anti-Semitism. As police commissioner, he opened the department to Jews and called the Jewish officers "Maccabees." One, Otto Raphael, said the traditional prayer for the dead over Roosevelt's body the night before he was buried.

performance, and Roosevelt's image as a reformer took firm hold on the public imagination. Lodge began to see the presidency as a possibility for his friend. "The day is not far distant when you will come into a large kingdom," he declared. Roosevelt himself was no doubt dazzled by the possibilities before him, but the uncertainties of politics preyed upon his mind. One day Riis and Steffens raised the question of whether he was considering running for president. In a rage, he leaped to his feet, ready to throttle Riis.

"Don't ask me that!" he shouted. "Don't you put such ideas in my head. No friend of mine would say a thing like that. Never, never, must either of you remind a man on a political job that he may be President. It almost always kills him politically. He loses his nerve; he can't do his work; he gives up the very traits that are making him a possibility."

Whether it was a result of his brightening prospects for the presidency or not, Roosevelt took an increasing interest in international affairs while on the Police Board. The Franco-Prussian War had awakened Europe from midcentury dreams of a peaceful concert of nations under the benign hegemony of Great Britain. A unified, powerful Germany now challenged British eminence and threatened the balance of power that had prevailed since the end of the Napoleonic Wars. Nationalism was in vogue, bringing with it harsh competition for new colonial empires and the right to exploit backward countries in Africa and East Asia.

In the United States, this new spirit was reflected in the reappearance of manifest destiny, which had lain dormant since the end of the Civil War. With the closing of the frontier, Latin America, the Caribbean, and the Far East were being eyed as areas where the great surpluses of America's factories and farms could be disposed of at a profit. The new imperialists did not envision colonizing these areas as the Europeans were doing, but wished to exploit them commercially to bring the benefits of Anglo-Saxon civilization and republican government to the benighted natives— "the lesser breeds without the law," in the words of Roosevelt's good friend Rudyard Kipling.

Always a strong nationalist concerned with the honor and prestige of the United States, Roosevelt readily adopted the views of the expansionists. Expansion to "masterful" people is "not a matter of regret," he insisted, "but of pride." Wishing to see European influence removed from the Western Hemisphere, he was willing to use force, if necessary. War would ennoble the American people by purging them of their concern with material gain. He was opposed to unjust wars, but found it impossible to conceive of the United States as acting unjustly—which led to the typically Rooseveltian view that all American actions abroad were moral and all opposition to them therefore immoral.

Roosevelt was among the first to recognize the importance of the teachings of Captain Alfred Thayer Mahan, the high priest of expansionism. In 1890, he wrote a glowing review in the *Atlantic Monthly* of Mahan's first and most important book, *The Influence of Sea Power Upon History, 1660–1783,* that called it to the attention of a much wider audience than it might have had. Some writers have contended that Roosevelt followed Mahan's theories, but in reality Mahan's ideas coincided with the basic thesis of his own first book, *The Naval War of 1812.* In that work, Roosevelt argued that America needed a strong navy to defend its interests abroad and to protect its coasts against enemy attack.

"America must begin to look outward," advised Mahan, and businessmen and politicians echoed him. Overseas trade was the key to national greatness, he maintained, citing the example of Britain, and command of the sea and a strong navy were vital to America's protection. To support the great fleet he envisioned, Mahan called for bases in the Caribbean and the Far East and for a canal across Central America—probably through Nicaragua—to facilitate commerce between East and West and the rapid movement of the American battle fleet from the Atlantic to the Pacific.

Roosevelt had no reluctance to speak his mind on foreign policy and military preparedness, even though the public positions he held were unrelated to these issues. When a Chilean mob attacked and killed two American sailors in 1892, he was eager for war, but was frustrated by President Harrison's decision merely to seek an apology. "For two nickels he would declare war—shut up the Civil Service Commission, and wage it sole," an amused John Hay had noted. Two years later, Roosevelt was demanding the annexation of the Hawaiian Islands after American residents had overthrown the regime of Queen Liliuokalani, and he talked of war with Britain after that nation became involved in a boundary dispute with Venezuela.

"Personally I rather hope the fight will come soon," Roosevelt wrote Lodge at the time of the Venezuelan crisis. "The clamor of the peace faction has convinced me that the country needs a war. Let the fight come if it must; I don't care whether our sea coast cities are bombarded or not; we would take Canada." He wrote Governor Levi P. Morton about a captaincy in the New York National Guard in case of war, and assailed the *Harvard Crimson* for opposing Cleveland's Venezuelan policy. "Nothing will tend more to preserve peace on this continent than the resolute assertion of the Monroe Doctrine," he declared. "Let us make this present case an object lesson, once and for all."

In dealing with backward nations, there was a certain sense of the superiority of Western civilization and institutions in Roosevelt's view, but it was seasoned by both idealism and realism. Although convinced that Americans were superior in terms of progress to Latin Americans or the Chinese or Filipinos, he was also certain these peoples would make advances in years to come. In the meantime, they should not be allowed to stand in the way of civilization. For example, although he admired the Boers, he felt it "in the interest of civilization that the English-speaking race should be dominant in South Africa, exactly as it is for the interest of civilization that the United States . . . should be dominant in the Western Hemisphere.*

"Every expansion of a great civilized power means a victory for law, order, and righteousness," he declared. "This has been the case in every instance of expansion during the present century, whether the expanding power was France, or England, Russia or America." Thus, it was the duty of the United States to annex Hawaii and intervene in Cuba, where a revolt had broken out against Spanish rule in 1895. "I wish our people would really interfere in Cuba," he wrote Bamie. "We ought to drive the Spaniards out of Cuba; it would be a good thing."

As a result of his ten- and twelve-hour days of paperwork, meetings, and hearings followed by nightly prowls about the city,

---

*By "race," Roosevelt meant an ethnic or national group, not a biological race in the sense in which the word is used today.

Roosevelt found little time to work on the fourth volume of *The Winning of the West* and lamented the fact that he saw "little more than a glimpse" of the children. During his infrequent weekends at Sagamore Hill, he tried to make time for a romp with Archie, a game with Ethel, man-to-man talks with Ted and Kermit, and chats with Alice, now as tall as her stepmother and showing signs of becoming a beauty. To save money, Edith dressed the boys in homemade overalls and tennis shoes. She felt a pang of nostalgia when she chaperoned Alice and Ted to Mr. Dodsworth's dancing classes, where she and her husband had learned the waltz and the polka under the same stern gaze years before.

Ted was nine when his father gave him his first rifle. One day the boy had been roaming about the farm and came home late to find the weapon and fell upon it at once. "I wanted to see it fired to make sure it was a real rifle," he recalled later, "but it would be too dark to shoot after supper." The elder Roosevelt picked up the weapon and slipped a cartridge in the chamber.

"You must promise not to tell Mother," he whispered conspiratorially.

The boy readily promised and his father fired into the ceiling. The report was slight, the smoke hardly noticeable, and the hole in the ceiling almost undetectable.

To save money, the boy's parents enrolled him in the one-room Cove Neck School, where he attended class with the children of gardeners, grooms, and coachmen from nearby estates. "Exceedingly active, normally grimy," and undersized for his age, he was always getting into fights. The rights and wrongs of these encounters were a subject of earnest discussion between father and son. "We always told him everything," Ted later wrote, "as we knew he would give us a real and sympathetic interest."

On Christmas Eve, Roosevelt went over to the school, where he urged the children to stand up for their rights, be kind to animals, and "do something worthwhile when they grew up." It was a speech he gave with minor changes every Christmas for the rest of his life. Following songs and recitations by the children, he distributed gifts of skates, toy trains, and dolls that he and Edith had purchased.

Christmas at Sagamore itself was an important occasion. Early in the morning, the Roosevelt children leaped from bed to burrow into their Christmas stockings, hailing each new treasure with squeals of delight. Then, after breakfast, everyone trooped into the library for the opening of gifts, where, as their father said, they "came as near as realizing the feelings of those who enter Paradise as they ever will on this earth." To Mrs. Winthrop Chanler, an old friend, this was the essence of Theodore Roosevelt. "Life was the unpacking of endless Christmas stockings."

The family was startled by a cable from Bamie announcing her engagement to William Sheffield Cowles, the American naval attaché in London. Cowles, a genial and portly man with a walrus moustache whom she called "Mr. Bear-o," was forty-nine; the bride was forty. Society at large might have been surprised at the sudden engagement of a semicrippled spinster to a worldly sailor, but those familiar with Bamie's charm and graciousness could understand how she had attracted him. "I have always felt it a shame that you, one of the two or three finest women whom I have met or known of, that you, a really noble woman, should not marry," her brother wrote. Unhappily, he was unable to attend the wedding,* saying it would be "dishonorable" for him to leave New York at the very time he had "plunged the Administration into a series of fights."

The most serious of these fights resulted from Roosevelt's decision to enforce the law making it illegal to serve alcoholic beverages on Sunday. He was no prohibitionist—in fact, he repeatedly stated that excessive use of alcohol could not be prevented by law. But he argued that the failure to enforce the law set a bad precedent that was tailor-made for under-the-counter payments to the police by saloonkeepers wanting to do business. If a wave of protests followed strict enforcement of the Sunday closing law, he reasoned that it would force the State Assembly to repeal it. Instead, to his surprise, the protests were directed against him.

---

*The wedding was followed by legal complications. Although Cowles was divorced, the decree was not recognized in New York State, and if the couple lived there after their return from England, he could be prosecuted for bigamy. Fortunately, the laws of Connecticut, where Cowles had a family home at Farmington, were more lenient, and they made it their legal domicile. In 1898, despite Mrs. Cowles's age and physical problems, they became parents of a healthy child, William Sheffield Cowles, Jr.

Thirsty New Yorkers charged that Roosevelt was instituting a dictatorship and called him "King Roosevelt I." Workingmen complained that they were being deprived of a few innocent beers on Sunday, while the rich could still get a drink in their clubs, including the Union League, of which Roosevelt was a member.* The large German-American population, whose family custom it was to gather on Sunday afternoons in the city's many beer gardens for refreshment, pinochle, and dominoes, were particularly angered. More incorruptible than sagacious, Roosevelt adopted the position that the law was the law—and would be enforced.

"I do not deal with public sentiment," he declared. "I deal with the law. I am going to see if we cannot break the license forthwith of any saloon keeper who sells on Sunday. This applies just as much to the biggest hotel as to the smallest grogshop. To allow a lax enforcement of the law means to allow it to be enforced just as far as the individual members of the police force are willing to wink at its evasion. Woe to the policeman who exposes himself to the taint of corruption."

Protest took many forms. Some people went over to New Jersey, where Sunday drinking was legal; others rowed out to excursion boats that floated offshore with taps flowing. The Germans announced a parade denouncing Roosevelt and dared him to attend. Picking up the challenge, he was on hand to join in the laughter when a coffin labeled "Teddyism" was carried by. Near the end of the parade marched an old German with thick eyeglasses who peered closely at the reviewing stand and loudly demanded:

"Wo ist der Roosevelt?"

"Hier bin ich!" the commissioner shouted back in his best Dresdener German, amid a wave of laughter.

Everyone agreed that Roosevelt had survived the ordeal, but not all the opposition was so good-humored. Once, a crudely made bomb arrived in the mail, and he was shadowed at night in a vain attempt to catch him in some indiscretion. Clearly, his days as police commissioner were numbered. By the end of the year, Roosevelt lamented to Lodge that "I have not one New York City newspaper or one New York City politician on my side" and the

---

*In fact, Roosevelt enforced the law against the rich as well as the poor, and the police raided Sherry's restaurant during a dinner attended by many prominent New Yorkers.

publisher of the *Tribune* had ordered that "I am not to be mentioned save to attack me unless it is unavoidable."

Tammany exploited the opportunity to discredit reform. Suddenly, long-forgotten "blue laws" were rigidly enforced by the police, who arrested operators of soda fountains, florist shops, and shoeshine parlors for doing business on Sunday. Rumors spread that ice could not be sold, and the *World* printed a tearful—but faked—story about a child who had died because ice was unavailable. Further complications were caused by the Assembly's passage of the Raines Law, which allowed Sunday drinking with meals in hotels and defined a hotel as a building with at least ten bedrooms. Within a few months, hundreds of new "hotels" had sprung up all over the city, many of them saloons that had partitioned off ten rooms on their upper floors,* and then, thumbing their nose at the law, used them for prostitution.

Enforcement of the Sunday closing law created strains that soon had the reformers at each other's throats. In a desperate attempt to hold on to the German-American vote, Mayor Strong indicated that he would fire Roosevelt unless he eased up on his unpopular campaign of law enforcement, while Roosevelt resolutely held his ground. There were reports that an attempt would be made in Albany to legislate him out of the job. Meanwhile, the reformers grumbled and demanded fewer political compromises and stricter enforcement. Parkhurst accused Roosevelt of making deals. Roosevelt retaliated by calling the preacher "a goose" and "an idiot." As any Tammany ward heeler could have foreseen, the air had quickly gone out of the reformers' balloon.

Roosevelt was hardly to blame for the failure of reform. From the beginning, he had not even been able to count upon united support from his own party. The fight among Republicans over the presidential nomination in 1896 was reflected on the local scene. Boss Tom Platt favored Governor Levi Morton; Strong supported William McKinley, while Roosevelt backed Tom Reed. "With proper power," Roosevelt had written Lodge, "I could make this Department of the first rank from top to bottom." But the division of authority on the Police Board, where any of the

---

*One enterprising furniture store advertised furnishings for all ten rooms for exactly $81.20.

four members held a veto over the decisions of the majority, made it difficult to administer the department. Roosevelt was indeed sometimes heavy-handed, but a weaker man or a more cautious man would have accomplished little. Within the limits in which he was forced to work, he brought a semblance of honesty and efficiency to the police force.

Tension was especially high on the Police Board, where Andrew Parker's distaste for Roosevelt and his reforms had grown by leaps and bounds. At first merely uncooperative and then rebellious, he persuaded Fred Grant to join him in blocking a series of promotions backed by Roosevelt and Andrews. Parker found a ready ally in Platt, who was angered by Roosevelt's dismissal of some of his minions and his support of Reed. Parker conspired so actively against Roosevelt that the board was virtually paralyzed by the internal dissension and deadlock.

To Lodge, Roosevelt deplored the fact that "I cannot shoot [Parker] or engage in a rough-and-tumble with him, and I hardly know what course to follow as he is utterly unabashed by exposure and repeats lie after lie with brazen effrontery." Parker was "mendacious, treacherous, capable of double dealing and exercising a bad influence on the force." Finally, with the support of Andrews and Parkhurst, he forced Mayor Strong to bring charges against Parker and hold a public hearing. But even as Roosevelt was leading the fight against Parker, he was already looking for a way out of the debacle.

In the summer of 1896, just a few weeks after William McKinley had won the Republican presidential nomination, Roosevelt invited an old friend, Mrs. Maria Longworth Storer, who also knew McKinley well, to Sagamore Hill. One day, he took her out rowing on the placid waters of Oyster Bay, and as the oars rose and fell, he confided a secret to her. "There is one thing I would like to have, but there is no chance of my getting it. McKinley will never give it to me. I should like to be Assistant Secretary of the Navy."

There is no greater contrast in American politics than between Theodore Roosevelt and William McKinley. Roosevelt was exuberant, emotional, and unpredictable; McKinley was a stolid, unimaginative Republican party wheelhorse who reminded some observers of a statue in search of a pedestal. William Allen White, the Kansas editor, called him "a kindly, dull gentleman . . . honest enough, brave enough, intelligent enough for politics—and no more." Another observer said that he had "the art . . . of throwing a moral gloss over policies which were dubious if not actually immoral." But McKinley had one thing going for him—and that was his good friend Mark Hanna.

Hanna was a Cleveland industrialist who had made a fortune in coal, iron, banking, and Great Lakes shipping, but had wider ambitions. He craved power and sought it in politics—not by making speeches or running for office himself but through the efficient management of others. Practical, shrewd, and cynical, he shouldered his way into the American political scene by becoming an unrivaled fund-raiser and the closest thing to a national political boss the country has ever seen.

Few political figures have had a worse image than Hanna, however. Cartoonists gleefully portrayed him as a sinister, toadlike figure wearing a flashy suit spangled with dollar signs. Editorial writers lashed out at "Dollar Mark" as "the evil genius of the plutocracy." In reality, he was comparatively honest in his personal dealings and far more enlightened than the average tycoon or politician of the day. Hanna paid his workers high wages, made provision for those who became sick or were injured, and condemned employers who fought unionization. And he was the first American political figure fully to grasp the implications of industrialization. "Politics are one form of business and must be treated strictly as a business," he declared. In place of the slippery spoilsmen, he envisioned a system in which business and politics were closely aligned. The great mission of the Republican party, as Hanna saw it, was to run the country for the benefit of business and spread the benefits of prosperity among all Americans.

"I love McKinley," Hanna once remarked and found in him the pitchman of the new politics. An Ohio country boy who had taught school and volunteered for the Union Army, McKinley emerged from the war a major—in fact, friends always referred to him as "the Major"—and after a year's study he was admitted to the bar. Amiable and eager to be liked, McKinley prospered as an attorney and married a young lady named Ida Saxton. They had two daughters but, tragically, both died in childhood, and their loss sent Ida into a lifelong decline that was broken by regular epileptic seizures.*

With such a background, McKinley naturally turned to politics, and his starched dignity, square face, and heavily tufted eyebrows impressed the voters of Canton, Ohio, and its environs. He was a devout Methodist who attended church regularly, read the Bible to his ailing wife, and would hide his cigar when his picture was being taken—lest he corrupt American youth. Elected to Congress in 1876, he remained a congressman until the public outcry against the McKinley Tariff resulted in his defeat in 1890. Two years later, with Hanna behind him, he overcame this setback to win election as governor. When McKinley was faced with financial ruin, Hanna formed a syndicate of business leaders who put up $100,000 to bail him out.†

Over the next four years, Hanna watched carefully over the Major's interests, plotting the strategy that would make him president. Ohio was a pivotal state and anyone who controlled it had a leg up on the election. While McKinley had a serene faith that destiny meant him to be president, Hanna was unwilling to trust to fate. Retiring from business, he devoted his full attention to McKinley's candidacy. A year before the Republican convention was gaveled to order in St. Louis, he already had most of the notoriously venal southern delegates safely in his pocket.

Theodore Roosevelt backed Tom Reed for the nomination and regarded him "as in every way superior to McKinley." As an early Reed supporter, he could count on a Cabinet seat if the Speaker won the presidency, but Reed showed a distressing reluctance to court support by the usual methods. He refused to promise jobs to prospective supporters or buy his share of southern delegates, and people were put off by his sardonic, overbearing manner. Roosevelt himself was uneasy about Reed's opposition to expansionism and his reluctance to speak out on the currency question, and soon realized he could not be nominated. "Upon my word,"

---

*When Ida suffered seizures in public, her husband draped a napkin or handkerchief across her face and blandly acted as if nothing had happened.

†Bellamy Storer, an Ohio businessman and husband of Maria Storer, had contributed $10,000 to the fund to rescue McKinley from financial disaster. In return, he hoped to be made ambassador to London or Paris; instead he got the second-rank post of minister to Belgium.

he complained to Cabot Lodge, "I do think Reed ought to pay some heed to the wishes of you and myself."

Roosevelt was unable to attend the Republican convention because of the row with Andrew Parker. "McKinley, whose firmness I utterly distrust will be nominated; and this . . . I much regret," he wrote Bamie. This prophecy proved correct. Hanna had done his work well and McKinley was chosen on the first ballot. Reed, who was nominated by Lodge, received only a handful of votes. Always quick to react to new political developments, Roosevelt wore a large McKinley button at Police Headquarters the next morning.

Roosevelt found Mark Hanna, who had been named chairman of the Republican National Committee, to be "a good natured, well meaning coarse man, shrewd and hard-headed, but neither very farsighted nor very broad-minded. He has a resolute, imperious mind; he will have to be handled with some care." Handling Hanna with care included the decision to keep his own future ambitions to himself while accepting Hanna's invitation to join in the crusade against the Democratic candidate, Nebraska's William Jennings Bryan.

Three weeks after the Republicans nominated McKinley, the Democrats gathered in Chicago, where they repudiated party conservatives and entered into an alliance with the Populists and western silver mining magnates. The platform condemned the gold standard as "anti-American," and the Democrats demanded "free and unlimited coinage" of silver at a ratio of sixteen to one for gold. In Bryan, the thirty-six-year-old "Boy Orator of the Platte," the free-silverites had found their Moses. Black eyes flashing, he strode to the rostrum to denounce the eastern bankers and spoke with a revivalist fervor that had the Chicago Coliseum ringing with cheers. Reaching the end of his speech, he declared:

Having behind us the producing masses of this nation and the world supported by the commercial interests, the laboring interests and the toilers everywhere, we will answer their demand for a gold standard by saying to them: "You shall not press down upon the brow of labor this crown of thorns, you shall not crucify mankind upon a cross of gold!"

Twenty thousand spectators leaped to their feet as one and roared approval. Weeping, chanting, shouting, they lifted Bryan to their shoulders and carried him about the hall in triumph. Cries of "Bryan! Bryan! Bryan!" echoed throughout the hall and his nomination, which came on the fourth ballot, was a foregone conclusion. Not everyone was carried away, however. Illinois Governor John Peter Altgeld turned with a quizzical smile to Clarence Darrow, the lawyer who had defended Eugene Debs, and said, "I have been thinking over Bryan's speech. What did he say anyhow?"

Theodore Roosevelt, for one, had no doubt about the meaning of Bryan's words. In them he heard the rumbling of the tumbrils that had carried the victims of the French Revolution to the guillotine. "Not since the Civil War has there been a Presidential election fraught with so much consequence to the country," he told Bamie. "The silver craze surpasses belief. The populists, populist-democrats, and silver-or populist Republicans who are behind Bryan are impelled by a wave of genuine fanaticism; not only do they wish to repudiate their debts, but they really believe that somehow they are executing righteous justice on the moneyed oppressor. . . ."

The campaign of 1896 roused the country to extremes of emotion and hatred. It was Silver against Gold; the People against the Interests; Agriculture against Industry; the common man against the banker and the speculator. Bryan swept across the nation like a prairie tornado. Sometimes he would make as many as thirty-six speeches in a single day, and as many as five million people heard him speak. Breaking with tradition, he crisscrossed the country from New England to the Pacific, riding hot, dirty day coaches, appealing to laborers and farmers, liberals and progressives. The only effective support he had in the East was from the newspapers of William Randolph Hearst, who had inherited the fortune his father had made in western mines.

Republican and Democratic conservatives feared that the election of Bryan meant bloody revolution and the end of the capitalist system. "Gold" Democrats, including Grover Cleveland, supported McKinley. The conservative Senator David B. Hill of New York spoke for many of them. Upon his return from the convention, he was asked if he were still a Democrat. "Yes," he replied, "I am a Democrat still, very still."

To combat Bryan's energy and dramatic appeal to the masses, Mark Hanna dubbed McKinley the "Advance Agent of Prosperity" and created the nation's first sophisticated political organization. The land was papered with copies of McKinley's speeches, the mails deluged with McKinley posters, tracts, badges, and buttons. McKinley was packaged and merchandised, Roosevelt is reported to have said, "as if he were a patent medicine." An admiring John Hay called Hanna "a born political general." Money spoke louder than oratory in Hanna's opinion, and the Republican campaign chest totaled $7 million—the largest until after World War I. Insurance companies and banks were assessed one quarter of a percent of their assets; J. P. Morgan and Standard Oil each contributed $250,000 to the McKinley campaign. Workers were told that if Bryan won, factory whistles would not blow the day after the election. In contrast, the Democrats were able to gather only about a half-million dollars for their campaign.

In those days presidential candidates did not campaign actively and McKinley remained home and conducted a traditional "front porch campaign." If the candidate would not go to the voters, then the voters should be brought to the candidate, reasoned Hanna. The railroads, fearful of a Bryan victory, provided free passes, and Republican delegations were brought to Canton from all over the country. Once there, they were shepherded through the flag-decked streets to McKinley's home. The candidate greeted them with a few carefully rehearsed platitudes, then shook hands with his dazzled visitors.

In McKinley's place, Hanna dispatched a flood of speakers— Theodore Roosevelt among them—to stump the country and to paint a dire picture of the consequences of a Democratic victory. Whistle-stopping furiously through the West, Roosevelt lashed out at Bryan, comparing him to "the leaders of the Terror of France." He oversimplified the silver controversy, juggling two loaves of bread, one larger than the other. "See this big one," he would say. "This is an eight-cent loaf when the cents count on a gold basis. Now look at this small one . . . on a silver basis it would sell for nine cents. . . ."

By midsummer, Bryan's great organ of a voice was reduced to a hoarse whisper and Hanna's relentless barrage of propaganda had tarnished his theme of "free silver." McKinley won easily, scoring the greatest electoral victory since before the Civil War, and the scalps of the free-silverites dangled from Hanna's belt. "God's in his heaven, all's right with the world!" he triumphantly telegraphed the Major on Election Night. Roosevelt also felt cause to celebrate. "You may easily imagine our relief over the election," he wrote Bamie. McKinley's election had flashed the green light for his own campaign to escape from New York and return to Washington as assistant secretary of the navy.

Roosevelt's friends—Cabot Lodge, the Storers, Tom Reed, and others—emphasized to Republican leaders that his past record of party loyalty, increasing popularity, and unstinting efforts in the campaign were deserving of recognition. While Hanna was understood to look with favor upon the idea, the president-elect was noncommittal. McKinley wished for peace and stability in his administration, and was troubled by Roosevelt's reputation for trouble. Following a personal visit to McKinley, Lodge reported: "He spoke of you with great regard for your character and your services and he would like to have you in Washington. The only question he asked me was this, which I give; 'I hope that he has no preconceived plans which he would wish to drive through the moment he got in.' I replied that he need not give himself the slightest uneasiness on that score."

Mrs. Storer cornered McKinley after dinner one evening and brought up Roosevelt's appointment. "I want peace," the president-elect answered, "and I am told that your friend Theodore—whom I know only slightly—is always getting into rows with everybody. I am afraid he is too pugnacious."

"Give him a chance to prove that he can be peaceful," she replied.

Throughout that winter Roosevelt fretted over the appointment, while his disgust at conditions on the Police Board deepened. Parker was dismissed by Mayor Strong, but Governor Morton, nudged by Tom Platt, refused to approve the mayor's order and Parker was still plaguing him. If McKinley thought him "hot-headed and harum-scarum," there was little he could do about it, Roosevelt told Mrs. Storer. To Lodge, he wrote despondently: "I think I could do honorable work as Assistant Secretary. If I am not offered it, then I shall try to do honorable work here as long as I can, and then turn to any work that turns up."

Yet, he continued to prepare for the hoped-for job by working over a revised edition of *The Naval War of 1812,* met several times with Captain Mahan to discuss naval policy, and spoke at the Naval Academy in Annapolis. Stress soon took its toll, however. He had a violent attack of asthma while playing with the children; a heavy bough fell on his arm as he was chopping wood and left a large bruise; he cut his forehead on the mantel while placing a log on the fire; and took a tumble while skiing and pulled a shoulder muscle.

Tom Platt was as much a stumbling block to Roosevelt's ambitions as Roosevelt's own reputation as a loose cannon. Although known as the "Easy Boss" because he was always ready for consultation and compromise, one McKinley aide noted, "he hates Roosevelt like poison." Platt was planning to return to the U.S. Senate after a lengthy absence, and with Republican control of the Senate resting on a single vote, the president was reluctant to make an appointment likely to arouse Platt's anger. The Boss was particularly concerned that Roosevelt might interfere with the rich patronage at the Brooklyn Navy Yard. Lodge urged Roosevelt to swallow his pride and make peace with the old man. "I shall write Platt at once, to get an appointment to see him," Roosevelt replied. Following the meeting, he reported that Platt was "exceedingly polite"—but noncommittal.

Lodge also dealt with reports that the new secretary of the navy, John D. Long, a former Massachusetts governor and congressman, was worried that Roosevelt, if appointed his assistant, would try to brush him aside and dominate the Navy Department. Once again at Lodge's suggestion, Roosevelt provided assurances that he would be a loyal subordinate. Tom Platt gave in, after his lieutenants persuaded him that sending Roosevelt to Washington was the best way to get him out of their hair. Roosevelt had to pass a test of loyalty, however. He was expected to support Platt's senatorial nomination. This created a wrenching dilemma for Roosevelt because Platt's major rival was Joseph Choate, an old family friend who had encouraged his first try for political office. Practical politics and ambition won out over loyalty. When Choate's friends asked Roosevelt to speak for their candidate, he refused. Instead, he attended a "harmony" dinner given by party regulars to endorse Tom Platt's candidacy.

Not long thereafter, on April 5, 1897, Lodge informed Roosevelt that the president was sending his name to the Senate for confirmation as assistant secretary of the navy. "Sinbad had evidently landed the old man of the sea," a triumphant Roosevelt wired back.

Chapter 12

"The Supreme Triumphs of War"

Plumes of smoke rising from tall funnels and white hull gleaming in the sun, the battleship *Iowa* steamed majestically past the Virginia Capes on September 7, 1897, with Assistant Secretary of the Navy Theodore Roosevelt on board. Speedy little torpedo boats dashed about the great ship, sometimes crossing her wake, like shepherd dogs nipping at the heels of their charge. The *Iowa*, the most powerful vessel in the U.S. Navy's battle line, was so new that she had never fired her big guns, and Roosevelt wished to observe as she engaged in target practice for the first time.

Gongs clanged all over the ship as general quarters was sounded. Roosevelt accompanied Captain William T. Sampson to the bridge, sailors scampered to battle stations, loose gear was stowed, and hatches were securely dogged down. Word was passed that the assistant secretary wanted to see how quickly the "enemy" could be destroyed. A surgeon distributed cotton earplugs to the party of visitors who accompanied Roosevelt. "Open your mouth, stand on your toes, and let your body hang loosely," when the guns were fired, he cautioned. Slowly and deliberately, the guns began to search out their target, a tiny speck of wood and canvas bobbing on the glassy ocean.

"Two thousand seven hundred yards," called out a naval cadet monitoring the range. "Two thousand six hundred yards . . . two thousand five hundred yards. . . ." Suddenly, there was a flash of fire and smoke from the eight-inch secondary battery, followed by a thunderous report. The first salvo fell fifty feet short of the target; a second was off to the left. Now, it was the turn of the *Iowa*'s main battery of twelve-inch guns. Long sheets of flame leaped from their muzzles followed by a roar that shook the ship. A small boat was stove in by the blast, several watertight doors were ruptured, and two visitors who had not followed instructions were thrown violently into the air. In the meantime, the assistant secretary scanned the sea for the target. Huge black water spouts obscured it, but when they subsided it was clear that if it had been an enemy warship it would have been on the way to the bottom.

"Oh, Lord!" Roosevelt enthusiastically exclaimed to a friend upon his return to Washington. "If only the people who are ignorant about our Navy could see these great warships in all their majesty and beauty, and how well they are handled, and how well fitted to uphold the honor of America, I don't think we would encounter such opposition in building up our Navy to its proper standing."

"Building up our Navy to its proper standing" was the task Roosevelt had set for himself upon assuming his duties at the Navy Department on April 19, 1897. No man worked harder to make the navy an effective instrument of war and diplomacy. One of his first moves symbolized the active role he intended to play. Rooting about in a basement storage room of the State, War and Navy Building* for a suitable desk, he found the one used during the Civil War by Gustavus V. Fox, the first assistant secretary of the navy. The Stars and Stripes were carved into the mahogany front, and cannon protruded fiercely from its sides.

President McKinley and Mark Hanna, soon to be senator from Ohio, were little concerned with naval affairs. Their paramount goal was the restoration of prosperity and the maintenance of peace; they wished to avoid any foreign adventures that might interfere with these goals. Under their benevolent gaze, railroad

---

*Now the Old Executive Office Building

mergers, industrial combinations, communities of interest, and trusts flourished. Tariffs were set at ridiculously high levels and business was guaranteed protection against hostile legislation. The alliance between politics and business in the making since the end of the Civil War was now complete.

In contrast, Roosevelt's head was buzzing with thoughts of nationalism, imperialism, and naval expansion. Within two weeks of his arrival in Washington, he outlined to Captain Mahan, now retired, a bold plan to make the United States a great power in both the Atlantic and Pacific oceans. He called for the immediate annexation of Hawaii or the establishment of a protectorate in the islands, for a dozen new battleships and an isthmian canal to facilitate their passage from one ocean to another. To control this waterway, the United States must dominate the Caribbean, which required the ouster of Spain from Cuba. "Until we definitely turn Spain out of the island (and if I had my way that would be done tomorrow) we will always be menaced by trouble there," he added. This letter was carefully marked "Personal and Private."

Roosevelt was pleased to be back in Washington, but with the exception of Henry Adams and the Lodges—with whom he stayed because Edith was pregnant again and had remained at Sagamore—most of his old circle had scattered. John Hay had been named ambassador to London; Lizzie Cameron also was living in Europe and Cecil Spring Rice was serving at the British embassy in Berlin. "As you will see from the heading of this letter, I am now Assistant Secretary of the Navy," he wrote Spring Rice. "My chief, Secretary Long, is a perfect dear."

Short, plump, and even-tempered, Long was something of a hypochondriac who gave the impression of being older than his fifty-eight years. Originally, he had been worried about the truculence of his assistant, but this concern quickly faded to be replaced by an avuncular indulgence. "Best man for the job," he recorded in his diary after their first meeting. Roosevelt and Long complemented each other. Long made no pretense of knowing the difference between a bowline and a bollard, but he was a capable administrator, kept the navy relatively free of patronage, and had an open mind. He saw no necessity to master the details of warship design, engines, armor plate, guns, or dry docks. "What is the need of my making a dropsical tub of any lobe of my brain," he noted, "when I have right at hand a man possessed with more knowledge than I could acquire."

Roosevelt was perfectly content to assume this burden—and the influence that went with it. He had a smattering of knowledge about the navy's affairs and the enthusiasm of the dedicated amateur. He was delighted to spend long hours discussing the intricacies of ship construction and armament with his staff and wrote articles and gave speeches propagandizing for a strong navy.* Before long, he was popping up everywhere to inspect ships and store stations, and the hulking figure with the flashing teeth became familiar to everyone. "He broke the record for asking questions," said one observer. Gossip had it that officers waited until Long was away—which was often—and brought problems to Assistant Secretary Roosevelt for final resolution.

With a boyish interest in the latest technical discoveries, he recommended that an experimental submarine be purchased and suggested that the navy look into the prospects for manned flight. Taking note of the experiments that Professor Samuel P. Langley of the Smithsonian Institution was conducting with a steam-driven "aerodrome," he wrote Secretary Long: "The machine has worked. It seems to me worth while for the government to try whether or not it will work on a large enough scale to be of use in case of war. . . . I think this is well worth doing."

The fleet itself consisted of four first-class battleships, two of the second class,† and forty-eight other vessels of various types, ranging from armored cruisers to torpedo boats. Five additional battleships, slightly larger than those in commission, were under construction, but work on three was at a standstill because the armor plate manufacturers refused to bid on contracts at the price established by Congress. This force ranked well behind the navies of Britain, France, Germany, and Russia, but many Americans regarded it as adequate for the nation's defense.

Critics opposed naval expansion on grounds that it would involve the United States in foreign wars and argued for a fleet that would be geared to coastal defense and, harking back to priva-

---

*Among them an anthology of quotations from various presidents on the need for a strong navy. Needless to say, Thomas Jefferson was absent from this honor roll.
†In reality, these ships, the *Maine* and *Texas,* were large armored cruisers.

teering in the age of sail, the destruction of an enemy's commerce. "These battleships are unsafe," charged Representative William Vandiver of California. "Set such ships upon mid-ocean and the first discharge of a broadside from one of their formidable guns will sink them to the bottom of the seas."

Such talk set Roosevelt's teeth on edge. Like Mahan, he believed that an adequate navy meant battleships that were able to meet an enemy fleet on the high seas and defeat it in battle. Coastal fortifications constituted a passive defense that allowed an enemy to strike where he wished. Commerce raiding* was also less attractive as a national strategy in an age in which the telegraph and fast, well-armed steam-driven cruisers made the long-time survivability of a raider problematic. "We need a large navy . . . containing . . . a full proportion of powerful battleships, able to meet those of any other nation," he declared. "It is not economy—it is niggardly and foolish shortsightedness—to cramp our naval expenditures, while squandering money right and left on everything else, from pensions to public buildings."

Once established in the Navy Department, Roosevelt embarked on an aggressive campaign to expand the fleet. He lobbied for an additional six battleships, six armored cruisers, and seventy-five more torpedo boats. Four of the battleships were for the Atlantic, where he foresaw trouble with Germany; two were for the Pacific, where Japan, which had just defeated China, was showing an untoward interest in Hawaii. Lukewarm to the idea of expansion, Long eventually settled for only one battleship, but Congress refused to appropriate funds for even that vessel.

Roosevelt also established himself as the leader of a coterie of like-minded imperialists: senators, congressmen, ranking military officers (both active and retired), writers, and prominent citizens, who met regularly at the Metropolitan Club† and plotted ways to influence American policy. One member of this group, Captain

---

*Submarines were just coming on line in the world's navies in the 1890s, and Mahan and his disciples failed to see the new dimension they added to commerce raiding. The German U-boat campaign against Allied commerce in World Wars I and II, which nearly proved decisive, and the spectacular successes scored by American submarines against the Japanese merchant fleet in the Second World War proved they were completely wrong about the importance of commerce raiding in modern warfare.

†TR had a passionate attachment to the double lamb chops at the club.

Leonard Wood, an army surgeon who had won the Medal of Honor while fighting the Apaches, became one of his closest friends. Two years younger than Roosevelt, Wood was a graduate of the Harvard Medical School and a natural-born soldier. He was, Roosevelt acknowledged, one of the few men in Washington who "walked me off my legs."

In May 1897, Roosevelt assigned the Naval War College the following problem and to make plans to deal with it:

Japan makes demands on Hawaiian Islands.
This country intervenes.
What force will be necessary to uphold the intervention, and how should it be employed?
Keeping in mind possible complications with another power on the Atlantic Coast (Cuba).

Two weeks later, on June 2, 1897, he appeared in person at the War College in Newport, Rhode Island, to make a speech that was a clarion call for naval preparedness. Taking as his text George Washington's "forgotten maxim"—"To prepare for war is the most effectual means to promote peace"—he preached a sermon that bristled with martial ardor. There had been much talk of arbitration since the settlement of the Venezuelan border dispute, Roosevelt noted. Although "arbitration is an excellent thing," those who sincerely desired peace "will be wise if they place reliance upon a first-class fleet of battleships rather than on any arbitration treaty which the wit of man can devise. . . . Peace is a goddess only when she comes with sword girt on thigh."

In the future, Roosevelt pointed out, the United States would not have the luxury of being able to prepare for war once a conflict had begun. Technological innovations in naval warfare made it impossible to continue to be isolated behind the oceans. National greatness, national welfare, national survival were linked to the existence of a powerful navy. "Diplomacy is utterly useless where there is no force behind it; the diplomat is the servant, not the master, of the soldier." Finally, his high, staccato voice reached the peroration, the words striking with the force of a trip-hammer:

No triumph of peace is quite so great as the supreme triumphs of war. The courage of the soldier, the courage of the statesman who has to meet storms which can only be quelled by soldierly virtue— this stands higher than any qualities called out merely in times of peace. . . . It may be that at some time in the dim future of the race the need for war will vanish; but that time is yet ages distant. As yet no nation can hold its place in the world, or can do any work really worth doing, unless it stands ready to guard its rights with an armed hand.

The speech had the impact of a shell from one of the *Iowa*'s twelve-inch guns. "Well done, nobly spoken!" said *The Washington Post.* The *New York Sun,* which was agitating for the liberation of Cuba, described his words as "manly, patriotic, intelligent and convincing." But anti-imperialist journals, such as *Harper's Weekly,* criticized Roosevelt for his "bellicose fervor." Secretary Long was upset by the speech, too, but decided not to make an issue of it. Most important, however, was President McKinley's reaction. "I suspect Roosevelt is right," he told an aide, "and the only difference between him and me is the greater responsibility."

Roosevelt is often accused of seeking war for its own sake. William James, his former professor at Harvard, thought he regarded "one foe . . . as good as another." Roosevelt did not regard himself as a militarist, but wanted the nation to be willing and able to fight if necessary. He believed the materialism of the "banker, the broker, the mere manufacturer and mere merchant" would leave the United States as defenseless as China had been against Japan. There is no simple, correct answer to the question of whether he was a man of war or a man of peace. Roosevelt was too complex. In truth, he was both.

Roosevelt's aggressiveness is often linked to his boyhood struggle against illness, to his father's failure to fight in the Civil War, and the tales of derring-do related by his Confederate uncles. Yet, other men of his time with psychological makeups far different from Roosevelt's held similarly bellicose views. Cabot Lodge, John Hay, Senator Albert J. Beveridge, and Henry Adams's brilliant and eccentric younger brother, Brooks, all shared Roosevelt's admiration for the soldierly virtues. Roosevelt differed from them only in being less circumspect in his remarks.

Well educated, cultivated, and patrician in outlook, these men

viewed themselves as pragmatic idealists. Their code was aristocratic, one in which their superiority was taken for granted but
must be paid for in the coinage of duty. Although they lived in
a commercial world, they regarded the standards of the marketplace with contempt. They lived in democracy but were not democratic by instinct. Having tried their hands at civic reform, they
found it futile and turned to expansionism.

One of the major reasons for the appeal of far-off bugles and
streaming banners to this generation of Americans was rooted in
boredom with a society increasingly dominated by the machine
and commercialism. With the exception of the Indian wars, which
were fought by a handful of hard-bitten professional soldiers out
of the eye of the general public, the nation had been at peace for
more than three decades. The bloodbath of the Civil War had
faded from memory and had been replaced by sentimentalized
myths of cavalry charges and the "Lost Cause." War seemed
romantic to men who had no experience of the privation, diseases,
and mass slaughter of modern combat.

Early in July, Roosevelt was able to spend some time at Oyster
Bay with Edith and the children. Washington was all very well,
but "nothing could be lovelier than Sagamore." Few sights were
more pleasing to him than his tow-headed children and their myriad cousins at play, and he counted "sixteen small Roosevelts" at
Oyster Bay that summer. Fifteen-year-old Franklin Roosevelt was
among them. A few weeks before, the elder Roosevelt had visited
Groton, where Franklin was a student, and gave "a splendid talk
on his adventures when he was on the Police Board," the boy
enthusiastically informed his parents. "He kept the whole room
in an uproar for over an hour, by telling killing stories about
policemen and their doings in New York." Franklin adored "Cousin Theodore," and when he was invited to visit Sagamore, he
eagerly accepted.

Upon one or two occasions, Eleanor Roosevelt, Elliott's orphaned daughter, was also allowed by her grandmother to visit
Uncle Ted and Aunt Edith. A tall, gawky girl with prominent
teeth, she lived in a dream world in which she was the heroine
and her dead father the hero. Uncle Ted's affection for her was
gargantuan. When she arrived, he "pounced" upon the girl like
a bear, hugging her to his chest "with such vigor that he tore all

the gathers out of Eleanor's frock and both button holes out of her petticoat." Edith saw something in Eleanor that escaped the others. "Poor little soul, she is very plain," she wrote Bamie. "Her mouth and teeth have no future. But the ugly duckling may turn out to be a swan."

Following this respite, Roosevelt departed on a tour of naval militia stations in the Great Lakes area, where he created another uproar. Several weeks earlier, McKinley had asked the Senate to ratify a treaty annexing the Americanized republic of Hawaii. Japan immediately protested on grounds that annexation would jeopardize the interest of the 25,000 Japanese living in the islands. "The United States," declared Roosevelt in a speech at Sandusky, Ohio, "is not in a position which requires her to ask Japan, or any other foreign power, what territory it shall or shall not acquire."

Secretary Long, who had barely restrained himself after the Newport speech, exploded when he heard about his assistant's latest outburst. "The headlines and comment . . . nearly threw the Secretary into a fit," Roosevelt told Lodge, "and he gave me as heavy a wigging as his invariable courtesy and kindness would permit." Profuse apologies and promises not to repeat the indiscretion smoothed the incident over, and Long departed for a monthlong vacation at his farm at Hingham Harbor, Massachusetts. Not long afterward, Roosevelt exulted to Bellamy Storer: "The Secretary is away, and I am having immense fun running the Navy."

Early in August, Edith and the children returned to Washington, where Roosevelt had rented a house at 1810 N Street, opposite the monumentally ugly British embassy. The place was furnished "largely from the wreck of Edith's fore-father's houses sixty years back, with an occasional relic of my own family thrown in—all of the mesozoic or horse-hair furniture stage," he sardonically wrote the vacationing Lodge. "We have come across some lovely momentos of a bygone civilization; including especially a number of stereopticon plates—'The Wedding Breakfast,' 'Dressing the Bride' . . . varied with views of the family tombs in Greenwood cemetery; which our ancestors always deemed highly edifying."

Throughout the summer, he sent letters off to Long urging him "to have an entire rest," to stay on vacation after Labor Day, and giving assurances that he would "do nothing . . . excepting as you directed," while doing his best to place the navy in readiness for a war with Spain. "The liveliest spot in Washington at present is the Navy Department," observed the *New York Sun* in mid-August. "The decks are cleared for action. Acting Secretary Roosevelt, in the absence of Governor Long, has the whole Navy bordering on a war footing. It remains only to sand down the decks and pipe to quarters for action."

As the crisis over Hawaii evaporated, the full attention of the expansionists turned upon Cuba. In the two years since the rebellion against Spain had first flared on the island, the struggle had become increasing barbarous. Both sides laid waste to the countryside and massacred prisoners. In the fall of 1896, to deprive the guerrillas of support, the Spanish commander, General Valeriano Weyler, herded large numbers of women and children into concentration camps, where hundreds died from disease and starvation.*

American sympathy for the rebels mounted. Although Yankee investments on the island were significant—amounting to about $100 million—the major reason for this interest was not imperialistic greed. The expansionists talked of expelling the decadent Spaniards from the Western Hemisphere and of dominating the Caribbean, but most Americans were motivated by a humane, if misguided, desire to end the suffering of the Cuban people and help them win their independence. In fact, businessmen were opposed to war because it would retard the return of prosperity and dislocate the economy, which accounts for Roosevelt's complaint that they were sordid and money-grubbing.

The United States was, however, far from being an innocent bystander in the Cuban crisis. The Wilson-Gorman Tariff of 1894, which levied new duties on Cuban sugar for the protection of Gulf Coast sugar growers, raised havoc with the island's economy and helped precipitate Cuban protests against Spanish rule. Economic and political unrest were met with repressive measures, which, in turn, led to open insurrection. Half a world away, a

---

*Following President McKinley's death, it was discovered that he had privately donated $5,000 to relieve the plight of the *reconcentrados*.

separate revolution against Spanish rule had also broken out in the Philippines.

These events coincided with a journalistic revolution in the United States—the rise of the popular press. Until the coming of William Randolph Hearst and Joseph Pulitzer, newspapers had been more or less aimed at a limited audience of educated readers. Hearst and Pulitzer created giant-circulation newspapers directed at the newly literate masses. The introduction of the Linotype, the web press, and photoengraving made it possible to print huge numbers of papers with actual pictures of events—papers designed to entertain as well as to inform their readers. Looking about for sensations to increase circulation, the "yellow" press exploited the unrest in Cuba. Cuban patriots and propagandists had an open line to Hearst's *New York Journal* and Pulitzer's *World,* and filled their columns with lurid tales of alleged atrocities by "Butcher" Weyler and other Spanish officials.*

"Blood on the roadsides, blood in the fields, blood on the doorsteps, blood, blood, blood!" one *World* correspondent reported in typical fashion. "The old, the young, the weak, the crippled—all are butchered without mercy. . . . Is there no nation wise enough, brave enough, and strong enough to restore peace in this bloodsmitten land?"

Acting Secretary Roosevelt had the opportunity to discuss the Cuban situation with President McKinley on three occasions in mid-September. On the fourteenth, the president invited Roosevelt for a drive in his carriage, and McKinley told him that he had read his pamphlet containing the pro-navy statements of previous chief executives and was "exceedingly glad that I had put it out." Much to Roosevelt's astonishment, McKinley also remarked that "I was quite right" in criticizing Japanese reactions to the annexation of Hawaii. And then he "expressed great satisfaction" with the way he had run the Navy Department for the

---

*For example, when Richard Harding Davis, the leading war correspondent of the day who had been sent to Cuba by Hearst, reported that Spaniards had boarded an American ship in Havana harbor and searched three Cuban women for treasonable documents, the *Journal* ran the story under a banner headline: DOES OUR FLAG PROTECT WOMEN? Hearst ordered Frederic Remington, the artist, to prepare a drawing showing the naked women being leered at by villainous-looking Spanish officials. In fact, the women had been examined by a police matron in the privacy of a cabin. All the other touches were products of Mr. Hearst's vivid imagination.

past seven weeks. "Of course the President is a bit of a jollier," Roosevelt told Lodge, "but I think his words did represent a substratum of satisfaction."

With the president in an affable mood, Roosevelt brought up the subject of Cuba. McKinley said he was "by no means sure that we shall not have trouble with either Spain or Japan," although he wished to avoid it. If war came, Roosevelt said, he would resign and enter the army. The president asked what Mrs. Roosevelt would think and Roosevelt replied that as much as she and Cabot Lodge would regret it, "this was one case where I would consult neither." McKinley laughed and assured him that he would have an opportunity to serve if war came.

McKinley had obviously enjoyed the outing, for Roosevelt received an invitation to dine at the White House three days later, and three days after that, he went for another carriage ride with the president. Emboldened by McKinley's interest, he presented him with a strategic plan outlining operations for a war against Spain that had been developed by the Navy Department. The main battle fleet should be based upon Key West, which would permit it to establish a blockade of Cuba within forty-eight hours. A squadron of fast armored cruisers should be dispatched to harass the Spanish coast. The Asiatic Squadron should blockade, and if possible, destroy the Spanish squadron based in the Philippines and take Manila. If the army quickly landed an expeditionary force in Cuba, he thought, the "acute phase" of the war would be over in six weeks. There is no record, however, of McKinley's reaction to this plan—which, to a remarkable extent, was followed by the United States when war with Spain finally broke out the following year.

The Asiatic Squadron required a commander who could act on his own initiative, and Roosevelt found the right man without difficulty; getting him the billet was another matter. Commodore George Dewey matched Roosevelt's own combative temperament and was a member in good standing of the expansionist coterie. But he lacked seniority, and some politicians were supporting Commodore John A. Howell for the post. In Roosevelt's opinion, Howell was "irresolute" and "afraid of responsibility." Worried that Secretary Long, who was due back in Washington momentarily, might appoint Howell, he summoned Dewey to his office and suggested that he contact Senator Redfield Proctor of his home state of Vermont, who was close to the president. Proctor put in a word for Dewey with McKinley and as Roosevelt later said, "in a fortunate hour for the Nation, Dewey was given command of the Asiatic squadron."

"Very unexpectedly, Quentin Roosevelt appeared two hours ago," Roosevelt informed Bamie early on the morning of November 19, 1897. "Edith is doing well. By the aid of my bicycle I just got to the Doctor & Nurse in time! We are very glad, and much relieved."* The boy was immediately put down for Groton, then his father was off to the Navy Department to beat the drum for war with Spain.

Throughout that winter, Roosevelt implored Long to place the fleet on alert and to base the North Atlantic Squadron at Key West. One can almost visualize the gentle Long, eyes closed and head in his hands, while for the tenth or hundredth time, his voluble assistant launches into a tirade about the need to be prepared for war before it is too late. Rather than thinking of warlike measures, however, the administration was doing its best to find a way out of the Cuban maze.

The Spanish government seemed willing to cooperate. Unwilling to get into a war with the United States that they knew their country could not win, the Queen Regent, María Cristina, and a newly installed moderate prime minister, Práxedes Mateo Sagasta, recalled General Weyler and offered to grant Cuba limited home rule. Sagasta's counciliatory program provided President McKinley with a breathing spell. In his annual message to Congress, he counseled patience and assured the world the United States had no intention of annexing Cuba. Everything now hinged on the success or failure of the reform plan.

But Cuban autonomy had no future. The rebels, sensing victory in the offing, refused to settle for anything short of independence, while loyalists considered the offer a betrayal. Rioting broke out in Havana on January 12, 1898, and Spanish officers wrecked the

---

*The boy was named in honor of Edith's grandfather, Isaac Quentin Carow. To the amusement of Quentin's father, Cabot Lodge agreed to attend the christening but "with gloomy reluctance, as it is against his principles to sanction anything so anti-malthusian as a sixth child." *Letters,* Vol. I, p. 745.

presses of several newspapers critical of the army. Alarmed for the safety of American residents, Fitzhugh Lee,* the U.S. consul general, requested assistance, and the battleship *Maine,* based at Key West, was placed on alert. However, Lee had misjudged the danger and the rioting quickly subsided.

Next morning, Roosevelt came to Secretary Long's office, shut the door and, according to Long's diary, "began in his usual emphatic and dead-in-earnest manner. After referring sensibly to two or three matters of business, he told me that, in case of war with Spain, he intends to abandon everything and go to the front. He bores me with his plans of naval and military movements, and the necessity of having some scheme of attack arranged for instant execution in case of an emergency. By tomorrow morning, he will have got half a dozen heads of bureaus together and have spoiled twenty pages of good writing paper, and lain awake half the night."

Long knew his man. Although Roosevelt did not believe the "flurry in Havana" would lead to war, he worked out his frustration by requesting the assistance of C. Whitney Tillinghast II, New York's adjutant general, in obtaining a commission as a major or lieutenant colonel in one of the National Guard regiments. And he spent the night "spoiling twenty pages of good writing paper" on a lengthy memorandum to the secretary all but directly demanding that the navy be placed on a war footing.

"Certain things should be done at once if there is any reasonable chance of trouble with Spain during the next six months," he declared. Ships on foreign stations should be concentrated; Dewey's Asiatic Squadron, although adequate to deal with Spanish forces in the Philippines, should be reinforced; a flying squadron should be established in the Canaries, where it would be held ready to raid Barcelona and Cádiz; recruiting should be stepped up, ammunition and coal stockpiled. "If we drift into [war], if we do not prepare in advance, and suddenly have to go into hostilities without taking the necessary steps beforehand, we may have to encounter one or two bitter humiliations. . . ."

Roosevelt's fine writing paper was not altogether wasted. McKinley was now worried that the rioting in Havana meant that

---

*Lee, a former Confederate officer, was General Robert E. Lee's nephew.

the Spanish authorities were losing control of the situation, and he ordered a redeployment of naval units. Squadron commanders were instructed to retain all men whose enlistments were about to expire; the South Atlantic Squadron, composed of small vessels, was told to proceed northward; and ships in European waters were shifted to Lisbon, where an eye could be kept on Spanish naval movements. The four battleships of the North Atlantic Squadron, the *Massachusetts, Iowa, Indiana,* and *Texas,* had already been ordered to join the *Maine* at Key West for winter exercises. On January 24, the *Maine* was sent to Havana, on a "courtesy visit." She arrived there the next day and anchored at a buoy in the harbor—a potent representative of the power of the United States.

Whatever satisfaction Roosevelt could take from these developments was moderated by domestic strains. Although Edith's recovery from Quentin's birth was swift, she was stricken with what seemed to be grippe soon after the turn of the year. She appeared weaker than her husband had ever seen her, and he suspected typhoid. A trained nurse was called in and the younger children were shipped off to friends, while Alice and Ted remained at home. The boy had his own problems, suffering from nervous headaches for which five Washington doctors were unable to find a cause.

Edith's condition steadily worsened. She developed acute neuralgia pains followed by high fever and sciatica, and she was unable to sleep. Friends thought she might be suffering from exhaustion, but her illness was much more serious. For four weeks her temperature remained above 101 degrees. Roosevelt was in a turmoil of apprehension, not knowing whether she would ever recover. Meanwhile, fourteen-year-old Alice was proving to be a handful. With Edith too sick to keep her in check, she was "running the streets uncontrolled with every boy in town." On one day's notice, she was packed off to join Bamie in New York, amid howls of protest. Ted later joined her there.

While Roosevelt was undergoing these torments, the *Maine* rocked peacefully at anchor on the tropical swells of Havana harbor. Captain Charles D. Sigsbee, an officer of judgment and diplomacy, maintained tight security, steam was kept up, and ammunition had been brought up for the guns. The Spaniards were cool but correct, and life seemed to be returning to normal in the Cuban capital. But there was an undercurrent of misgivings about the ship's presence there. "You might as well send a lighted candle on a visit to an open cask of gunpowder," observed Mrs. Richard Wainwright, wife of the *Maine's* executive officer.

This uneasy calm was broken by the February 9 publication in the *New York Journal* of a private letter written by the Spanish minister in Washington, Enrique Depuy de Lome, to a friend in Havana, which had been purloined by rebel agents. The letter not only revealed bad faith by de Lome during the current negotiations but contained an unflattering portrait of the U.S. president: ". . . McKinley is weak and a bidder for the admiration of the crowd, besides being a would-be politician who tries to leave a door open behind himself while keeping on good terms with the jingoes of his party." The yellow press had a field day with this letter, and Depuy de Lome, his usefulness at an end, resigned. Although the Spanish government promptly apologized, the American public, already inflamed, was slow to forget.

That night, Roosevelt encountered Mark Hanna and two other senators at a reception. In his excitement over the de Lome letter, he launched into a typical arm-flailing harangue that backed Hanna's guest, a Frenchwoman named Henriette Adler newly arrived from Paris, into the wall, and his arm swept increasingly closer to the bodice of her gown. Finally, his elbow ripped off a silk rose and some gauze from her shoulder, and she uttered a shocked *"Mon Dieu!"* Roosevelt immediately showered her with apologies in French. Nannie Lodge came to the rescue with a safety pin and, with the senators providing a screen, she repaired the damage to Mlle. Adler's dress.

Now joining in the heated conversation, Mlle. Adler suggested that France and Germany would hardly stand by and let Spain be despoiled of her possessions in the New World. She said she had heard such reports in Paris only two weeks earlier. Roosevelt quickly dismissed these objections. "I hope to see the Spanish flag and the English flag gone from the map of North America before I'm sixty!" Hanna, who had been staring impassively at him, with his chin resting on his white tie, growled, "You're crazy, Roosevelt! What's wrong with Canada?"

On the carriage ride home later that evening, Mrs. Hanna tried to reassure Mlle. Adler that Mr. Roosevelt, contrary to all appearances, was "really amusing" but "he did get violent about things." The senator had a different opinion. He thanked God that Roosevelt had not been put in the State Department. "We'd be fighting half the world."

Six nights later, on the evening of February 15, 1898, Captain Sigsbee was in his cabin on the *Maine* writing to his wife when the marine bugler sounded taps. "I laid down my pen to listen to the notes of the bugle, which were singularly beautiful in the oppressive stillness of the night," he recalled later. "The echoes floated back to the ship repeating the strains of the bugle fully and exactly." Sigsbee looked at his watch and noted that it was 9:40 P.M.

I was enclosing my letter in its envelope when the explosion came. . . . It was a bursting, rending and crashing roar of immense volume . . . there was a trembling and lurching motion . . . a list to port, and a movement of subsidence. The electric lights went out. Then there was intense blackness and smoke.

Groping his way on deck, Sigsbee found the forward part of his ship was shattered and rapidly sinking into the mud. Casualties were heavy—266 of the 354 officers and men on board were killed. In the gray light of dawn, a naval officer brought news of the disaster to the White House, and President McKinley was awakened. He had some difficulty comprehending the report. "The *Maine* blown up!" he mumbled slowly over and over again as he paced back and forth. "The *Maine* blown up!"

Washington was engulfed by a sense of black calamity not seen since President Garfield's assassination in 1881. Flags were lowered to half-staff; official functions were canceled. Silent crowds gathered before the White House and the adjoining State, War and Navy Building. In the hallway outside Secretary Long's office, a group of people watched silently as workmen opened a glass case containing a model of the *Maine,* removed the tiny ensign, and replaced it with one at half-staff. For Long, "The saddest thing of all is the constant coming of telegrams from some sailor's humble home or kinspeople inquiring as to whether or not he is saved. . . ."

Who had destroyed the ship?* Was it the Spaniards? The Cuban revolutionaries hoping to provoke a war with Spain? Or was it an accident? Long himself was inclined to the accident theory. The forward magazines of the vessel were crammed with munitions, which could have been set off by defective electrical wiring or a chance spark. Captain Sigsbee had cautioned that "public opinion should be suspended until further report," but the yellow press, spoiling for a war with Spain, provided its own answer. THE WAR-SHIP MAINE WAS SPLIT IN TWO BY AN ENEMY'S SECRET INFERNAL MACHINE blared the New York *Journal*. DESTRUCTION OF THE MAINE BY FOUL PLAY bannered the *World*.

War fever swept the nation. In Buffalo, mass meetings were held calling upon McKinley to declare war on Spain. Lehigh University students began drilling and parading under a banner reading TO HELL WITH SPAIN! Students in Omaha and a mob in Chicago burned the Spanish flag. In the face of a tidal wave of public emotion, the president urged Americans to be calm and keep an open mind. Unlike most of the jingoists, McKinley knew the horrors of combat at first hand and sincerely wished to avoid war. "I have been through one war," he declared. "I have seen the dead piled up, and I do not want to see another." The president ordered a naval board of inquiry to investigate the explosion and tried to keep from being swept off his feet. "My duty is plain," he contended. "We must learn the truth and endeavor, if possible, to fix the responsibility. The country can afford to withhold its judgement and not strike an avenging blow until the truth is known."

Assistant Secretary of the Navy Roosevelt had no doubt where the responsibility lay. In his official correspondence, he carefully called the disaster an accident, but left no doubt of his private opinion. "I would give anything if President McKinley would order the fleet to Havana tomorrow," he raged to a Harvard

---

*The cause of the explosion has never been conclusively determined. In 1976, Admiral Hyman G. Rickover theorized, on the basis of modern technical studies, that a fire resulting from spontaneous combustion in a forward coal bunker—a not uncommon occurrence at the time—set off ammunition in an adjoining magazine. *How the Maine Was Destroyed* (Washington, D.C.: Naval History Division, 1976).

classmate the morning after the disaster. "The *Maine* was sunk by an act of dirty treachery on the part of the Spaniards. . . ."

War with Spain—only a possibility before the sinking of the *Maine*—now appeared probable even as McKinley continued to seek a diplomatic solution. Outraged by the president's failure to avenge the *Maine,* Roosevelt is reputed to have described him as having "no more backbone than a chocolate eclair" and feverishly bombarded Long with suggestions for war preparations. To add to his highly emotional state, Edith's health took a turn for the worse. In mid-February, she noticed an ominous swelling in her abdomen near her pelvis, and her alarmed husband was "extremely anxious" about her. On February 25, he called in Sir William Osler, a world-famous specialist on abdominal tumors at Johns Hopkins Hospital in Baltimore.

That same day, Long, weary after ten days of tension and emergency meetings, took the afternoon off, leaving Roosevelt in charge of the Navy Department, even though he had misgivings about him. "He is so enthusiastic and loyal that he is, in certain respects invaluable, yet I lack confidence in his good judgement and discretion," Long confided in his diary. "He goes off very impulsively. . . ."

No sooner was the secretary out of the way than Roosevelt issued a flurry of peremptory orders. Ships were redeployed, ammunition ordered, coal requisitioned, and vessels earmarked for purchase as auxiliary cruisers. Messages were sent to members of Congress requesting immediate legislation to authorize the enlistment of unlimited numbers of men; guns were ordered, and officers were shifted about so the best men were assigned the right ships. The most far-reaching of these orders was a cable to Commodore Dewey ordering him to concentrate the ships of the Asiatic Squadron at Hong Kong:

KEEP FULL OF COAL. IN THE EVENT DECLARATION OF WAR SPAIN, YOUR DUTY WILL BE TO SEE THAT THE SPANISH SQUADRON WILL NOT LEAVE THE ASIATIC COAST AND THEN OFFENSIVE OPERATIONS IN PHILIPPINE ISLANDS.

Dewey immediately took steps to make his ships battle ready. Bunkers were filled with coal, engines were repaired, magazines were restocked, and white hulls were painted slate gray. One ship was dry-docked, scraped, and painted within twenty-four hours.

Returning to the office the next day, Long was thunderstruck. He angrily declared that Roosevelt had "gone at things like a bull in a China shop" and "has come very near causing more of an explosion than happened to the *Maine*. . . . The very devil seemed to possess him yesterday afternoon." Long concluded that Edith's and Ted's serious illnesses had "accentuated . . . his natural nervousness," and he implied that Roosevelt's temporary authority had gone to his head.

Roosevelt's actions were not impulsive; they were the result of a deliberate plan worked out over some time with Lodge, Mahan, and other members of his expansionist coterie. In fact, Lodge was present when Roosevelt prepared the cable to Dewey. The secretary's absence merely provided an opportunity to set in motion plans that already existed.

Yet, despite his anger at Roosevelt's audacity, Long made no effort to countermand his orders; even the controversial order to Commodore Dewey was allowed to stand. He may have been more angered by his assistant's decision to act on his own than by the actions themselves. For his part, Dewey credited Roosevelt's exercise of authority with making possible the subsequent victory at Manila Bay. As one prominent historian* has written: "The Assistant Secretary had seized the opportunity given by Long's absence to insure our grabbing the Philippines without a decision to do so by either Congress or the President, or at least of all the people. Thus was important history made not by economic forces or democratic decisions but through the grasping of chance authority by a man with daring and a program."

Shocked into action by Roosevelt's cyclonic performance, the Navy Department now entered a period of intensified activity. More ships were placed in commission and others were reassigned. Most important, the battleship *Oregon,* fitting out on the West Coast, was ordered to join the North Atlantic Squadron. She raced around Cape Horn to Key West, a distance of some thirteen thousand miles, in sixty-one days, a feat that dramatized the need for a Central American canal connecting the Atlantic

---

*Beale, *Theodore Roosevelt and the Rise of America to World Power,* p. 63.

and Pacific. In the final analysis, because of Roosevelt's persistence, the navy was far better prepared for war than the army.

That same day, Roosevelt had also sent a "strictly confidential" letter to Adjutant General Tillinghast of the New York National Guard warning him that "conditions are sufficiently threatening" to warrant preparations for general mobilization. "Pray remember that in some shape I want to go," he added. "I was three years in the National Guard, and have had a good deal of experience in handling men. . . ."

In the meantime, Dr. Osler had examined Edith, found that she was critically ill with a noncancerous inflammatory abdominal tumor and recommended an immediate operation. But Roosevelt, relying upon what Winthrop Chanler called "a lot of perfectly incompetent doctors, taxidermists and veterinarians, sportsmen and excellent athletes," hoped his wife would recover without surgery. Yet Edith grew steadily weaker. Finally, on March 5, he called in a gynecologist, who confirmed Osler's diagnosis and operated the next day. "Everything went well; but of course it was a severe operation and her convalescence may be a matter of months," Roosevelt told Bamie. "She behaved heroically; quiet, and even laughing, while I held her hand until the ghastly preparations had been made."

For weeks Edith lingered between life and death, but in mid-March Roosevelt said she was slowly "crawling back to life." Ted's illness was also diagnosed at last by Dr. Alexander Lambert, a family friend. Lambert attributed the boy's nervous condition to the senior Roosevelt having driven him too hard physically and mentally. "Hereafter I shall never press Ted either in body or mind," the chastened father declared. "The fact is that the little fellow, who is particularly dear to me, has bidden fair to be all the things I would like to have been and wasn't, and it has been a great temptation to push him." Fortunately, as Edith recalled, in later years he "was too busy to exert the same pressure on the others!"*

*Throughout his life, Theodore Roosevelt, Jr., was bedeviled by invidious comparisons with his father. "Poor T.R., Jr.," Alice once said. "Everytime he crosses the street, someone has something to say because he doesn't do it as his father would. And if he navigates nicely, they say it was just as T.R. would have done."

*   *   *

Pending the outcome of the naval board's inquiry into the destruction of the *Maine,* the nation hung suspended between peace and war. As Americans anxiously awaited the board's findings, the *New York Sun* quoted an anonymous New Yorker as saying that if the assistant secretary of the navy had been placed in charge of the investigation, it would have been over by now. "Teddy Roosevelt is capable of going down to Havana, and going down in a diving-bell to see whether she was stove in or stove out." Roosevelt himself, consumed by a passion for war, continued to chivy Tillinghast for a commission in the New York National Guard while pushing preparations for war.

"Have you and Theodore declared war yet?" McKinley jokingly asked Leonard Wood, one of the White House physicians, on several occasions.

"No, Mr. President, but we think you should," Wood shot back.

On March 20, the board of inquiry confidentially informed the president that the *Maine* had been the victim of a submarine mine, although it was "unable to obtain evidence fixing responsibility." Roosevelt must have learned of these findings from Sheffield Cowles, a member of the board, and vented his anger in a letter to Brooks Adams:

> In the name of humanity and of national self-interest alike, we should have interfered in Cuba three years ago. . . . The craven fear and brutal selfishness of the mere money-getters, have combined to prevent us from doing our duty. The blood of the Cubans, the blood of women and children who have perished by the hundred thousand in hideous misery lies at our door; and the blood of the murdered men of the *Maine* calls not for indemnity but for the full measure of atonement which can only come by driving the Spaniard from the New World. I have said this to the President before his Cabinet; I have said it to Judge Day, the real head of the State Department; and to my own chief. I cannot say it publicly, for I am of course merely a minor official in the administration.

Not long afterward, however, Roosevelt managed to make his feelings public. On the evening of March 26, both he and Mark Hanna were present at the annual Gridiron Dinner attended by Washington's political and journalistic elite. The club's president took note of the split in the administration regarding the war and, noting Hanna's opposition, turned to him and said: "Senator Hanna, can we have this war?" Rather than responding with a joke, Hanna made a brief, serious speech against war. When he sat down, the president introduced the next speaker. "At least we have one man connected with this Administration who is not afraid to fight—Theodore Roosevelt, Assistant Secretary of the Navy."

"We will have this war for the freedom of Cuba, Senator Hanna," Roosevelt declared, glaring at Hanna, "in spite of the timidity of the commerical interests."

Two days later, McKinley made public the report of the board of inquiry. The American people, whipped to a frenzy by the press, were convinced that an "external explosion" meant that the Spaniards had deliberately blown up the *Maine*. Restraint was tossed to the winds, and the slogan of the hour was:

To Hell with Spain!
Remember the Maine!

McKinley was so worn out by the strain that he had to use drugs to sleep, but he still hoped to avoid war. First, he thought of trying to buy Cuba, but Spain would not sell the troubled island. Then, he issued an ultimatum demanding an armistice and American artitration of the conflict. To the Spaniards, this meant the independence of Cuba. Spanish pride forbade acceptance of what was looked upon as abject surrender and anti-American rioting broke out in several Spanish cities. The interventionists regarded McKinley's diplomatic efforts as temporizing. "Do you know what that white-faced cur up there has done?" Roosevelt reportedly raged to a friend after a White House meeting. "He has prepared two messages, one for war and one for peace, and he doesn't know which one to send in!"

Roosevelt did have the consolation of Edith's full recovery from her illness. She felt so well on April 4 that she came downstairs after he had left for the office and decided to go for a carriage drive in the balmy spring sunshine. On impulse, she told the driver to take her to the Metropolitan Club with the intention of surprising her husband, who lunched there every day. Seeing Leonard Wood, she asked him to go into the club and tell the unsuspecting Theodore there was a lady outside who wished to speak with him. Upon seeing his wife, Roosevelt ran toward her. "I wish you could have seen his face of surprise and delight," Edith told Ted.

In the end, McKinley, pushed along by the popular tide, sent a war message to Congress on April 11, seeking authorization to use the armed forces to end hostilities in Cuba. Two days *before,* Spain had agreed to the proposed armistice, but the president took the position that the Spanish government had not lived up to previous promises and could not be trusted. An able politician, McKinley fully realized that any attempt to thwart the people's will could jeopardize reelection in 1900. William Jennings Bryan, his likely opponent, was arousing support across the country with tub-thumping speeches for an independent Cuba.

Eight days later, on April 19, at three o'clock in the morning Congress approved a joint resolution that was tantamount to a declaration of war. Northerners, southerners, easterners, and westerners were all carried along on the flood tide. Groups of members sang "The Battle Hymn of the Republic"; others pounded each other on the back. A few sat dejectedly in their seats. The formal declaration followed six days later—and Theodore Roosevelt had his war.*

A naval blockade was immediately imposed upon Cuba and a call issued for 125,000 volunteers to flesh out the regular army of 28,000 men. The order included provision for three regiments of mounted riflemen—the work of Judge Jay L. Torrey, a prominent Wyoming political figure and Civil War veteran, who had gone directly to the White House with the proposal. Secretary of War Russell A. Alger had little difficulty in finding a commanding officer for the first of these regiments. In fact, Theodore Roosevelt had been beating on his door for weeks requesting an army commission. The regiment sounded very much like the troop of "harum-scarum rough riders" that he had talked about leading into battle in case of war with Mexico back in 1886.

Roosevelt was delighted by Alger's offer, but he declined on

---

*Spain declared war on the United States on April 24, and the formal American declaration of war was made retroactive to April 21.

grounds that three years' service in the New York National Guard was hardly enough experience to command a regiment in wartime. "I believed I could learn to command the regiment in a month," he said, but that month might mean missing the chance of seeing action. However, he would be delighted to serve as lieutenant colonel of the regiment if the battle-tested Leonard Wood were named its colonel. Alger accepted the proposal, and Wood and Roosevelt were assigned command of the 1st United States Volunteer Cavalry.*

Family, friends, and superiors all implored Roosevelt to remain in the post in which he had done so much to prepare the navy for war. Edith, still frail from illness, was despondent but knew that he was determined to go and loyally supported his decision. President McKinley and Secretary Long both made personal appeals for him to stay in Washington. Almost forty years old, nearly blind without his glasses, and with six children and a wife recuperating from a near-fatal illness, Roosevelt should remain where he was—and where he could accomplish the most. "I really think he is going mad," said Winthrop Chanler. Long also doubted Roosevelt's sanity. "He has lost his head . . . and running off to ride a horse and, probably brush mosquitoes from his neck on the Florida sands," Long wrote in his diary. "He is acting like a fool." Then he had second thoughts. "How absurd all this will sound if, by some turn of fortune, he should accomplish some great thing and strike a very high mark."†

Roosevelt was determined to get into the fight. "It was my one chance to do something for my country and for my family and my one chance to cut my little notch on the stick that stands as a measuring-rod in every family," he told a friend several years later. "I know now that I should have turned from my wife's deathbed to have answered that call." Yet, there was more to Roosevelt's decision than the beckoning of military glory. Writing to Dr. Lambert, he said: "If I am to be any use in politics it is

---

*The other two regiments were organized by Torrey and Melvin B. Grigsby, attorney general of South Dakota and also a Civil War veteran. Roosevelt's regiment was the only one to see action.

†Several years later, Long had occasion to look back upon this page of his diary and scrawled: "Roosevelt was right and we his friends were wrong. His going into the army led straight to the Presidency."

because I am supposed to be a man who does not preach what he fears to practice. . . . For the last year I have preached war with Spain. I should feel distinctly ashamed . . . if I now failed to practice what I have preached."

As soon as the newspapers reported that Wood and Roosevelt were raising a regiment, they were deluged with applications not only from the West but from all over the nation. "We could have raised a brigade or even a division," Roosevelt said. The regiment was also being given alliterative nicknames—"Teddy's Terrors," "Wood's Wild Westerners," "Roosevelt's Rangers," "Cavalry Cowpunchers," and "Roosevelt's Rough Riders." The last was the most popular name and it stuck. Wood departed immediately for San Antonio, where the regiment was to be mustered and undergo training while Roosevelt remained in Washington to clear his desk at the Navy Department.

One bit of unfinished business was the Spanish flotilla in the Philippines. Orders were sent to Dewey to launch an attack immediately, and on April 27, his ships stood out into the China Sea with the band of the flagship, the cruiser *Olympia,* playing John Philip Sousa's *El Capitan.* In the squadron's wake bobbed wooden chairs, tables, chests, waste and paint cans—anything that might feed a fire.

For more than a week, the nation waited for news from Dewey, but there was only silence. It was as if the squadron had sailed over the rim of the world, and everyone was in a fever of anxiety. The silence was finally broken on the morning of May 7, when a dispatch vessel arrived at Hong Kong with Dewey's official report. When the encoded message reached Washington, it was given to Secretary Long, who stared at it as if by sheer willpower he could extract some meaning from the jumble of letters, which began: CRAQUIEREZ REFRANAMS VUFVOE . . . He handed it to a cryptographic officer, who disappeared into the cipher room.

Assistant Secretary Roosevelt emerged about a half hour later to confront a crowd of frantic newsmen. Six days before, Roosevelt informed them, Dewey had crushed the Spanish squadron in Manila Bay. Every one of the enemy ships had been destroyed or captured without the loss of a single American life. Cheers resounded throughout the room—and shortly across the nation— even though most Americans could have joined President

McKinley in confessing that he "could not have told where those darned islands were within two thousand miles."

Roosevelt turned in his resignation on this note of triumph and, after packing a uniform ordered from Brooks Brothers and twelve pairs of spectacles, left for Texas to join his regiment. "Father went to war last Thursday," Kermit informed an aunt. "I sted up untill he left which was at 10."

Chapter 13

"A Body of Cowboy Cavalry"

Lieutenant Colonel Theodore Roosevelt, wearing a new khaki uniform with yellow cavalry trim, arrived in San Antonio, Texas, on May 15, 1898, to assume his place with the First Volunteer Cavalry. The camp where the regiment was being mustered was named for Leonard Wood, but the sign at the railroad station read: This Way to Camp of Roosevelt's Rough Riders. Piling his baggage onto a buckboard, Roosevelt promptly headed for the camp at the fairgrounds just outside town. He was immediately surrounded by men eager to shake his hand. Some, who had never met him before, were surprised by his thick glasses and squeaky voice.

Over the previous ten days, cowboys, ranchmen, miners, gamblers, Indians, lawmen, and hard-bitten men who had tangled with the law had drifted into Camp Wood, where they mixed with a smattering of sportsmen, soldiers of fortune, Ivy League athletes, society clubmen, New York City policemen, actors, and musicians. In all, the regiment numbered about a thousand men and was described as "a society page, a financial column, a sports section, and a Wild West show all rolled into one." It quickly captured the public imagination, from which it was never to be dislodged.

276

Roosevelt found that Colonel Wood had performed a miracle of organization. The camp was humming with activity. Men had been detailed to squadrons, troops, and squads. Somehow, horses, uniforms, supplies, weapons had been pried out of a Quartermaster Department that was near collapse from the unexpected strains placed upon it and reluctant to divert scarce matériel to a volunteer regiment. Wood had even obtained new Krag-Jörgenson magazine carbines rather than the single-shot Springfields left over from the Indian wars that were issued to many units. As a result of his efforts, the Rough Riders were among the army's best-equipped regiments.

The roster included William "Bucky" O'Neill, onetime sheriff and mayor of Prescott, Arizona; Ben Daniels, marshal of Dodge City; Bob Wrenn and Bill Larned, the nation's two top tennis players; Hamilton Fish, Jr., captain of the Columbia crew and grandson of another Hamilton Fish who, as secretary of state under President Grant, had prevented the United States from becoming embroiled in a war in Cuba twenty years before; George M. Dunn, master of the Chevy Chase hounds; Thomas P. Ledwidge, who had fought with the Cuban guerrillas; Dudley Dean, said to be Harvard's best quarterback ever; Captain Allyn K. Capron, Jr., a highly regarded Regular officer; and Emilio Cassi, the regimental trumpeter, who had served with the Chasseurs d'Afrique in Algeria.

With rough humor, the men gave one another derisive nicknames. An eastern clubman was known as "Tough Ike"; his tentmate, a rough cowboy, was christened "The Dude." One cowpuncher was called "Metropolitan Bill"; a young Jew accepted with equanimity the name "Pork Chop," while a huge red-haired Irishman became "Sheeny Solomon." A quiet man was dubbed "Hell Roarer," and another, whose language and conduct were completely opposite, became "Prayerful James."

By the spit-and-polish standards of the Regular Army, the Rough Riders were an ill-disciplined and motley-looking lot. Most of the men wore flannel shirts with loosely knotted blue polka-dot handkerchiefs around their necks—already the badge of the regiment—and the canvas trousers of the cavalry fatigue uniform stuffed into boots or leggings, rather than the heavy blue uniforms that had been supplied the rest of the army, which were more suitable for campaigning in Alaska than the tropics. Government-issue felt hats were discarded for sombreros. Sabers were considered useless and so were not issued, but each man packed a long-barreled Colt revolver.* Altogether, observed Roosevelt fondly, the regiment looked "exactly as a body of cowboy cavalry should look."

Only the most patient—or blasphemous—drill sergeant could get such a varied group of men to advance, wheel, and turn in some semblance of military precision either on horseback or on foot. Sentries greeted officers making the rounds of regimental outposts with a cheerful "good evening" instead of a salute. A cook summoned the officers to mess with the call "If you fellas don't come soon, everything'll get cold." But the troopers were tough, good marksmen and excellent horsemen. And they were eager to fight. The Rough Riders existed as a regiment for only 133 days, but one in three who landed in Cuba were killed, wounded, or stricken by disease, the highest casualty rate for any unit in the war with Spain.

Before they developed a mutual respect for each other, the cowboys were much amused by the "Fifth Avenue Boys" and devised an elaborate plot to embarrass one of them. The man had gotten permission to go to San Antonio and the cowboys selected the meanest mount they could find. They stood about waiting for the horse to explode into volcanic action the moment the city slicker put a foot into a stirrup. But he gathered up the reins and swung into the saddle with practiced ease. The animal tensed as though ready to lunge into the air, but the rider quickly pulled his head up with a sure hand. Instantly, the horse recognized his master and quieted down. Solemnly tipping his hat to the slack-jawed spectators, the horseman moved off at a gentle canter. Not until later did the cowboys learn that their intended victim was Craig Wadsworth, the country's top steeplechase rider.

Training consisted primarily of long rides on the trails leading out of San Antonio in heat worthy of a blast furnace. Even a light wind sent dust and mesquite swirling in heavy clouds, caking the troopers with dirt and sweat. On the way back from one such march, Roosevelt's squadron passed a saloon and he called a halt. Turning his horse about, he faced the men and announced:

---

*Roosevelt carried a pistol that had been salvaged from the *Maine*.

"Captains will let the men go in and drink all the beer they want, and I will pay for it."

Then, he shook his fist at the line, gritted his teeth fiercely, and added: "But if any man drinks more than is good for him, I will cinch him."

Wood heard about the incident and promptly called his second-in-command on the carpet. Any officer who drank with his men was quite unfit to hold a commission, he declared. Roosevelt saluted without a word and retreated to his tent. Later that night, he turned up at headquarters again. "I wish to tell you," he said, "that I took the troops out without thinking of this question of officers drinking with their men and gave them all a schooner of beer. I wish to say, sir, that I consider myself the damnedest ass within ten miles of the camp. Good night."

From Edith, he received letters that strained to provide a light-hearted account of domestic life at Sagamore. "Ted hopes there will be one battle so that you can be in it, but come out safe. Not every boy has a father who has seen a battle, he says." But try as she might, her loneliness shone through. "Always I have the longing and missing in my heart, but I shall not write about it for it makes me cry."

In a remarkably brief time, the regiment's disparate elements were welded together into a fighting force, and Roosevelt's enthusiasm for it was boundless. "This is just a line to tell you we are in fine shape," he wrote President McKinley. "Wood is a dandy Colonel, and I think the rank and file of this regiment are better than you would find in any other regiment anywhere. In fact, in all the world there is not a regiment I would so soon belong to. The men are picking up the drill wonderfully. They are very intelligent, and rather to my surprise they are very orderly—and they mean business."

While Roosevelt and his Rough Riders were chafing to be off to Cuba, the navy was engaged in a game of blindman's buff with a Spanish squadron of four armored cruisers and several smaller ships that had sailed from the Cape Verde Islands at the end of April and disappeared into the Atlantic. On paper, these vessels were the equal of their Yankee counterparts, but they were in poor condition and some had sailed without all their guns. In fact, the Spanish commander, Admiral Pascual Cervera y Topete, had no hope of victory. "Nothing can be expected . . . except the total destruction of the fleet or its hasty and demoralized return," he gloomily declared after a survey of his ships.

Nevertheless, the threat of Cervera's squadron loose upon the high seas sent panic racing up and down the Atlantic coast of the United States like fire in a ship's rigging. Newspapers proclaimed that "the galleons of Spain" were off Boston and New York, and the Navy Department was inundated with demands for protection. Rear Admiral William T. Sampson, commander of the main battle fleet, calculating that Cervera would head for San Juan, Puerto Rico, to refuel and steamed there to intercept him. But the Spanish ships were nowhere to be found.

While Sampson was on his way back to base at Key West, word was received that Cervera was at Martinique. Once their ships were refueled, Sampson thought the Spaniards would now make for Cienfuegos on the southern coast of Cuba. But Cervera outfoxed the Americans. Having been denied fuel by the French, he sailed to Curaçao, where the Dutch were willing to oblige. After partially replenishing his bunkers, he steamed directly across the Caribbean to Santiago, near the southeastern tip of Cuba, and took shelter behind the batteries and minefields guarding the narrow passage into the harbor.

The chagrined Sampson established a blockade of Santiago, determined to prevent the Spaniards from escaping again. Each night one of his battleships edged in closer, played its searchlights on the harbor mouth, and blindly shelled the Spanish ships to little avail. The navy was unable to penetrate the harbor because of the minefields. As a result, the presence of Cervera's squadron at Santiago determined the strategy of the war.

Earlier on, there had been talk of landing an American expeditionary force near Havana or Cienfuegos. Now, the navy wanted the army to capture the fortifications guarding Santiago so the passage into the harbor could be swept and its ships could go in after the Spanish vessels. It was probably the first instance in which a fleet called upon an army to help capture another fleet. The War Department, eager for a share of the glory and headlines being harvested by the navy, accepted the challenge. On May 30, Major General William R. Shafter, commander of the Fifth Army Corps at Tampa, Florida, the city closest to Cuba with rail and port facilities that could handle an army, was ordered to prepare 25,000 troops for an immediate landing near Santiago.

The men nearest the headquarters tent of the Rough Riders were aroused from a heat-induced torpor by the sound of shouting. They ran out of their tents to find Lieutenant Colonel Roosevelt dancing a little war dance—like the one he had performed when he shot his first buffalo—in front of Wood, his hat in one hand, a telegram from the War Department in the other. The message inquired when the regiment would be ready to leave San Antonio. Wood promptly replied: AT ONCE. Soon, orders were received to entrain for Tampa.

The San Antonio city fathers gave the Rough Riders a gala send-off that included a concert by Professor Carl Beck and his band, reputed to be the best in Texas. The highlight of the evening was to be *The Cavalry Charge* and, to add a touch of realism, Beck provided a few troopers with blank cartridges and instructed them to fire their pistols into the air when he gave the signal. All went well in the beginning. The music swelled into a pounding rhythm that simulated the exhilaration of the charge and the regiment stamped its feet in excitement. As a crescendo was reached, the Professor arched his back, rose on his toes, and thrust his baton at the pistol detail standing at the side of the platform. There was a flurry of shots and then the evening exploded into gunfire and cowboy whoops as the entire regiment joined in the firing. Some two thousand rounds were fired—and bandsmen and civilian spectators stampeded for cover. Someone cut the electric cable, and in the darkness the troopers vanished into town to continue their merrymaking.

For four days, the long train carrying the Rough Riders and their horses rolled eastward toward Florida. Roosevelt rode with his men in "a dirty old ramshackle coach," having relinquished his place in the Pullman car reserved for officers to an ailing trooper. The regiment received an enthusiastic reception all across the South. It was as if the war with Spain had reunited a nation pulled asunder by an earlier war. Pretty girls turned out to bring the troops flowers, watermelon, and pails of milk, and bands serenaded them. "Everywhere we saw the Stars and Stripes," Roosevelt reported, "and everywhere we were told, half-laughing by grizzled ex-Confederates that they had never dreamed in the bygone days of bitterness of greeting the old flag as they were now greeting it, and to send their sons, as they were now sending them, to fight and die under it."

Roosevelt rapidly learned that war is mostly confusion. No one was on hand to meet the Rough Riders when they arrived in Tampa on June 3; there was no one to tell them where to bivouac, no one to issue food or forage. With some difficulty, Wood found General Shafter's headquarters and was assigned a campground, where the cavalry division was commanded by Major General Joseph Wheeler. Short and bandy-legged, "Fighting Joe" Wheeler looked like a bewhiskered gnome, but he had been a dashing Confederate cavalry leader and McKinley had offered him a commission to show the nation was indeed united. For several days, Wood and Roosevelt spent their own money to buy food for the men and forage for the horses. They were brigaded together with the Ninth and Tenth cavalries, two black Regular regiments that had won their spurs on the plains.

Tampa, miserably situated in a blistering waste of sand hills and scrub pine, was composed mostly of derelict wooden shacks at the end of the railroad line that was the sole reason for its existence. The port itself was nine miles away and the only connection to the solitary pier was a single-track railroad. Trains loaded with men and supplies had been pouring into the town for a month and there were some thirty thousand troops on hand, mostly Regulars with a few volunteer units scattered among them. More than a thousand freight cars stood on the sidings, stretching all the way back to Columbia, South Carolina.

Units drilled in civilian clothes while fifteen cars loaded with uniforms sat unnoticed on a nearby siding. Thousands of rifles, badly needed by the troops, were out there somewhere. Invoices and bills of lading had not been shipped, and officers, directing men with crowbars, hunted from car to car for weapons, ammunition, clothing, horse equipment, and artillery pieces. "No words can paint the confusion," fumed Roosevelt in the diary he had taken up again. "No head, a breakdown of both the railroad and military systems of the country."

Under Colonel Wood's watchful eye, the Rough Riders erected their tents on the treeless sand flats, established picket lines, drilled in the full glare of the Florida sun—and asked over and over again when they would get into combat. Rumors swirled about, each contradicted by the next: the army was going to invade Spain; no, it was not. New York had been attacked and the troops were going north; no, it was not and they were not. Sampson's squadron had been attacked and destroyed; no, it was the Spaniards who had been defeated. On payday—privates received $13 a month—the men thronged the saloons and houses of ill fame that had sprung up on "Last Chance Street." White soldiers also learned how to play craps from the black cavalrymen.

Off-duty officers lounged on the broad veranda of the Tampa Bay Hotel, a bizarre Moorish confection of ornamental brick and silver minarets and domes that loomed like a mirage above the low palmettoes. The hotel became General Shafter's headquarters, as well as for newspapermen, sightseers, foreign military attachés, and not a few suspiciously overdressed women. While waiting for orders to move out, everyone sank into the veranda's rocking chairs. Richard Harding Davis of the *New York Herald* called these days the "rocking chair period" of the war. Another newsman, iced drink at his elbow and Cuban cigar in hand, remarked to his fellows on the veranda: "Gentlemen, as General Sherman truly said, "War is hell."

General Shafter was himself one of the foremost members of the rocking-chair brigade. A three-hundred-pound Civil War veteran, Shafter had spent the intervening years fighting Indians and he was a bluff, gruff old soldier. In his youth, he had won the Medal of Honor, but owed his command of the Fifth Corps to the fact that Secretary of War Alger found him agreeable. Gouty and dropsical, Shafter hardly inspired confidence as he was helped up the grand stairway of the hotel with sweat pouring from his face. He had no experience in commanding such a sizable force—neither had any American general—and absolutely no sense of public relations. He regarded the horde of journalists who had descended upon the camp as an unnecessary evil, and their stories were already beginning to reflect disdain for him.

Once the Rough Riders were settled in, Edith Roosevelt arrived in Tampa to spend a few days with her husband. She found him "thin but rugged and well." Wood gave him permission to stay overnight at the hotel with his wife for the duration of her visit, although he had to return to camp for reveille at 4:00 A.M. Roosevelt invited two family friends, both sergeants in the Rough Riders, to join them for dinner one evening, much to the scandal of the Regular officers. During the day, Edith was squired about by the handsome and debonair Dick Davis.

The swashbuckling beau ideal of all ambitious young newspapermen, Davis had almost singlehandedly transformed journalism from a rather seedy calling into a glamorous profession. He had already covered several wars while tossing off popular short stories and best-selling novels. His square-jawed good looks were familiar to everyone, for he was also the model for the handsome escort of the coolly elegant Gibson Girl, the feminine ideal of the day, drawn by his friend Charles Dana Gibson. Initially, Davis and Roosevelt did not get along; after an encounter in Washington several years earlier, Roosevelt had described Davis as "an everlasting cad." But Davis recognized a good story when he saw it and attached himself to the Rough Riders. He and Roosevelt, sensing a mutual need for each other, soon became good friends. Roosevelt, said Davis, had the "energy and enthusiasm to inspire a whole regiment."

On June 6, Edith and the military attachés were guests at a review of the cavalry brigade. "It was a wonderful sight to see two thousand of these men advancing through the palmettoes, the red and white guidons fluttering at the fore, and the horses sweeping onward in a succession of waves, as though they were being driven forward by the wind," Davis observed. "It was a fine spectacle, and it was due to such occasional spectacles in and around the camps that the rocking-chair life was rendered bearable." Unknown to everyone, however, the curtain was already coming down on the rocking-chair period of the war.

Under pressure from Washington, Shafter had issued orders that morning for the embarkation of the Fifth Corps for a "destination unknown." But a double bombshell was dropped on the Rough Riders. Only two thirds of the troopers and the horses of the senior officers were to go, because there was a shortage of space on the transports. It was the worst time, so far, for Wood and Roosevelt. With considerable sadness, they went over the muster rolls and broke the news to the men who were to be left behind. They consoled the troopers by promising them they and the horses would rejoin the regiment after it reached its designation. Then began a mad scramble to the transports at Port Tampa.

The one-track railroad quickly collapsed under the demands placed upon it, and the train that was supposed to carry the regiment to the pier failed to show up. "We were up the entire night standing by the railway track . . . hoping for a train that did not come," reported Roosevelt. At dawn, they were shifted to another track but no train appeared there, either. When a line of empty coal cars hove into view, Roosevelt commandeered them and the men climbed in with their equipment. Upon their arrival at the pier, everything was in a "higgledly-piggledly" state. Roosevelt and Wood searched through "this swarming ant-heap of humanity" for Shafter's chief quartermaster, who told them the regiment was to sail in the *Yucatan,* an old coastal steamer, still out in midstream.

Wood hijacked a launch, went out to the *Yucatan,* and ordered the vessel to come up to the pier. In the meantime, Roosevelt, who had learned that the ship had been allotted to two other units, the Seventy-first New York Volunteers and the Second Regular Infantry, ran at full speed back to the coal cars and marched his grimy men double-quick to the ship in time to beat the other regiments up the gangplank. When an officer of the Seventy-first tried to come on board, this interchange took place:

"Hello," said Roosevelt cheerily, "what can I do for you?"

"That's our ship."

"Well, we seem to have it."

Shaking their fists at the jeering Rough Riders who lined the rail of the transport, the other regiments angrily marched off to commandeer another transport.

On the morning of June 8, Shafter, who established his headquarters on the pier—using one packing crate as a desk and two others on which to rest his huge bulk—finally completed the loading of the largest military force that had ever left the shores of the United States. It numbered about sixteen thousand men and three thousand horses and mules, plus thirty-eight pieces of artillery, including four Gatling machine guns. But just as the general was boarding his headquarters ship, the *Seguranca,* he received orders that kept the transports in Tampa Bay. Several Spanish cruisers were reported to be at large in the Caribbean, and if they got in among the convoy, it would be disastrous. As it turned out, some American vessels had been misidentified as enemy ships, but for the next six days, the troops sweltered on the transports, rocking in the greasy swells off Tampa while the navy tracked down the "ghost squadron."

Roosevelt exploded at the mismanagement of the expedition. While a certain amount of confusion was to be tolerated in any such operation, "no words could describe to you the confusion and lack of system and the general mismanagement of affairs here," he told Lodge. The *Yucatan* was anchored in a "sewer" and was "crowded to suffocation" with double the number of men it could hold comfortably. Only a small number could come on deck at a time; the rest had to remain in the lower hold, which was "unpleasantly suggestive of the Black Hole of Calcutta." Although morale was high, Roosevelt was worried that the delay in getting to sea would adversely affect the health and efficiency of the regiment.

The navy eventually ascertained that all of Cervera's ships were safely bottled up in Santiago harbor, and at midday on June 14, the thirty-one transports and accompanying warships were given permission to sail. In a great din of whistle-blowing, cheering, and discordant music, the vessels slowly steamed out of Tampa Bay in three lines. Someone hung a sign over the side of the *Yucatan* reading: "Standing Room Only." Soon, the bands all harmonized in a high-stepping minstrel march, and the tune floated over the water, spreading from ship to ship, until thousands of voices took up the refrain: "Come along, come along, get you ready, wear your brand new gown . . . for there'll be a hot time in the old town tonight."

Over the next six days, the convoy steamed southeast through "a sapphire sea, wind rippled, under an almost cloudless sky," Roosevelt wrote home. Turning philosophical, he noted that "it is a great historical expedition, and I thrill to feel that I am part of it. If we fail, of course, we share the fate of all who fail, but if we are allowed to succeed (for we certainly shall succeed if allowed) we have scored the first great triumph of what will be a world movement."

The voyage was scheduled to take only three and a half days, but the convoy poked along at only four to seven knots—the speed of the slowest vessel—and the time stretched out to six days. For the men, packed into the holds of the "prison hulks," as they called the transports, it was agony. The water was foul, the canned beef served out as the main ration was spoiled, and the stench of manure from the horses and mules pervaded the ships. Civilian crewmen engaged in a brisk trade in whiskey of doubtful provenance. One Rough Rider acknowledged that he had "experimented with some pretty rough bug extract in Arizona," but had never "tackled any red liquor that would come up to the standard of the rat poison sold right here on board this government ship."

Only Shafter and his staff knew the destination of the armada, and as it plowed along with the trade winds blowing in their faces, the men debated whether the landing would be at Santiago or Puerto Rico. But as soon as the ships rounded the eastern tip of Cuba and turned to the southwest, everyone knew their objective was Santiago. At daybreak on June 20, the coast of Oriente Province loomed out of the blue mist, and the troops crowded the rails of the transports for their first sight of Cuba. For several hours, the convoy plodded along the coast, which was described by Roosevelt as "high barren looking mountains rising abruptly from the shore . . . looking much like those of Montana." The Morro Castle of Santiago and the semicircle of gray blockading warships surrounding the harbor entrance soon appeared in view.

Admiral Sampson came out to the *Seguranca* and, together with Shafter, attended a council of war with the Cuban leader, General Calixto García, in a palm-thatched hut on shore. The guerrilla chieftain was an imposing figure. Tall and vigorous, he had a flowing white moustache and a deep scar in his forehead, the result of trying to shoot himself when captured by the Spaniards ten years before. Sampson proposed that the army land on both sides of the entrance to Santiago harbor, scale the cliffs, and capture Morro Castle and the Socapa battery on the opposing headland. Once the way was opened, the navy would sweep the mines from the harbor entrance and dash in to attack Cervera's squadron.

Shafter had his own plans, however. On the slow voyage to

Cuba, he had read an account of a disastrous attack by the British admiral Edward Vernon* on Santiago in 1741, and had no desire to repeat the British example of trying to storm a stone fort perched on a 230-foot cliff. Instead, he decided to land his troops at Daiquirí, fifteen miles east of Santiago, and at nearby Siboney, even though they lacked adequate port facilities, and to strike quickly at the city from the rear. To confuse the Spaniards, a stretch of coast twenty miles long would be bombarded by the navy and feints would be made at various points. García agreed to this plan and promised to keep the enemy busy.

Lieutenant Colonel Roosevelt greeted the news that the Rough Riders were to be part of the landing force by whipping off his hat with a flourish and performing one of his little war dances accompanied by a song:

> Shout hurrah for Erin go Bragh
> And all the Yankee nation.

Replacing his hat, he took Captain Capron by the arm. "Come along, you old Quaker," Roosevelt said, "and let's go to supper."

No one slept during the night of June 21. Each man readied himself for battle, preparing the blanket roll, three days' field rations, and the one hundred rounds of ammunition he was to carry. Weapons were cleaned and oiled one last time. Boats were swung out at the davits. From the deck, the soldiers could see fires ashore at Daiquirí and in the mountains behind the village. Some hazarded the guess that these were signals lit by the Cubans; others surmised that they were the campfires of the waiting Spaniards. In the dull light of dawn, the men studied their objective. Daiquirí was little more than a notch in the jagged shoreline and consisted of a few corrugated iron huts and a railroad pier that had been built by an American mining company.

Specialized landing craft were four decades in the future, so the troops had to go ashore in ships' boats. The sea was rough and the heavily laden men had considerable trouble dropping

---

*Lawrence Washington, George Washington's half brother, served in the Caribbean under Vernon and named his estate on the Potomac after the admiral. George Washington later inherited Mount Vernon.

from the transports into the boats, which rose and fell several feet on the swells. As the boats bobbed alongside, the warships unleashed a bombardment. The cheering was constant as heavy shells shrieked overhead to land amid great clouds of dust and debris and broken tree limbs. When the guns fell silent, several fires were burning ashore and there were no signs of life except for a solitary Cuban energetically waving a white cloth from the end of the pier to signal that the Spaniards had left.

The boats were ready to begin the run to the beach, but the civilian transport skippers refused to hazard their vessels close inshore, so the boats had to cover a considerable distance through the pounding sea. Navy steam launches tried to round up as many as they could and tow them in. Some ran alongside the pier, but it was so high that the men had to leap from the boats at the exact moment they rose on a wave. If a man miscalculated, he would fall into the sea, to be dragged down by his equipment or crushed against the pilings. Other boats raced up to the beach, tumbling the men out into the surf. Had the Spaniards contested the landing, they could have turned it into a bloody shambles.

The Rough Riders were not scheduled to be among the first to land, but Roosevelt's navy contacts proved providential. A converted yacht, the *Vixen,* commanded by a Lieutenant Sharp, a former aide, volunteered to lead the *Yucatan* in closer to shore than the other transports. This enabled the regiment to outdistance other units and to be among the first to land. Once ashore, officers and sergeants scurried about reorganizing their units and rounding up stragglers before moving inland to form a defensive perimeter. In all, some seven thousand soldiers were landed with the loss of only two men.* Another force of about the same size landed at Siboney.

Horses and mules were pushed overboard from the transports to swim to the beaches. Confused, wild-eyed, and snorting with terror, some of the animals headed for the open sea as the soldiers looked on helplessly. Suddenly, a quick-thinking bugler sounded the call "right wheel," and they came around and headed for the safety of the shore.

---

*The casualties were both black troopers from the Tenth Cavalry who slipped while trying to jump from their boat onto the pier. Major Bucky O'Neill of the Rough Riders dived in to try to save them, but was unable to do so.

Roosevelt returned to the *Yucatan* to supervise the unloading of his own two horses. As the first, Rain-in-the-Face, was swung over the side with a band under its belly, a huge breaker smashed against the ship and the animal drowned. Snorting like a bull and stamping back and forth, Roosevelt "split the air with one blasphemy after another." The sailors took such care in lowering Texas, the second horse, that the frightened mare seemed to hang in the air until the outraged Roosevelt roared: "Stop that goddamned animal torture!" This time, there was no slipup and the horse swam safely to shore.

None of the army's baggage was landed—Roosevelt had only a yellow slicker and a toothbrush—and exhausted and wet, the soldiers slept on the ground during their first night in Cuba with their weapons within easy reach. But most regarded this as better than the holds of the transports. From the warships, searchlights gleamed, casting prying beams along the mountainsides and over the forms of the huddled men. Before going to sleep, Roosevelt made an entry into his diary that was a record for brevity: "June 22—landed."

Two days later, General Shafter launched his campaign to capture Santiago. A pair of Regular infantry regiments was ordered to probe the enemy's defenses while the dismounted cavalry, including the Rough Riders, who were regarded as of a somewhat dubious quality, were to take up defensive positions until the beachhead was consolidated. But Shafter, who remained on board the *Seguranca* to oversee the unloading of equipment and supplies, failed to take account of the intense rivalries between the various units of his army.

Fighting Joe Wheeler, the senior officer ashore, had his own ideas on how the campaign should be conducted. During the War Between the States, foot soldiers had never scouted ahead of cavalry—not even dismounted cavalry—and he saw no reason for it now. Cuban irregulars had reported that a force of some two thousand Spaniards had established a well-fortified position about three miles inland from Siboney, and Wheeler was determined to strike the first blow against the enemy. Slyly, he pushed the cavalry, which had the previous day been force-marched up to Siboney, to the front.

Although Wheeler may have launched his campaign with comic-opera overtones, his reasoning was sound. Having reconnoitered the area in person, he realized that the beachhead was dangerously exposed to a Spanish counterattack. The enemy also controlled a gap in the hills through which the invaders must pass as they advanced upon Santiago. If the Americans did not press forward immediately, he reasoned, the enemy might array their army there, where from trenches at the top of a 250-foot ridge they would command the approaches and pour fire down upon an attacking force. If the Fifth Corps were pinned down on the beachhead for any length of time in the rainy season, yellow fever, typhoid, and dysentery would create havoc.

There were two roads winding over the foothills from Siboney to Santiago: One was the main highway, or Camino Real, little more than a wagon track, and the other was a narrow trail about a half mile to the east. Like an inverted V, the roads came together at Las Guasimas, about twelve miles from Santiago. It was not a town, but merely the site of a clump of guasimas—trees with low spreading branches that bore nuts Cuban farmers fed to their pigs. The Cuban scouts also gave Wheeler a more macabre landmark. When the Americans reached the body of a dead man lying sprawled across the trail, they would know that the Spaniards were in the vicinity.

Having outsmarted the infantry, Wheeler ordered Colonel Wood to take five hundred of his men down the rugged path to the east while an equal number of troopers from the First and Tenth cavalries were assigned to the Camino Real. The Regulars had a pair of light Hotchkiss field guns with them. Once the columns had joined at Las Guasimas, they would deploy, and while the black troopers made a direct assault on the enemy-held ridge, the Rough Riders would attack the flank.

Wood sent out an advance guard headed by Captain Capron, who in turn sent out five men headed by Sergeant Hamilton Fish to serve as point for the column. The night before, Roosevelt later recalled, he had observed Capron and Fish conferring in the flickering light of a campfire. "Their frames seemed of steel, to withstand all fatigue; they were flushed with health; in the eyes shone high resolve and fiery desire. . . . Within twelve hours they both were dead."

Chapter 14

"My Crowded Hour"

Reveille was at three o'clock the following morning. No bugles sounded, but the sergeants moved about in the darkness among the huddled forms, shaking the men out of their blankets. Shortly after daybreak, following a breakfast of rancid bacon, hardtack, and bitter coffee, the Rough Riders moved out of Siboney in a column of fours. As the dense underbrush closed in, the troopers were forced to narrow down to two, and finally to single file. A thick curtain of cactus, vines, and low trees lined both sides of the trail, hiding small, secret places covered with undulating tall grass. Soon, the underbrush was so thick that no trace of the column of Regulars on the right could be seen or heard except for a faint bugle call now and then, although they were less than a half mile away.

To Theodore Roosevelt, it seemed the perfect way to go to war. Riding near the head of the regiment, he thought the expedition more like a hunting trip than the beginning of the conquest of Cuba. At the crest of a hill, he found that the tropic sun had burned away the morning mist, and he momentarily reined in his horse to gaze across the seemingly peaceful valley that lay beneath him. Royal palms reached into the sky and here and there he saw trees covered with masses of brilliant scarlet flowers. "It seemed hard to believe we were about to go into a sharp and bloody little fight," Roosevelt later recalled.

Only the senior officers were mounted, so the men—now wryly calling themselves "Wood's Weary Walkers"—trudged uphill with heavy packs, their flannel shirts black with sweat. Land crabs scuttled about in the jungle and vultures wheeled ominously overhead. The trail led over the spur of a low mountain range and then twisted for more than a dozen miles down to the Spanish stronghold at Santiago. Every tree might conceal an enemy sniper, but the troopers were, according to Roosevelt, "filled with eager longing to show their mettle." In spite of the heat and swarms of flies and gnats that swirled about them, they laughed and joked.

"Damn!" shouted one trooper amid cries of approval. "Wouldn't a cold glass of beer taste good?"

New to the tropics and always an enthusiastic naturalist, Roosevelt was fascinated by everything he saw. He peppered Dick Davis, who was riding behind him on a mule and knew the island well, with questions about unfamiliar birds and flowers. It looked like good deer country, he observed. Indeed, it had been, Davis replied, before being devastated by war. The cooing of a brush cuckoo caught and held Roosevelt's attention. Not until later did he learn that the Spaniards imitated the bird's call to signal the Americans' approach.

The regiment kept up a brisk pace and there were few stragglers even though the cowboys were unused to marching. Toward late morning, Captain Capron sent back word he had come upon the corpse mentioned by the guerrillas. Colonel Wood ordered a halt where the trail dropped sharply into a deep ravine, and then turned upward to a long ridge topped by the crumbling ruins of a large ranch house. The officers dismounted and the men were ordered to load their carbines.

Roosevelt was telling Edward Marshall of the *New York Journal* a funny story about his employer, William Randolph Hearst, when his eye fell on some barbed wire curling out from a fence on the left side of the trail. He reached for a strand and looked it over with the eye of an experienced ranchman. "My God!" he declared. "This wire has been cut today."

"What makes you think so?" asked Marshall.

"The end is bright, and there has been enough dew, even since sunrise, to put a light rust on it."

Suddenly, off to the right the two field guns with the Regulars boomed out. Whether intended or not, these two shots were the signal for the opening of a firefight. Bullets kicked up the ground at the feet of the startled Rough Riders. Under fire for the first time, they hesitated in the open. "Deploy!" shouted Wood. "Take cover!"

Springing to life, the men darted into the dense bush. They used the butts of their carbines to bat down frantically the almost impenetrable growth. "It was like forcing the walls of a maze," reported Davis. "If each trooper had not kept in touch with the man on either hand, he would have been lost in the thicket. At one moment the underbrush seemed swarming with troopers, and the next, except that you heard the twigs breaking, and the heavy breathing of the men, or a crash as a vine pulled someone down, there was not a sign of a human being anywhere."

Roosevelt later recalled that he had no time to feel fear, but experienced a tremor of self-doubt. Wood ordered him to take three troops to the right and, if possible, link up with the Regulars on the main road. "In theory this was excellent," he recalled, "but as the jungle was very dense the first troop that deployed to the right vanished forthwith, and I never saw it again until the fight was over—having a frightful feeling meanwhile that I might be courtmartialed for losing it."

The remaining men were deployed in column to keep them from losing touch with each other. "I had an awful time trying to get into the fight and trying to do what was right when in it," Roosevelt remembered. "All the while I was thinking that I was the only man who did not know what I was about. . . ."

In a few minutes, the soldiers broke out into a small patch of high grass in the underbrush. Some threw themselves onto the ground, while others crouched in the grass desperately looking for some sign of the enemy. The Spanish were using smokeless powder and were invisible. And they had sighted in the trails over which the Americans were advancing. The only sounds were the deadly *whit-whit* and shrill *z-e-e-u-u* of Mauser bullets followed by a *chug* if a slug hit home. The enemy's fire was heavy, and their aim was low. The dense underbrush up ahead seemed to spout bullets. Within three minutes, nine men lay crumpled in the tall grass, among them Edward Marshall. Hit in the spine, it was thought the wound was fatal but he survived.

Pinned down by an invisible enemy, Roosevelt and a few officers searched the jungle with field glasses for some trace of the enemy so the Rough Riders' fire could be delivered with more effect. Suddenly, he heard a voice shouting for his attention.

"There they are, Colonel! Look over there! I can see their hats near that glade!"

Roosevelt turned to see Dick Davis, glasses to his eyes, pointing across the ravine. Looking in that direction, he spotted the hats and pointed them out to three or four marksmen, who concentrated their fire upon the indicated place. At first, there was no indication of results, but after several volleys, the Spaniards leaped from the underbrush and sought safety elsewhere.

Having at last spotted the enemy, the Rough Riders began to move forward. The advances were made in quick desperate rushes—half a troop would rise and race forward and then burrow deep in the hot grass and fire. The heat was intense, and the men discarded all their equipment except for their carbines, canteens, and cartridge belts. They caught only fleeting glimpses of the Spaniards and men were killed and wounded by an all-but-invisible enemy. Roosevelt took cover behind a palm. Just as he stuck his head out—"very fortunately," he later observed—to look around it, a bullet passed through the tree, filling his eyes and ears with tiny splinters.

By this time, the line of skirmishers had reached the place where the regiment's point had been ambushed. The grass and rocks on either side of the trail were spattered with blood, and discarded blanket rolls, haversacks, and carbines were strewn all around. Fifty feet away lay the body of Captain Capron. Beyond a turn in the trail, at the farthest point of advance, Roosevelt found Sergeant Fish, the first man to be killed.

The fight had now lasted about an hour, and the Rough Riders had fought their way into more open country, where the land sloped up toward a ruined farmhouse that was being used as a blockhouse by the Spaniards. Both Wood and Roosevelt, out of sight of each other and at opposite ends of the line, decided to drive the enemy from the building. Roosevelt picked up a carbine from a wounded man, and joined in the charge. The advance upon the ruined building was made in stubborn, short rushes, sometimes in silence, and sometimes with the troopers firing as they ran. Yelling at the top of their lungs, they swept up the hill. At almost the same moment, the First and Tenth regiments, which had also been heavily engaged, made contact with them and joined in the assault. "Come on!" shouted Fighting Joe Wheeler. "We've got the damn Yankees on the run!"

Following this skirmish, the Rough Riders camped for the next six days in a pleasant glade on the far side of the ridge they had captured, close to a clear, cool stream. The regiment's spirits were high and most newspapers had given them credit for driving off the Spaniards, all but ignoring the Regulars. Roosevelt occupied a prominent place in these accounts, and he was pleased with his performance. He had shown coolness and bravery under fire and had earned the respect of his men. Some of the newspapers were mentioning him for Congress or the governorship of New York. "Well, whatever comes I shall feel contented with having left the Navy Department to go into the army," he wrote Lodge. "For our regiment has been in the first fight on land, and has done well."

Casualties were heavy for a brief skirmish, however. In all, the attacking force lost sixteen killed and fifteen wounded; of these, eight Rough Riders were killed and thirty-four were wounded. Regular officers charged that the fight had been unnecessary because the Spaniards had already decided to withdraw. They said that the amateur soldiers, in their eagerness to press on, had been ambushed. Wood and Roosevelt were irate at this suggestion and argued that it was quite different to be attacked by an enemy known to be lying in wait and blundering into an ambush.*

Following simple ceremonies, the dead were buried in a trench beside the trail where they had fallen. There was nothing romantic about the aftermath of the fight. "The vultures were wheeling overhead by the hundreds," Roosevelt grimly reported. "They

---

*In his original dispatch to the *New York Herald,* Richard Harding Davis wrote that there had been an ambush but, undoubtedly influenced by his friends Wood and Roosevelt, he altered his views in later accounts to match theirs. See Davis, *The Cuban and Puerto Rican Campaigns,* pp. 133–134.

plucked out the eyes and tore the faces and the wounds of the dead Spaniards before we got to them, and even one of our own men who lay in the open."

The Fifth Corps was now tantalizingly close to Santiago, only about seven or eight miles away, and from the corps' forward position on El Pozo, a conical hill flanking the Camino Real, the red and blue rooftops of the city could be seen. The troops could also see men stringing barbed wire and making yellow slashes in the ground as they dug trenches along a series of hills known as the San Juan Heights. One, San Juan Hill, was topped by a blockhouse with an arching roof that looked like a Chinese pagoda, more quaint than threatening. There were other blockhouses along the Spanish lines and a stone fort at the village of El Caney on the right of the heights. From these positions, the Spaniards dominated the Camino Real and the trails leading out of the jungle into the open ground over which the Americans would have to advance.

Throughout this period, General Shafter remained on the *Seguranca,* and the army made no attempt to advance. Despite the urging of his commanders and the open criticism of the correspondents, he failed to issue orders to reconnoiter the ground ahead, to cut additional trails through the brush in preparation for an attack, or to shell the trench diggers and wire-stringing details. Without interference from American artillery, the yellow slashes grew longer and deeper and the wire entanglements more complex.

The rainy season began. Every afternoon the skies opened and rain poured down in torrents for an hour. Some of the men went about naked in an effort to keep their uniforms dry because they had received no fresh clothing since the landing. The trail back to the beachhead at Siboney was a slippery gutter of mud on which even the pack mules had trouble keeping their footing. Only a trickle of supplies reached the forward positions, and the men foraged for provisions in the pouches of dead Spanish mules.

Rumors of a stockpile of supplies on the beach reached Roosevelt, who took a detail of thirty or forty men with pack mules to see if they could obtain some for the regiment. They scrounged around and found some sacks containing about eleven hundred pounds of beans, but a commissary officer refused to let them have them. Producing a well-thumbed book of regulations, he pointed to a subsection stating that beans were to be issued only to an officers' mess. Roosevelt went away and as he later related, " 'studied on it' as Br'er Rabbit would say and came back with a request for eleven hundred pounds of beans for the officers' mess."

"Why, Colonel, your officers can't eat eleven hundred pounds of beans," the officer protested.

"You don't know what appetites my officers have," replied Roosevelt as he ordered the sacks to be loaded upon the mules.

The military bureaucrat insisted that a requisition would have to be sent off to Washington. Roosevelt responded that he didn't care as long as he could take the beans back to his regiment. As he signed the requisition amid warnings that the cost would probably be deducted from his pay, the sacks were loaded on the mules. "Oh! what a feast we had, and how we enjoyed it," Roosevelt told his family.

On the morning of June 30, General Shafter, having finally come ashore and briefly examined the enemy's positions, summoned his senior commanders to lay out his plans for an attack the next day. "There was no strategy at all," Shafter later related, "and no attempt at turning their flank. It was simply going straight at them." Reports had reached him that the Spanish garrison at Santiago, numbering some thirteen thousand men, was about to be reinforced by another eight thousand troops—actually about half that number—from Manzanillo, approximately forty-five miles away, and he was anxious to strike the enemy before they arrived.

The attack was to begin at daybreak with an assault on El Caney and its stone blockhouse by a Regular infantry division under Brigadier General H. W. Lawton, supported by artillery commanded by Captain Allyn K. Capron, Sr., father of the Rough Rider killed at Las Guasimas. In the meantime, a battery under the command of Captain George Grimes would shell enemy positions on the San Juan Heights from El Pozo, while the rest of the infantry and dismounted cavalry took up positions in the jungle in front of the San Juan Heights. As soon as Lawton's men had taken El Caney, they would join in the main frontal attack on the ridge. Shafter intended to storm the San Juan Heights, rout the enemy, and capture Santiago all in one fell swoop.

This plan was faulty in several respects. The ridges and hills, augmented by the Spanish trenches and blockhouses that anchored them, were naturally suited for defense. Once past the line of departure, the attacking troops would be on open ground, without cover and exposed to a withering fire from modern weapons. Admiral Sampson, for one, was surprised when informed of Shafter's intentions. Believing that the harbor forts, rather than Santiago itself, were the principal objectives, he had expected Shafter to advance along a railroad line that ran up the coast to within four miles of Morro Castle and then turn inland to storm the fortifications from the rear.

Before the conference broke up, there was one further bit of business. Several senior officers, including Brigadier General S. B. Young, had come down with fever, and Wood was promoted to brigadier general and given command of Young's brigade. Command of the Rough Riders passed to Theodore Roosevelt, and the regiment was Roosevelt's Rough Riders in fact as well as in the popular imagination. "I was very glad, for such experience as we had had is a quick teacher," he later noted. "By this time the men and I knew one another, and I felt able to make them do themselves justice in march or battle."

By late afternoon, the army was on the move. The confusion that ensued while some twelve thousand men jostled for position on the narrow, muddy road leading up to their jumping-off points was indescribable. It took the Rough Riders nearly eight hours to cover the three miles to their position near El Pozo. There was not much talk that night as they bedded down on the muddy ground on their ponchos. Like Shakespeare's Henry V before Agincourt, Roosevelt moved among his men, lending encouragement and checking on the sentries. "Above us," reported Dick Davis, "the tropical moon hung white and clear in the dark purple sky, pierced with millions of white stars. . . . Before the moon rose again, every sixth man who had slept in the mist that night was either killed or wounded."

For as long as he lived, Theodore Roosevelt regarded July 1, 1898, as "the great day of my life." He was up before daybreak, calmly shaved, and after having breakfast—the usual fat bacon washed down with black coffee—joined Wood on top of El Pozo to watch the brigade take its position. "It was a very lovely morning," he later recalled, "the sky of cloudless blue, while the level simmering rays from the just-risen sun brought into fine relief the splendid palms which here and there towered over the lower growth."

This reverie was broken by the dull boom of Capron's guns as they opened fire on El Caney. The noise sent clouds of birds screeching into the air and their cries resounded across four miles of jungle to El Pozo. Captain Grimes's battery was directed to create a diversion by shelling the pagodalike blockhouse on San Juan Hill. Little damage was done by the obsolescent American artillery, and the black powder supplied the gunners created thick clouds of smoke that exposed their positions to counterbattery fire. Wood remarked that he wished that for safety's sake he could move his troops elsewhere.

No sooner were these words out of his mouth than "there was a peculiar whistling, singing sound in the air, and immediately afterwards the noise of something exploding over our heads," Roosevelt noted. It was Spanish shrapnel. The two officers leaped to their horses, but a piece of metal struck Roosevelt on the wrist, hardly breaking the skin yet raising a bump as big as a hickory nut. Several men were killed or wounded before he got his regiment under the cover of the thick underbrush.

Lawton's attack on El Caney, which began at 7:00 A.M. and would be over within two hours, quickly bogged down in the face of a stubborn defense by some five hundred Spanish troops. They poured fusillades of Mauser fire into the ranks of the attacking Americans and held off a force ten times their number for eight hours without artillery support. Instead of breaking off the engagement or continuing with a small force, while moving the bulk of his division to attack the San Juan Heights, the main objective of the battle, Lawton persisted in a full-scale assault on Caney, and the fort was not taken until 4:00 P.M.

Shafter, overcome during the night by his physical exertions and gout, directed operations from his cot three miles in the rear, and issued orders through his adjutant, Colonel Edward J. McClernand. Realizing that Lawton was not going to take El

Caney on schedule or withdraw, McClernand ordered the rest of the army to get into position to attack the San Juan Heights as planned. The actual assault awaited direct orders from Shafter. McClernand later recalled giving the order to prepare for attack directly to Roosevelt, "which seemed to please him."

Infantry and dismounted cavalry plunged into a maze of underbrush and trees with only the Camino Real and a few trails running through. The Spaniards, entrenched a half mile away on the sloping hills, had previously sighted their guns along the trails and the border of the woods. Under merciless artillery and rifle fire, the Americans bunched up and the area was soon filled with dead and wounded and the torn carcasses of horses and mules. Snipers also took a heavy toll of medical corpsmen and surgeons as they worked over the wounded in makeshift dressing stations.

At the end of about a mile, the country opened up at the point where the Camino Real crossed a stream called the San Juan River. Having made their way through the underbrush, the Rough Riders halted at the stream to lend support to the Regular cavalry regiments. They took whatever cover they could find and awaited further orders. Directly above them was a knob that became known as Kettle Hill because of a huge iron kettle found there that was used in refining sugar. Behind it and to the left rose San Juan Hill and its blockhouse. Some of the men were deployed along the right side of the stream in the tall grass while others crouched under the bank of the far side.

A Signal Corps observation balloon was sent aloft to search out the enemy positions and was, in Dick Davis's angry words, "an invitation to kill everything beneath it. And the enemy responded to the invitation." For a long hour and a half the troops endured a trial by fire and intense heat. A battalion of the Seventy-first New York was caught in an iron hail of shrapnel and, in action for the first time, panicked. The men either ran away or threw themselves on the ground in terror. Their officers managed to prevent the panic from spreading, and other units cursed and heaped ridicule on the New Yorkers as they moved up to the front.

Roosevelt moved up and down the line of his regiment making certain the men had as much cover as possible. The heat was intense and many of them were already showing signs of exhaustion. Volleys from the Spanish trenches "sputtered and rattled and the bullets sang continuously like the wind through the rigging in a gale, shrapnel whined and broke, and still no order came from Shafter," Davis related. "The situation was desperate. Our troops could not retreat as the trail for two miles behind them was wedged with men. They could not remain where they were for they were being shot to pieces. . . ."

Casualties among the Rough Riders were heavy—as heavy as if the regiment had been on the attack, Roosevelt observed. And they could not return the fire. A West Point cadet named Ernest Haskell, in Cuba to obtain combat experience during his summer vacation, was hit in the stomach while talking with Roosevelt. William Saunders, the colonel's orderly, collapsed from the heat, and Roosevelt detailed another trooper to take his place. Shortly after, as Roosevelt directed him to go back and find a general— any general—to get permission to attack, the man was hit in the head and fell dying across Roosevelt's knees. Farther down the line, Bucky O'Neill strolled up and down in front of his troopers calmly smoking a cigarette. He thought officers should never take cover because it set a bad example for the men.

"Captain, a bullet is sure to hit you!" one of his sergeants warned over the din of battle.

"Sergeant, the Spanish bullet isn't made that will kill me," O'Neill turned and shouted back.

Just as he turned toward the ridge again, a Mauser bullet struck him in the mouth and tore through the back of his head. He was dead before he hit the ground.

By now it was 1:00 P.M. and Roosevelt was anxious to get his men in a position to return the enemy fire. He was about to take matters into his own hands and advance without orders when the Rough Riders were directed to support the Regular cavalry in an assault on Kettle Hill. "The instant I received my orders I sprang on my horse and then my 'crowded hour' began," he recalled. Back and forth, he galloped down the line, shouting orders to advance. Section by section, the troopers rose, moved forward in open skirmishing order, and dropped down to fire. Instantly they were followed by another section.

Roosevelt spotted a trooper lying on the ground who refused to advance. "Are you afraid to stand up when I am on horseback?" he shouted. Before the man could move, he fell flat on his face. He had been hit in the head by a bullet that went through his entire body lengthwise.

Soon, Roosevelt had ridden through his line of skirmishers and was closing in on the Ninth Cavalry. Finding its senior officers hesitant to advance without orders, he shouted: "If you don't want to go forward, let my men pass!" The regiment's junior officers and black troopers sprang into line with the Rough Riders. "I waved my hat and we went up the hill with a rush." Dick Davis described the assault in graphic terms:

> They had no glittering bayonets, they were not massed in regular array. There were a few men in advance bunched together and creeping up a steep, sunny hill, the top of which roared and flashed with flame. The men held their guns pressed against their breasts and stepped heavily as they climbed. Behind these first few, spreading out like a fan were single lines of men, slipping and scrambling in the smooth grass, moving forward with difficulty, as though they were wading waist high through water, moving slowly, carefully, with strenuous effort. It was much more wonderful than any swinging charge could have been. They walked to greet death at every step, many of them, as they advanced, sinking suddenly or pitching forward and disappearing in the high grass, but the others waded on, stubbornly, forming a thin blue line that kept creeping higher and higher up the hill. It was inevitable as the rising tide. It was a miracle of self-sacrifice, a triumph of bulldog courage, which one watched with breathless wonder. . . .

Roosevelt was at the forefront of the charge, and the blue polka-dot handkerchief he wore on the back of his hat to keep off the sun streamed straight out behind him like a guidon. "By this time," he said, "we were all in the spirit of the thing and greatly excited by the charge, the men cheering and running forward between shots" as Mauser bullets zipped through the air. Partway up the hill, they encountered a barbed-wire fence, and men frantically raced at it with knives and bayonets. Jumping off his horse, Roosevelt turned the animal loose and continued to advance on foot. The Spaniards fired a few last shots at the fast-moving wave that was sweeping over them and retreated across a wide valley to the next line of hills. Soon, the crest of Kettle Hill was swarming with Rough Riders and black troopers. Years later, they were still debating who got there first.

To the left, Roosevelt saw a line of infantry slowly advancing up San Juan Hill in the face of heavy fire from the Spanish trenches and blockhouse. He directed his men to open fire on the enemy to help out the attacking force. Suddenly, he heard "a particular drumming sound" above the cracking of the Krag carbines. "The Spanish machine guns!" one of the men cried with a touch of panic. Listening, Roosevelt determined that the firing was coming from the flat ground to the left and jumped to his feet. "It's the Gatlings, men, our Gatlings!" Cheering spread all along the line and with the support of the machine guns, the infantry swept forward and drove the enemy from their positions.

Once his men had caught their wind, Roosevelt decided to charge the next line of Spanish trenches on a spur of San Juan Hill to the front of the Rough Riders from which steady fire was coming. Believing that the regiment was following him, he leaped over a barbed-wire fence and ran forward into a little valley before the Spanish position. But the men either did not hear or see him. Roosevelt had advanced about a hundred yards when he turned around to discover that he had only five men with him and that bullets were ripping through the air all around them. Telling the men to take cover, he ran back and angrily shouted and signaled for the regiment to follow. The troopers nearest him understood.

"We didn't hear you, Colonel!" they shouted. "We didn't see you go! Lead on and we'll follow!"

They came on with a rush, white and black troopers, Rough Riders and Regulars all mixed together. Sweat ran down their faces into their eyes and they could not see the sights of their weapons, but they rose and drove ahead. Just as the Americans leaped into the enemy trenches, most of the Spaniards fled, but two men bounded up and fired at Roosevelt. As they turned to run, he emptied his revolver at them, missing the first man but killing the second. He "doubled up . . . like a jackrabbit," the Colonel told a friend.* Upon reaching the crest of the hill, the

---

*Roosevelt said the enemy trenches were filled "with dead bodies in the light blue and white uniform of the Spanish regular army. . . . Most of the fallen had little holes in their heads from which their brains were oozing." There is some question about the accuracy of his observations, however. Another witness claimed there were no trenches in this area

victors drove the yellow silk flags of the cavalry into the soft earth—and unfurled them within sight of Santiago.

The Fifth Corps had the Spaniards backed up against their last line of fortifications, only seven hundred yards away, but the American hold on the captured ridges was tenuous. Casualties had been heavy—205 killed and 1,180 wounded, or about 10 percent of those engaged—and the remainder were disorganized, hungry, and exhausted. Santiago had also been reinforced by about 3,500 men from Manzanillo. "We are within measurable distance of a terrible military disaster," Roosevelt scribbled to Lodge two days after the battle. "Tell President McKinley for Heavens' sake to send us every regiment and above all every battery possible."

As the Rough Riders dug in, he made quick tally of the regiment's strength. It had gone into action with about four hundred men: Eighty-six had been killed or wounded, six were missing, and nearly forty were down with heat prostration. Some troops were now commanded by second lieutenants and sergeants. There was no food except for what they found in the blockhouses, the water was bad, and they slept in the rain-soaked, muddy trenches without cover.

Personally, Roosevelt was proud of his own performance under fire. "For three days I have been at the extreme front of the firing line," he wrote home. "How I escaped I know not." He was pleased when General Wheeler recommended him for the Medal of Honor. While he very much wanted the decoration, "it doesn't make much difference," he said, "for nothing can take away the fact that for ten great days of its life I commanded the regiment, and led it victoriously in a hard fought battle. I never expected to come through! I am as strong as a bull moose. . . ."

In the meantime, Shafter, shaken by the army's casualties, had sent an urgent message to Admiral Sampson: "Terrible fight yesterday. . . . I urge you to make every effort immediately to force the [harbor] entrance to avoid future losses among my men, which are already very heavy. You can operate with less loss of life than

---

and tartly observed: "These trenches being then, imaginary, it is fair to argue that they were filled with imaginary, dead Spaniards." Trask, *The War with Spain in 1898*, p. 243.

I can." To Sampson's surprise, the navy was now being asked to come to the rescue of the army, which had originally been sent to Cuba to assist the navy. Early on the morning of July 3, he sailed eastward along the coast of Cuba in his flagship, the *New York,* to confer with Shafter about the grim realities of sweeping mines under enemy fire.

It was a Sunday, so the routine of another day on the blockade was broken by shipboard church services and the monthly reading of the Articles of War. Suddenly, a gun boomed out from the *Iowa* and Signal 250 was hoisted: "Enemy Ships Coming Out!" Admiral Cervera had been ordered to break out of Santiago by the authorities in Havana, who were convinced the city was about to fall, and told him to try to escape to Cienfuegos or Havana. The sortie took the American squadron by surprise; some of its ships were refueling at Guantánamo, a captured base about forty-five miles to the east, while others did not have steam up. Had Cervera made his dash for freedom at night, he might have gotten away.

The Spanish flagship, the *Infanta Maria Teresa,* smoke belching from her funnels and the crimson and gold flag of Spain snapping at her masthead, led the way with three other cruisers and two destroyers in her wake. Once out of the passage, Cervera's ships dashed at full speed along the coast to the west with the American squadron in hot pursuit. The Spanish vessels were in poor repair, however, and one by one, they came within range of the American guns. Within three hours, every enemy ship had been destroyed or beached by its crew. Cervera and his entire squadron was the U.S. Navy's Fourth of July present to the nation.

The destruction of Cervera's squadron meant that for all practical purposes the land campaign in Cuba was over, but the siege of Santiago dragged on for another two weeks. The Spaniards refused to surrender, saying military honor prevented them from doing so while they had the power to resist. Day by day, Shafter tightened his grip upon Santiago, emplacing artillery, making his trenches more formidable, gambling that he could win the city by siege before disease decimated his army. Typhoid, malaria, dysentery, and the first cases of yellow fever were already exacting a greater toll among the besiegers than Mauser bullets. Flies and mosquitoes made the men's lives miserable, their uniforms were in tatters, and the stench of rotting mule carcasses and half-buried corpses uncovered by the persistent rains was everywhere.

Through it all, Colonel Roosevelt—the rank was now official—hurled jeremiads at Shafter with his usual fury. "Not since the campaign of Crassus against the Parthians has there been so criminally incompetent a general as Shafter. . . . It is bitter to see the misery and suffering, and think that nothing but incompetency in administering the nation's enormous resources caused it." At the same time, Roosevelt worked tirelessly for his troopers. "He tried to feed them," Stephen Crane wrote later in the *World*. "He helped build latrines. He cursed the quartermasters and the 'dogs' on the transports, to get quinine and grub for them. Let him be a politician if he likes. He was a gentleman down there."

One day, a mud-stained figure appeared at a Red Cross supply depot in Siboney. "I have some sick men with my regiment who do not wish to go to the hospital but are unable to eat army rations," he told a Dr. Gardner, who was in charge. "Can you sell me some of the things you are issuing here?"

"Not for a million dollars, Colonel Roosevelt," Gardner replied, recognizing the visitor.

"But, Doctor, you have the things I need for my men. I think a great deal of my men. If you will not sell them, how can I get them?"

"I suppose you might ask for them, Colonel."

"Then I ask for them."

"All right. Make out a list of things you need and send for them."

"Give me some of them now. I'll take them myself."

Gardner filled a sack with rolled oats, condensed milk, rice, dried fruit, and other items and Roosevelt tossed it over his shoulder. "I'm proud of my men," he said as he walked back into the jungle.

Honor satisfied, the Spaniards finally gave up on July 17. Shafter formally accepted the surrender in the main plaza of Santiago, and the troops on the six-mile ring of entrenchments about the city cheered as the flag of Spain, which had floated over the Governor's Palace for nearly four centuries, was replaced by the Stars and Stripes. A battery fired a twenty-one-gun salute, a band crashed into "The Star-Spangled Banner," and an honor guard slapped their Krags and presented arms. Within days, another expeditionary force was sent to Puerto Rico that quickly mopped up that island's defenders.

In the meantime, Roosevelt—certified by the newspapers as one of the war's heroes—was receiving pleas from New York Republicans that he return immediately and run for governor. "You are most kind," he responded to one such communication, "but . . . I would not be willing to leave the regiment while the war is on for even so great an office as that of Governor of New York."

Nevertheless, events kept him on the front pages. Cuba had been conquered, but the Fifth Corps remained on the island, its ranks ravaged by malaria. More than four thousand men were on the sick list—Roosevelt calculated that fewer than half the six hundred Rough Riders who had landed four weeks before were fit for duty. Shafter implored the War Department to bring the men home. Secretary of War Alger, who believed the troops were infected with yellow fever and was afraid they would bring it home with them, replied that the army must remain in Cuba until the sickness had run its course.

Shafter called a meeting of his ranking officers on August 3, with Roosevelt, now an acting brigade commander, among them. They were unanimous in urging the withdrawal of troops from Cuba. To keep the men on the island much longer would cause the death of thousands. But having reached this agreement, no one was certain how the American people could be informed of the Fifth Corps' situation. Shafter proposed "some authoritative publication which would make the War Department take action before it was too late to avert the ruin of the army." Regular officers were chary, however, of risking their careers by openly offending President McKinley and Secretary Alger.

Roosevelt, who would return to civilian life and thereby could freely criticize the War Department, was the obvious choice to issue such a statement. He agreed to give an interview to the press, but Leonard Wood thought it better to send Shafter a round-robin letter, signed by all those present, which stated: "The persons responsible for preventing such a move will be responsible for the unnecessary loss of many thousands of lives." This was accompanied by an even stronger personal letter from Roosevelt to Shafter, and both were leaked to the press. "To keep us here," Roosevelt wrote, "in the opinion of every officer commanding a division or a brigade will simply involve the destruction of thousands."

McKinley and Alger were angered by the publication of the letters—in fact, the president only learned of the round robin when he read about it in the newspapers—and the War Department emitted veiled threats of a court-martial for Roosevelt. In revenge, Alger saw to it that he never got the Medal of Honor that he thought he deserved. It was, as Edith wrote later, "one of the bitterest disappointments of his life." Undoubtedly one of the contributing factors to Alger's ire was that, unknown to the public, the War Department had already planned to bring the troops home and had ordered that a camp be set up at Montauk Point, on the tip of Long Island, to receive them. Thus, when Alger issued orders three days later for the Fifth Corps to return to the United States, Theodore Roosevelt got the credit for bringing the boys home.

The transport *Miami,* carrying the Rough Riders, eased against a pier at Montauk on August 15, 1898. The scattered spectators, many of them members of the regiment that had been left behind at Tampa, searched her sides and cheered when they spotted a burly, bronzed figure in the well-worn uniform of a colonel of cavalry on the bridge with the vessel's captain. The brim of his campaign hat was turned up on the side and the sun glinted off his glasses. Roosevelt practically ran down the last few steps of the gangway, with his pistol jouncing at his side. Next came the troopers, some limping, some so weak they had to be helped along, some on stretchers. "I feel positively ashamed of my appearance when I see how badly some of my brave fellows are," Roosevelt declared. "Oh, but we have had a bully fight!"

Edith had been informed that her husband was to return in mid-August, but she did not know the exact date. As soon as he could, Roosevelt telephoned Oyster Bay to inform his wife of his arrival and asked her to come out to Montauk at once. Sagamore Hill had no telephone, so the message was relayed from the village by a boy on a bicycle. By the time Edith reached the camp, the regiment had been placed under a strict, five-day quarantine. But Roosevelt had made special arrangements to see her clandestinely. She was met by a young officer, who smuggled her husband out for the rendezvous. They had an hour together, and all that Edith would say was, "Theodore looks well but thin."

Three days before the Rough Riders had arrived at Montauk, the war with Spain was formally declared at an end—less than four months after it had begun. "It has been a splendid little war," John Hay wrote Roosevelt. Cuba had won its independence, and Guam, Puerto Rico, and the Philippines were ceded to the United States. For Roosevelt, the war had been a vindication of his physical courage and his abilities as a leader, and it thrust him into the national limelight. He found himself one of the most famous men in the nation, and everyone rushed to pay him honor. Newspapers speculated about his future, and political leaders were attracted to Montauk like iron filings to a magnet.

John Jay Chapman, essayist, poet, fellow Porcellian, and head of the tiny Independent party, was the first to make the pilgrimage. Chapman had a dream—he believed that with Roosevelt at the head of its ticket, the party could not only win the governorship, but lesser state offices as well. Roosevelt's strong sense of party loyalty and canny practicality inclined him to be noncommittal, about Chapman's offer of his party's nomination but Chapman was convinced of his approval by the Rough Rider's failure to decline it outright.

Congressman Lemuel E. Quigg, a prominent Republican and one of Tom Platt's chief lieutenants, was the next to make an appearance at Roosevelt's tent. Platt, who controlled some 700 of the 971 delegates to the state convention scheduled for Saratoga late in September, was flatly opposed to giving Roosevelt the gubernatorial nomination, but he was in a quandary. He needed someone to replace Governor Frank S. Black, whose administration had been blackened by insurance frauds and scandals in the reconstruction of the Erie Canal, and Quigg emphasized that Roosevelt was the only Republican who could win in November.

Chauncey M. Depew, president of the New York Central and a leading Republican orator, put it succinctly. If Black were renominated, he told Platt, it would be difficult to deal with the corruption issue. But if Colonel Roosevelt were the candidate, he could say with conviction that if he were elected the voters could be assured that " 'every thief will be caught and punished, and every dollar that can be found will be restored to the public treasury.' Then I will follow the colonel leading his Rough Riders up San Juan Hill and ask the band to play the 'The Star Spangled Banner.' "

Platt saw the point, but his distaste for Roosevelt had not mellowed. In addition to an ingrained animosity toward him, he thought Roosevelt "a little loose" on the issue of trusts and wanted assurances that once he was in office he would not "make war" on the Republican machine. He was also reluctant to give Roosevelt's career a further boost. "If he becomes Governor of New York, sooner or later, with his personality, he will have to be President of the United States," the old man declared. "I am afraid to start that thing going." Nevertheless, Quigg persuaded Platt to allow him to sound out Roosevelt.

Quigg wasted no time in asking the Colonel for "a plain statement" about whether or not he was interested in the governorship, and if elected, would he "make war" on Platt and his organization. Roosevelt's reply was forthright. Indeed, he would like to have the nomination, and he gave assurances that he would "not make war on Mr. Platt or anybody else if war could be avoided." As a good Republican he would make every effort to work with the organization with "the sincere hope" there might be "harmony of opinion and purpose." He reserved the right to consult with anyone he pleased, however, and "to act finally as my own judgement and conscience directed." Quigg replied that this was the answer he had expected and would immediately consult with Platt about the next move.

Other visitors came to Montauk as the Rough Riders recuperated and awaited demobilization. President McKinley reviewed the troops, and it was noted that he got out of his carriage to greet Roosevelt personally. Edith brought Alice, Ted, and Kermit for an overnight visit. Round-eyed with wonder, the boys listened to tales told by the troopers, inspected everything, and slept with their father in his tent. Pretty Alice, every inch the "Colonel's daughter" at fourteen and a half, was a hit with the younger officers.

Roosevelt was writing at his desk on September 13 when several men ducked in and asked him to come out for a brief ceremony. Blinking in the sunlight, he found the regiment drawn up in a hollow square; in the center was a rough plank table bearing a mysterious object covered by a horse blanket. On the fringe of the gathering were the black troopers of the Ninth and Tenth cavalries, doctors and nurses from nearby hospitals, and assorted visitors. Trooper Will Murphy of M Troop, who was a judge back home in the Indian Territory, stepped forward to announce that the regiment's enlisted men wished to present their Colonel with "a slight token of admiration, love and esteem. . . ."

Whipping away the blanket, Murphy revealed a two-foot-high casting of Frederic Remington's dramatic statue of the "Bronco Buster," a cowboy mounted on a rearing wild pony, one hand firmly fixed in the animal's mane, the other uplifted, grasping a rawhide quirt. It was the personification of the Rough Rider.

Plainly moved, Roosevelt's tanned face took on a deeper hue and behind the thick lenses of his glasses moisture welled up in his eyes. "I am proud of this regiment beyond measure," he declared, voice faltering. "It is primarily an American regiment, and it is American because it is composed of all the races which have made America their country. . . . To have such a gift from this peculiarly American regiment touches me more than I can say. This is something I shall hand down to my children." Roosevelt also payed tribute to the black soldiers who had been on the regiment's flanks at Las Guasimas and San Juan Hill. "Between you and the other cavalry regiments there is a tie which we trust will never be broken." And then he asked the Rough Riders to file by, one by one, so he could personally shake the hand of every man and officer.

One great adventure was over; another was about to begin.

Thee. Theodore Roosevelt, Sr. THEODORE ROOSEVELT COLLECTION,
HARVARD COLLEGE LIBRARY

Mittie. Martha Bulloch Roosevelt. THEODORE ROOSEVELT COLLECTION,
HARVARD COLLEGE LIBRARY

The house at 28 East Twentieth Street in New York City where Theodore Roosevelt was born THEODORE ROOSEVELT COLLECTION, HARVARD COLLEGE LIBRARY

"Teedie." Theodore Roosevelt's first picture, taken when he was nearly two. THEODORE ROOSEVELT COLLECTION, HARVARD COLLEGE LIBRARY

Theodore Roosevelt at the age of five THEODORE ROOSEVELT COLLECTION, HARVARD COLLEGE LIBRARY

The funeral procession of Abraham Lincoln passing the home of Cornelius V. S. Roosevelt at Union Square and Broadway on April 25, 1865. Theodore and Elliott Roosevelt are at the second-floor side window.
THEODORE ROOSEVELT COLLECTION, HARVARD COLLEGE LIBRARY

Theodore Roosevelt (left) *at the age of seventeen with his brother Elliott and sister Corinne with Edith Kermit Carow* (seated on the ground) THEODORE ROOSEVELT COLLECTION, HARVARD COLLEGE LIBRARY

A souvenir of the "tintype spree." Theodore Roosevelt and Alice Hath-away Lee during their courtship. Rose Saltonstall is with them. THEODORE ROOSEVELT COLLECTION, HARVARD COLLEGE LIBRARY

"Sunshine." Alice Lee about the time of her marriage to Theodore Roosevelt. THEODORE ROOSEVELT COLLECTION, HARVARD COLLEGE LIBRARY

Theodore Roosevelt's favorite picture of his first wife, Alice, taken when she was fourteen. This is the first general publication of this photograph. THEODORE ROOSEVELT COLLECTION, HARVARD COLLEGE LIBRARY

Ranchman. Theodore Roosevelt in the Bad Lands. THEODORE ROOSEVELT COLLECTION, HARVARD COLLEGE LIBRARY

Edith Carow about the time of her marriage to Theodore Roosevelt THEODORE ROOSEVELT COLLECTION, HARVARD COLLEGE LIBRARY

Theodore Roosevelt and his family at play (a photo taken by Edith in 1894) THEODORE ROOSEVELT COLLECTION, HARVARD COLLEGE LIBRARY

Sagamore Hill, the house at Oyster Bay THEODORE ROOSEVELT COLLECTION, HARVARD COLLEGE LIBRARY

The conquerors of San Juan Hill. Colonel Theodore Roosevelt and his Rough Riders after the battle. THEODORE ROOSEVELT COLLECTION, HARVARD COLLEGE LIBRARY

Bamie. Anna Roosevelt Cowles, Theodore Roosevelt's older sister. FRANKLIN D. ROOSEVELT LIBRARY

Cabot. Henry Cabot Lodge, Roosevelt's closest friend. A portrait by John Singer Sargent. NATIONAL PORTRAIT GALLERY—SMITHSONIAN INSTITUTION

The Roosevelt family in 1907. Left to right: *Kermit, Archie, President Roosevelt, Ethel, Edith, Quentin, and Ted. Alice had married Nicholas Longworth.* THEODORE ROOSEVELT COLLECTION, HARVARD COLLEGE LIBRARY

Return from the jungle. Exhausted from the ordeal of his South American trip and having lost considerable weight, Roosevelt manages a smile for a shipboard photographer. LIBRARY OF CONGRESS

Princess Alice, Alice Roosevelt Longworth FRANKLIN D. ROOSEVELT LIBRARY

Champion campaigner. Roosevelt greets admirers from the rear of a train. THEODORE ROOSEVELT COLLECTION, HARVARD COLLEGE LIBRARY

Theodore Roosevelt and his successor, William Howard Taft, at the White House on March 4, 1909, before Taft's inauguration THEODORE ROOSEVELT COLLECTION, HARVARD COLLEGE LIBRARY

On safari in Africa. TR with a fallen rhino. LIBRARY OF CONGRESS

TR meets with Kaiser Wilhelm II during German Army maneuvers in 1910. One of a series, the photograph is inscribed by the Kaiser with the comment "The Colonel of the Rough Riders lecturing the Chief of the German Army." THEODORE ROOSEVELT COLLECTION, HARVARD COLLEGE LIBRARY

The eternal TR. How many Americans saw him. LIBRARY OF CONGRESS

Theodore and Franklin Roosevelt (right) during a break in the Barnes libel trial in Syracuse in 1914. Recently discovered, this is the only *known photograph of them together. The man in the center is one of TR's attorneys.* FRANKLIN D. ROOSEVELT LIBRARY

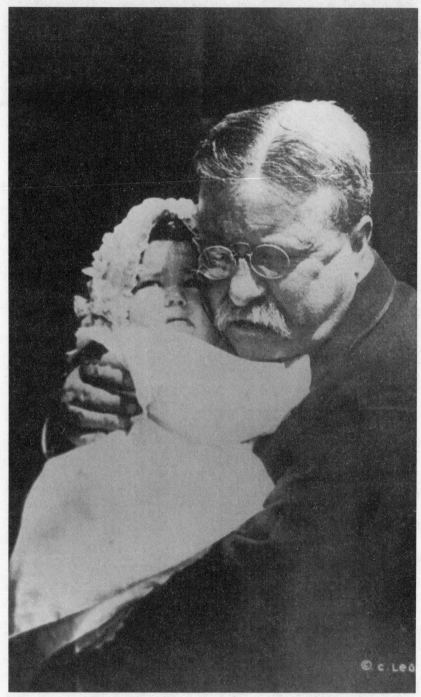

TR and his granddaughter Edith Roosevelt Darby. One of his last photographs. THEODORE ROOSEVELT COLLECTION, HARVARD COLLEGE LIBRARY

Chapter 15

"A Legacy of Work Well Done"

The Roosevelt Special pulled out of Weehawken, New Jersey, on the morning of October 17, 1898, in a gray cloud of steam. As the flag-bedecked train whistle-stopped up the Hudson Valley, Theodore Roosevelt made as many as twenty speeches a day. Small knots of farmers and their families clustered beside the tracks to greet the Republican gubernatorial nominee as he spoke from the rear platform. Perspiring crowds gathered at fairgrounds and in town squares to pump his hand. They were captivated by the flashing grin, the determination in the blue eyes squinting behind the thick spectacles, and the vigor in the staccato flood of words driven home by the steady pounding of fist on palm.

Billy O'Neil, his friend and ally of Albany days, provided the fullest appraisal of Roosevelt's impact as he crisscrossed the state. At Carthage in Jefferson County, some three thousand people waited patiently in the rain and mud to hear the Rough Rider:

He spoke for about ten minutes—the speech was nothing, but the man's presence was everything. It was electrical, magnetic—I looked in the faces of hundreds and saw only pleasure and satisfaction—when the train moved away scores of men and women

ran after [it] waving hats and handkerchiefs and cheering trying
to keep him in sight as long as possible. . . .

The Colonel, as he now preferred to be called, seemed to
O'Neil to project "sincerity 'six feet high.' " A born showman,
he understood that his audience wished to see "not an etching
but a poster" and exuded "streaks of blue, yellow, and red to
catch the eye." As the Roosevelt Special pulled into each town,
Bugler Emilio Cassi, one of the Rough Riders accompanying the
candidate, sounded the charge. "You have heard the trumpet that
sounded to bring you here," the Colonel would begin his speech.
"I have heard it tear the tropic dawn when it summoned us to
fight at Santiago!"

Roosevelt's reentry into politics had not been without considerable pain and embarrassment, however. Following his return
from Cuba, he had conducted a month-long flirtation with John
Chapman and the Independents, who had offered him the gubernatorial nomination—a flirtation that angered Tom Platt and
the Republican Old Guard. The latter insisted that if he accepted
the Republican nomination, he must run as a Republican only.
To resolve the matter, a meeting was arranged with the Easy Boss
on September 17, 1898, a few days after Roosevelt had returned
to civilian life. News of the meeting leaked out and spectators
milled about the Fifth Avenue Hotel, Platt's headquarters, to
catch a glimpse of the hero of San Juan Hill. Shortly before three
o'clock, the Colonel, somber in civilian black and gray but wearing
a jaunty military-style hat, slipped through a side door with Le-
mual Quigg. They went to a third-floor room reserved by Benjamin Odell, chairman of the Republican State Committee.

Two hours later, Roosevelt was spotted coming down the grand
staircase alone. He paused for dramatic effect and the crowd in
the lobby surged toward him.

"I had a very pleasant conversation with Senator Platt and Mr.
Odell . . . ," he began, but was cut off by an impatient reporter.

"Will you accept the nomination for Governor?"

"Of course I will!" he replied with a great show of teeth. "What
do you think I am here for?"

With these words, Roosevelt proclaimed himself both a regular
Republican and a gubernatorial candidate. Platt's laying on of

hands ended all doubt about his intentions. "We buried past differences," Platt commented later. Roosevelt had agreed to "consult with me and other party leaders about appointments and legislation in case he was elected." Not long after the meeting, Chapman received a "Dear Jack" letter. "I do not see how I can accept the independent nomination and keep in good faith with the other men on my ticket," Roosevelt declared. "It has been a thing that has worried me greatly. . . ."

The Independents, having convinced themselves of Roosevelt's commitment to their ticket, charged that he had strung them along and then dumped them after receiving the Republican nomination. Precisely because of the lofty image he cultivated, they felt betrayed. "By accepting Platt he becomes the standard bearer of corruption and demoralization," charged the *Evening Post*. "The matter is a question of honor," Chapman told his wife. Reverend Parkhurst and Carl Schurz angrily declared that he had sold out the cause of reform. The affair was so distressing, Roosevelt told Lodge, he "hardly had been able to eat or sleep" for several days. Significantly, he made no mention of his dealings with the Independents or their unhappy outcome in his *Autobiography*.

It was 1884 all over again. Then, Roosevelt shocked the mugwumps by refusing to bolt the party after Jim Blaine's nomination for the presidency. Now, they accused him of political expediency and of making a deal with Platt. As always when his motives were questioned, Roosevelt took this as a personal insult and gave as good as he got. He angrily called Parkhurst "a goose" and linked Schurz and E. L. Godkin of *The Nation* to the "idiot variety of 'Goo-Goos.' " To cover himself, he made certain his own version got on the record. An account appeared in the New York *Commercial Advertiser,* the paper on which Lincoln Steffens was a reporter, stating that the Colonel had decided to reject the Independent nomination before his meeting with Platt—an account undoubtedly planted by Roosevelt.

Had the Independents understood Roosevelt better, they would not have been led astray. While he believed he was as much a reformer as they, he also saw himself as a practical politician. Reforms should be made from within the Republican party, and he had no taste for insurgency—until 1912, that is. To make reforms, he believed, power must be matched with power, and he was willing to accept the support of Tom Platt to obtain it. If a man "goes into politics," he had written, "he must go into practical politics, in order to make his influence felt."

Edith Roosevelt observed her husband's new prominence with mixed feelings. For the first time in their twelve-year marriage, the financial clouds that had hung over them were lifting. *Scribner's Magazine* had contracted for six articles on his experiences in Cuba at $1,000 each and to print them as a book;* the Red Cross had reimbursed him for money he had spent on supplies for the regiment and he had been invited to give the Lowell Lectures at Harvard for a $1,600 fee. If Theodore were elected governor, she noted, he would be paid $10,000 a year and the family would live at state expense in the rambling old Executive Mansion at Albany.

Yet, at the same time, Edith was irritated by the avalanche of newsmen, photographers, politicians, and curiosity seekers that descended upon Sagamore Hill and intruded upon the family's privacy. Returning from a swim one day, she was shocked to find some "camera fiends" sitting on the fence taking snapshots. "It is something horrid," she declared.

Even the children were pestered by reporters. "Where is the Colonel?" one asked Archie.

"I don't know where the Colonel is," the boy replied, "but Father is taking a bath."

Roosevelt had just gotten over the charges of double-dealing hurled at him by the Independents when he faced a new crisis. Three days before the Republican convention was to open in Saratoga on September 27, newspaper headlines proclaimed he was ineligible to run for the governor's office. New York law required a gubernatorial candidate to be a resident of the state for five years preceding nomination, and the papers published copies of an affidavit filed in March 1898, while he was assistant secretary of the navy, that named Washington as his legal residence. Tammany supporters had filched the embarrassing document from the city files and slipped it to Governor Black's campaign managers.

---

*Published in 1899 as *The Rough Riders*

An unhappy Roosevelt explained that in August 1897, in reaction to the doubling of his personal property taxes in Oyster Bay, he had moved his legal address to Bamie's house in New York City. Only a few months later, the city slapped him with a heavy tax assessment and, following the advice of his uncle James Alfred Roosevelt and his cousin and attorney John Roosevelt, he had switched his domicile to Washington. But upon being told by his cousin that he would lose his right to vote under this arrangement, he had changed his mind and issued instructions for the taxes to be paid at Oyster Bay. In the muddle surrounding Edith's illness, his departure for war, and James Alfred Roosevelt's death, no one had taken care of the matter. Roosevelt produced a letter to John Roosevelt supporting his story,* but cynics snickered at his discomfiture.

Platt decided to brazen the matter out. Roosevelt was instructed to lie low at Sagamore Hill while two leading Republican attorneys, Elihu Root and Joseph Choate, were detailed to damage control. Choate, undoubtedly smarting from Roosevelt's refusal to support his own candidacy for the Senate in 1896, was "unwilling to put himself on record sustaining" the candidate's eligibility, but Root plowed ahead. He argued that Roosevelt's domicile in Washington was temporary because of his official duties and, besides, his letter to John Roosevelt plainly indicated he intended to maintain his legal residence at Oyster Bay. Had the case gone to the courts, these flimsy arguments might not have held up—but the judge and jury were the delegates to the Republican convention who wanted to back a winner.

Keeping up the appearance of being unruffled by the charges, Roosevelt remained at Sagamore rather than going to Saratoga and he declined to have a telephone installed in his house. As the convention was gaveled to order, reporters found him "wearing white flannels, calmly sauntering over the lawn with his wife, seemingly unconcerned about the doings in Saratoga." But the attack upon his integrity had cut deeply, and he anxiously awaited the parade of messenger boys from Oyster Bay with the latest news from the convention. Privately, as usual in moments of stress, he was forecasting his own imminent political demise.

---

*Someone once observed that if the mutilated remains of Roosevelt's grandmother were discovered in his cellar, he would immediately produce documents proving he was somewhere else at the time. See Einstein, *Roosevelt: His Mind in Action*, p. 104.

As soon as the Colonel's name was placed in nomination, Root strode to the platform and declared that Roosevelt wished the delegates to know the full facts of the eligibility question before casting their votes. "I mixed my argument with a lot of ballyhoo and it went over with a bang," he later declared. Even Black's supporters accepted Root's explanations and Roosevelt was nominated by a vote of 753 to 218.* Not long afterward, state Republican leaders—with one conspicuous exception—came out to Long Island to notify him formally of his nomination. Pleading an attack of arthritis, Tom Platt sent his regrets, making it clear that political power was still centered in the "Amen Corner," the niche off the lobby of the Fifth Avenue Hotel where the Easy Boss held court for suitors and supplicants.

Roosevelt was under no illusions about his chances for the governorship. "There is great enthusiasm for me, but it may prove to be mere froth, and the drift of events is against the Party in New York this year," he told Lodge. For one thing, the Democratic candidate, Judge Augustus Van Wyck of Brooklyn, matched him in probity if not in popular renown. The Republicans were also tarred with the scandals of the Black administration, the Democrats were certain to keep the spotlight on Roosevelt's inept effort at tax dodging, the unions considered him antilabor, and the German community had neither forgiven nor forgotten his efforts to enforce the Sunday closing law while police commissioner. The election was expected to be close, and wiseacres joked that the only certainty was that the next governor of New York would be a Dutchman.

To offset his problems on the state level, Roosevelt decided to divert the attention of the voters to the issue of expansionism. Since the end of the war with Spain, the nation had been debating whether to absorb the conquered islands. While Roosevelt advocated independence for Cuba, he believed the United States should hold on to the Philippines until its people were ready to govern themselves. Businessmen were also looking at the islands as a jumping-off point for East Asian trade. The decision to refuse

---

*Once the nomination was his, Roosevelt quietly paid back taxes of $1,005 in New York City to ensure his right to vote. The books at Oyster Bay had already been closed.

the islands' independence touched off a nasty little guerrilla war when the Filipinos, who had been fighting the Spaniards for their freedom, resisted the American takeover.

Flanked by several uniformed Rough Riders, Roosevelt opened his campaign at Carnegie Hall on October 5 with a ringing demand for annexation of the islands. "We cannot avoid facing the fact that we occupy a new place among the people of the world," he declared. "The guns of our warships in the tropic seas of the West and the remote East have wakened us to the knowledge of new duties. . . ."

Anti-imperialists such as John Chapman sneered that Roosevelt "really believes that he is the American flag." Even more important, Platt, who had opposed the war, let it be known that he thought a Rough Rider campaign would cost more votes than it would win. Roosevelt was told to remain at Sagamore Hill while party leaders took over the campaign. For a week or so, the Colonel seethed with anger while Van Wyck gained ground. What he desperately needed was a state issue on which he could take the offensive—and the Democrats unexpectedly supplied it.

Tammany boss Richard Croker refused to renominate Judge Joseph F. Daly to another term on the state supreme court because he declined to name a Tammany hack as court clerk and Roosevelt leaped upon it. From his experience in the Assembly, he knew the one thing New Yorkers would not stand for was tampering with the judiciary. Croker's action, Roosevelt observed, enabled him to "fix the contest in the public mind as one between himself and myself." All but ignoring Van Wyck, he made Croker the major issue of the campaign.

Roosevelt hastily put together a campaign train and mobilizing a handful of Rough Riders and reporters, whistle-stopped his way across the state. Glossing over the Republican scandals, he noted that Van Wyck was an honorable man, but told audiences that a vote for the Democratic candidate was a vote for Croker and Tammany control of the state. The Roosevelt Special chugged through the brilliant autumn foliage along the Hudson River and across the prosperous Mohawk Valley to the gritty industrial towns of the northern tier, as the crowds grew in size and enthusiasm.

For the first time, Roosevelt became aware of his ability to move large numbers of people. He talked of the simple virtues of "courage," an "upright judiciary," and "honesty in government." In truth, he had no program, and as he told James Bryce, the British political commentator, all he planned to do if elected was to conduct an "honest administration." He lauded the Irish and the Germans; the Catholics and the Jews; labor and capital. In Kingston, he promised a searching investigation of the canal scandals. He assured Haverstraw brickmakers and the ironworkers of Cornwall of his newfound sympathy for labor's problems. And he urged his audiences to ratify the results of the war with Spain by voting for a Republican congress—and a Republican governor.

At Port Jervis, ex-Sergeant Buck Taylor entertained the crowd. "I want to talk to you about mah Colonel," he declared. "He kept ev'y promise he made to us and he will to you. When he took us to Cuba he told us . . . we would have to lie out in the trenches with the rifle bullets climbing over us, and we done it. . . . He told us we might meet wounds and death and we done it, but he was thar in the midst of us, and when it came to the great day he led us up San Juan Hill like sheep to the slaughter and so he will lead you." This "hardly seemed a tribute to my military skill," Roosevelt observed, "but it delighted the crowd."

Edith did not join her husband on the campaign trail, but remained at Sagamore fretting over his safety. In the dozen years since the Haymarket Riot, the anarchist movement had targeted several prominent political figures and she feared something might happen to him. Empress Elizabeth of Austria was fatally stabbed that year by an anarchist, and two years earlier President Sadi Carnot of France had met a similar fate. "No sooner . . . had she [seen] him by Scylla than Charybdis loomed up," noted a sympathetic Jacob Riis. In Cuba, the danger had been open and acknowledged; now it lurked in the shadows. To keep occupied, Edith, helped by Alice and a secretary, managed her husband's voluminous correspondence. She sorted out the most important letters for his personal attention, dictated answers to others, and served as his surrogate—and did so for the rest of his political career.

Roosevelt's whirlwind swing through the state helped turn around a situation that had looked unpromising only a few weeks before. "Theodore Roosevelt has grown mightily in the public estimation since he appeared in person in the campaign," noted the *Troy Times,* which had originally been cool to his candidacy. When the votes were tallied on November 8, Roosevelt narrowly defeated Van Wyck by 17,794 votes out of some 1.3 million cast, and became governor of New York at the age of forty. "No man besides Roosevelt could have accomplished that feat in 1898," Platt observed.* Roosevelt himself put it all down to luck. "I have played it with bull luck this summer," he wrote Cecil Spring Rice. "First to get into the war; then to get out of it; then to get elected. . . ."

Traditionally, New York governors are sworn in on New Year's Day, but in 1899 the holiday fell on a Sunday, so the formal ceremony took place the following day.† The temperature in Albany was six below zero and streets were crusted with a half foot of new snow. Roosevelt and his family were taken in horse-drawn sleighs to the Capitol for the inaugural ceremony. The trumpets of the military escort froze and only drum taps sounded on the thin, still air. Edith and Alice, looking grown up in pink silk and sables, were seated on the dais in the Assembly chamber, while the smaller children were in the gallery with Mame as their father took the oath as New York's thirty-sixth governor. Upon his arrival at the rostrum, he blew them a kiss.

The new governor's first annual message was a typically Rooseveltian performance. He supported the rights of labor to organize, an improved civil service law, governmental economy, tax reform, reform of the municipal government in New York City, and biennial sessions of the legislature. He also appealed for practical morality and manly virtue in government. "It is absolutely impossible for a Republic long to endure if it becomes either corrupt or cowardly," he declared. Edith described it as "a most solemn and impressive ceremony. I could not look at Theodore or even listen closely or I should have broken down."

Once the swearing in was over, the Roosevelts braced them-

---

*The victory was assisted by a last-minute transfusion of $60,000 in campaign funds that Platt had wrung out of Wall Street, reputedly including $10,000 from J. P. Morgan.
†There was a small private ceremony on Saturday, December 31, 1898.

selves for the ordeal of receiving about six thousand well-wishers who had been lining up at the doors of the Executive Mansion since early morning. Smiling yet reserved, Edith carried a bouquet of flowers, making it clear she intended to avoid the tiring and painful ritual of shaking hands; her husband shook the hand of each guest with evident gusto. A select group of relatives and friends was invited to stay on for tea. Already forecasting doom, Roosevelt told Fanny Smith that with the governorship he had "shot his last bolt" but would be satisfied if he could leave the children "a legacy of work well done."

Roosevelt proved to be a good governor—in his own estimate "better either than Cleveland or Tilden"—and in Albany he honed the skills that would become useful to him in the White House. As chief executive of the most important state in the Union, he had to deal with the complex problems of an urban-industrial society and exhibited a mixture of imagination, pragmatism, and shrewdness. In the speeches and legislation of this period lay the roots and objectives of the Square Deal. Under Roosevelt, New York assumed some of the early characteristics of a progressive state government. Fearing radicalism on the one hand and the excesses of the great corporations and trusts on the other, Roosevelt saw himself as a mediator, or honest broker, between these contending forces who had the interests of all Americans in mind.

Roosevelt's first great challenge, however, was to declare his independence of Tom Platt, while working with him until he had time to create a political base of his own. Not long after he assumed the governorship, a magazine cartoon summed up the relationship between the two men. Both are shown grinning broadly and shaking hands—but each holds a well-sharpened dagger behind his back. It was expected that they would quickly fall upon each other, but the clash occurred far sooner than anyone expected.

Among Roosevelt's major campaign promises had been to conduct a thorough investigation of the canal scandals, and to appoint a new superintendent of public works. But even before he had settled into office, Platt not only named his own man to the post, but handed Roosevelt his telegram of acceptance. Such arrogance was unacceptable. "It was necessary to have it understood at the outset that the Administration was my Administration and no one else's but mine," Roosevelt declared. "So I told the Senator very politely that I was sorry but that I could not appoint his man. This produced an explosion. . . ."

While Platt looked on maliciously, the governor had difficulty in finding a first-class man for the post because no one dared to risk the Boss's displeasure. In the end, Roosevelt drew up a list of four acceptable candidates and suggested that Platt select the one he preferred. The man chosen, Colonel John N. Partridge, an able engineer and administrator, was endorsed by both machine and antimachine elements in his home borough of Brooklyn, and turned out to be an excellent selection. Both men could claim victory, and the affair made it clear that while Roosevelt would not be Platt's captive, he did not intend to be a foe.

Throughout the remainder of his term, Roosevelt usually followed the same procedure of submitting lists of competent candidates to Platt when it came to major appointments. The result was a minimum of friction with the Easy Boss and a generally high level of appointees. Cannily, Roosevelt usually allowed the machine's candidate to prevail when it came to minor offices so as to reduce strife and to maintain party unity and morale. These were compromises, not surrenders. As a result, Roosevelt obtained party support for many of the reform laws that were passed during his governorship.*

Roosevelt and Platt developed a prickly cordiality. On the one hand, Roosevelt observed that Platt "did not use his political position to advance his private fortunes—therein differing absolutely from many other political bosses." For his part, Platt acknowledged that Roosevelt "religiously fulfilled" the pledge he had made to consult with him on appointments, although, he added bitingly, he then "did just what he pleased."

Usually they met over breakfast on Saturday mornings in the

---

*Roosevelt avoided nepotism and cronyism in all but two cases: Joe Murray, his original political benefactor, was named deputy superintendent of public buildings, and Avery D. Andrews, his former colleague on the Police Board and a West Point graduate, was appointed adjutant general. In the two years that he was governor, he took more than $1,000 from his own pocket and gave it to needy acquaintances rather than put them on the state payroll.

Amen Corner of the Fifth Avenue Hotel. These visits were vigorously criticized by reformers as additional evidence of Roosevelt's subservience to Platt, but he shrugged off the criticism. Had
they taken the trouble to check, he said, "they would have seen
that any series of breakfasts with Platt always meant that I was
going to do something he did not like, and that I was trying,
courteously and frankly, to reconcile him to it. My object was to
make it as easy as possible for him to come with me. As long as
there was no clash between us there was no object in my seeing
him; it was only when the clash came or was imminent that I went
to see him. A series of breakfasts was always the prelude to some
active warfare."

While the new governor was growing accustomed to his office,
Edith set about settling the family in at Albany. First, she saw to
the children's education. Alice and Ethel were placed in the hands
of a proper English governess, a Miss Young; Ted and Kermit
were enrolled at the Albany Military Academy, and Archie and
Quentin remained under Mame's strict supervision. Next, she
turned to the shabby and gloomy Executive Mansion, which her
husband likened "in appearance and furnishing" to a less-than-
prosperous Chicago hotel. Some of the "unspeakable" old paintings were replaced by works done by her favorite artists, including
John LaFarge, a fellow member of Henry Adams's old circle. She
gave prominent places to the few good pieces of furniture and
reunited matching tables and chairs that had strayed from one
room to another over the years. Space was found for a schoolroom
downstairs, and a third-floor billiard room was converted into a
gymnasium.

The latter was a necessity because Roosevelt loved to eat and
his waistline was expanding accordingly. On his visits to New York
to meet Platt, the governor usually stayed with his sister Corinne
Robinson. While his wife had breakfast in her room, he joined
Corinne and her friends, where out from under Edith's eye, he
consumed huge amounts of food and cup after cup of heavily
sweetened coffee. "I'm not hungry," he would say, "but thank
God I'm greedy." With an expression of feigned guilt on his face,
he would point to the ceiling and say, "I feel Edie's stern disapproval trickling down from the third floor."

In Albany, Edith entertained often* if not lavishly, and the younger children were sometimes allowed to sit at the top of the stairs to watch the guests and listen to the music. Upon one occasion, they slipped down a drainpipe from their bedrooms in their pajamas and were discovered throwing snowballs at each other. They also kept a menagerie of pet rabbits, guinea pigs, and hamsters in the cellar, all named after real people without respect to sex. One day Archie burst in upon his parents and their guests, shouting excitedly: "Father! Father! Bishop Doane has had twins!"

Gifford Pinchot once arrived for a meeting with Roosevelt to find the Executive Mansion "under ferocious attack from a band of invisible Indians, and the Governor of the Empire State was helping a houseful of children to escape by lowering them out of the second-story window on a rope." After he had joined in "saving" the children, Pinchot was invited to the gym, where he knocked Roosevelt "off his very solid pins" in a sparring match. Once this was out of the way, they sat down to a discussion of forestry and conservation.

While the other children enjoyed life in Albany, Alice found it insufferably dull. "She cares neither for athletics nor good works," Edith wrote, "the two resources of youth in this town." Plans were discussed to send her to Miss Spence's School in New York City, but the prospect filled the girl with horror. "I had seen Miss Spence's scholars marching two by two in their daily walks, and the thought of becoming one of them shrivelled me," Alice recalled. "I practically went on strike. I said that I would not go—I said that if the family insisted, and sent me, I should do something disgraceful." Just before her departure, her parents relented.

The same thing happened when it became time for Alice to be confirmed. She had already read through the Bible—twice, in

---

*Among the guests were Lord Minto, the governor-general of Canada, whose party included a young newspaper correspondent named Winston S. Churchill. The Roosevelts thought Churchill insufferably rude because he slumped in his chair, puffed on a cigar all through dinner, and refused to rise when ladies came into the room. Alice Roosevelt Longworth in the *Saturday Evening Post,* December 4, 1965.

Richard Harding Davis agreed with this assessment. When young Churchill announced his intention of marrying Davis's ward, the actress Ethel Barrymore, he urged her to reject him because Churchill was "a cad."

fact—she had no interest in organized religion, and considered Christian dogma "sheer voodoo." When forced to attend church, she would read a book or practice her "one-sided nose wrinkle." Once, she heard a woman remark: "Poor Governor Roosevelt, his daughter has a twitch." For the most part, Alice absorbed her education by being "let loose" in the family library. Her father's only requirement was that she learn something from her reading every day and tell him what it was the next morning.

Every day in Albany, Roosevelt briskly walked the few blocks from the Executive Mansion to the Capitol and ignoring the elevator, took the 150 steps up to his office on the second floor two at a time. An intellectual in politics, he instituted the practice of seeking information and advice from college professors and experts on such subjects as taxation, canal improvements, education, labor, and conservation. Among his visitors was Woodrow Wilson, a rising professor of political science at Princeton who spent a weekend at Sagamore. Roosevelt also inaugurated weekly "cabinet" meetings attended by all the top state officials.

The most revolutionary change from past practice was Roosevelt's treatment of the press. Under his predecessors, the Capitol correspondents were tolerated as little more than an unavoidable evil. "At that time," Roosevelt observed, "neither the parties nor the public had any realization that publicity was necessary, or any adequate understanding of the dangers of the 'indivisible empire' which throve by what was done in secrecy." Having already learned to use the press both to further his career and to make the public aware of issues, Roosevelt turned the press corps into invaluable allies.

Twice daily, the governor summoned reporters for a fifteen-minute meeting in his office. Perched on the edge of his desk, he briefed them on the events of the day and his reaction to them, all spiced with the latest gossip and stories about backstage maneuvering. There was little time for questions because, loquacious as always, he took up most of the time. Sometimes he would issue a formal statement, and then, on an off-the-record basis, let the newsmen in on the background behind it. He used the press to push his agenda, float ideas, and shoot down unwanted legislation. "He has torn down the curtain that shut in the Governor and taken the public into his confidence," observed *The New York Times*.

Among the journalists Roosevelt became friendly with at this time was the humorist Finley Peter Dunne, creator of "Mr. Dooley," a genial Irish saloonkeeper-philosopher, whose observations on life and politics were popular. "I regret to state that my family and intimate friends are delighted with your review of my book," he wrote Dunne after Mr. Dooley had suggested that the title of *The Rough Riders* be changed to *Alone in Cuba*.

It was at one of Roosevelt's press conferences that he unveiled his support for the most important single piece of reform legislation of his governorship—and set the stage for a bitter clash with Boss Platt. Public agitation over large corporations and their political influence had not vanished with the defeat of William Jennings Bryan in 1896, and New Yorkers complained of a heavy tax burden while public service corporations paid no taxes on their franchises. On March 18, 1899, Roosevelt offhandedly told reporters that he looked "with favor" upon such a levy. A bill treating utility franchises as real property had been introduced two months before by Senator John Ford, a New York City Democrat, and was languishing in committee. Although Roosevelt had reservations* about certain aspects of the Ford bill, his statement was enough to jar it loose out onto the floor of the Assembly.

"I was hardly prepared for the storm of protest and anger which my proposal aroused," Roosevelt later wrote. The utilities were paying Platt to protect them against such legislation, and the Easy Boss quickly announced his opposition to it. As a compromise, he suggested that a joint legislative committee be appointed to study the whole question of tax reform and report back during the 1900 session. The governor, convinced that the Ford bill could not be passed in the face of Platt's opposition, agreed to put off the issue until the following year.

Reformers immediately pounced on this decision as kowtowing

---

*Roosevelt found particularly obnoxious a provision that taxes on utilities were to be levied by local governments rather than by the state. This, he felt, would leave the companies at the mercy of Tammany Hall.

to Platt and the corporations. Within a few weeks, Roosevelt, conscience apparently stung, reversed himself and announced that if the Ford bill were approved he would sign it. To everyone's surprise, the Senate immediately passed the bill and sent it over to the Assembly. Obviously, Platt had decided to oppose the measure—and Roosevelt as well—in the Assembly, confident that the bill could be tied up there. This seemed to be the case because it was buried in the Rules Committee, and a rival bill, rife with loopholes favorable to the corporations, appeared in its place. The Senate, for its part, refused to consider the substitute and both chambers were deadlocked.

With the legislative session due to end only a few days later, on April 28, Roosevelt lobbied individual assemblymen and launched a press campaign to bring the Ford bill to the floor for a vote. In the meantime, Platt and the lobbyists for the franchise holders threatened him with political extinction unless he retreated. Platt reminded Roosevelt that at the time his nomination was being considered, some Republican leaders had expressed concern "at what has been called your 'impulsive nature' " and thought him "a little loose on the relations between capital and labor, on trusts and combinations," but this had been overlooked. Now his stand had caused businessmen to "wonder how far the notions of Populism, as laid down in Kansas and Nebraska have taken hold upon the Republican party of the State of New York. . . ." In effect, Platt accused him of closet Bryanism.

Roosevelt's reply reflected his emerging philosophy on the issue of government and business. He argued that measures such as the franchise tax would strengthen the Republican party among the mass of Americans by convincing them it stood squarely for the people, not the corporations. "I do not believe that it is wise or safe for us as a party to take refuge in mere negation and to say that there are no evils to be corrected," he wrote. "Our attitude should be one of correcting the evil and thereby showing that . . . we Republicans hold the just balance and set ourselves as resolutely against improper corporate influence on the one hand as against [the] demagoguery and mob rule" of populism and socialism on the other.

With this, Roosevelt accepted the consequences of opposing Platt, and as the Boss later observed, "clinched his fist and gritted his teeth and drove the franchise tax law through the legislature." Informal polls on April 27 showed a majority for the bill, and the governor played his trump card. Under the rules, he could send a special emergency message to the Assembly forcing a bill out of committee, and that afternoon he took advantage of this procedure. But the Speaker, S. Fred Nixon, who was ordered by Platt to prevent the bill from coming to a vote, tore up the message without reading it to the Assembly.

A furious Roosevelt learned of this challenge early the next morning, the final day of the session. Without delay, he fired off another message to the Assembly that demanded immediate passage of the bill. And he warned that if it were not promptly read, he would come to the chamber and read it himself. The opposition faded before Roosevelt's determination and the Ford Franchise Tax Bill was approved by an overwhelming vote of 109 to 35. "Well, I suppose I have ended my political career today," Roosevelt remarked to Assemblyman Nathaniel Elsberg. "You're mistaken, Governor," Elsberg replied. "This is only the beginning."*

"All together, I am pretty well satisfied with what I have accomplished," Roosevelt wrote Lodge following the close of the legislative session. In addition to the franchise tax law, he won approval of a clutch of reform measures. Among these were a new civil service law, and legislation strengthening the state forest, fish, and game commissions, preserving the Palisades against development, and preventing the dumping of sawmill waste into the streams of the Adirondacks and Catskills. Believing that a strong public school system was the best tool for the Americanization of New York's polyglot population, he supported salary increases for teachers and signed a bill banning local-option segregation of the public schools by race. "My children sit in the same school with colored children," he declared during the debate on the measure, and said he saw nothing wrong with it.

Surprisingly enough for a man looked upon by organized labor

---

*The following month, Roosevelt called the legislature into special session to approve amendments to the Ford bill that placed the taxing authority in the hands of state rather than local officials—a move which somewhat assuaged Platt's anger.

with suspicion, some of Roosevelt's most significant accomplishments were in labor relations. He conferred with union officials more than any of his predecessors and supported bills to create an eight-hour-day law for state employees, increase the number of factory inspectors, and establish more stringent regulation of working conditions in the tenements where clothing and cigars were made. This was followed by a measure placing greater liability upon management in case of industrial accidents. The unions were angered, however, because he had ordered out the National Guard in several strikes.

Once these bills were signed into law, Roosevelt embarked on a wide-ranging speaking tour climaxed by the first reunion of the Rough Riders in Las Vegas, Nevada, at the end of June.* One speech, "The Strenuous Life," delivered at the Hamilton Club in Chicago in April, gave a title to his way of life.

The speech was an attack on materialism and the commercial spirit, a denunciation of "the doctrine of ignoble ease," and a glorification of "the doctrine of the strenuous life, the life of toil and effort, of labor and strife." Essentially, it was a defense of expansionism combined with the familiar appeal to national greatness. "We cannot avoid the responsibilities that confront us in Hawaii, Cuba, Porto Rico, and the Philippines," Roosevelt declared. "The guns that thundered off Manila and Santiago left us echoes of glory but they have also left us a legacy of duty. . . ."

But to what end? In this floodtide of oratory, Roosevelt's message was easy to overlook. He pleaded with Americans to adopt "the strenuous life" not for its own sake but so that the United States might advance the cause of civilization in backward lands. With its call to patriotism, overtones of Social Darwinism, preachments of the value of character and the need for a rule of law, this speech perfectly reflected Roosevelt's innermost beliefs and convictions.

Traveling westward, Roosevelt was embarrassed to find he was being greeted everywhere by cheering crowds as if he were a presidential candidate. In Kansas, William Allen White, the ed-

---

*Reunions were held annually until 1969, when a single member of the regiment attended. There were two other survivors, but both were too ill to attend.

itor of the *Emporia Gazette,* was working to win him the Republican nomination in 1900. Understanding full well that any challenge to President McKinley could destroy his political future, Roosevelt implored White to cease and desist and issued a statement supporting McKinley for reelection. Once Roosevelt was gone, the editor merely moved the target date forward by four years. The governor of New York "is more than a presidential possibility in 1904, he is a presidential probability," wrote White. "He is the coming American of the twentieth century."

Upon his return from the West, Roosevelt, "feeling completely tired out," settled in at Oyster Bay for "six weeks of practically solid rest." It was his first lengthy period of comparative inactivity since becoming civil service commissioner in 1889. Roosevelt's idea of "solid rest" was to embark on a biography of Oliver Cromwell. Each morning, reported William Loeb, his secretary, he came down to his study with a sheaf of notes and a few reference books, and began to dictate the text of the book with "hardly a pause." Colonel Arthur Lee, a British military attaché who was visiting Sagamore that summer, noted that Roosevelt sometimes called in another stenographer and dictated alternate paragraphs of gubernatorial correspondence and the book while he was being shaved.

Roosevelt, who regarded *Oliver Cromwell* as "a labor of love," may have been attracted to Cromwell by the resemblances between them. Both Cromwell and Roosevelt were fighting men, both in politics and in the field, and while they were sincere believers in constitutional liberty and parliamentary institutions, they took shortcuts to achieve their ends. Roosevelt's description of Cromwell's cavalry regiment at the start of the English Civil War could well have served for the Rough Riders. It was "from the beginning, utterly different from most of the Parliamentary cavalry; it was composed of his own neighbors; yeomen and small farmers, hard serious men whose grim natures were thrilled by the intense earthiness of their leader, and whom he steadily drilled into good horsemanship and swordsmanship."

Although he realized that the book had its limits, Roosevelt thought he had done a good job. "The more I have studied Cromwell, the more I have grown to admire him," he said, "and yet the more I have felt that his making himself a dictator was unnecessary and destroyed the possibility of making the effects of that particular revolution permanent." The book has endured less well than any of Roosevelt's other biographies, however. One friend, probably Elihu Root, described it as "a fine, imaginative study of Cromwell's qualifications for the governorship of New York."

Chapter 16

"That's an Acceptance Hat"

"Theodore Roosevelt," observed James Bryce, "is the hope of American politics." But the subject of the British political commentator's encomium looked to the future with uncertainty in the summer of 1899, for he was at a political crossroads. As governor of New York, he was automatically a prospect for a future Republican presidential nomination. In five of the seven campaigns since the Civil War, one or the other of the major parties had nominated a governor or former governor of the Empire State.* Obviously, he could not run against President McKinley in 1900 and was not at all sure that he could keep his name alive until 1904.

Unimpressed by the cheers he had received in the West, Roosevelt told Lodge he might be forgotten by then. "I have never yet known a hurrah to endure five years." In order to keep his name before the public, he thought about running for governor again in 1900, and then holding himself in readiness for presidential lightning to strike. What he really wanted to be was sec-

---

*Additionally, in 1872, New York City editor Horace Greeley was the Democratic candidate.

retary of war—"How I would like to have a hand in remodeling our army!"—or governor general of the Philippines. But when McKinley finally got rid of Russell Alger, Elihu Root was chosen to head the War Department, and the president chose Judge William Howard Taft of Ohio to be proconsul in the Philippines.

Shortly after Roosevelt's return from the western tour, Lodge urged his friend to run for vice president in 1900 as a stepping-stone to the White House. Initially, the idea of a return to the center of power in Washington seemed attractive, but Roosevelt quickly brushed it aside. First of all, the incumbent vice president, Garret A. Hobart, showed no signs of stepping aside. Besides, he thought the vice presidency a dead end. "I am a comparatively young man yet and I like to work," he told Lodge. "I do not like to be a figurehead. It would not entertain me to preside in the Senate." And he was influenced by his wife's financial concerns. "Even to live simply as Vice President, would be a serious strain upon me and . . . would cause Edith continual anxiety about money."*

Hobart's unexpected death on November 21, 1899, radically altered Roosevelt's prospects. Public pressure immediately began building upon him to become McKinley's running mate in 1900. But the more he considered the vice presidency, the more he was convinced it had no future. While Lodge repeatedly assured him that it would mark him as McKinley's natural successor in the White House, Roosevelt knew full well that since 1800 only one vice president, Martin Van Buren, had been elected directly to the presidency, and he resolved to run again for governor.

The story of Theodore Roosevelt's campaign against the vice presidency is the story of a man overwhelmed by the sweep of events. It also underscores the futility of planning in politics. Some authorities have doubted the sincerity of Roosevelt's efforts to avoid the nomination, but an examination of his attempts to prick the vice-presidential balloon shows that he really tried to keep from being nominated. He failed because of his popularity, especially among westerners in whose hearts he had staked a claim, and because Tom Platt, heartily sick of Roosevelt and reform,

---

*The vice presidency paid only $8,000 a year compared to the $10,000 he was getting as governor and the Roosevelts would have to supply their own housing.

wanted him out of the governorship. "I want to get rid of the bastard," the Boss told a crony. "I don't want him raising hell in my state any longer. I want to bury him."

The cold war between the two men heated up over the issue of control of large corporations during Roosevelt's second year in Albany. Unlike most Republican leaders, Roosevelt was sensitive to the public's concern over the rapid growth of trusts and monopolies. The number of trusts grew by leaps and bounds in the closing years of the nineteenth century as businessmen and industrialists both sought to restrict competition and generate profits through stock manipulation.*

The Sherman Anti-Trust Act of 1890 was designed to deal with trusts by forbidding combinations in restraint of trade, without any distinction as to "good" or "bad" trusts. Bigness not badness was the sin. Enforcement of the Sherman Act was ineffective, and it was known as the "Swiss Cheese Act" because of its numerous loopholes. When the American Sugar Refinery Company absorbed 98 percent of the industry's output in 1895, the Supreme Court saw nothing illegal in it. More often than not the Sherman Act was used to block workers from organizing unions, on grounds that they were in restraint of trade.

Republicans like Tom Platt also saw nothing illegal in these machinations and supported "the right of a man to run his own business in his own way with due respect of course to the Ten Commandments and the Penal Code." Looking for an issue to replace free silver, the Democrats denounced the Republicans, calling them defenders of the plutocracy. Although Roosevelt was convinced of the inevitability of business consolidation, he condemned corporate malpractices. Furthermore, he believed the Republicans should present sensible remedies to the trust problem to prevent more radical solutions being imposed.

"Our laws should be so drawn as to protect and encourage

---

*Trusts were usually formed by exchanging stock in the participating firms for shares in the new combination—usually at highly inflated terms. When Consolidated Steel and Wire merged with several other companies to form American Steel and Wire, every $100 share of Consolidated stock was exchanged for $350 in American shares. When American was merged into an even larger combination, these shares were exchanged for $490 in stock. Finally, when this giant was absorbed by J. P. Morgan's new U.S. Steel Corporation, the lucky owners got $564.37 in U.S. Steel stock. Thus, as soon as the insiders unloaded their shares of "watered," or overly inflated, stock on others, they had made a 450 percent profit.

corporations which do their honest duty by the public and to discriminate sharply against those organized in a spirit of mere greed, for improper speculative purpose," Roosevelt declared in his annual message for 1900. He advocated full publicity for corporation profits so the public could determine if they were excessive, the right of the state to intervene against monopoly, and the need to ensure that corporations paid their fair share of taxes. Reminiscent of his denunciation of the "wealthy criminal class" during his Assembly days, Roosevelt's criticism of the trusts was, to a man of Boss Platt's conservatism, the equivalent of waving a red flag in front of an angry bull.

Under normal circumstances, the Easy Boss might have been content with merely blocking the governor's legislative program, but Roosevelt compounded the transgression with a refusal to reappoint Louis F. Payn,* one of Platt's more tarnished henchmen, as state superintendent of insurance. Roosevelt had a keen distaste for Payn because he was responsible for the publication of the embarrassing tax affidavit during the gubernatorial campaign. He had also learned that Payn had "peculiarly intimate relations" with some of the insurance companies that he was supposed to regulate—not enough evidence to warrant criminal charges but sufficient to bar his reappointment. With the hope of bringing Platt around to his view, Roosevelt offered to appoint another organization man in Payn's place, but Platt, under pressure from the insurance companies that wanted Payn, would have none of it.

Fortunately for Roosevelt, a disgruntled stockholder in the State Trust Company in New York City provided evidence that the firm's directors, who had looted its treasury, had "loaned" Payn $435,000, although he had a yearly salary of only $7,000. Armed with this information, the governor carried the fight to the public. "When I go to war," he declared, "I try to arrange it so that all the shooting is not on one side."

Meeting with Platt and Benjamin Odell on January 20, Roosevelt gave Platt three days to accept one of a list of prospective

---

*Once accused of "voting tombstones" during an election in his bailiwick in Columbia County, Payn replied that he had cast the ballots the same way the deceased would have if alive. "We always respect a man's convictions," he said.

nominees to replace Payn. If Platt failed to do so, he would choose his own man and ram him through the State Senate. An ugly, open battle seemed in the offing. On the evening of January 23, Odell again met with the governor at the Union League Club. Neither side showed any sign of yielding. Odell declared that Platt would not retreat under any circumstances and was certain to win the fight and bring Roosevelt's political career to "a lamentable smash-up."

If that was the case, Roosevelt replied, nothing more could be gained by further talk and he got up to leave.

"You have made up your mind?" asked Odell.

"I have," Roosevelt answered.

"You know it means your ruin?"

"Well, we'll see about that," Roosevelt said as he headed toward the door.

"You understand, the fight will begin tomorrow and will be carried on to the bitter end."

"Yes," declared Roosevelt as he reached the door. "Good night."

When Roosevelt opened the door, Odell suddenly called out: "Hold on! We accept. Send in Hendrick's name.* The Senator is very sorry. He will make no further opposition!"

Looking back upon his handling of the incident, Roosevelt thought he "never saw a bluff carried more resolutely through to the final limit." And writing to a friend a few days later, he observed: "I have always been fond of the West African proverb: 'Speak softly and carry a big stick; you will go far.' "

Roosevelt had scored a hard-won victory, but he soon found the ground giving way under his feet. The big insurance companies and corporate magnates were incensed by his interference and pressed Platt to rid them of the Rough Rider. In the beginning, the governor could not understand Platt's enthusiasm for his running for the vice presidency. And then the truth dawned upon him. "All the high monied interests that make campaign contributions of large size and feel that they should have favors in return are extremely anxious to get me out of the State," he told Lodge. Informants brought word that Platt, fearing that another two years

*One of the men on Roosevelt's list

of Roosevelt would wreck the state Republican party, intended to kick him upstairs to a place on the national ticket. If he resisted, Platt would secretly support his Democratic opponent for governor.

Platt was also manufacturing sentiment for Roosevelt's nomination as vice president by planting newspaper reports that "recent events have made Governor Roosevelt the logical candidate of the party for Vice-President." In reply, Roosevelt issued a formal statement that "under no circumstances could I or would I accept the nomination for the vice-presidency." And he told Platt, "The more I have thought it over the more I have felt that I would a great deal rather be anything, say a professor of history, than vice-president." Smiling thinly, the Easy Boss brushed off these protestations. "Why Roosevelt might as well stand under Niagara Falls and try to spit the water back as to stop his nomination," he declared.

In April, Platt engineered the selection of Roosevelt as one of New York's four delegates-at-large to the Republican national convention due to open in mid-June in Philadelphia, calculating that his appearance would stampede the delegates into nominating him. Lodge warned his friend that if he went to the convention, he would be nominated and, once nominated, could not refuse to run. Roosevelt insisted, however, that he had to attend the convention to make it clear he did not want to be vice president. "I should be looked upon as rather a coward if I did not go."

Roosevelt's campaign against the vice presidency continued up to the moment when the convention was gaveled to order. He wrote his western supporters, asking them to desist in their efforts to nominate him. In April, he told reporters in Chicago that he "would rather be in private life than be Vice-President. I believe I can be of more service to my country as Governor of the State of New York." Observers noted, however, that Roosevelt's words fell far short of his previous "under no circumstances" statement. Obviously, he was becoming aware that a Shermanlike stand might result in his elimination from politics altogether. That night, an audience cheered him for fifteen minutes and chanted: "We want you, Teddy! Yes we do!"

Nevertheless, two weeks later, he went down to Washington to inform President McKinley and Mark Hanna of his objections to the vice presidency. Much to his surprise—and undoubted chagrin—

they told him he was not being considered for the office. "Teddy has been here. . . ," Secretary of State John Hay wrote a friend with amused malice. "It has been more fun than a goat. He came down with a somber resolution thrown on his strenuous brow to let McKinley and Hanna know once and for all that he would not be Vice President, and found to his stupefaction that nobody in Washington except Platt had ever dreamed of such a thing."

Both the president and his chief adviser had neither forgotten nor forgiven Roosevelt's criticism of the administration's conduct of the war with Spain and considered him too impetuous and erratic to be on the ticket. Officially, McKinley took a statesman-in-bronze neutral position, but made it clear that he did not want Roosevelt as a running mate. However, Platt and his ally, Matthew Quay, the cynical Pennsylvania boss who had an old score to settle with Hanna, were conspiring to stampede the convention for the Rough Rider.

In spite of her opposition to her husband running for vice president, Edith Roosevelt intended to accompany him to the convention. "As I have never attended a National Convention," she told Judge Alton B. Parker,* a dinner guest in Albany on a June evening not long before she was to leave for Philadelphia, "I am expecting to have a good time."

"Oh!" replied Parker, "you will have the most wonderful time of your life. . . . Just a bit late, you will see your handsome husband come in and bedlam will at once break loose, and he will receive such a demonstration of applause from the thousands of delegates and guests as no one else will receive. And being a devoted wife you will be very proud and happy. . . . You will see your husband unanimously nominated for the office of Vice-President of the Unites States. . . ."

"You disagreeable thing," interrupted Edith with mock horror. "I don't want to see him nominated for the vice-presidency."

Parker proved to be uncannily clairvoyant. On June 19, as the convention was gaveled to order by Hanna, the Republican national chairman, Roosevelt staged one of his dramatic entrances. With his jaw determinedly set and wearing his wide-brimmed

---

*The Democratic presidential candidate in 1904

Rough Rider-style hat, he strode down the aisle and took his seat in the New York delegation as the cheering crowd chanted, "We want Teddy! We want Teddy!" "Gentlemen," quipped one observer, "that's an acceptance hat."

While Roosevelt continued to deny any interest in the vice presidency, his resistance to the nomination seemed to be crumbling. He issued a statement alleging he could best help the national ticket by running for governor of New York, and appealed to his friends to respect his wishes, but observers thought the statement wishy-washy. Some professed to see a surer augury of future developments in the fact that "Buttons Bim" Bimberg, the leading supplier of campaign souvenirs, had already laid in a large stock of McKinley-Roosevelt buttons.*

The night before, Platt and Roosevelt had met for a showdown. Having suffered a fall and cracked a rib shortly before coming to Philadelphia, the old man was conducting business from a hotel bed. Roosevelt insisted once again that he did not want to be vice president and if nominated, would decline. "I can serve better as governor than as vice-president," he declared. Platt flatly told him he would not be renominated for governor and pointing to Benjamin Odell, said, "Your successor is in this room." Roosevelt replied that he would not yield to threats and that if Platt wanted war he could have it.

Stalking out angrily, Roosevelt went directly to a caucus of the New York delegation to warn them that if he were nominated for vice president, he would openly denounce Platt and his conspiracy on the floor of the convention. "There was great confusion," Roosevelt noted, "and one of Platt's lieutenants came to me and begged me not to say anything for a minute or two until he could communicate with the Senator." Faced with Roosevelt's open challenge, Platt, in seeming acquiescence, agreed to back Lieutenant Governor Timothy Woodruff as a "favorite son" for vice president and to give Roosevelt another term as governor. In exchange, Roosevelt agreed to abide by the choice of the convention and if nominated for vice president, he agreed he would accept. With that, Platt gleefully left the actual nomina-

---

*"If it isn't Roosevelt," Bimberg said, "there will be a dent in the Delaware River caused by Bim committing suicide."

tion of the Rough Rider in the capable hands of Matt Quay.

Mark Hanna was, however, still determined to keep Roosevelt off the ticket. To do this, he faced several insurmountable hurdles. While he had lined up the southern delegates against Roosevelt's nomination, Roosevelt was sweeping the rest of the country. Hanna considered Navy Secretary John D. Long or Representative Jonathan Dolliver of Iowa as candidates, but couldn't get McKinley to support anyone openly—and found it was impossible to beat somebody with nobody. "Don't you realize that there's only one life between that madman and the White House?" he raged at Roosevelt's supporters. Meanwhile, western delegates stamped up and down the hall outside Hanna's hotel room chanting, "We want Teddy! We want Teddy!"

Quay made his move on the second day of the convention. Almost offhandedly, he rose to propose a change in the rules to base the number of votes allowed each delegation on the number of Republican votes cast in that state instead of the total population. Platt's men were momentarily mystified. What did this have to do with Roosevelt's nomination? The howls of outrage from the southern delegates soon made everything clear. Republican votes were scarce in the South, and Quay's amendment would drastically reduce the power of these delegations, which were the foundation of Hanna's influence. In the solidly Democratic South, the Republican party was merely a mechanism for the loaves-and-fishes division of federal patronage.

Privately, Quay informed Hanna that he would withdraw the motion if Hanna would drop his opposition to Roosevelt's nomination. Realizing that he had been outmaneuvered, Hanna caved in that night with a statement that "since President McKinley is to be nominated without a dissenting voice, it is my judgment that Governor Roosevelt should be nominated with the same unanimity." The only thing left was to persuade Roosevelt to accept. Without his name on the ticket, he was told, the West might be lost to William Jennings Bryan,* the certain Democratic nominee, and Roosevelt reluctantly agreed to lead the fight.

---

*When the Democrats arrived in Kansas City to hold their convention, the superintendent of the city morgue, a Republican, draped a banner across the front reading WELCOME.

Everyone's eye was on Roosevelt when the convention reconvened. Once again, the hall was rocked by an outburst of cheers as the Rough Rider made his way to the rostrum to give a speech forcibly seconding the renomination of the president. It was, in effect, his own nomination speech. As he gazed out over the crowd, Roosevelt, as he later told Lodge, thought "back sixteen years when you and I sat in the Blaine convention on the beaten side, while the mugwumps foretold our ruin. . . ."

Once the president had been unanimously renominated, a hush fell over the hall. The chairman of the Iowa delegation rose to withdraw the name of favorite son Jonathan Dolliver and he placed Theodore Roosevelt in nomination for vice president. The crowd again exploded into a cheering frenzy. Roosevelt banners and placards suddenly appeared everywhere. Delegates stood on their chairs, tossed their hats in the air, waved state standards, and shouted themselves hoarse while the band swirled into "There'll Be a Hot Time in the Old Town Tonight." Looking up into the row of boxes, Roosevelt caught a glimpse of Edith and waved. "With just a little gasp of regret," noted a reporter for the *New York World,* "Mrs. Roosevelt's face broke into smiles." Roosevelt was nominated with every vote except one— his own.

"I am glad that we had our way," Tom Platt declared after the convention had adjourned. And then, realizing the enormity of this slip, he hastily added, "The people, I mean, had their way." Mark Hanna was haunted by darker thoughts. To McKinley, he wrote fervently, "Your *duty* to the Country is to *live* for four years from next March."

Theodore Roosevelt accepted the nomination philosophically. "I am completely reconciled and I believe it all for the best," he told Lodge. "I should be a conceited fool if I was discontented with the nomination when it came in such a fashion. . . . Edith is [also] becoming somewhat reconciled." For the rest of his life, he believed that the nomination had been spontaneous; to say otherwise was to admit that he had been bested by Platt and the bosses. With that, he plunged into the campaign. "I am as strong as a bull moose and you can use me up to the limit," he told Hanna. Following tradition as he had in 1896, McKinley remained

at home in Canton, Ohio, conducting a benign front-porch campaign, while Roosevelt stumped the country, taking on Bryan
directly.*

McKinley had a good word for everyone who came to visit him.
"With education and integrity, every pathway of fame and favor
is open to all of you," he told a delegation of Chicago bricklayers.
Alabama blacks were advised to "cultivate homes, make them
pure and sweet, elevate them, and other good things will follow."
And southern whites, proud of the role Confederate veterans had
played in the war with Spain, were told, "Everybody is talking
about General Wheeler, one of the bravest of the brave. . . ." In
the meantime, Roosevelt went for Bryan's jugular. "What a thorough-paced hypocrite and demagogue he is, and what a small
man!" he wrote Lodge.

Anti-imperialism replaced free silver as the Democrats' key
issue, and the problems of labor and capital took a back seat to
the annexation of the Philippines. The voters heard thousands
of speeches condemning alleged atrocities by the American
occupation forces and demanding troop withdrawals. The Republicans eagerly accepted the challenge and "Don't Haul Down
the Flag!" joined the "Full Dinner Pail" as their slogan. Reports
of gallant American conduct in Peking during the siege of foreign
legations by the Boxers, a Chinese secret society, that summer,
along with American participation in the international relief column, also created support for the nation's enlarged role in world
affairs.

The Republicans held all the trump cards and played them well.
Roosevelt proclaimed that McKinley had won the war with Spain,
made the nation respected in the world, and had brought prosperity. He campaigned so vigorously that many voters thought
the Rough Rider was the presidential candidate. He spiritedly
denounced Bryan's proposals for an inheritance tax, graduated
income tax, reduced tariff, and expanded money supply while
preaching the gospel of Republicanism and responsibility—leav-
ened with cautionary words about the trusts. Based upon his

---

*Bryan's running mate was Adlai E. Stevenson of Illinois, vice president during Grover
Cleveland's second term and grandfather of the Democratic presidential nominee in 1952
and 1956. He was all but ignored by Roosevelt.

experience as governor, he recommended publicity for corporate profits, taxation of corporations, and "the unsparing excision of all unhealthy, destructive and anti-social elements."

In all, Roosevelt gave hundreds of speeches and covered some 21,000 miles. People in many of the states heard him speak for the first time, and his gestures, the way in which his pent-up thoughts seemed almost to strangle him, his smile showing white rows of teeth, his swift flashes of humor, his fist clenched as if to strike an invisible adversary, all became familiar to millions of his countrymen. The story of the campaign is chronicled in newspaper headlines patiently clipped by Edith and pasted in her scrapbooks:

RANGE GREETS ROOSEVELT
WYOMING IS STIRRED UP
ROOSEVELT ROUSES BUTTE
ENTHUSIASM AROUSED IN UTAH

"Tis Tiddy alone that's r-runnin', and he ain't runnin', he's gallopin'," observed Mr. Dooley.

There was a poignant moment when Roosevelt's campaign train pulled into Medora in the Dakota Bad Lands. "The romance of my life began here," he declared as he gazed toward the familiar buttes that stretched gray and bleak in every direction. Sadly, he shook his head. "It does not seem right that I should come here and not stay."

The outcome of the election was never in doubt. When the campaign began, the betting was two to one on the McKinley-Roosevelt ticket; by the close it was five to one. The enterprising Hanna had no difficulty in raising a huge campaign chest, more than five times that of the Democrats. In the end, the Republicans won their greatest victory since 1872. McKinley received 292 electoral votes to 155 for Bryan and a plurality of nearly a million popular votes. Outside the Solid South, the Democrats carried only four silver-mining states and Bryan even lost his native Nebraska. The voters had ratified America's new role as world power. Roosevelt could take credit for the decisiveness of the victory—and had a leg up on the presidential nomination in 1904.

* * *

The first presidential inauguration of the twentieth century was also the centenary of the first inauguration held in Washington. On the cloudy morning of March 4, 1901, Americans everywhere were conscious of the splendid contrast between the infant republic over which Thomas Jefferson had presided and the bustling modern world power. In January, a great oil boom had begun in Texas when a gusher came in near Beaumont. A recent government report stated there were more than ten thousand automobiles in the country—although President McKinley declined to ride in one that morning out of fear of being ignominiously stalled on Pennsylvania Avenue. One family in six now had a telephone. And on the very eve of the inaugural, J. P. Morgan and Company announced the formation of a monster industrial combine to be known as the U.S. Steel Corporation with the unprecedented capitalization of $1.4 billion.

The Senate gallery was already crowded when Edith, the children, and a dozen family members took their places as Theodore Roosevelt was sworn in as vice president. In preparation for the move to Washington, Edith had rented Bellamy Storer's large house on Connecticut Avenue for $3,000 a year and was already worrying about how they would live on the vice-presidential salary of $8,000. The children nearly fell over the railing as their father entered the chamber, shoulders thrown back and muscles rippling under his tightly buttoned frock coat. "He was quiet and dignified," Edith noted. "Spoke in a low voice and yet so distinctly that not a word was lost."

Once more, Roosevelt talked of the "great privileges and great powers" of America and the responsibilities that went with them. "We belong to a young nation, already of giant strength, yet whose political strength is but a forecast of the power that is to come. We stand supreme in a continent, in a hemisphere. East and West we look across two great oceans toward the larger world life in which, whether we will or not, we must take an ever-increasing share."

Tom Platt and Governor Odell were in the audience, having, as the Easy Boss mischievously put it, come down "to see Teddy take the veil." There was another interested spectator as the inaugural parade proceeded down Pennsylvania Avenue to the

White House. Alice, now a vivacious seventeen, watching the parade from a window above Madame Payne's Manicure Shop on Fifteenth Street, noted the stark contrast between the pale, constrained president and her exuberant father. "As I looked at President McKinley," she later recalled, "I wondered, in the terminology of the insurance companies, what sort of 'risk' he was."

The duties of the vice president have been described as presiding over the Senate and inquiring each morning about the health of the president. Theodore Roosevelt's service in the active part of this role lasted only four days. The Senate dealt with a sheaf of presidential appointments and then adjourned until December. Uncertain about his knowledge of Senate procedure, the vice president kept a clerk at his side during the deliberations to prompt him. Inadvertently, he revealed his partisanship when dealing with one of his first motions. "All in favor will say Aye," he called, nodding in the direction of the Republicans. And then, amid chuckles, he turned toward the Democrats and said, "All those opposed say No."

The brevity of the session suited Roosevelt because he found it boring and exasperating to preside over debates in which he could not take part. "I shall get fearfully tired in the future . . . and of course I should like a more active position," he told Cecil Spring Rice. Frustrated and feeling unused, he regarded the vice presidency as a comedown after the governorship. "The vice president . . . is really a fifth wheel to the coach," he declared. "It is not a stepping stone to anything but oblivion."

To fill time, Roosevelt contemplated studying law again under the supervision of Supreme Court Justice Edward D. White, lecturing on American history, or writing a multivolume history of the United States. "I intend studying law with a view to seeing if I cannot go into practice as a lawyer when my term as Vice-President ends," he told Leonard Wood. "Of course I may go on in public life, but equally of course it is unlikely. . . . What I have seen of the careers of public men has given me an absolute horror of the condition of the politician whose day is past . . . and then haunts the fields of his former activity as a pale shadow of what he once was. . . ."

Returning to Sagamore, he spent the summer relaxing as best as his restless spirit would permit, surrounded by all the family except Ted, now enrolled at Groton, who would join them later. "Our cup of happiness was full," recalled his daughter Ethel, because his arrival meant "there would be expeditions and adventures." Not long after, Roosevelt noted that he was "rather ashamed to say that I am enjoying the perfect ease of my life at present. I am just living out in the country, doing nothing but ride and row with Mrs. Roosevelt, and walk and play with the children; chop trees in the afternoon, and read books by a wood fire in the evening." When Ted came home from school, they went for a week's shooting in the marshlands of Long Island. There was leisure for teaching Alice to ride bareback, for reading to Kermit and Ethel; for romps with Archie and Quentin, the latter a "small boisterous person . . . in fearful disgrace . . . having flung a block at his mother's head."

Yet, even as Roosevelt relaxed, his supporters quietly laid plans for a run for the presidency in 1904. They had to be careful, however, not to antagonize McKinley and Hanna. Campaigning for a presidential nomination is a lot like tiptoeing through a minefield; one false step can be fatal. Nevertheless, Governor Henry T. Gage of California virtually endorsed the vice president's candidacy and suggested that Roosevelt make a national tour the following year. William Allen White was also beating the drums out on the high wheatlands of Kansas. Roosevelt was "greatly astonished" at the western crowds that greeted him during his annual trip to the Rough Riders' reunion. A swing through the South was planned for the autumn during which he would speak at Tuskegee Institute, in Alabama, at the invitation of Booker T. Washington, the black educator.

At the end of the summer, both the president and the vice president embarked on tours of the country, letting themselves be seen by the American people. Roosevelt began a series of speaking engagements that took him, on September 6, to a meeting of the Vermont Fish and Game League on Isle La Motte in Lake Champlain. That same day, McKinley was winding up a two-day visit to the Pan American Exposition at Buffalo, a lavish international festival designed to dramatize progress in the Western Hemisphere. The vice president had already toured the exhibition, but the president had postponed his appearance because of his wife's poor health. Following a summer in Canton, the first lady now felt better, and he had happily brought her with him.

While Mrs. McKinley was resting at the home of John G. Milburn, the exposition's president, McKinley attended a public reception at the ornate Temple of Music. Anyone who wished could shake the hand of the president as he stood in a bower of palm trees, and hundreds of people were waiting in line for the doors to open. Worried about the lack of security, George B. Cortelyou, the crisply efficient presidential secretary, tried to persuade McKinley to call off the reception, but the president refused. "Why should I?" he asked. "No one would want to hurt me."

The day was warm and as the line flowed along to the strains of an organ playing a Bach cantata, some people wiped their brows with handkerchiefs. With only minutes to go before the doors were shut, a young Polish-American named Leon Czolgosz approached the president. His right hand was wrapped in a handkerchief as though it were injured, and his left hand was extended instead. McKinley reached for it, but they never clasped hands. Czolgosz brought up his "bandaged" right hand, pressed it almost against the president's vest, and fired two shots from a concealed revolver.

Smoke curled from the "bandage," and the president sagged into the arms of shocked officials. Shouts and screams blended into a rapid blur of violence. The assassin was wrestled to the ground and pummeled by a falling avalanche of spectators and guards. "Go easy on him, boys," pleaded the president before lapsing into unconsciousness.

Vice President Roosevelt was preparing to attend a reception on Isle La Motte when the telephone rang. He was told the president had been shot by a man identified as an anarchist, and exploratory surgery was under way.* No one yet knew the severity of McKinley's wounds, but the vice president was requested to come to Buffalo immediately. Roosevelt hurried to Burlington,

---

*Czolgosz told police he had shot the president because "I don't believe one man should have so much service and another man should have none." After McKinley's death he was quickly tried for murder, convicted, and executed.

where a special train was waiting. On the trip across the lake, someone remarked that he might at any moment become president of the United States. Roosevelt quickly rebuked the man by saying all thoughts ought to be of the stricken chief executive.

In Buffalo, Roosevelt learned that one of Czolgosz's bullets had ripped through the president's stomach, damaging the liver and pancreas before lodging in his back. The doctors were worried about gangrene, but there was no sign of infection. McKinley, who had been taken to the Milburn home, rallied following surgery and optimistic bulletins were issued about his condition. "The President is coming along splendidly," Roosevelt reported to his sister Bamie. "Awful though this crime was against the President it was a thousand-fold worse crime against this Republic and against free government all over the world."

On September 10, four days after the shooting, McKinley seemed to be out of danger, and Roosevelt was told he need not remain by the presidential bedside. In fact, aides believed the vice president's departure would help assure the nation that the crisis was over. Mark Hanna called McKinley's condition "just glorious" and went home to Cleveland. Roosevelt joined Edith and the children at a camp in the Adirondacks, near Mount Tahawus, where she had taken them in hopes that the cooler air would speed their recovery from various illnesses.

Three days later, the president unexpectedly took a turn for the worse and began drifting in and out of consciousness. "Is Mark there?" he asked several times, but Hanna had not yet returned to Buffalo. George Cortelyou dispatched a telegram to the vice president informing him of the alarming change in McKinley's condition. This was followed by messages urging him to come to Buffalo immediately. Occasionally, the stricken man regained consciousness long enough to murmur a few disconnected lines of his favorite hymn, "Nearer My God to Thee," and snatches of prayer. Hanna arrived at last and rushed to the side of his dying friend. "Mr. President, Mr. President," he called out to the pale figure. There was no sign of recognition in the glazed eyes. In his anguish, Hanna dropped all formality. "William! William!" he cried out. "Don't you know me?"

The president's breathing grew labored and finally at 2:45 in the morning of September 14, it ceased entirely. One of the attending physicians, who had been listening to McKinley's heart, straightened up and quietly announced, "The President is dead."

For twelve suspenseful hours, the nation had no president. Unaware of the crisis, Roosevelt had spent September 13 with friends climbing Mount Marcy, the highest peak in the Adirondacks. The climb was exhausting, the clouds and rain so thick the climbers could not see ten feet ahead of them, and the rocks were slippery. When they reached the summit, the mist suddenly cleared, bathing them in sunshine. "Beautiful country!" Roosevelt declared as he gazed across the expanse of trees, lakes, and mountains that stretched to the horizon. "Beautiful country!" Just as quickly as it had cleared, however, the sky darkened and heavy clouds covered the vista like the sea.

On the way down, the party stopped to eat at a shelf of land where there was a little lake named Tear-of-the-Clouds. Roosevelt was just about to bite into a sandwich when he looked up and saw a guide on the trail leading up from below. "I had had a bully tramp and was looking forward to dinner with the interest only an appetite worked up in the woods gives you," he recalled. "When I saw the runner I instinctively knew he had bad news— the worst news in the world."

Upon reading Cortelyou's telegram, Roosevelt scrambled down the slope and hiked as fast as he could to the nearest telephone, located at the Tahawus Club, about a dozen miles away. Further reports confirmed the president's deteriorating condition, but curiously he did not leave for Buffalo for another four and a half hours. Why the delay? Edith supplied an answer several years later. "When the party came down from Mount Marcy, my husband came to me and he said, 'I'm not going unless I am really needed,' " she explained. " 'I have been there once and that shows how I feel. But I will not go to stand beside those people who are suffering and anxious. I am going to wait here.' "

Word was received about ten o'clock that night that McKinley was dying. One of the children, who knew the president had been shot, began to cry because he was afraid his father would be shot, too, if he became president. A buckboard was hitched up, and Roosevelt and the driver set off on a wild ride to North Creek, forty miles away, where William Loeb, his secretary, was standing by with a special train. Bouncing against rocks, and sliding in the mud, they raced down a twisting wilderness trail where a wrong turn meant a plunge over a precipice.

"Faster!" Roosevelt cried to the driver. "Faster!" The buckboard was reined in only long enough to change horses and drivers, and then they were off again. Dawn was breaking as Roosevelt clattered up to the North Creek station. He jumped down from the mud-splattered vehicle to learn that McKinley had died during the night. By the flickering light of a kerosene lamp, he read the telegram informing him he was now president of the United States.

Throughout the trip across the breadth of New York State, the normally exuberant Roosevelt kept his own counsel. Perhaps he was thinking how many major crises of his life—the death of his father, the passing of Alice Lee—had found him in similar circumstances. Newsmen mobbed the train at Albany, Utica, Rome, and Syracuse, but he remained in seclusion. Earlier, he had sent a brief telegram to Edith whose formality spoke volumes:

> PRESIDENT MCKINLEY DIED AT 2:15 THIS MORNING.
> THEODORE ROOSEVELT

The special train reached Buffalo at 1:34 P.M. Wearing a top hat that Loeb had found somewhere, Roosevelt went to the Milburn house to pay his respects to the dead president and his family. Then he was driven to the home of Ansley Wilcox to await the swearing-in. A Buffalo policeman, Anthony J. Gavin, thrust his head into the carriage window. "Mr. Roosevelt," he asked, "will you shake hands with me?" Lost in thought, Roosevelt glanced up and then brightened. "Why, hello, Tony," he replied, "I'm glad to see you." There is no record indicating where they met, but the incident was a forecast of the astonishing range of presidential acquaintances that were soon to fascinate the nation.

As the senior member of the Cabinet present,* Secretary of War Elihu Root was in charge of swearing in the new president. No one was certain of the proper procedure, so an aide was sent to the city library to consult newspaper accounts of the induction

---

*Secretary of State John Hay and Lyman J. Gates, the Treasury secretary, were unable to reach Buffalo in time for the ceremony.

of Chester Arthur following the death of President Garfield. Shortly after three o'clock, Cabinet members, a few guests, and a handful of reporters gathered in the somber shadows of the Wilcox library, amid the ghostly shapes of furniture still draped with summer dustcloths.

Roosevelt's eye was caught by a small bird perched on the windowsill, chirping and fluttering its wings, before turning his attention to Root. Voice choked with emotion, the war secretary suggested that he take the oath of office without delay. Roosevelt, his face stern and stiff, signaled his acquiescence with a slight bow and responded with words that were designed to allay national uncertainty and tensions: "The administration of the government will not falter in spite of the terrible blow. . . . It shall be my aim to continue, absolutely, unbroken, the policy of President McKinley for the peace, the prosperity, and the honor of our beloved country."

With his right arm held up in the air like a schoolboy who wished to be recognized, Roosevelt took the oath of office as twenty-sixth president of the United States. In repeating the words after Judge John R. Hazel, his high-pitched voice showed signs of nervousness at first, but by the time he reached a final "And so I swear"—a touch he added himself—the words rang out loud and clear.

Later he and Root went for a short walk. Just as they returned, a carriage rolled up to the Wilcox house. From it emerged Mark Hanna, bent with rheumatism and with his spirit crushed. William McKinley had been more than a friend and political ally, he had been his life. Long before Hanna reached the door, Roosevelt bounded down the steps, hand outstretched to greet him. For a moment, Hanna lost his composure and then, pulling himself together, said: "Mr. President, I wish you success and a prosperous administration. I trust you will command me if I can be of any service."

At forty-two, Theodore Roosevelt was now the youngest president in American history.

Chapter 17

"I Acted for the Common Well-being. . . ."

Theodore Roosevelt regarded it as an omen. As he signed letters and documents on this, his first day in the White House—September 23, 1901—he noted that it was his father's birthday. The initial shock of President McKinley's assassination had worn off, but the scars lay unhealed across the land. Roosevelt himself had just returned from his predecessor's funeral in Canton; flags everywhere flew at half-staff, and the official stationery on his desk was edged with stark, black mourning borders. Yet he felt the comforting presence of his father, invisible yet strangely palpable, at his side as he took up the most important task of his life.

Edith was at Sagamore, supervising the move of the Roosevelt brood to Washington, so the new president, unwilling to dine alone his first night in the White House, asked his sisters and their husbands to join him for dinner. Corinne noticed that her usually talkative brother was in a reflective mood. The meal was nearing its end when he asked his sisters if they remembered that this would have been their father's seventieth birthday.

"I have realized it as I signed various papers all day long, and I feel it is a good omen that I begin my duties in this house on

353

this day," he said. "I feel my father's hand on my shoulder, as if there were a special blessing over the life I am to lead here."

Coffee was served just as he finished. It was the custom at the White House in those days to present each male guest with a boutonnière along with the coffee, and as a yellow saffonia rose was placed before the president, his face suddenly flushed. "Isn't that strange!" he exclaimed. "This is the rose we all connect with our father."

Both Bamie and Corinne excitedly recalled that, indeed, they had often seen Father pruning with special care a saffonia rose-bush, which grew in the garden behind the Twentieth Street house. "He always picked one for his buttonhole from that bush, and whenever we gave him a rose, we gave him one of these," said Bamie.

"I think there is a blessing connected with this," observed the president thoughtfully.

Roosevelt eagerly took up the reins of office. Undoubtedly writing for the record—as he often did—he expressed the mournful sentiments expected of him in a letter to Lodge earlier that day. "It is a dreadful thing to come into the Presidency this way. But," he added with a typically Rooseveltian rationalization, "it would be a far worse thing to be morbid about it." He had no intention of allowing the manner in which he had inherited the office he had so long coveted to rob him of his pleasure. "Here is the task, and I have got to do it to the best of my ability; and that is all there is about it."

Roosevelt's ascendancy to power coincided with the coming of age of the United States both industrially and as a world power. In the end, he became the leader of the generation that reached its political maturity with the beginning of the new century. "I did not care a rap for the mere form and show of power," he later declared. "I care immensely for the use that could be made of the substance." But in the beginning, he had, as an "accidental president," to proceed cautiously. His first task was the restoration of public confidence in the stability of the government following McKinley's murder.

Conservatives awaited the new president's first moves with nervous trepidation. Mark Hanna, for one, was "intensely bitter" about Roosevelt's rise, according to H. H. Kohlsaat, editor of the *Chicago Times-Herald* and a Republican insider. He later told of riding on the McKinley funeral train with Hanna, who "damned Roosevelt and said, 'I told William McKinley it was a mistake to nominate that wild man at Philadelphia. I asked him if he realized what would happen if he should die. Now look, that damned cowboy is President of the United States.' "*

Businessmen were worried about Roosevelt's unpredictability. The cowboy, the impetuous Rough Rider, the boss-defying governor, was not considered altogether "safe." Even before McKinley's death, Douglas Robinson, Roosevelt's brother-in-law, urged him to assure a shaken business community that if he became president, he would follow his predecessor's policies. If "you will give . . . the feeling that things are not to be changed and that you are going to be conservative . . . it will take a weight off the public mind."

The conservative alarm was premature. Numerous elements combined to keep Roosevelt "safe." He was neither a "madman" nor a "radical." Nor was he given to making quixotic gestures despite the boisterous image that delighted the cartoonists. For all his "wild talk," as Henry Adams noted, Roosevelt was "exceedingly conservative." By temperament, he was middle-of-the-road, pragmatic, and too skillful a politician to allow dogma to get in the way of exercising power. Well aware that the vice presidents who had been elevated to the presidency by the death of their predecessors had quickly alienated their parties' leadership and been denied renomination, he had no intention of repeating the error.

In the beginning, Roosevelt spent much effort in consolidating his authority, not only within the nation but within his own party. He met with Hanna, who urged him to "go slow," and he assured the senator deferentially that he would do nothing to rock the boat. He declared that McKinley's policies were his policies, McKinley's Cabinet was his Cabinet. He would lean upon his associates for advice—men like John Hay† and Elihu Root—who

---

*There is some question about the accuracy of some of Kohlsaat's recollections, however. For example, he also stated that Woodrow Wilson visited Roosevelt in Buffalo at the time of McKinley's death. Wilson biographers make no mention of such a visit.

†Hay regarded Roosevelt as if he, Hay, were a benevolent and amused uncle. "Teddy said the other day, 'I am not going to be a slave of the tradition that forbids Presidents from seeing their friends,' " Hay reported. " 'I am going to dine with you and Henry Adams and Cabot whenever I like. But' (here the shadow of the crown sobered him a

had close ties to Wall Street. To complete the image of continuity, he kept George Cortelyou, McKinley's secretary, at his side. On such matters as the tariff and currency—in which he had little interest anyway—he took counsel with such champions of the *status quo* as Senators Nelson W. Aldrich of Rhode Island,* Orville H. Platt of Connecticut, William B. Allison of Iowa, and John C. Spooner of Wisconsin. Roosevelt courted these representatives of the conservative oligarchy, known as the "Senate Four," out of necessity, yet tried to remain true to his character as a moderate reformer. Aware of this conflict, he told Chauncey Depew: "*How* I wish I wish I *wasn't* a reformer, oh, Senator! But I suppose I must live up to my part, like the Negro minstrel who blacked himself all over!"

With these assurances of Roosevelt's good behavior in hand, conservatives breathed easier. "Businessmen have confidence in the administration of President Roosevelt," the *Boston Herald* declared, and added that "the President has proved his conservatism." For his part, Hanna rumbled approval: "Mr. Roosevelt is an entirely different man today from what he was a few weeks since. He has now acquired all that is needed to round out his character—equipoise and conservatism." But this did not mean that he would support Roosevelt for the Republican presidential nomination in 1904. He had that honor in mind for himself.

Hanna may have seen Roosevelt as merely a White House caretaker, but Roosevelt's concept of the presidency was too active and vigorous for him to play such a subordinate role for long. Unlike the complaisant McKinley and all his predecessors since Abraham Lincoln who had abdicated leadership to Congress, Roosevelt believed that the president should be the ultimate authority in government. Upon becoming president, Roosevelt set three goals for himself: to become the preeminent leader of the Republican party in order to ensure his election to the presidency in his own right in 1904; to transform the presidency into the most important position in the federal government; and to make the

---

little) 'of course I must preserve the prerogative of the initiative.' " *Letters of Henry Adams,* Vol. II, p. 367.

*Aldrich had married the daughter of oil tycoon John D. Rockefeller and was the grandfather of Vice President Nelson Aldrich Rockefeller.

federal government the most important and decisive influence in public affairs.

"My view," he later wrote in a telling passage in his *Autobiography*, "was that every executive officer, and above all every executive officer in high position, was a steward of the people bound actively and affirmatively to do all he could for the people, and not to content himself with the negative merit of keeping his talents undamaged in a napkin. . . . I did and caused to be done many things not previously done by the President and the heads of the departments. I did not usurp power, but I did greatly broaden the use of executive power. . . . I acted for the common well-being of all our people whenever and in whatever manner was necessary, unless prevented by direct constitutional or legislative prohibition."

Roosevelt's elevation to the presidency coincided with a revival of the crusading reform spirit that, under different names, has periodically swept the nation. The "Progressive Movement," as it came to be called, had its roots in the 1880s and 1890s with the Social Gospel Movement—rural populists and urban settlement-house workers, municipal reformers and antimachine politicians—and grew stronger with each passing year. Social Gospel clergymen thundered from pulpits; muckraking* journalists exposed the corrupt link between business and politics, and reform politicians battled the Old Guard bosses. Midwestern insurgents, like Robert M. La Follette of Wisconsin and Albert B. Cummins of Iowa, both of whom captured governorships from Old Guard Republicans, and reform mayors, like Democrat Tom Johnson of Cleveland and the independent Samuel "Golden Rule" Jones, began to appear on the scene.

Popular writers quickly took up the flag of reform. In October 1902, *McClure's* magazine published Lincoln Steffens's "Tweed Days in St. Louis," the first of a series of articles called *The Shame of the Cities* that exposed municipal corruption across the nation. The following month, Ida Tarbell began her investigative report on the Standard Oil Company's ruthless road to domination of the oil business. This was followed by exposures of the meat-packing and insurance industries among others.

---

*A term coined by Roosevelt himself

In contrast to the Populists, the Progressives were generally urban, middle class, educated, and articulate. Believing that the country had grown too big and that the people had lost their voice in directing its affairs, the Progressives popularized numerous "direct democracy" electoral and legislative reforms: the direct primary to replace party caucuses in selecting candidates for office; the election of U.S. senators directly by the voters rather than by state legislatures; the initiatives through which citizens could enact laws; and the recall for removing unsatisfactory public officials.

For some Progressives, social reform and labor unions were suspect and radical, and they believed that all that was needed was clean government, democracy, and trust busting. Other Progressives championed thoroughgoing economic and social reform, from top to bottom, in American society. Whatever their disagreements, it was clear that the conservatives in both parties, and most of the leaders of big business, were their enemies. In order to cleanse politics and to guarantee that small business be free of monopolistic control, the Progressives found it necessary to regulate big business. Their target was the interlocking relationship between business and government, as exemplified by the trusts that had consolidated industrial production and distribution in the hands of a few corporations. Utilities, shipping, steel production, railroads, food processing, and other basic industries were all coming under the control of such combinations. A reformer as far back as the 1880s, a reform governor like La Follette and Cummins, Theodore Roosevelt became, in time, the first Progressive president.

Roosevelt's arrival in the White House, a name he quickly made official by executive order, was like a blast of fresh and bracing air in the fetid atmosphere of Washington. Informal, energetic, and—to the delight of reporters—as outspoken as ever, he saw dozens of people every day, listened attentively to what they had to say, and sent them on their way with machine-gun rapidity. His exuberance was infectious. "You go into Roosevelt's presence, you feel his eyes upon you, you listen to him, and you go home and wring the personality out of your clothes," said one visitor. "Every day or two," noted the *Detroit Press,* "he rattles the dry bones of precedent and causes sedate Senators and heads of departments to look over their spectacles in consternation."

Lincoln Steffens came down to Washington to see his friend and painted a vivid portrait of the new president in action. "His offices were crowded with people, mostly reformers, all day long, and the President did his work among them with little privacy and much rejoicing. He strode triumphant around among us, talking and shaking hands, dictating and signing letters, and laughing. Washington, the whole country, was in mourning, and no doubt the President felt he should hold himself down; he didn't; he tried to, but his joy showed in every word and movement. I think that he thought he was suppressing his feelings and yearned for release, which he seized when he could. . . . With his feet, his fists, his face and his free words he laughed at his luck. He laughed at the rage of Boss Platt and at the tragic disappointment of Mark Hanna. . . . And he laughed with glee at the power and place that had come to him."

Musty and run down, the White House itself was hopelessly inadequate for a family with six active children. "Edie says it's like living over the store," the president told William Allen White soon after the family moved in. A century of makeshift repairs and a hodgepodge of presidential tastes had reduced the old mansion to a gloomy shambles of smoke-darkened frescoed ceilings and heavy plush furniture. Except for a small private dining room, the entire ground floor was devoted to public rooms, and the space over the East Room on the second floor was taken up by offices. Only a glass screen separated these rooms from the family living quarters, which consisted of a large and drafty hall, six bedrooms, two antiquated bathrooms, a library, and a maid's room.

The gloom of her new home and the shadow of McKinley's assassination oppressed the new first lady. "I suppose in a short time I shall adjust myself," Edith wrote, "but the horror of it hangs over me and I am never without fear for Theodore. The secret service men follow him everywhere. I try to comfort myself with the line of the old hymn, 'Brought safely by His hand thus far, why should we now give place to fear?' "

Edith's first move was to figuratively and literally throw open the windows of the White House and let pleasant Indian summer air into the stuffy rooms. She directed the shifting of furniture, consigning horsehair monstrosities from the Hayes era to the attic, and weeding out dusty volumes in the black walnut cabinets that served as bookcases, restocking them with books that she and the president loved. Painters, carpet layers, and curtain makers trooped in to redecorate the family's private quarters. Edith filled the house with flowers, and sometimes went snooping for bargains for the private living quarters in the city's antique and junk shops.

The first lady allocated the large southwest bedroom with a dressing room and bath to herself and the president. From its windows they could see the Washington Monument, the Tidal Basin, and the Potomac. Ethel was given the room across the hall on the northwest corner facing her father's old office in the State, War and Navy Building. Alice had the room next door overlooking Lafayette Park. Kermit shared a bedroom with Ted, when the latter was home from Groton. Archie and Quentin were placed together in the room adjoining their parents' bedroom, where an eye could be kept on them. Edith appropriated the library next to the president's office as her private sitting room. There was a connecting door between the two rooms, allowing them to consult with each other between appointments.

With the exception of Alice, the children soon made themselves at home in Washington. Evading the servant assigned to watch them, Kermit and Ethel explored the streets of the capital by bicycle, while the younger children found the cavernous public rooms of the White House ideal for stilt walking, and they roller-skated in the cellars. Kermit and Archie were enrolled in a nearby public school, Ethel was sent to the private Cathedral School, and Alice was to join her there. But as usual, Alice had ideas of her own. She stubbornly refused to be enticed to Washington, and remained with Bamie at the Cowles estate at Farmington, Connecticut, for several months. None of Alice's social set did anything so public as hold office, and she considered all the publicity surrounding her father "rather vulgar."

Slowly, the shadows lifted and Edith began to enjoy her new position. The $50,000-a-year presidential salary and allowances for household expenses and upkeep had finally relieved her persistent worries about finances. Because of the period of official mourning following McKinley's death, the Roosevelts limited their entertaining to teas and small dinners—attended by whoever interesting happened to be in Washington at the time. One of these dinners, at which Booker T. Washington became the first black to dine at the White House, plunged Roosevelt into the first controversy of his presidency.

The president had known Booker Washington for some time and like many prominent whites, was attracted to the black educator because he preached an evolutionary policy that called upon blacks to improve their condition through self-help rather than by agitation or violence.* Washington was well acquainted with southerners of both races and he was a good judge of men, so Roosevelt frequently called upon him for advice in selecting candidates, both black and white, to fill federal positions in the South as part of his campaign to wrest political control from Mark Hanna. For his part, Washington felt that Roosevelt "wanted to help not only the Negro, but the whole South."

To ensure his nomination in 1904, Roosevelt began building up a party apparatus loyal to him, especially in the South, where many Republican jobholders were venal and owed their political allegiance to Hanna. In the years since Reconstruction, the Republican party had abdicated its championship of the rights of blacks and existed in the South merely as a means of distributing federal patronage. "In the South Atlantic and Gulf States there has really been no Republican Party," the president told Lodge. There was "simply a set of black and white scalawags . . . who are concerned purely in getting the Federal offices and sending to the national convention delegations whose venality makes them a menace to the whole party."

"If I can't find [good] Republicans I am going to appoint Democrats," Roosevelt told a shocked group of Republican congressmen soon after moving into the White House. True to his word, he made Thomas Goode Jones, a Cleveland Democrat and onetime Confederate officer, a federal judge in Alabama, and followed it up with similarly controversial appointments

---

*Washington's views were being challenged by the younger and more militant W.E.B. Du Bois, who argued that it was folly to believe that blacks could win economic security without political power or achieve self-respect as long as they acquiesced in their own inferiority.

in other states. Mr. Dooley proposed "an emergency hospital f'r office-holders an' politicians acrost th' sthreet fr'm the White House" in the wake of the president's actions. Gradually, however, Roosevelt managed to shift the balance of power away from the Old Guard to a party structure that was loyal to him.

While still vice president, Roosevelt had planned a visit to Booker Washington's vocational school at Tuskegee in Alabama, but had postponed it following the death of McKinley. Instead, he urged Washington to come to the White House "as soon as possible" in order "to talk over the question of possible future appointments in the south exactly on the lines of our last conversation together." This meeting took place on October 1, 1901, but the president wanted to discuss further the patronage situation, and when Washington returned to the capital on October 16, he was invited to dinner at the White House that evening. "We talked a considerable length concerning plans about the South," Washington later recalled.

Efforts were made to avoid publicity, but news that a black man had dined with Roosevelt and his wife found its way into the newspapers when a reporter noticed Washington's name in a routine list of official callers. These reports touched off a wave of protest from southern whites, who accused the president of encouraging racial mixing and social equality for blacks. The *Memphis Scimitar* termed the dinner "the most damnable outrage ever," and a Mississippi demagogue, James K. Vardaman, claimed the White House had become "so saturated with the odor of the nigger that the rats have taken refuge in the stable." Blacks, on the other hand, sized up the dinner as a fragment of hope amid a rising tide of discrimination.

Roosevelt was surprised and embarrassed by the hostile southern reaction and attributed it to the irrationality of the South on the question of race. "I had no thought whatever of anything save of having a chance of showing some little respect to a man whom I cordially esteem as a good citizen and good American," he explained to a friend. "The outburst of feeling in the South about it is to me literally inexplicable. It does not anger me. As far as I am personally concerned I regard their attacks with the most contemptuous indifference, but I am very melancholy that such feeling should exist in such bitterly aggravated form in any part of the country."

But Roosevelt made no effort to defend his action in public, obviously hoping the affair would blow over without affecting his southern strategy. The Democrats, on the other hand, made good use of it to prevent defections to the Republicans. A few weeks later, the president invited Finley Peter Dunne to dinner with the rueful observation, ". . . You need *not* black your face." While he continued to consult Booker Washington on patronage matters and invited black officials to White House receptions, he never again issued a dinner invitation to Washington or any other black, nor did he mention the incident in his *Autobiography*. The president was secure enough in his own position to have a black as a guest, but the politician in him quickly grasped the tactlessness of offending the prejudices of a large section of the country.*

Roosevelt's personal attitude toward blacks was contradictory, a result of the constant war between his social consciousness and sense of noblesse oblige on the one hand, and his politician's sense of the immensity of the problem and poor prospects for success on the other. Like most white Americans of the day, he accepted the view that as a group blacks, in terms of education and social attainments, were inferior to whites, but felt that discrimination, with its attendant lynchings, antiblack riots, and poverty, was morally wrong. And his belief in biological evolution convinced him that if the environment in which blacks lived was improved, they would be uplifted. He also conceded that there were blacks, among them Booker Washington, who were superior to individual whites, and he felt strongly that they should have full opportunity to prove their merit. In summary, Roosevelt was no bigot, but he took a paternalistic view of blacks and their

---

*Not long afterward, Washington sent the president a clipping from the *Baltimore Herald,* which he said recounted a true story of his encounter with an elderly southern colonel.

"Suh, I am glad to meet you," the colonel said. "Always wanted to shake your hand, suh. I think, suh, you're the greatest man in America."

Washington modestly replied that he thought President Roosevelt was the greatest man in America.

"No suh!" roared the old man. "Not by a jugful: I used to think so, but since he invited you to dinner, I think he's a———scoundrel."

The president was vastly amused by the story. "I think that it is one of the most delightful things I have ever read," he told Washington. "It is almost too good to believe."

problems rather than leading a fight for the rights due them as American citizens. Even if he had, however, there was little he could have accomplished in the area of civil rights given the rulings of the Supreme Court since Reconstruction.

On the evening of February 18, 1902, J. P. Morgan was enjoying dinner with several favored associates from his great banking house. Fierce and forbidding, he was regarded as the "Jupiter" of finance and hurled his thunderbolts from an office at 23 Wall Street. Everything about him signified solidity and permanence: his bulky figure, his plug hat, and the heavy stick he brandished at unwary photographers. Most remarkable was his nose, a huge, glaring, ruby-colored deformity that commanded attention. Upon one occasion, he wryly commented that it was "part of the American business structure." Morgan's forte lay in the creation of enormously powerful industrial combinations, such as U.S. Steel, General Electric, and International Harvester, to ease competition and to increase profits for the insiders.

The day had been pleasantly uneventful. The market had closed higher and the latest example of the old man's organizational genius, the Northern Securities Company, a giant holding company that controlled the major western railroads, was proceeding successfully. Midway through dinner the telephone rang with an urgent call. A newspaperman was on the line with a bulletin from Washington: President Roosevelt had ordered the Justice Department to proceed against Northern Securities for violating the Sherman Anti-Trust Act, on grounds that it was in restraint of trade.

Shocked, Morgan returned to the table. He should have been warned, he told his guests. He should have had a chance to fix things up, or at least to dissolve the combination voluntarily. To attack him in this fashion was not only wrong but ungentlemanly. After all, he and Roosevelt were from the same class. Hadn't he and the president's father both been founders of the American Museum of Natural History? Hadn't they sat on the same boards together? The attitude of James J. Hill, Morgan's ally in organizing Northern Securities, was blunter. "It really seems hard," the railway magnate declared, "that we should be compelled to fight for our lives against the political adventurers who have never done anything but pose and draw a salary."

The stock market immediately plunged on the news of the suit against Northern Securities, but Roosevelt's action was popular with ordinary Americans. Public outrage over the latest of Morgan's monopolies was widespread. "Wall Street is paralyzed at the thought that a president of the United States would sink so low as to try and enforce the law," said one newspaper, mocking the gloom of the financial community. "The moral effect of this injunction is in the highest degree important," observed the *Literary Digest* with approval. "It will encourage other efforts to restrain the greed of combinations."

Had Morgan and his fellow Wall Streeters understood Roosevelt better, they would not have been shocked by the suit. Although he had promised to follow in McKinley's footsteps, he had no sympathy for speculators and financial manipulators, whom he was later to call "malefactors of great wealth."* "Of all forms of tyranny," he declared in his *Autobiography,* "the least attractive and the most vulgar is the tyranny of mere wealth." He was opposed to the trusts on grounds of both political faith and personal ambition. He believed it was dangerous to allow these enormous economic powers to remain unresponsive to the people, while at the same time he reinforced his reputation as a forceful leader by making war on them.

More sophisticated than most Progressives, Roosevelt recognized that the large corporation was a fact of modern economic life and that a general rise in the standard of living depended far more on increased productivity than on radical redistribution of existing wealth. Rather than follow the common Progressive line that by their nature all trusts were inherently evil, he divided them into "good" and "bad" trusts. If a trust was charging fair prices and offering good service, he favored allowing it to continue. If it was restraining trade and jacking up prices, then it was evil. The line was to be drawn on conduct, not bigness. Roosevelt and Roosevelt alone decided which trust was good and which was bad, however. Northern Securities, he decided, was bad—and should be broken up. Roosevelt also advocated selective prosecutions because the government did not have enough lawyers to prosecute all monopolies and because he realized that the con-

---

*A phrase first used in a speech at Provincetown, Massachusetts, on August 20, 1907

servatism of the courts would make all prosecutions difficult.*

Upon taking office, Roosevelt had been irritated, alarmed, and finally moved to strike against the "tyranny of mere wealth" by the arrogance of men like old Morgan, who acted as if their power were greater than that of the national government. When he became president, he declared, "the total absence of governmental control had led to a portentous growth . . . of . . . corporations. In no other country in the world was such power held by men who had gained these fortunes. . . . The power of the mighty industrial overlords of the country had increased with giant strides, while the methods of controlling them, or checking abuses . . . remained . . . practically impotent. . . ."

For those who looked carefully, Roosevelt's first annual message to Congress had contained a storm warning. Sent on December 1, 1901, it proposed a moderate program of balanced reforms, but was not as cautious as it seemed. In particular, Roosevelt drew attention to the "real and grave" evils inherent in the rapid expansion of trusts, and while noting that large corporations were a "natural" outgrowth of the process of industrialization, he called for federal government supervision and regulation of all corporations engaged in interstate commerce, especially in all cases where there was "some monopolistic element or tendency in its business." In essence, prohibition was not necessary but supervision and control were.

Specifically, Roosevelt called for the establishment of a Bureau of Corporations in a Cabinet-level Department of Commerce to determine if regulation was warranted—a program he had advocated as governor of New York. The bureau would provide the president with information on the trusts, which he could at his own discretion make public and present to the Justice Department as a basis for indictments.

The equivocal stop-and-go aspects of the message inspired one of Mr. Dooley's most acidulous comments: "The thrusts are heejous monsthers built up by th' inlightened intherprise iv th' men that have done much to advance progress in our beloved counthry,

---

*The Justice Department had five antitrust lawyers during Theodore Roosevelt's administration. In contrast, there were 190 during the New Deal years. Leuchtenberg, *Franklin D. Roosevelt and the New Deal*, p. 259.

he [Roosevelt] says. 'On wan hand I wud stamp them under fut; on the other hand, not so fast.' " Perhaps Roosevelt failed to grasp fully the problems facing the nation. More likely, however, because of his ambiguous political position in the early months of his presidency, he thought a conservative Congress would ignore anything stronger. If so, he was correct, for his proposals were given short shrift by the lawmakers.

In fact, Roosevelt had resisted efforts by Wall Street to persuade him to present an even blander program. Hanna, Root, and Attorney General Philander C. Knox recommended changes in the message, which were rebuffed. Robert Bacon, his Harvard classmate, and another Morgan partner, George W. Perkins, had been dispatched by the financier to urge him to "go slow." They argued like an "attorney for a bad case," the president told Douglas Robinson. They wanted him "to go back on my messages to the New York Legislature and on my letter of acceptance of the nomination for the Vice-Presidency . . . I intend to be most conservative, but in the interests of the big corporations themselves and above all in the interest of the country I intend to pursue, cautiously but steadily, the course to which I have been publicly committed again and again, and which I am certain is the right course."

The Northern Securities Company was an outgrowth of an attempt by two railroad barons, E. H. Harriman of the Union Pacific and James J. Hill of the Great Northern, to win a route into Chicago. The ideal connection for each was the Chicago, Burlington and Quincy Railroad. Hill, who also owned the Northern Pacific and was working with Morgan and the Rockefeller interests, purchased control of the Burlington. Not to be outdone, Harriman, with Jacob Schiff of Kuhn, Loeb as his agent, attempted to buy enough Northern Pacific shares from under Hill's nose to give him ownership of both railroads. Within a few days, the price of Northern Pacific stock soared from $100 a share to $1,000 a share—precipitating a disastrous financial panic as short sellers hurriedly dumped all their other stocks to obtain cash to cover positions in Northern Pacific.

This was too much for Jupiter Morgan. The clash between Hill and Harriman violated his sense of order and he brought the two men together to create a community of interest. The Northern

Securities Company, which resulted from this agreement, was launched on November 12, 1901, and held control of the Great Northern, the Northern Pacific, and the Burlington railroads. Harriman was given representation on the board to bring him into the deal. The new company was capitalized at $400 million, about a third of it thought to be watered stock. High-priced legal talent assured the participants that the Sherman Act was no obstacle. Since the U.S. Supreme Court had thrown out the suit against the American Sugar Refining Company in 1895—known as the Knight case—not a single antitrust indictment had been brought under the law by the Justice Department in Washington, although there were a few cases in the states.* It was used instead to break strikes.

But even Mark Hanna thought Morgan had gone too far this time. The creation of another huge holding company so soon after U.S. Steel had alarmed large numbers of Americans already frightened by the power of the trusts. The general outcry against Northern Securities provided Roosevelt with the opportunity to wield his Big Stick against the trusts. Several states had already planned or initiated suits against Northern Securities, but he preempted them with dramatic abruptness.

Without discussing it with his Cabinet, he secretly ordered Attorney General Knox to bring suit to dissolve the company. In reply to criticism of the cloak-and-dagger atmosphere surrounding the filing of the suit, the president claimed it had been done to "prevent violent fluctuations and disaster in the market." In fact, Roosevelt probably resorted to secrecy because he did not trust himself to withstand demands from Root and others to shelve the suit if they had known about it.

Shortly afterward, the irate Morgan, chaperoned by Hanna, paid a visit to the White House. It was a classic confrontation. On one side was the imperious lord of finance, whose fiery nose and fixed glare reminded one observer of the headlights of an onrushing locomotive. On the other was the young and debonair president of the United States, who had an undisguised contempt

---

*In the Knight case, the Supreme Court held that the Sherman Act did not prohibit a corporation from acquiring all the stock of other corporations through the exchange of its stock for theirs even if it permitted the first corporation to control the entire production of a commodity.

for "men who seek gain, not by genuine work . . . but by gambling."

Morgan began with a typically blunt proposal. "If we have done anything wrong, send your man [the Attorney General] to my man [Morgan's lawyer] and they can fix it up."

Roosevelt replied that this was impossible. He did not intend to "fix it up" but to "stop it."

"Are you going to attack my other interests, the Steel Trust and the others?" Morgan asked. He added that the primary reason he had come to the White House was to get the answer to this question.

"Certainly not," replied the president, "unless we find out that . . . they have done something that we regard as wrong."

Roosevelt thought Morgan completely typified the Wall Street viewpoint. "Mr. Morgan could not help regarding me as a big rival operator, who either intended to ruin all his interests or else could be induced to come to an agreement to ruin none," he told Knox after the meeting.

Two years passed before the courts unraveled the Northern Securities case. Most observers felt that when it finally wended its way to the Supreme Court, the justices would rebuff the president and reaffirm their ruling in the Knight case. But in March 1904, by a five-to-four vote,* the Court upheld a lower-court ruling dissolving the company—and Theodore Roosevelt's reputation as a trustbuster was made. A typical approving comment was made by Joseph Pulitzer, who opposed the president on almost every other issue: "The greatest breeder of discontent and socialism is the . . . popular belief that the law is one thing for the rich and another for the poor."

The Northern Securities case marked the high-water mark of Roosevelt's campaign against the trusts during his first term. Except for filing a successful suit against Swift and Company, the so-called Beef Trust, he laid aside his Big Stick until after he won election on his own in 1904. Instead, he embarked on a succession

---

*Roosevelt was disconcerted by the dissenting vote filed by Justice Oliver Wendell Holmes, Jr., whom he had appointed to the Supreme Court in 1902 under the impression they held similar views. Holmes took the position that an exchange of stock was not in itself an act of commerce. "I could carve a better judge out of a banana," the president reportedly exploded upon hearing the news.

of speaking tours in which he emphasized the need for some sort of federal control over corporate evils. Wall Street accused him of demagoguery, but Roosevelt was unconcerned because this type of abuse endeared him to ordinary Americans. Big business had been regarded by them as all-powerful, and so the moral fervor of Roosevelt's challenge was thrilling and electrifying.

"The great corporations which we have grown to speak of rather loosely as trusts are the creatures of the State," he declared in a speech at Providence, "and the State not only has the right to control them, but it is in duty bound to control them wherever the need of such control is shown." And in Boston, he told an enthusiastic audience that he was "acting in the most conservative sense of property's interests. . . . Because when you can make it evident that all men, big and small alike, have to obey the law, you are putting the safeguard of law around all men." As the 1902 midterm congressional elections approached, it was clear that Roosevelt's policy and popularity would be confirmed by the voters except for one obstacle—a coal strike that had shut down the anthracite fields of eastern Pennsylvania in May 1902.

Trouble had been brewing in the minefields for some time. Miners worked twelve-hour shifts, six days a week, for an average yearly wage of $560, from which rent, oil for their lamps, and a compulsory charge for the company doctor were deducted. Working conditions in the mines were hazardous, child labor endemic, and there was a long history of violence. No matter how hard a man worked, he was constantly in debt to the company store. Investigators discovered a twelve-year-old boy, whose forty-cent-a-day pay was regularly credited against a debt left by his father, who had been killed four years earlier in a mine accident.

The strike was called by the United Mine Workers after the operators, mostly coal-carrying railroads, rejected the miners' demands for a 10 to 20 percent increase in pay, an eight-hour day, a fairer system of weighing coal, and union recognition. Two years before, the operators had made minor concessions under pressure from Mark Hanna, who not only wished to avoid a strike in an election year but had a profound contempt for employers who exploited their workers. Resentful of the intervention by outside interests in 1900, the operators were determined not to recognize the UMW and to "restore discipline" to the mines. George F. Baer, president of the Reading Railroad, a Morgan property, and the industry's chief spokesman, cavalierly brushed off the miners' complaints. "They don't suffer," he declared. "Why, they can't even speak English."

John Mitchell, the president of the UMW, offered to submit the miners' grievances to binding arbitration by a commission of three clergymen, but Baer flatly refused to compromise and 150,000 men struck the mines. Public opinion was on the side of the miners, although the operators had supporters, among them Woodrow Wilson, who had just become president of Princeton University. The strike, Wilson said, was not about wages and working conditions but constituted an effort by the union "to win more power." Baer proclaimed arrogantly: "The rights and interests of the laboring man will be protected and cared for—not by the labor agitators—but by the Christian men to whom God in His infinite wisdom has given the control of the property interests of the country. . . ."

The operators plainly counted upon a coal shortage and violence and sabotage in the minefields to force President Roosevelt to intervene to put down the strike, as Grover Cleveland had done in the 1894 Pullman strike. They held most of the cards, and not even Mark Hanna could persuade them to settle this time. The strike dragged on for the next several months. Mitchell's appeals to miners not to resort to violence and to the humanity and sense of fair play of the American people won public approval, especially when compared with the harsh, unyielding intransigence of the mine operators.

In the beginning, the president took no formal notice of the strike. There was some talk that he should intervene to break up the coal trust, but Roosevelt informed George Perkins that he did not contemplate action because the strike constituted no threat to the general welfare. But as winter approached, the stockpile of coal dwindled, bringing the nation face to face with the prospect of freezing homes and a complete shutdown of industry and rail transport. The price of coal soared from $5 per ton in New York to $30, and was heading higher. Schools started to close for lack of fuel, and in some areas mobs seized coal cars. In Rochester, New York, enterprising citizens chopped down telegraph poles for firewood. Lodge warned Roosevelt about the possibility of rioting in Boston if coal did not start moving again, and he expressed fear of the effect of a continuing strike upon the elections in November. Even sane, conservative men were suggesting that the government seize the coal fields, he reported. "Is there nothing we can *appear* to do?" Lodge asked plaintively.*

The president was acutely aware of the effect of a coal famine on Republican political hopes in the off-year elections, and by late summer he was seeking a way to intervene in the strike. "What is the reason we cannot proceed against the coal operators as being engaged in a trust?" he inquired of Attorney General Knox in August. "I ask because it is a question continually being asked of me." Knox advised him that under the Constitution the president lacked such power. The whole affair was further proof to Roosevelt of the need for government supervision of corporations.

While trying to find a way to resolve the coal strike, Roosevelt had a close brush with death. In Pittsfield, Massachusetts, on September 3, his carriage was struck by a trolley car. A Secret Service man riding with the coachman was killed, George Cortelyou sustained a head injury, and the president himself was cut on the head and left leg. Brushing off the injury, Roosevelt continued his speaking tour, but his leg began to swell. While in Indianapolis to address a meeting of Spanish War veterans on September 23, he experienced considerable pain. The examining physicians found an abscess and suggested an immediate operation to drain the wound to prevent blood poisoning.

Always deeply interested in any novel experience, the president was a keen observer as the doctors prepared him for the operation. He was given a local anesthetic and kept up a bantering conversation with the operating team throughout the procedure. Through gritted teeth, he told them that if they required expert opinion on what the operation felt like, "he could inform them that something was happening in the vicinity of his shinbone."

---

*Italics mine. Some authors (see Hofstadter, *The Age of Reform*, p. 233) have interpreted Lodge's words to indicate that most of Roosevelt's past actions were symbolic. Instead, it was the plea of an alarmed politician for the president to do something—anything—to convince the electorate that the administration was trying to deal with the problem.

That evening Roosevelt was taken to a special train for the return to Washington and put to bed with his leg propped up on pillows. A reporter noted that he immediately took out a book and became engrossed in it.

The White House was undergoing a long-deserved remodeling* and the rest of the family was at Oyster Bay, so the president took up temporary quarters at 22 Jackson Place, across from the mansion, where Edith joined him. Four days of rest provided no relief, however, and the doctors had to operate again. Using cocaine as a sedative, they cut a two-inch gash in his leg and scraped the shinbone. "As soon as it was over," Edith wrote Kermit, who had just left home for Groton, "he was lifted back into his wheeled chair and came into the sitting room where he is now reading."

During the period of recuperation ordered by the doctors Roosevelt read everything from Dickens's *Our Mutual Friend* to the latest study of European expansion into Asia, whirling about in his wheelchair at great peril to his staff. Within a few weeks he was up and about, but he never completely recovered from the injury. "The shock," he observed years later, "permanently damaged the bone." When he became tired, or his leg was jarred, it ached, although he carefully concealed his discomfort in public.

The coal strike challenged Roosevelt's belief that the general welfare of the American people must take precedence over the demands of both labor and capital. Although eager to take a hand in the crisis, he was frustrated in his efforts to find a role. Pronouncing himself "at wit's end," he eventually decided to take up Massachusetts Governor W. Murray Crane's suggestion to appeal to the reason of both sides. Telegrams were dispatched to all the parties requesting them to come to Washington for a meeting on October 3. Roosevelt acknowledged that he had "no legal or constitutional duty—and therefore no legal or constitutional right in the matter"—but he was convinced he had to deal with the emergency.

In general, the press and public favored the meeting. It would be the first time an American president had intervened in a labor dispute except to issue an antiunion injunction. Writing to Bamie,

---

*To be discussed in Chapter 19.

Roosevelt explained why he had acted in this case. "There would be no warrant in interfering under similar conditions in a strike of iron workers. Iron is not a necessity. But I could not see misery and death come to the great masses of the people in our large cities and sit by idly because under ordinary conditions a strike is not the subject of interference by the President. . . ."

The representatives of the coal operators arrived in the capital in special railroad cars and then moved in a procession of glistening coaches to the temporary White House on Jackson Place. In contrast, Mitchell and his aides arrived by streetcar. Wearing a gray dressing gown, and a very vigorous invalid despite his wheelchair, Roosevelt explained that while he had no legal right or duty to call the meeting, he represented a third party to the strike—the public—which was suffering the most from the shutdown, and he appealed to everyone's patriotism to end the dispute. From across the street, reporters could see a small section of the second-floor conference room, and periodically a wheelchair with a vigorously gesticulating occupant would flash by the window.

Although snubbed by Baer, Mitchell offered to accept the findings of an arbitration committee appointed by the president. The mine operators, angry at the presidential intervention and its implied recognition of the UMW, flatly rejected the proposal. "We object to being called here to meet a criminal, even by the president of the United States," snarled Baer. The operators demanded an injunction ordering the miners back to work, using federal troops if necessary as in the Pullman strike, and the meeting broke up on this grim note.

Behind closed doors, the president railed against the "gross blindness" of the mine operators. Mitchell, he observed, had been the only man in the room who behaved like a gentleman—including himself, for he had lost his temper. Roosevelt had been so angered by Baer's insolence, he later declared, that "if it wasn't for the high office I hold, I would have taken him by the seat of the breeches and the nape of the neck and chucked him out of that window."

Now convinced that the operators bore greater responsibility for the strike than the miners, Roosevelt told Mitchell that if the men went back to work, he would appoint a commission of inquiry and use his personal influence to convince the mine operators to accept its findings. Mitchell was wary of the proposal because he didn't believe the operators could be persuaded to accept the commission's findings.

Nevertheless, Roosevelt had gotten an idea from the October 3 meeting. Troops would be sent into Pennsylvania, all right, but they would seize the mines and run them under his authority to maintain public order if requested to do so by the local authorities. Word was passed to Governor William A. Strong of Pennsylvania of the president's intentions. General John M. Schofield was ordered to stand by with ten thousand federal troops to take over the mines. Roosevelt may have been bluffing, but the operators could not be certain. Schofield, described by the president as "a most respectable-looking old boy, with side whiskers and a black skull cap," was told "to act in a purely military capacity under me as Commander-in-Chief, paying no heed to any authority, judicial or otherwise, except mine."

"What about the Constitution of the United States?" Congressman James E. Watson protested to the president. "What about using private property for public purposes without due process of law?"

Roosevelt stopped suddenly, took a firm grip on Watson's lapels and looking him squarely in the eye, fairly shouted: "The Constitution was made for the people and not the people for the Constitution!"

Before this drastic action could be taken, Elihu Root met secretly on October 11 with J. P. Morgan in New York at his own initiative—but with the president's approval—on the financier's yacht *Corsair*. Undoubtedly assisted by Roosevelt's threat to seize the mines, Root convinced Morgan to accept arbitration by a five-member commission after the miners agreed to go back to work. Prodded by Morgan, the mine operators caved in. To save face, they stipulated that the commission should include a businessman familiar with the coal industry, a mining engineer, a federal judge, an army or navy officer, and an eminent sociologist. Significantly, there was no union representative on the list.

Mitchell objected that the commission had been "packed" by the operators and insisted that a representative of organized labor be included. Roosevelt agreed, but the operators were adamant.

Throughout the night of October 15, the president, Root, George Perkins, and Robert Bacon tried to reach some agreement. Perkins and Bacon were constantly in touch by telephone with Morgan and the operators. They were "nearly wild," Roosevelt observed, and declared that even if "anarchy and social war" occurred, the operators would under no circumstances accept a labor representative.

"Suddenly," Roosevelt later recalled, "it dawned on me that they were not objecting to the thing but the name. I found that they did not mind my appointing any man, whether he was a labor man or not, so long as he was not appointed *as* a labor man, or *as* a representative of labor." Having made this startling discovery, the president named E. E. Clark, the head of the Brotherhood of Railway Conductors, to the place reserved for the "eminent sociologist"—and the appointment was accepted by the mine operators.

"I shall never forget the mixture of relief and amusement I felt when I thoroughly grasped the fact that they would heroically submit to anarchy rather than have Tweedledum, yet if I would call it Tweedledee they would accept with rapture," Roosevelt later recalled sardonically. "It gave me an illuminating glimpse into one corner of the mighty brains of these 'captains of industry.' In order to carry the great and vital point and secure agreement by both parties, all that was necessary for me to do was to commit a technical and nominal absurdity with a solemn face. This I gladly did."

The strikers returned to work while the commission held its hearings. The miners' case was presented by Clarence Darrow, while that of the operators was not helped by Baer's poor sense of public relations. Ultimately, the miners won a nine-hour day, a 10 percent wage hike, and the presence of their own checkers at the weighing of coal, but they failed to win recognition of the union. The commission also recommended a 10 percent increase in the price of coal—and the mine operators wasted no time in taking advantage of this. Nevertheless, the great coal strike had been peacefully settled, in the public's mind largely as a result of the tireless energy of the president.

In dealing with the coal crisis, Roosevelt had not only demonstrated his mastery of the presidency but had strengthened his position as head of the Republican party. He had also established numerous precedents. For the first time a president had intervened to bring about a negotiated settlement of a labor dispute; for the first time a president had proposed binding arbitration, and for the first time a president had threatened to use troops to seize a strike-bound industry. The federal government had moved clearly into the role of broker between capital and labor that Roosevelt had envisioned for it.

Later Roosevelt claimed that his efforts to settle the coal strike were aimed at ensuring "a square deal" for both sides—a phrase that caught the fancy of the public and gave its name to his administration.* The episode also opened the way for the federal government to become a third party in major labor disputes. Ironically, young Franklin Roosevelt, whose father had invested heavily in coal mines, was critical of his idol's action. "The President's tendency to make the executive power stronger than the Houses of Congress is bound to be a bad thing," he declared, "especially when a man of weaker personality succeeds him in office."

Having revealed a personal delight in the exercise of power, sometimes with an authoritarian disregard of legal restraints, Roosevelt saw his role quite differently. To him, the coal crisis was a test of what he termed the "Jackson-Lincoln" theory of the activist presidency. "Occasionally great national crises arise which call for immediate and vigorous executive action, and in such cases it is the duty of the President to act upon the theory that he is the steward of the people" who "has the legal right to do whatever the needs of the people demand unless the Constitution or the laws explicitly forbid him to do it," he declared.

In the remaining years of his presidency, Roosevelt's relations with organized labor varied. While he firmly believed in labor's right to organize, he saw "nothing sacred in the name of labor itself." In 1903, he ordered federal troops into Arizona when violence flared during a copper miners' strike, only to withdraw them when they were used to cow the workers. On the other hand, he refused to intervene that same year when Colorado mine

---

*Roosevelt first used the term "square deal" in a speech at Grand Canyon on May 6, 1903. *Theodore Roosevelt Association Journal,* Summer 1991.

owners used the state militia to suppress a strike called by the radical Western Federation of Miners. He also favored the "open shop," much to the distress of union leaders, and refused to condone the dismissal of William Miller, a foreman in the Government Printing Office bookbindery, after he had been expelled by his union.

Basically, Roosevelt's aim was to prevent radical change. In November 1904, writing to Philander Knox, who had just been elected to the Senate from Pennsylvania, Roosevelt urged him to reject the views of the reactionary wing of the Republican party lest "we shall some day go down before a radical and extreme democracy with a crash which will be disastrous for the Nation. We must not only do justice, but be able to show the wage workers that we are doing justice. . . . The friends of property must realize that the surest way to provoke an explosion of wrong and injustice is to be shortsighted, narrow-minded, greedy and arrogant, and to fail to show in actual work that here in this republic it is particularly incumbent upon the man with whom things have prospered to be in a certain sense the keeper of his brother with whom life had gone hard."

Roosevelt's vigorous action in the Northern Securities case and his personal intervention in the coal strike were overwhelmingly endorsed by the voters in the 1902 congressional elections. Traditionally, the party that controls the White House loses seats in off-year elections, but Roosevelt's coattails were broad enough to carry additional Republicans into the House. Buttressed by this victory, he again pressed—with considerably more vigor and clout this time—for the legislative program that Congress had ignored the previous year.

The agenda was once more topped by the demand for a Department of Commerce with a Bureau of Corporations within it. "I believe that monopolies, unjust discriminations, which prevent or cripple competition, fraudulent overcapitalization and other evils in trust organizations and practice which injuriously affect interstate trade can be prevented under the power of the Congress to 'regulate commerce . . . among the several states,' " Roosevelt declared, despite those who argued that he could not take such steps under the Constitution. And if such proved to be the case, he warned, "we should not shrink from amending the Constitution so as to secure beyond peradventure the power sought."

The other measures sent to Congress included an act to expedite antitrust prosecutions by hiring more prosecutors, a bill barring railroad freight-rate rebates, tariff reciprocity with Cuba and the Philippines, subsidies for the merchant marine, the construction of a canal across Central America, naval expansion, reform of the army's command structure, and conservation. This program received a mixed reception. The so-called Expedition Act, which authorized the appointment of two additional assistants to the attorney general to expedite the trying of suits brought under the Sherman Act, met with little opposition. The Elkins Act barring freight rebates was approved with the blessings of the railroads, which no longer wished to make such payments.

The president's efforts at trust regulation ran into difficulty, however. While almost everyone supported the creation of a Department of Commerce—now expanded to include a Department of Labor with the support of congressmen from labor constituencies—there was strong opposition to the Bureau of Corporations, especially to a provision giving it the power of subpoena and to compel attendance and testimony. The opposition substituted its own bill, drawing the teeth of the presidential version, while reformers, led by Congressman Charles E. Littlefield of Maine, supported a more drastic measure providing for punative action against corporations.

"I am having a terrific time trying to get various things through Congress," the frustrated president wrote Kermit, "and I pass my days in a state of exasperation, first with the fools who do not want to do any of the things that ought to be done, and second, with the equally obnoxious fools who insist upon so much that they cannot get anything." Although he sometimes despaired of getting any bill at all, Roosevelt was in no mood to accept defeat.

The modified bill was in a House-Senate conference committee, where differences between the two versions were being ironed out, and Roosevelt promptly set about persuading the committee to amend the bill to restore full investigatory powers to the Bureau of Corporations. Unless his proposal was accepted, the president cannily implied that he would support the radical Littlefield proposal. He also told the press that unless Congress passed his version, he would call a special session as soon as the current one ended in March 1903. And in a public relations masterstroke, he leaked telegrams that indicated that John D. Rockefeller of Standard Oil was leading the opposition to his version of the bill.

In actuality, Rockefeller paid little attention to politics, and the telegrams, which Roosevelt had not seen, were sent by his attorney, John D. Archbold. Roosevelt, with his firm grasp of public psychology, realized that Archbold's name meant nothing to the public, while Rockefeller personified the evils of the trusts, and a storm of protest arose in Congress following the release of the telegrams. Under public pressure, opposition to the president's bill collapsed. The House passed it on February 10, by 251 to 10; the Senate shouted its approval the following day. "I got the bill through by publishing those telegrams and concentrating the public attention on the bill," Roosevelt subsequently declared.*

The tariff was regarded by most antitrust zealots as "the mother of trusts," but by and large Roosevelt sidestepped the issue. If one major issue had separated Republicans and Democrats in the years since the Civil War, it had been the tariff, but as a result of the country's relative prosperity during Roosevelt's years in office, public interest in tariff reform was low. On several occasions, Roosevelt spoke of the necessity of tariff reform—using it as a club to keep the Old Guard in line—but in general he avoided the issue because of its potential for divisiveness. He came out strongly for reciprocity on Cuban sugar, however, arguing that the United States had a moral duty to assist the ailing economy of the liberated island, but western sugar-beet growers aligned with eastern refiners blocked such a program until November 1903. A similar tariff reciprocity measure for the Philippines was defeated.

Roosevelt's passionate interest in the environment, which went back to the organization of the Boone & Crockett Club, was reflected in his advocacy of a full-scale federal conservation program. "It is as right for the National Government to make the streams and rivers of the arid regions useful by engineering works for water storage as to make useful rivers and harbors of the humid

---

*George Cortelyou was named the first secretary of commerce. The Department of Commerce and Labor was split into two separate Cabinet-level agencies in 1913.

regions by works of another kind," he stated in his first message to Congress. Reservoirs, reclamation projects, forest reserves, and wildlife protection were all part of his program.

The Newlands Act, which called for part of the receipts from the sale of public lands in the western states and territories to be earmarked for dams and reclamation projects, had a high presidential priority. When House Speaker Joseph G. Cannon, expressing the views of eastern and midwestern agricultural interests, opposed the measure on grounds that it would result in an increase of farm products already in oversupply, Roosevelt personally took a hand in defending the measure. "I do not believe that I have ever before written to an individual legislator in favor of an individual bill," he told Cannon, "yet I feel from acquaintance with the far West that it would be a genuine and ranking injustice . . . to kill this measure." Western Republicans and Democratic lawmakers combined to pass the measure, and it was signed into law by the president—the most significant piece of American environmental legislation approved to that date. Twenty-one federal irrigation projects were established in fourteen states.

In little more than a year in office, Roosevelt had captured the imagination of the American people. He had launched the Square Deal, brandished his Big Stick against the trusts, personally settled the coal strike and won the support of labor and the public, enlarged the power and prestige of his office, and emerged as the leader of the Republican party. He had shown the will and skill to capitalize on the opportunities that came his way. Now, having made himself the arbiter of American policy at home, Theodore Roosevelt was ready to turn his attention to the world scene.

Chapter 18

"... The Proper Policing of the World"

Unlike most Americans of his day, Theodore Roosevelt was fascinated by international affairs and realized that the war with Spain had transformed the United States from a provincial nation on the fringes of global affairs into a world power. He also possessed a keen understanding of the intricate network of international relationships and the fragile balance of power upon which peace largely depended—an equilibrium disrupted at the beginning of the new century by the emergence of Germany as the leading military power in Europe and the contemporaneous rise of Japan in Asia. "Whether we desire it or not," he warned Congress in his first annual message, "we must henceforth recognize that we have international duties no less than international rights."

In the beginning, Roosevelt carried out the broad outline of the foreign policy of the previous administration. The process of creating an independent Cuba was fulfilled, but with sufficient restrictions on the sovereignty of the new republic to make the island an American protectorate when the United States withdrew in 1902. The pacification of the Philippines continued, an effective civilian administration was put in place, and the Filipinos were placed under tutelage until they were deemed ready to govern themselves.

However, Roosevelt was also determined to put his own brand on the nation's foreign policy just as he had upon every other aspect of government, and so turned his attention to several items of unfinished business: to secure a site for the long-dreamed-of canal across Central America, to settle a vexing dispute over the boundary between Canada and Alaska, and to reinvigorate the Monroe Doctrine in order to block European intervention in certain Latin American nations that were tempting targets for imperialistic expansion.

Isolationism was strong, however, and most Americans opposed further foreign adventures. Roosevelt was convinced that his fellow citizens harbored two illusions: one, that the United States, insulated by two oceans and a tradition of nonentanglement with other nations, could escape involvement in foreign affairs; the other, that peace was the normal international condition. Rapid changes in transportation, communications, and weaponry, Roosevelt believed, had created both a growing interdependence of nations and a potentiality for global turmoil in which only power and the willingness to use force, if necessary, could ensure a peaceful international order.

To defend its new empire in this increasingly unstable climate, Roosevelt argued that it was essential that the United States preserve stability in the Caribbean, take a strong position in the Pacific basin, and immediately begin constructing a canal across Central America to permit the efficient transfer of American naval power from the Atlantic to the Pacific. "I wish to see the United States the dominant power on the shores of the Pacific," he had declared in 1900. Little more than a year later, he stated that no single objective was more important to the American people than the building of an interocean canal.

The capstone of this ambitious structure was the strong battle fleet he had urged when assistant secretary of the navy. "The American people must either build and maintain an adequate Navy or else make up their minds definitely to accept a secondary position in international affairs," the president declared. To bring the public under the spell of the navy, Roosevelt gave his approval to colorful reviews and ceremonies, such as the interment of the body of John Paul Jones at the U.S. Naval Academy in Annapolis. Fleet maneuvers were held and attempts to improve naval marksmanship were made.* In every year of Roosevelt's administration, Congress authorized at least one new battleship, with five such vessels being laid down in 1903 alone. When Roosevelt became president, the United States ranked fifth among naval powers. By 1907, it had twenty battleships at sea, and its fleet was second only to that of the Royal Navy. Bases were also established at Subic Bay in the Philippines and at Guantánamo in Cuba.

Roosevelt met the challenge to his foreign initiatives posed by the isolationist tradition of the American people by exercising strong leadership. "In domestic policy, Congress in the long run is apt to do what is right," he noted. "It is in foreign politics, and in preparing the army and navy that we are apt to have the most difficulty, because these are just the subjects as to which the average American citizen does not take the trouble to think carefully or deeply." He had little regard for Congress in conducting foreign affairs. He consulted with it when legally obligated to do so, but in many cases—such as intervening in the Dominican Republic and taking the swath of land across Panama for the canal—he pressed executive power to the limit.

Strategic necessity influenced Roosevelt's foreign policies more than economic interests. He had no interest in colonies and, seeing the Philippines as a potential Achilles' heel, looked forward to withdrawing from those islands as soon as the Filipino people could stand alone. Colonialism was intended as a step on the difficult road to self-government. Unlike Britain in Egypt and Japan in Korea, as he repeatedly pointed out, the United States had withdrawn from Cuba as it had said it would. Later on, he denied any intent to annex the Dominican Republic, saying he had as much desire to do so "as a gorged boa constrictor might have to swallow a porcupine wrong-end to." Nor did Roosevelt think in terms of trade or markets. Instead, his aim was to protect the United States and the nations of the Western Hemisphere from those that might threaten their security.

Roosevelt brought to the world stage the same attitudes that

---

*The navy's standard of marksmanship had proved abysmal during the Spanish war, despite Roosevelt's efforts at improving it when he was assistant secretary of the navy.

had motivated him as a young political reformer in Albany, New York City, and Washington. Then, he had fought with righteous zeal against corrupt politicians and businessmen. Now, he projected a mixture of nationalism and practical idealism into global affairs. America, he believed, had the moral obligation to overawe international bullies, maintain order, and uplift backward peoples. His policies were the precursor of what would, in time, become the key issue of American foreign policy in the twentieth century: How far should the United States go in seeking freedom and equality for others? "More and more," he declared, "the increasing interdependence and complexity of international and political and economic relations render it incumbent on all civilized and orderly powers to insist on the proper policing of the world."

Such bold words, especially his widely quoted aphorism, "Speak softly and carry a big stick"—first used to describe his attitude toward Boss Platt and the New York Republican machine—have given Roosevelt a reputation for impetuosity and pugnacity. Critics have charged that he rarely spoke softly and wielded his Big Stick all too readily. In fact, the Big Stick was more of a useful image than a reality. Roosevelt was a cautious and skillful diplomat who negotiated the peaceful settlement of numerous serious disputes. Frederick W. Marks III, a student of his foreign policy, observes that Roosevelt "stands in a class by himself for that blend of statesmanlike qualities which might best be described as velvet on iron."*

Taking a direct hand in helping end the Russo-Japanese War, he not only became the first American to win the Nobel Peace Prize, but also averted the possible widening of the conflict. In Europe, he patiently laid the groundwork for the Algeciras Conference of 1906 to resolve the dispute over Morocco, which may have put off the outbreak of World War I by nearly a decade. He mediated disputes between several Central American countries and was the first world leader to submit a dispute involving his own country to the Permanent Court of International Justice at The Hague.

Some historians have tried to bridge the gap between Theodore Roosevelt's supposed brag and bluster and his efforts as a peace-

---

*Marks, *Velvet on Iron: The Diplomacy of Theodore Roosevelt,* 1982.

maker by attributing his success to the restraining counsel of John Hay and Elihu Root, who became secretary of state after Hay's death in 1905. In fact, however, he kept control of the foreign policy decision-making process in his own hands, often acting on his own initiative and responsibility. Following Hay's death, Roosevelt publicly praised Hay's statesmanlike qualities, but in a letter to Cabot Lodge he characterized Hay as a "fine figurehead." In times of crisis, he maintained, "rarely did I consult Hay." And inasmuch as Roosevelt served for the most part as his own secretary of state, Root usually implemented policy instead of creating it.

"When I left the presidency, I finished seven and a half years of administration, during which not a shot had been fired against a foreign foe," Roosevelt proudly declared in his *Autobiography*. "We were at absolute peace, and there was no nation in the world . . . whom we had wronged, or from whom we had anything to fear."*

Having little patience with the formalities of diplomacy, Roosevelt, like many of his successors, had slight regard for the State Department and many American diplomatic representatives abroad. "The trouble with our Ambassadors in stations of real importance is that they totally fail to give real information, and seem to think that the real life work of an Ambassador is a kind of glorified pink tea party," he declared. He sought information outside official channels from his own network of informants— among them Arthur Lee and John St. Loe Strachey, editor of the *Spectator*—and on important occasions had confidential messages transmitted by private envoys, with the State Department serving as a mere post office.

Personal likes and dislikes governed his relations with some foreign envoys to Washington. In fact, he lobbied for the appointment of two old friends, Cecil Spring Rice and Hermann Speck von Sternberg, as British and German ambassadors, in the case of the latter, successfully. Jean Jules Jusserand, the French ambassador, was another presidential confidant, while Count Arturo Cassini, the Russian envoy, was viewed as untrustworthy.

---

*Roosevelt rationalized the bloody Philippine Insurrection, which he had inherited from McKinley, as a civil war on American territory rather than a struggle against a "foreign foe."

Preferring directness even in diplomacy, Roosevelt was irked by the absurd formalities of protocol. He was particularly upset by the title "Excellency" by which foreign diplomats had traditionally addressed the president. "Let them call me 'The President,' or 'Mr. President,' or 'Sir,' but not a title to which I have no right," he told the State Department. To maintain the dignity of his office, he refused to address kings and emperors as "Your Majesty," but wrote them directly in the second person as an equal. When he held a simple family dinner at the White House for Prince Henry, the visiting brother of German Kaiser Wilhelm II, he was exasperated by the details of precedence. "Will the Prince take Mrs. Roosevelt [into dinner], while I walk in solemn state ahead by myself?" he asked Hay. "I am quite clear that I ought not to walk in with my wife on one arm and my daughter on the other and the Prince somewhat alongside. . . ."

Whenever he was able, the president tried to break through the stiff, formal protocol procedures. After receiving Speck von Sternberg's credentials, he took him shooting, tramping, and riding at Oyster Bay. Archie Butt, his military aide, recalled that after the usual formalities were out of the way, Roosevelt took a new Dutch minister aside to assure him that he knew some Dutch, including a few nursery songs, and immediately launched into one for the delighted diplomat. Once, Roosevelt invited Butt to "draw up a chair" and join in a chat with a special Japanese envoy, Baron Kentaro Kaneko, who had been one of the president's contemporaries at Harvard. And Roosevelt once interrupted a formal White House luncheon to demonstrate a judo hold by throwing the Swiss minister to the floor several times, to the vast amusement of the other guests.

Roosevelt's presidency coincided with a trend toward greater collaboration between the United States and Great Britain along with a corresponding estrangement from Germany. In the years immediately following the Civil War, relations between London and Washington had been tense because of American anger over British aid to the Confederacy—including the building of the *Alabama* under Uncle Jimmy Bulloch's direction—and what the British perceived as American designs on Canada. Tensions subsided after the British supported the United States during the war with Spain, but the key element in the drawing together of the two countries was the challenge of Imperial Germany. Faced with the provocative policies of Kaiser Wilhelm II and the German threat to their command of the seas, the British found it advisable to concentrate their fleet in home waters and cultivate ties with the Americans. The resulting vacuum in the Caribbean was filled by the United States, while the Japanese, who signed a treaty of alliance with Britain in 1902, extended their influence over Far Eastern waters.

Technological obstacles, let alone distance, lay in the way of any German incursion into the Western Hemisphere, but antagonism between the United States and Germany had been brewing for some time. In 1889, there had been friction over a coaling station in Samoa that almost led to war. This incident was followed by German attempts to interfere with Commodore Dewey's squadron at Manila in 1898. Repeated reports that the Germans were scheming to pick up naval bases in the Caribbean—in Santo Domingo or on an island off the Venezuelan coast—along with an increased German commercial presence in the Caribbean also aroused alarm.

While many Americans, including Hay, Root, and Cabot Lodge, believed that the German menace was substantive, Roosevelt had mixed feelings about Germany. Ever since his boyhood as a student in Dresden, he had appreciated German culture and admired the qualities of thrift and hard work brought to the United States by German immigrants. Where Germany's imperial ambitions did not conflict with American interests, he sympathized with them. Nevertheless, Roosevelt was worried about the future. In 1901, while he was still vice president, he had told Spring Rice, "For a generation or two, probably for the present century, Germany may show herself an even more formidable rival [than Russia] of the English-speaking peoples; perhaps in warlike and almost certainly in industrial competition."

Much of the American distrust and suspicion of Germany was focused on the kaiser, who was regarded as a theatrical, bombastic, yet dangerous figure. His "me and Gott" attitude and fierce, spiked moustache made him a favorite target of cartoonists. Keenly aware of the unpopularity of his nation in the United States, Wilhelm tried to ingratiate himself with the American people and the president. He struck a medal in Roosevelt's honor—John Hay dryly observed that it was worth about thirty-five cents and of questionable artistic merit—sent his brother, Prince Henry, to visit, had a yacht built in an American shipyard,* and presented a bronze statue of Frederick the Great to the United States. In view of what was seen as the kaiser's menacing designs upon the Western Hemisphere, some observers suggested that the United States might reciprocate the gift with a statue of James Monroe.

Numerous observers professed to see a similarity between the kaiser and the president. Perhaps the most commented-on trait was their common capacity to interest and charm people whom they met personally. Nicholas Murray Butler, the president of Columbia University who knew both men well, found them alike "in physical build . . . the same strong stocky structure of bone and muscle, the same well-tanned complexion from much exposure to the open air. Their eyes were much alike, too, especially in the alert way in which they travelled about and the intent expression they took on when deeply interested." In their youth, both had conquered physical handicaps, doodled warships, and dreamed of the "hammering guns that beat out mastery of the high seas." Wilhelm openly admired Roosevelt, but the president had mixed feelings about the German ruler. "I wish to Heaven our excellent friend, the Kaiser, was not so jumpy and did not have so many pipe dreams," he once told William Howard Taft. Nevertheless, upon more than one occasion he took steps to prevent the kaiser from the embarrassing consequences of his own follies.

Whatever gains the German emperor may have made from his courtship of the United States and its president were thrown away by a misadventure in the Caribbean. The Germans, as well as several other European nations, had lent large sums of money to revolution-wracked Venezuela. In 1901, the foreign creditors wanted payment on the bonds, of which the current Venezuelan dictator, Cipriano Castro, refused to pay even the overdue inter-

---

*Alice Roosevelt was chosen to launch the vessel and to make a brief speech. She asked her father what she should say, and the president later observed: "The only motto sufficiently epigrammatic that came to my mind was 'Damn the Dutch!' " *Letters,* Vol. III, p. 219.

est. Germany offered to accept arbitration of its claims, but Castro, described by Roosevelt as an "unspeakably villainous little monkey," brushed off the offer, and Britain and Germany undertook joint collection at the cannon's mouth.

Nothing in the Monroe Doctrine as it was then understood forbade such attempts at debt collection. In fact, Roosevelt felt that irresponsible Latin American nations should not be permitted to hide behind the skirts of the United States. "If any South American country misbehaves toward any European country, let the European country spank it," the then vice president had told Speck von Sternberg. To Roosevelt, self-reliance was as important to a nation as to an individual. No people deserved to be independent until they could maintain stability at home and fulfill their international obligations.

Chastisement for stepping out of line could conceivably include a blockade, the seizure of customs receipts or shipping to satisfy unpaid obligations, even the bombardment of coastal cities. There was one limitation, however. The Monroe Doctrine, as Roosevelt explained in his first annual message to Congress, did "not guarantee any State against punishment if it misconducts itself, provided that punishment does not take the form of the acquisition of territory."

Theodor von Holleben, the German envoy to Washington, informed Secretary Hay of the Anglo-German intentions to blockade the Venezuelan coast in November 1902, but he also assured Hay that the Germans had absolutely no intention of acquiring or permanently occupying any Venezuelan territory. Referring to the president's statement, the secretary indicated that the United States had no objections to such action. Within days, an Anglo-German naval squadron blockaded five Venezuelan ports, captured several gunboats, landed troops at La Guaira, and bombarded Puerto Cabello.

Public opinion in both the United States and Britain was aroused by this display of the mailed fist despite Castro's sordid record of misbehavior. For example, the *Literary Digest* worried that "England and Germany will overstep the limits prescribed by the Monroe Doctrine." Nevertheless, the bombardment achieved its purpose, for the Venezuelans now eagerly made it known through the United States that they would accept arbitration. Britain and Germany agreed, according to the official version of the incident, and asked Roosevelt to arbitrate. The president was inclined to accept, but decided that the matter should be entrusted to the Permanent Court of International Justice at The Hague, which had recently been organized to settle such disputes.

Several weeks of maneuvering followed during which the public outcry intensified, especially after the Germans twice bombarded Venezuelan territory in January 1903. The blockade was lifted a month later, and the dispute submitted to the Hague Tribunal, which eventually found for the claimants. As a result of the kaiser's poor image in the United States, much of the blame for the intervention fell upon Germany. On the British side, Prime Minister Arthur Balfour, worried about the effect of the episode on Anglo-American relations, calmed the troubled waters by publicly denying any intention of acquiring additional territory in the hemisphere, and accepted the Monroe Doctrine as international law. Such deference to the United States by a European power was a clean break with the past.

Although the president's public role in the resolution of "the Venezuelan business" did not appear overtly to have been crucial to the settlement, Roosevelt placed it among the leading accomplishments of his first term—indicating there may have been more to his actions than first met the eye. "The striking enforcement of the Monroe Doctrine and its acquiescence in it by [the] great foreign powers, while at the same time a great step forward was taken in the cause of international peace by securing The Hague Tribunal as its arbitrator," he noted as he prepared to run for election in 1904.

Thirteen years later, during the height of World War I, Roosevelt startled the nation with a far more dramatic account of the imbroglio. The ex-president revealed that it was not the Venezuelans but the Germans who had dragged their feet on arbitration. In answer to inquiries from William Roscoe Thayer, a Harvard classmate working on a biography of John Hay, Roosevelt recalled that he had "become convinced that Germany intended to seize some Venezuelan harbor and turn it into a strongly fortified place . . . with a view to exercising some measure of control of the future Isthmian Canal and over South American affairs generally."

In December 1903, Roosevelt continued, he had warned von

Holleben that if the kaiser failed to accept arbitration within ten days, a powerful squadron under the command of Admiral Dewey, and already conducting maneuvers in the Caribbean, would be ordered to the Venezuelan coast to prevent seizure of territory. While Roosevelt acknowledged that he may have confused some of the details with the passage of time, he recalled that the envoy had expressed "very grave concern and asked me if I realized the serious consequences that would follow such action. . . . I answered that I had thoroughly counted the cost before I decided on the step" and suggested that von Holleben look at a map. "A glance would show him that there was no spot in the world where Germany in the event of a conflict with the United States would be at a greater disadvantage than in the Caribbean sea."

A week later, Roosevelt said, he again saw the ambassador and asked if the kaiser had issued a reply to his ultimatum. When von Holleben replied in the negative, the president told the shocked diplomat that he would not wait any longer and advanced the deadline by twenty-four hours. Less than a day later, von Holleben returned with the news that the kaiser had agreed to arbitrate. It was, noted Roosevelt, a fine example of "how I applied the policy of 'speak softly and carry a big stick.' " Nothing was said publicly about the incident to keep from embarrassing the kaiser, and the president went out of his way to praise him as a friend of arbitration.

In his own biography of Roosevelt, Thayer stated he had discussed the reasons for the kaiser's sudden backdown with the president, and the conclusion was that von Holleben had either not communicated the ultimatum to Berlin or had described it as a bluff. Following the second meeting, the now thoroughly shaken envoy begged for advice from Karl Bunz, the German consul general in New York—who knew Roosevelt well—and he "assured von Holleben that the President was not bluffing. So Holleben sent a hot cablegram to Berlin, and Berlin understood that only an immediate answer would do."

Over the ensuing years, a debate raged among historians about the veracity of this story. Hostile writers pointed out there was no documentary evidence to support it and claimed Roosevelt had fabricated the account under the influence of his anti-German fervor prior to the American entry into World War I.* If the Germans had indeed backed down, the debunkers said, it was because public opinion had been aroused against the coercion of Venezuela and because the Germans feared the loss of American goodwill rather than Roosevelt's Big Stick. Modern researchers have, on the other hand, made a convincing defense of Roosevelt's claim to have issued the ultimatum. In fact, Howard Beale argued it was the president's wish to help the kaiser save face by keeping the episode secret—his concern "with substance, not with personal prestige"—that created the controversy in the first place.†

Whatever the ultimate truth about the Venezuelan crisis, the incident alerted Roosevelt to the need to deal with similar foreign interventions in the future. While he had originally supported efforts to force the Latin Americans to adhere to an acceptable level of international behavior—as long as there was no territorial aggrandizement—the conduct of the British and Germans had given him second thoughts. What had once been considered permissible was now perceived as a danger to hemispheric security. The alternative was obvious. The United States, garbed in the robes of the Monroe Doctrine, must be prepared to step in to rescue the "wretched governments" of the Caribbean and Latin America, whether they wished it or not.

Roosevelt regarded the Monroe Doctrine as an expression of

---

*These attacks upon Roosevelt's veracity helped create a picture of the ex-president as boastful and somewhat of a fraud, a man who invented incidents and doctored stories to enhance his reputation. All this contributed to the precipitate decline of his historical reputation during the 1930s. But when Roosevelt's voluminous correspondence became available to researchers after World War II, at least fifteen instances were discovered prior to World War I—dating back to 1905—in which he had privately imparted to friends partial details of the incident. Even Henry Pringle, his most critical biographer, was inclined to accept the ultimatum story after an interview with William Loeb, the president's private secretary, who said he was present at the first conversation with von Holleben.

"You gave that Dutchman something to think about," Loeb recalled telling Roosevelt following the meeting. Pringle, op. cit., p. 288.

†Beale, *Theodore Roosevelt and the Rise of America to World Power,* p. 407. Frederick Marks strengthened Beale's position. On the basis of an in-depth study of the relevant American and foreign archives, he determined that "a body of material was destroyed for diplomatic reasons." The obvious gaps, he stated, cannot be coincidental, but indicate a gentleman's agreement among the American, British, and German governments to suppress all details of the kaiser's humiliation. See Marks, op. cit., ch. 2. For a study of the historiography and a thorough examination of the evidence, see Morris, "A Few Pregnant Days . . ."

principle, not as law or an instrument of diplomacy. "It is a question of policy," he had written as far back as 1896. "To argue that it cannot be recognized as a principle of international law is a mere waste of breath. Nobody cares whether it is or is not so recognized, any more than one cares whether the Declaration of Independence and Washington's Farewell Address are recognized." In essence, the Monroe Doctrine was significant because it was an expression of an American nationalism.

The catalyst for what became known as the "Roosevelt Corollary" to the Monroe Doctrine was the troubled financial affairs of the Dominican Republic. Germany, France, and Italy were talking about intervening to collect debts owed their nationals. By early 1904, the country was in a state of chaos and Dominican officials were pleading with Washington to make it a virtual protectorate to save them from European creditors. While Roosevelt instructed the navy to protect American lives and property in case of violence, he was reluctant to intervene in an election year. "I have been hoping and praying for three months that Santo Domingans would behave so that I would not have to act in any way," he wrote a friend. "I want to do nothing but what a policeman has to do in Santo Domingo."

Events finally forced the president's hand. As conditions in the Dominican Republic deteriorated, the specter of European intervention increased and Roosevelt felt it necessary to assume the responsibility he had hoped to avoid. On May 20, 1904, he sent Secretary of War Root an open letter with the suggestion that it be read that night at a dinner marking the second anniversary of the establishment of the Republic of Cuba. It spelled out the Roosevelt Corollary to the Monroe Doctrine.

"If a nation shows that it knows how to act with decency in industrial and political matters, if it keeps order and pays its obligations, then it need fear no interference from the United States," Roosevelt declared. But, he added pointedly, "brutal wrongdoing, or an impotence which results in a general loosening of the ties of civilized society, may finally require intervention by some civilized nation, and in the Western Hemisphere the United States cannot ignore this duty. . . ." He saw no irony in the fact that the Monroe Doctrine, originally designed to prevent intervention in the hemisphere by the European powers, would now be used to justify intervention by the United States.

This pronouncement was greeted with a mixture of praise and condemnation. "Rather amused at the yell," Roosevelt told Root. He insisted that "what I wrote is the simplest common sense, and only the fool or the coward can treat it as aught else. . . . What a queer set of evil-minded creatures, mixed with honest people of preposterous shortness of vision, our opponents are!" In his annual message that December, the president emphasized the United States had no designs upon the territory of its neighbors to the south and he was expanding the Monroe Doctrine for their welfare.

Not long afterward, Roosevelt found it necessary to act under his corollary. Chaos now reigned in the Dominican Republic, and an Italian cruiser appeared off the island, apparently prepared to press the claims of Italy's creditors. The French were also pressing claims. Toward the end of January 1905, Washington persuaded the Dominicans to request U.S. assistance in the collection of customs receipts, a request that had been made twice previously without American prompting. Under terms of a protocol signed by both parties, an American would be placed in charge of these collections, with 55 percent of the receipts earmarked for discharge of the foreign debts and the rest to be retained by the local government for its operations.

With anti-imperialist Democrats such as Senators John Morgan of Alabama and Augustus O. Bacon of Georgia protesting that the protocol would lead to an American protectorate in the Caribbean, the Senate adjourned after an angry debate without approving the protocol. To the president's disgust, Senator Morgan proposed that the European nations be encouraged to collect the debts on their own—a step that Roosevelt was bending every effort to prevent. "I do not much admire the Senate because it is such a helpless body when efficient work for good is to be done," he fumed. "Creatures like Bacon, Morgan, et cetera, backed by the average yahoo among the Democratic Senators, are wholly indifferent to national honor or national welfare. They are primarily concerned with getting a little cheap reputation among ignorant people. . . ."

Ignoring the Senate, Roosevelt entered into an informal *modus vivendi* with the Dominicans for American oversight over the customs revenues. Although of questionable legality, this agreement remained in effect for twenty-eight months until February

1907, when the Senate finally formalized the relationship. The results demonstrated the president's practicality if not his respect for the Constitution. With the Dominican customs houses in the hands of the Americans, conditions on the island improved. Corruption was reduced if not wiped out, roads were built, schools opened, hygiene improved, and the foreign creditors satisfied. "The affairs of the island are on a better basis than they have been for a century—indeed it would not be an overstatement to say than they have ever been before," Roosevelt declared.

Despite the angry protests of the anti-imperialists, the Roosevelt Corollary was popular with most Americans, who saw it as a reflection of the nation's growing importance in the world. And contrary to most assumptions, the republics of Latin America showed remarkably little concern over the president's efforts to play hemispheric policeman. In fact, American prestige in Latin America was exceptionally high under Roosevelt, and when he visited the area in 1913, he was enthusiastically welcomed. Not until later, when the corollary was used to justify the landings of marines in Central America and the Caribbean, did Latin Americans come to view the United States with mistrust and alarm.

Imperial Germany and the hapless Latin American republics were not alone in feeling the weight of Roosevelt's Big Stick. He also waved it under the nose of the British lion to bring an end to the troublesome dispute over the boundary between Canada and Alaska. The problem began following the discovery of gold in the Klondike region of northwest Canada in 1896, when it was found that the best water approaches to the gold fields lay through the Alaska panhandle running along the Pacific coast. Canadian leaders now scrutinized the rather ambiguous Anglo-Russian Treaty of 1825 that had established the boundary, and built a trumped-up case that it gave them control of the inlets closest to the gold fields. If the Canadian claim held, the United States would lose a large area it had possessed without question since the purchase of Alaska from the Russians in 1867.

In 1899, to prevent this irritant from jeopardizing the warming friendship with Britain, Secretary Hay had reached a temporary understanding to "let sleeping dogs lie" without prejudice to the claims of either party. This agreement, although unpopular in the

United States, permitted tempers to cool. Upon taking office, Roosevelt, unwilling to cause problems for the British as long as the Boer War lasted, decided to go along with it. But in 1902, the president reexamined the controversy and was convinced that the Canadian case had no foundation because the boundary had been accepted by the Russians, the British, and Americans for sixty years. "I think the Canadian contention is an outrage pure and simple," he told Hay. "To pay them anything where they are entitled to nothing would in a case like this [be] dangerously near blackmail." To back up the American claim, Roosevelt ostentatiously ordered eight hundred troops sent to Alaska.*

Unwilling to risk alienating the United States over a secondary issue, the British privately assured Roosevelt that if Washington agreed to arbitration, the final decision would approximate the American claim. Reluctantly, Roosevelt accepted arbitration, while insisting there was nothing to arbitrate. The agreement provided for "six impartial jurists of repute"—three chosen by the president and three by the British government—who were to meet in London and decide the boundary issue by majority vote.

The three "impartial jurists" named by Roosevelt shocked the British and Canadians. They were neither "impartial" nor "jurists": Elihu Root, Cabot Lodge, and former Senator George Turner of Washington State. Americans greeted the names with derisive laughter, with the *Brooklyn Eagle* suggesting that the Canadians had as much chance of winning as of "a blizzard in Hades." Swallowing their pride, the British named two Canadians and Lord Alverstone, the Lord Chief Justice of England, as their representatives on the tribunal.

With Roosevelt's parting instructions ringing in the ears of the American commissioners—"You are not to yield any territory whatever"—the tribunal began deliberations in London on September 20, 1903. Since the Canadians were just as committed to upholding their country's claim as the Americans were to theirs, Lord Alverstone held the balance. Roosevelt pressed every diplomatic and political weapon into service, including interviews

---

*Although it is often assumed that these troops were sent to threaten the British, they were actually intended to prevent disturbances from breaking out among American and Canadian miners after rumors spread that gold had been discovered in the disputed territory.

with American and British newspapers and private communications, to make certain the British understood that he meant business. Using Justice Oliver Wendell Holmes, who was visiting London, as an intermediary, the president informed British officials that if the commission failed to reach a decision, he would send troops to occupy the disputed area and "run the boundary on my own hook."

In the end, Lord Alverstone sided with the Americans, and on October 20, 1903, the tribunal sustained the American position by four votes to two. As a compromise, the United States received a narrower strip than it claimed, and Canada was awarded two inconsequential islands. The Canadians angrily charged that their interests had been sacrificed on the altar of Britain's need for improved relations with the United States, but Lord Alverstone maintained that his decision had been based solely upon the evidence before the tribunal. Whether or not he was influenced, consciously or unconsciously, by Roosevelt's threats, it was the correct decision to nurture the Anglo-American alliance. As the president later told Admiral Mahan, the resolution of the Alaskan boundary question "settled the last serious trouble between the British Empire and ourselves."

REVOLUTION IMMINENT.

So read the cable that clattered into the Department of State from the American consul at Colón, on the Atlantic side of the Isthmus of Panama, on the afternoon of November 3, 1903. Acting Secretary Francis B. Loomis restrained himself for an hour and five minutes. Then he fired off a cable to the consul at Panama City, on the Pacific slope: UPRISING ON ISTHMUS REPORTED. KEEP DEPARTMENT PROMPTLY AND FULLY INFORMED. Four hours later, the Panama City consulate responded: NO UPRISING YET. REPORTED WILL BE IN THE NIGHT. SITUATION IS CRITICAL.

As evening fell, Loomis was frantic because there was no sign of the expected revolution. The conspirators were to proclaim Panama independent of Colombia, be quickly recognized by the United States, and make available a swath of land for constructing a long-desired canal across the Isthmus that the Colombians had been blocking. But a message intended for the captain of the U.S. Navy gunboat *Nashville,* which had been sent to the region to prevent Colombian government troops from landing, had miscarried. Hurriedly, Loomis drafted instructions for the consuls at Panama and Colón. "Act promptly," he directed. They were also told to convey an order to the skipper of the *Nashville:* "In the interests of peace make every effort to prevent Government troops at Colón from proceeding to Panama."

Having ordered direct American intervention in a revolution against a friendly government—a revolution that to his knowledge had not yet even begun—Loomis was left to agonize for another hour. Finally, a new cable arrived reading: UPRISING OCCURRED TONIGHT, NO BLOODSHED. GOVERNMENT WILL BE ORGANIZED TONIGHT. Undoubtedly Loomis—and Theodore Roosevelt as well—breathed deep sighs of relief. Within hours, Washington recognized the newly independent Republic of Panama, and a treaty was signed giving the United States control of a ten-mile zone across the Isthmus where the canal was to be built. Years after the event, Roosevelt proudly boasted: "I took Panama."*

No action during Roosevelt's presidency aroused greater controversy among his contemporaries and later generations than the methods used to "take" Panama—or reveals more about the workings of his mind. Personal ambition, intense patriotism, and firm belief in the leadership of the "superior" nations guided him throughout the adventure. He viewed the construction of the waterway as an act of high moral principle—"in the vital interests of civilization"—and this messianic conviction had much to do with his actions. Roosevelt regarded it as one of the most important accomplishments of his presidency, comparable to the Louisiana Purchase and the annexation of Texas.

The green light for the waterway project was given by the abrogation of the Clayton-Bulwer Treaty between Britain and the United States as part of the ongoing reduction of tensions between the two nations. Under this agreement, originally signed in 1850, the canal was to be a joint operation, with neither nation having exclusive control, which made it less attractive to Americans. In the meantime, a French stock company had obtained a concession from Colombia to construct a canal across Panama. In 1878, dig-

---

*Abbott, *Impressions of Theodore Roosevelt,* pp. 61–62.

ging began under the direction of Ferdinand de Lesseps, builder of the Suez Canal. Hampered by engineering problems, financial mismanagement, and the ravages of yellow fever, which killed some 22,000 workers, the project was abandoned in 1889 amid a miasma of scandal.

Increasingly apprehensive about foreign control of a strategic waterway so close to its borders, the United States sought to end the Clayton-Bulwer Treaty. Following protracted negotiations between Secretary Hay and Lord Pauncefote, the British ambassador to the United States, Britain bowed to the inevitable and agreed in 1901 to replace the old treaty with a new one giving the United States a free hand to build, control, and fortify such a canal.* President Roosevelt pronounced himself "*dee*lighted!"

The Hay-Pauncefote Treaty cleared away diplomatic roadblocks to constructing the canal, but the next question concerned a choice between routes across Panama and Nicaragua. Southern Democrats, in particular, favored the Nicaraguan route because it was eight hundred miles closer to New Orleans.† In November 1901, the Isthmian Canal Commission, which had conducted a two-year study, stated that a route across Panama would be shorter, but recommended a Nicaraguan waterway because of the lower cost. The Panama route had been made prohibitively expensive by a new French company, which had taken over the now defunct de Lesseps organization. Instead of digging the canal, the company offered to sell the concession and the machinery rusting in the Panamanian jungle to the Americans for $109 million. With Roosevelt apparently supporting a Nicaraguan route, the House of Representatives voted all but unanimously—308 to 2—on January 9, 1902, in favor of this route.

The French company's American lobbyist, William Nelson Cromwell, founding partner of the prestigious Wall Street law firm of Sullivan & Cromwell, now swung into action. White-haired and pink-cheeked, Cromwell projected an air of affable geniality,

---

*An earlier version of the Hay-Pauncefote Treaty, permitting the United States to build the canal but not to fortify it, was defeated in the Senate in 1900 with Roosevelt, still governor of New York, one of its most vociferous opponents. "Better have no canal at all than not be given the power to control it in time of war," he declared. *Letters*, Vol. II, p. 1187.

†Earlier, James Roosevelt, FDR's father, had helped organize the Maritime Canal Company, which proposed to build an interocean canal through Nicaragua. Before work could begin, the country was wracked by an economic crisis and funds dried up.

but one observer branded him "the most dangerous man the country has produced since the days of Aaron Burr." Together with Philippe Bunau-Varilla, the swashbuckling chief engineer of the original canal company and a large stockholder in the new enterprise, Cromwell launched a pro-Panama lobbying campaign aimed principally at Roosevelt, Hay, Mark Hanna, and other influential Republican senators. Hanna's support was assured by a $60,000 contribution to the Republican campaign chest in 1900.*

Under the advice of Cromwell, the directors of the French company slashed their price to $40 million, the value placed by the Isthmian Canal Commission upon its assets. Upon learning of this change, the president summoned the commissioners to the White House to review the situation. Based upon the technological data and reduced price, the commission reported that Panama offered "the most practicable and feasible" route for a canal. Before the month was out, Senator John C. Spooner, acting for the administration, moved to authorize the president to pay up to $40 million for the French concession and to build the canal in Panama.

Over the next five months, a "Battle of the Routes" raged in the Senate, a duel waged with charts, technological studies, and the testimony of expert witnesses. While Cromwell kept a close eye upon the proceedings from the gallery, Bunau-Varilla propagandized against the Nicaraguan route, raising the bogey of volcanic activity in that country while emphasizing that there were no volcanoes in Panama. This campaign had little success until May 1902, when Mount Pelée, on the island of Martinique, suddenly erupted and wiped out a town of forty thousand people. Not long after, a Nicaraguan volcano became active—the very same mountain shown on the country's postage stamps. The quick-witted Bunau-Varilla immediately bought every available Nicaraguan stamp.

"I was lucky enough to find . . . ninety stamps, that is one for every Senator, showing a beautiful volcano belching forth in magnificent eruption," he later wrote. "I hastened to paste my precious postage stamps on sheets of paper. . . . " Below the stamps were written the following words, which told the whole story:

---

*Cromwell charged this contribution to the French company, and later billed it a staggering $800,000 for his services.

"An official witness of the volcanic activity of Nicaragua." One of these stamps was placed in the hand of each senator—an inspired bit of lobbying.

The stamps may or may not have had an appreciable effect on the outcome of the debate, but on June 19, 1902, the Senate voted to approve a canal across Panama by 42 to 34 votes. The president was authorized to purchase the French concession for $40 million, provided the Colombians would cede sufficient territory across the Isthmus to construct the canal. If a satisfactory agreement could not be concluded within "a reasonable time," the canal was to be built across Nicaragua.

These dramatic events, with their overtones of backroom intrigue and the huge sums of money involved, created the impression that the Panama route was selected as a result of some corrupt deal. In point of fact, even though Cromwell probably had some influence with Hanna, Hay, and even the president,* there is no creditable evidence to indicate the final selection was made on any basis but that of expert engineering opinion.

Throughout the Senate debate over the canal route, Roosevelt had remained aloof, but once the selection was made, he assumed an active role in bringing the Panama Canal project to fruition. To hurry along the negotiations with Colombia, he urged Hay to take personal charge. "The great bit of work of my administration, and from the material and constructive standpoint one of the greatest bits of work that the twentieth century will see, is the Isthmian Canal," the president declared.

Everyone expected the negotiations to be concluded quickly because the Colombians had, over the years, repeatedly assured Washington that they wanted the canal built in Panama. But the talks dragged on for months, with the annual rent a stumbling

---

*Hanna is said to have introduced Cromwell to Roosevelt. "You want to be very careful, Theodore," the senator is supposed to have told the president. "This is a very ticklish business. You had better be guided by Cromwell; he knows all about the subject and all about those people down there."

Roosevelt replied that "the trouble with Cromwell is he overestimates his relation to the Cosmos."

"Cosmos?" said Hanna. "I don't know him—I don't know any of those South Americans; but Cromwell knows them all; you stick close to Cromwell." Sullivan, *Our Times,* Vol. II, p. 319.

block. "Why cannot we buy the Panama isthmus outright instead of leasing it from Colombia?" Roosevelt impatiently asked Hay. "I think they would change their constitution if we offered enough." Pressure was also building from Congress, which was insisting that an agreement be reached or negotiations opened for a Nicaraguan canal.

Finally, after Hay threatened—by order of the president—to break off talks and begin negotiations with Nicaragua, the Colombian chargé d'affaires, Dr. Tomás Herran, signed the agreement on January 22, 1903. In exchange for an initial payment of $10 million in gold and $250,000 annually, the United States was given control over a six-mile-wide strip of Panama for a hundred years, renewable at its option. Hay presented the pen with which he signed the treaty to the ubiquitous William Nelson Cromwell.

The U.S. Senate ratified the Hay-Herran Treaty a few weeks later, but the agreement struck a snag in Bogotá. In fact, three days after Herran had affixed his signature to the treaty, instructions reached him not to do so. Colombia was being ruled at the time by an octogenarian dictator,* José M. Marroquín, and Roosevelt and Hay confidently expected the old man to ratify the treaty by decree; then he and his cohorts would happily carve up the $10 million among themselves.

Marroquín surprised them, however. Rather than ratify the treaty by decree, he objected to the alleged paucity of the payment, and to infringements upon this nation's sovereignty, and summoned the Colombian congress, which had not met in five years, to pass upon the pact. The Colombians demanded an increase in the American payment and there was talk of sharing in the $40 million allotted to the French company. Cromwell, ever alert to the interests of his clients, persuaded Hay to issue instructions that Colombia was not to have a cent of these funds— thereby using the State Department to protect the interest of a foreign company in a dispute with another nation.

There were other signs of which way the wind was blowing. On June 9, Hay instructed the American minister in Bogotá to warn

---

*Roosevelt gleefully noted that Marroquín, then vice president, had led a coup in which the Colombian president was seized, and after imprisoning him outside Bogotá, "declared himself possessed of the executive power because of 'absence of the President.' " *Autobiography,* p. 533.

the Colombians that rejection of the treaty might compromise "the friendly understanding" between the two countries. Four days later, Cromwell came to the White House to meet with the president and then planted a story in the *New York World* that proved remarkably prophetic. If Colombia rejected the treaty, the paper reported, quoting unnamed sources, Panama would secede and "enter into a canal treaty with the United States." Moreover, "President Roosevelt is said to strongly favor this plan." Significantly, there was no denial from the White House.

Greed and nationalism overcoming their better judgment, the Colombians ignored these storm warnings, and sought to pocket the whole $40 million. With the French concession due to expire in October 1904, they reasoned, why not wait until the deadline had passed and sell it to the eager Yanquis? Why let the French walk away with $40 million? With this end in view, they adopted the risky game of playing for time. Thus, no one was much surprised when they unanimously rejected the treaty on August 12, 1903—but offered to reopen the negotiations.

For a brief moment, Roosevelt seemed willing to stay his hand. "To wait a few months, or even a year or two, is nothing compared with having things done right," he declared in September. But eager "to make dirt fly" before the 1904 elections, he began to view the Colombian action as a shakedown. Raging against "the blackmailers of Bogotá," he vowed that those "jack rabbits" would not be permitted to "bar one of the future highways of the civilization." American newspapers angrily suggested that the United States take the canal zone by right of international eminent domain. For his part, the president prepared the draft of a message to Congress suggesting that the Canal Zone be seized by force under terms of an 1846 treaty with New Grenada, Colombia's predecessor, which gave the United States the right to guarantee free transit across the Isthmus.

But there was no need to send the matter to Capitol Hill because on the Isthmus the pot was already boiling. Throughout most of its history as a province of Colombia, Panama had seethed with revolution—fifty-three uprisings in as many years, according to Roosevelt's count—and the United States had helped put some down as a favor to the Colombians or to preserve transit. Rumors of a new revolt were rife, and on August 31 *The New York Times* reported that an uprising appeared imminent. Panamanian businessmen and politicos, who had expected to reap great commercial benefits from the canal, were ready to proclaim an independent Panama and were seeking support from the representatives of the French canal company in both Panama and the United States.

The center of revolutionary activity was Room 1162 of the Waldorf-Astoria Hotel, in New York City, where the audacious Philippe Bunau-Varilla had set up shop.* (Later, he called it "the cradle of the Panamanian Republic.") There, he arranged for funds to pay recruits and bribe Colombian troops on the Isthmus, conferred with the other plotters, and prepared a do-it-yourself revolution kit that included a declaration of independence, a constitution, and a flag stitched by the "agile and discreet fingers" of his wife. Had he been permitted to take part, the president would have reveled in this cloak-and-dagger atmosphere.

The unresolved question facing the conspirators was Roosevelt's attitude toward the coup. Would he recognize the new government and make it the same $10 million offer that had been turned down by the Colombians? Bunau-Varilla went directly to the horse's mouth for his answer. On October 10, he was ushered into the president's office in the White House. In the conversation that followed, Roosevelt and his visitor were as circumspect as if tiptoeing through a minefield.

"Well, what do you think is going to be the outcome of the present situation?" Bunau-Varilla recalled the president asking. The Frenchman predicted a revolution and the "features of the President manifested profound surprise."

"A revolution?" murmured Roosevelt, according to Bunau-Varilla. "Would it be possible?"

Both men claimed later that no specific assurances of American support for the uprising were asked or given. Nevertheless, Bunau-Varilla came away from the White House certain of Roosevelt's backing. The president himself observed some months later: "I have no idea what Bunau-Varilla advised the revolutionists. . . . But he is a very able fellow, and it is his business to find

---

*Cromwell had gone to Paris to preserve "deniability."

out what he thought our Government would do. I have no doubt that he was able to make a very accurate guess and to advise his people accordingly. In fact, he would have been a very dull fellow had he been unable to make such a guess."

Not long after, Bunau-Varilla told Dr. Manuel Amador, the representative of the Panamanian junta, that $100,000 would be forthcoming to finance the uprising, and he was instructed to return to the Isthmus immediately to prepare for action on November 3. The date had been set after a meeting with Hay, in which the secretary, in giving his own informal assent* to the coup, informed Bunau-Varilla that three American naval vessels had been dispatched to the area. By calculating their steaming time, the Frenchman worked out a timetable for the coup. Early in the evening of November 2, the *Nashville* dropped anchor at Colón as scheduled, and the long-waited revolution was carried out the following night by a little army of railroad section hands, the local fire brigade, and some bribed Colombian troops. The only casualties were a Chinese bystander and his dog.

Panicked by the turn of events, the Colombians now feverishly promised to ratify the ill-fated Hay-Herran Treaty by executive decree if the United States would step in and put down the rebellion.† But it was already too late. Under the watchful eye of American marines and sailors, Panama proclaimed its independence on November 4. "The world is astounded at our heroism!" declared Amador, who had been named president. "Yesterday we were but the slaves of Colombia; today we are free. . . . President Roosevelt has made good." Several Latin American nations also extended immediate recognition to the new republic—refuting later claims that Roosevelt's actions in Panama inspired a wave of anti-Americanism throughout Latin America. The Latin Americans had more to gain economically from the canal than the United States.

.*Bunau-Varilla later related that when he pressed Hay for assurances of American support, the secretary talked instead about a novel he had been reading, *Captain Macklin*, by Richard Harding Davis. He took this as a positive sign because the book was about a revolution in Honduras led by a young American and a French soldier of fortune. Bunau-Varilla, *Panama . . .*, pp. 326–332.

†To Roosevelt, the sudden offer by the Colombians to approve the rejected Hay-Herran Treaty by decree was proof positive of their basic dishonesty during the whole episode. "It shows that the government which made the treaty really had absolute control over the situation," he told Congress, and could have ratified it earlier if it had so desired. *Works*, Vol. XV, p. 210.

If the president himself was anxious as these momentous events unfolded, he hid it well. He spent the day at Oyster Bay, where, surrounded by well-wishers, he cast his ballot in a local election. Returning to Washington, he was immediately caught up in the "boiling caldron [*sic*] on the Isthmus of Panama." But he was delighted by a telegram from Kermit informing his father of a Groton football victory. "Instantly I bolted into the next room to read it aloud to mother and sister, and we all cheered in unison when we came to the Rah! Rah! Rah! part of it."

For a half century, he told Kermit, the United States had been policing the Isthmus in the interest of "the little wildcat republic of Colombia," but did not intend to intervene on its side during this insurrection. "Any interference I undertake now will be in the interest of the United States and of the people of the Panama Isthmus themselves. There will be some lively times in carrying out this policy. Of course, I may encounter some checks, but I think I shall put it through all right. . . ."

In the wake of this momentous triumph, Bunau-Varilla emerged from the shadows in the role of Panama's representative in Washington. Only fifteen days after the coup, he negotiated a new pact that all but made Panama an American protectorate. The Hay–Bunau-Varilla Treaty conveyed to the United States in perpetuity the "use, occupancy and control" of a ten-mile-wide strip across the Isthmus, language that became a bone of contention for the next seventy years. In return, the Panamanians received the same terms the Colombians had so cavalierly rejected, and the United States undertook to guarantee the new nation's independence. The French stockholders pocketed $40 million; Colombia got nothing.

Most Americans cheered the prospect of a canal across Panama and had little inclination to examine the means used to achieve a goal that meant so much to the world. However, there was a vocal group that bitterly attacked Roosevelt for what they called a sordid land grab at the expense of a weak neighbor.* "This mad

---

*In 1908, the *New York World* charged that the president had protected the investors in the French company because friends and at least one relative, Douglas Robinson, were secret stockholders and the government was hiding records that would prove the truth of these charges. Roosevelt heatedly denied the allegations and had Henry L. Stimson, the federal attorney in New York City, bring libel charges against Joseph Pulitzer. There was

plunge of ours is simply and solely a vulgar and mercenary venture, without a rag to cover its sordidness and shame," declared the *New York Evening Post*. On the other hand, the *Detroit News,* with refreshing candor, simply stated: "Let us not be mealy-mouthed about this. We want Panama."

Incensed at the denunciations, the president asked Attorney General Philander Knox to help construct a defense. "Oh, Mr. President, do not let so great an achievement suffer from any taint of legality," the amused Knox purportedly replied. Having launched into a detailed account of his position at another Cabinet meeting, Roosevelt is supposed to have looked around the table and have caught the twinkling eye of Elihu Root. "Have I answered the charges?" he asked. "Have I defended myself?"

"You certainly have, Mr. President," replied Root. "You have shown that you were accused of seduction and you have conclusively proved that you were guilty of rape."

Over the years, Roosevelt steadfastly denied complicity in the uprising in Panama despite the numerous surreptitious winks and nods given the conspirators. "There had been innumerable revolutions in Panama prior to when I became President," he related. "While I was President I kept my foot down on these revolutions so that when the revolution . . . did occur, I did not have to foment it; I simply lifted my foot."

Roosevelt had no doubts about the righteousness of his conduct. In summary, he argued that the Colombians had upon several occasions agreed to provide a site for the canal and then had dishonorably reversed themselves; they had no right to block a work so vital to the interests of mankind; the national security of the United States would brook no further delay. "From the beginning to the end our course was straightforward and in absolute accord with the highest standards of international morality," Roosevelt declared. "To have acted otherwise than I did would have been on my part betrayal of the interests of the United States."

Following a flurry of opposition, the Senate, on February 23,

---

some question whether the government had jurisdiction, however, and the case petered out after Roosevelt left the White House. When the documents were disclosed, they showed nothing irregular about the transaction. Collin, *Theodore Roosevelt: Culture, Diplomacy and Expansionism,* p. 32.

1904, approved the Hay–Bunau-Varilla Treaty by a vote of 66 to 14. Work on the canal itself began in midyear, well in time for the presidential election—and capped Roosevelt's international activities with a triumph as great as any of his achievements in domestic policy.

Chapter 19

"The Bride at Every Wedding . . . "

The president was coming. Crowds had been gathering since early morning outside the brownstone just off Fifth Avenue, where Theodore Roosevelt was expected momentarily at the wedding of Eleanor and Franklin Roosevelt. It was March 17, 1905—Saint Patrick's Day—and the strains of "Oh Promise Me" were almost drowned out by "The Wearin' o' the Green." Men, women, and children who had raced over after the annual parade, which had just been reviewed by the president, surged about the arriving Astors, Burdens, and Winthrops. Shortly before 3:30 P.M., a shrill cry of "Hooray for Teddy!" was heard, and with a flash of famous teeth, the president dashed up the steps with a shamrock in his buttonhole to give the bride away.

"Well, Franklin," Uncle Ted told the groom when the ceremony was over, "there's nothing like keeping the name in the family."*

But the newlyweds soon found themselves almost alone. Like a pied piper, the president had swept most of the guests along with him into the library. "The room in which the President was holding forth was filled with people laughing gaily at his stories,

---

*FDR was a fifth cousin to the president and a fifth cousin once removed to Eleanor.

410

which were always amusing," Eleanor recalled. There was nothing else for the newlyweds to do but to follow in his wake.

"Father always wanted to be the bride at every wedding and the corpse at every funeral," one of Theodore Roosevelt's sons is supposed to have remarked. If so, he succeeded, for no president was more successful in holding the limelight. In fact, Roosevelt's zest for being at the head of the parade was the characteristic that most distinguished him from his predecessors. He saw the president as a symbol of a modern America, and brought to the White House a star quality not seen in decades. On the other hand, critics charged that by perverting democratic ideals and centralizing governmental power in the White House, Roosevelt laid the foundation for what became known as the "imperial presidency."

Numerous long-standing precedents about what a president should and should not do were rendered obsolete by Roosevelt. He changed the presidency, its power, its scope, and its possibilities. Before the Rough Rider's elevation to the White House, presidents sent messages to Congress only once a year or in times of national crisis; Roosevelt regularly went to the public to make his opinions and priorities known.* While custom dictated that presidents did not leave the continental United States during their term of office, he went to Panama in 1906 to inspect the progress of the Panama Canal.

Not the least of Roosevelt's contributions to the making of the modern presidency was his celebrity status. Already a dashing figure to many of America's seventy-eight million people because of his exploits in the West and Cuba, he turned the widespread curiosity about himself and his activities into a political asset. Having come to power with many political leaders opposed to him and his policies, Roosevelt reached over their heads to convince the great mass of the American people to follow his leadership. He had a way of "slapping the public on the back with a bright idea," said one editor. To be successful, Roosevelt believed that a president must project a personal force that inspired and influenced people.

---

*Woodrow Wilson revived the practice of presidents addressing Congress in person—which had been abandoned by Jefferson—and how Roosevelt must have wished that he had first thought of it.

Roosevelt's impulses were as contradictory as they were strong. As George Mowry has noted,* he was many things to many people because he was such a bewildering array of things to himself. He was likened at one time or another to Julius Caesar, Andrew Jackson, Napoleon, Kaiser Wilhelm II, and a combination of Simple Simon and Machiavelli. In the presence of conservatives, he denounced radicals and preached conservatism; to progressives, he attacked reactionaries and advocated reforms. In his own view, he was a practical man dealing in justice for all. Characteristically American, he wanted to make things work.

Despite Roosevelt's high moral tone, he possessed a streak of ruthlessness and at times broke his own rules for fairness and justice. Individuals were condemned without hearing, he rarely admitted to the possibility of error, and upon occasion employed the dangerous tactic of guilt by association. Opponents who could not be won over were dismissed as traitors or worse. Critics called him cunning, selfish, vindictive, melodramatic, megalomaniacal, dishonest, shallow, and cynical. Perhaps Roosevelt's greatest failure was his insistence on being both a political and moral leader. When he tried to justify political acts in moral terms, he sometimes cast himself in the role of insufferable hypocrite.

To mobilize support, Roosevelt used the White House as "a bully pulpit" from which to preach his version of the gospel, and created the idea of the president as an active participant—if not the dominating factor—in the making of national policy. This provided him an opportunity to dramatize issues and focus public attention upon them. He had a passionate sense of moral right and a fierce need to convey it to others. "Wherever he goes," one caustic critic remarked, "he sets up an impromptu pulpit, and his pious enunciations fall—like the rain and the sunshine, upon the just and the unjust—accompanied with a timely warning to the latter to look sharp!"

Although Roosevelt told William Roscoe Thayer that "one of the most wearing things about being President is the incessant publicity," he was himself a master of publicity and public relations. Every day, the newspapers were filled with stories and pictures of the president and his hyperactive family. Unlike many

---

*The Era of Theodore Roosevelt, 1962.

of his predecessors and successors, who have railed against the "splendid misery" of the presidency, Roosevelt gloried in his office. "I don't think any family has enjoyed the White House more than we have," the president wrote Kermit in the summer of 1904. "I was thinking about it just this morning when mother and I took breakfast on the [south] portico and afterwards walked about the lovely grounds and looked at the stately historic old house. It is a wonderful privilege to have been here and to have been given the chance to do this work. . . ."

Some observers wondered, however, how the "historic old house" was bearing up under the sustained assault of the Roosevelts. Life in the White House resembled nothing so much as a three-ring circus. People were always coming and going, and guests ranged from Henry Adams, who had taken to referring to the president as "Theodorus I, Czar Rooseveltoff," to John L. Sullivan, the prizefighter, and Bat Masterson, the frontier marshal. "Distinguished civilized men and charming civilized women came as a habit to the White House while Roosevelt was there," noted Owen Wister. "For once in our history we had an American *salon.*"

Novelists, sculptors, historians, philosophers, poets, artists, former Rough Riders, and old Harvard friends came to stay for a night or a week, mingle with the political regulars, and over lunch and dinner discuss everything from the day's events to Roman history, or one or more of the hundreds of subjects on which the president was eager to talk. Grudgingly, H. L. Mencken later acknowledged that Roosevelt was the "only . . . President since the birth of the Republic who . . . ever welcomed men of letters at the White House," even though he judged them "by their theological orthodoxy and the hair on their chests."

One companion from Roosevelt's days in Dakota who had been invited to dinner was cautioned: "Now, Jimmy, don't bring your gun along tonight. The British ambassador is going to dine, too, and it wouldn't do for you to pepper the floor round his feet with bullets, in order to see a tenderfoot dance."

"Ike" Hoover, the longtime chief usher, said the Roosevelt years were "the wildest scramble" in the history of the White House. Visitors were sometimes startled by a loud clatter as Ar-

chie or Quentin slid down the main staircase on metal trays.
Cabinet members recoiled as small boys—members of the "White
House Gang," as Quentin's buddies were known—popped out
of vases in the East Room. Quentin once took his Shetland pony,
Algonquin, upstairs in the elevator to visit Archie, who was sick
in bed. A family parrot screeched, "Hurrah for Roosevelt!" at
unsuspecting guests. When Speaker of the House Joe Cannon
tried to discuss weighty matters with the president, a kitten leaped
into his lap. The South Lawn became a baseball field for the
chums Quentin brought home from public school. The president
was often an interested spectator and cheered the batters. "Hit
it!" he would yell. "Hit it!" Sometimes he joined Ethel's friends
for games of hide-and-seek in the attic, usually insisting upon
being "it," one participant recalled years later.

It was not unusual for an important meeting to be interrupted
by a tap on the door of the presidential office and the appearance
of a small group of boys whose leader might shyly announce, "It's
after four."

"By Jove, so it is!" the president might say, and as he adjourned
the meeting, he would tell his visitor that he had promised to take
the boys walking at four o'clock. "I never keep boys waiting. It's
a hard trial for a boy to wait."

Evenings before dinner, he dropped in on Archie and Quentin
before they went to bed to read to them or for chats and pillow
fights. "The other night before the diplomatic dinner, having
about fifteen minutes to spare, I went into the nursery, where the
two small persons in pink tommies instantly raced for the bed
and threw themselves on it with ecstatic conviction that a romp
was going to begin," he reported to Kermit. "I did not have the
heart to disappoint them, and the result was that my shirt got so
mussed that I had to change it."

Although careful of the dignity of his office, Roosevelt did not
let it interfere with his own pursuit of the strenuous life. He
continued to box, even though he received a blow while sparring
with a military aide that eventually cost him the sight of his left
eye. Characteristically, he hid the injury, so as not to alarm the
young man. He also took up such pastimes as wrestling, judo,
and parrying with single sticks. He was the first president to go
down in a submarine, and after leaving the White House, was

the first to go up in an airplane. "I've had many a splendid day's fun in my life," he said after taking the controls of the submarine *Plunger* and practicing dives in Long Island Sound, "but I can't remember ever having crowded so much of it into such a few hours."* Of the flight, in 1910, he said, "By George, it was great!"

He went riding at every opportunity and disturbed by reports of the poor physical condition of senior army officers in Washington, decreed that they would have to pass a fitness test that included a ninety-mile ride in three days. To counteract grumbling by desk-bound officers, he personally led a party that rode to Warrenton, Virginia, and back, covering the hundred miles in a day. The last few miles were covered in the dark, with snow and sleet blowing in the faces of the riders.

Roosevelt organized his closest personal, political, and diplomatic friends into a "Tennis Cabinet," which joined him on the newly laid out court behind the White House or in rambles through Rock Creek Park, where streams were forded and cliffs were scaled. Jules Jusserand, the French ambassador, wrote a lighthearted account of an afternoon "promenade" with Roosevelt:

> I arrived at the White House punctually, in afternoon dress and silk hat, as if we were to stroll in the Tuilleries Garden. . . . To my surprise the President soon joined me in a tramping suit. . . . Two or three other gentlemen came and we started off at what seemed to me a breakneck pace, which soon brought us out of the city. On reaching the country, the President went pell-mell over the fields, following neither road nor path, always on, on, straight ahead! I was much winded, but I would not give in, nor ask him to slow up, because I had the honor of *la belle France* in my heart. At last we came to a bank of a stream, rather wide and too deep to be forded. I sighed relief, because I thought that now we had reached our goal and would rest a moment and catch our breath before turning homeward. But judge of my horror when I saw the President unbutton his clothes and heard him say, "We had better strip, so as not to wet our things in the Creek." Then I, too, for the honor of France removed by apparel, everything except my lavender kid gloves. The President cast an inquiring look at these as if they, too, must come off, but I quickly forestalled any remark

---

*Following his visit to the *Plunger,* Roosevelt ordered that enlisted men detailed to submarines be granted an extra $10 a month as hazardous-duty pay.

by saying, "With your permission, Mr. President, I will keep these on; otherwise it would be embarrassing if we should meet ladies." And so we jumped into the water and swam across.

The president's day usually began at 7:30 A.M. He bathed, shaved, and dressed in the clothes laid out for him by James Amos, his valet, who doubled as a White House waiter. Unlike his days as a "dude" at Harvard, Roosevelt cared little about what he wore except for shoes. He had small feet for a man of his bulk, and his shoes were specially made for him. He usually wore a frock coat, a light waistcoat, striped pants, a shirt with a turned-down collar, and a four-in-hand tie. At Oyster Bay, he favored knickers and was persuaded only with difficulty to change to more formal attire to receive visiting dignitaries. He always dressed for dinner both in Washington and at Sagamore, however, even when he and Edith dined alone.

Breakfast, usually served about 8:00 A.M., was a family affair in the Red Room. With Alice often visiting elsewhere, Ted and Kermit at Groton, and Ethel away at school during the week, those present included the president, his wife, Archie, Quentin, and whoever had spent the night at the White House. Roosevelt had a passion for peaches and cream, and when the fruit was in season he demolished a soup-bowl–sized serving every morning. He also drank cup after cup of heavily sweetened coffee, using a large cup that Ted described was "more in the nature of a bath-tub."

Following breakfast, the president took a short walk, then was at his desk by 9:30 every day except Sunday. In 1902, the White House underwent an extensive renovation, and the offices, which had adjoined the second-floor living quarters, were moved to a newly constructed West Wing. Inspection had revealed structural beams in danger of collapse, a welter of heating pipes and electrical wiring in the basement that were a menace to safety, and an infestation of vermin—all of which persuaded Congress to appropriate $540,000 for repairs and new construction.

Working with Charles F. McKim, a member of the distinguished architectural firm of McKim, Mead, and White, Edith directed the most extensive restoration of the old mansion in its history, returning it to the classic lines and decor of the Adams and Jef-

ferson eras. The old offices were converted to bedrooms and a
private study for the president. "The changes in the White House
have transformed it from a shabby likeness to the ground floor
of the Astor House into a simple and dignified dwelling for the
head of a great republic," Roosevelt declared.

Low and unobtrusive, the West Wing was attached to the main
house by a colonnade and contained ample space for the president
and his thirty-eight assistants, including several typists—called
"typewriters" in those days. The president's office itself was about
thirty feet square, with three large windows looking south to the
Potomac and the Virginia hills. The top half of the Washington
Monument was visible over the backscreen of the tennis court.*
Sliding doors opened to the Cabinet Room, where the heads of
the nine major departments of government met with the president
on Tuesdays and Fridays at eleven o'clock.

The presidential office was simply decorated with olive burlap
walls and matching curtains. There was a fireplace opposite the
large mahogany desk, which usually had a few papers and books
upon it. Over the mantel hung a portrait of Lincoln. Bookshelves,
a leather divan, an art nouveau lamp, and a few chairs completed
the decor. There was no flag or telephone. The only personal
touches were a large globe, a photograph of a bear, and a copy
of a sonnet, "Opportunity," signed by the author, Senator John
J. Ingalls of Kansas:

> Master of human destinies am I.
> Fame, love, and fortune my footsteps wait;
> Cities and fields I walk; I penetrate
> Deserts and seas remote, and passing by
> Hovel, the mart, and palace, soon or late
> I knock unbidden once at every gate!
> If sleeping wake—if feasting, rise before
> I turn away. It is the hour of fate,
> And they who follow me reach every state
> Mortals desire, and conquer every foe
> Save Death; but those who hesitate,
> Condemned to failure, penury and woe,

---

*The Oval Office was added during the presidency of William Howard Taft on the site
of the Roosevelt-era tennis court—which could be construed as an effort by the 350-
pound Taft to exorcise the strenuous spirit of his predecessor.

Seek me in vain, and uselessly implore,
I answer not, and I return no more.

Once Roosevelt was seated at his desk, the office hummed at full speed. William Loeb came in with a list of the day's appointments, and then the president began answering the most important and urgent of the five hundred letters received daily at the White House. Edith went over his mail first, culling out letters worthy of his personal attention. Usually answering letters within twenty-four hours of their receipt, he dictated in rapid-fire style, "often stopping, recasting a sentence, striking out and filling in, not disturbed by interruptions." As soon as one stenographer had filled his notebook, another took his place while the first batch of letters was being typed for the president's signature. Secretaries came and went with notes for the president without disturbing his train of thought. Upon one occasion, he dictated a thirty-thousand-word critique of a book he had just read and called for a fresh "shorthand man" to deal with government business while the first stumbled out of the office, exhausted.

Roosevelt also sifted through the pile of newspaper and magazine clippings gathered for his attention. He insisted that every news article, no matter how unfavorable to him or his administration, be shown to him, and a staff member was assigned to look through 350 newspapers each day and to clip items that reflected the mood of the nation. Edith also read four newspapers, as well as several magazines and journals, and pointed out items that he should be aware of.

When he had finished the mail, Roosevelt threw open the doors of the Cabinet Room and received visitors from about ten o'clock until 1:00 P.M. Congressmen and senators did not need an appointment if they came to see him from ten to noon, while members of the public could speak with the president if they made arrangements through William Loeb. William Bayard Hale, a *New York Times* reporter who spent considerable time observing Roosevelt, painted a vivid picture of the scene:

Sometimes a score of people will be in the Cabinet Room at one time, and the President goes from one to another, making the circle of the room half a dozen times in a morning, always speaking with great animation, gesturing freely, and in fact talking with his whole being, mouth, eyes, forehead, cheeks and neck all taking their mobile parts. . . . When the President sits, it may be on the divan or on the Cabinet table, he is very much at his ease, and half the time one foot is curled up under him. Curiously whenever he tucks one foot under him his visitor is very likely to do the same thing. . . . A hundred times a day the President will laugh, and, when he laughs he does it with the same energy with which he talks. It is usually a roar of laughter, and it comes nearly every five minutes. His face grows red with merriment, his eyes nearly close, his utterance becomes choked and sputtery and falsetto, and sometimes he doubles up in paroxysm. You don't smile with Mr. Roosevelt; you shout with laughter with him, and then you shout again while he tries to cork up more laugh[ter] and sputters: "Come gentlemen, let us be serious. . . ."

Once, when Senator Chauncey Depew came to the White House to lobby in behalf of a candidate for a diplomatic post, he tried to take the president aside to talk with him privately. "We have no secrets here; tell it all right out," Roosevelt declared—and after hearing him out in public, promptly declined to make the appointment. Upon another occasion, Depew entered Roosevelt's office just as another senator was leaving.

"Do you know that man?" the president asked.

"Yes, he is a colleague of mine in the Senate," Depew replied.

"Well, he's a crook," Roosevelt declared.

Each afternoon, at one o'clock, the president met the press. Usually the half dozen or so reporters assigned to the White House* gathered in a small room between his office and that of his secretary, while Roosevelt was shaved by a Treasury Department messenger who also served as a barber. Roosevelt would bustle in, coattails streaming behind him, eyeglass cord flying, teeth gleaming, and plop into an armchair, and as the barber did his work, the newsmen took turns throwing questions at him. The president, who usually filibustered when he spoke, had to learn

---

*The White House press corps is supposed to have been created one chilly, wet day in 1902 when Roosevelt happened to look out the window of his old office on the second floor and saw a cluster of sodden newsmen huddled on the North Portico waiting to interview visitors passing in and out of the building. Taking pity on them, he ordered that a small room adjoining his study on the first floor be set aside for the press. The reporters were given more permanent headquarters when the West Wing was opened. Juergens, *News from the White House*, p. 14.

to restrain himself because of the circumstances, but there were some close calls between the razor and the presidential throat.

"A more skillful barber never existed," recalled Louis Brownlow of the *Louisville Evening Post*. " 'Teddy' . . . would be lathered, and as the razor would descend toward his face, someone would ask a question. The President would wave both arms, jump up, speak excitedly, and then drop into the chair and grin at the barber, who would begin all over. Sometimes these explosions interrupted a shave ten or a dozen times. It was more fun to see than a circus. . . ."

Roosevelt also saw particular favorites on an individual basis, among them Oscar K. Davis, chief of *The New York Times* bureau. Known to the president by his nickname, "O.K.," Davis dropped by two or three times a week in the late afternoon while Roosevelt was signing letters and documents. The newsman marveled at the president's ability to keep a conversation going while scanning letters, making corrections and insertions. By and large, Roosevelt liked most of the men assigned to cover him, was on a first-name basis with them, and enjoyed the verbal fencing. There was no formal post of White House press secretary—that did not occur until the Hoover administration—but William Loeb fulfilled its functions.

Washington reporters were amazed at Roosevelt's openness. For a press corps accustomed to hearing presidential remarks no more revealing than a Fourth of July oration, the Roosevelt style was both exhilarating and unsettling. While he could not be quoted, he was free with his comments and sometimes so blunt in his characterizations of Washington personalities that even the reporters were concerned about the things he told them. But if Roosevelt was outspoken, he was also no fool. To be welcome in the White House, a man had to adhere to strict rules, rules that no reporter ever had the opportunity to violate twice. Punishment was swift and without appeal—banishment from the presidential inner circle. If Roosevelt said something that was not for publication, it remained a secret. Reporters had to have a personal sense of what was and was not intended for publication, for the president had no intention of seeing his most outrageous remarks in print.

While many major newspapers—among them the *Times* and

the *New York World*—opposed Roosevelt's policies, the president realized that by addressing the working press he could override the effect of the editorial pages. More than any other president before him, he understood that anything coming out of the White House was news and he capitalized on this fact. Realizing that newspapers faced the immense problem of filling their columns on Monday after a dull weekend, he timed his statements so they would appear in the Monday morning newspapers, when there was less competition for public attention. The *Times* index for 1907 has him beating the bishop of London at tennis, swimming in a freezing lake, dining on bear meat in the White House, and announcing that he will no longer use the expression "*dee*-lighted."

There was a mutual benefit in the relationship between the president and the press. If the reporters remained in the presidential good books, they were plugged into an unbeatable source. They were in an enviable position to learn about Roosevelt's plans and policies before they surfaced elsewhere, and information received on background might well transform something into an important story at a future date. But the process gave the president the whip hand. Reporters had to toe the line to remain insiders—which caused some editors to grouse that their men in the White House often seemed to be working for the president rather than for their newspapers.

Roosevelt was a past master of the leak, the trial balloon, and the unattributed source, and resorted to them to float policy. Often, he used off-the-record remarks to put the lid on a story that appeared likely to break at an inconvenient moment. If a proposal was shot down, he was free to deny ever having considered it. Sometimes, if events made it necessary, he was not above denouncing the reporter to whom he had leaked the story as a liar and fraud and to consign the hapless fellow to his Ananias Club, so named for the New Testament figure who was struck dead for lying.

Nevertheless, covering Roosevelt was an intoxicating experience. Excitement and controversy are the heartbeat of journalism, and the president was constantly embroiled in public squabbles and feuds that became public entertainments. He fell out with Bellamy Storer, whom he had appointed envoy to Vienna, after

Maria Storer involved the White House in intrigues with the Vatican to secure a cardinal's red hat for Archbishop John Ireland of St. Paul. The ensuing battle between the president and "Dear Maria" was waged in the newspapers, which gleefully printed heretofore secret letters that had been exchanged by the parties as well as the maledictions they now hurled at each other.

And there were the "nature-fakers." In his position as naturalist and big-game hunter, Roosevelt was angered by the tendency some writers had of crediting birds and animals with the ability to think and act like human beings. The major object of Roosevelt's ire was a clergyman named William J. Long, who wrote wildlife books featuring woodcocks that made casts out of mud for broken legs and wolves befriending lost children, and he broadened his attack to include "nature-fakers" in general. Three months of headlines, cartoons, and jokes followed, much to Reverend Long's distress. He would have been interested in Roosevelt's explanation of the origins of his attack on writers of "unnatural" history to John Burroughs: " . . . I ought not to do this; but I was having an awful time toward the end of the session and I felt I simply had to permit myself some diversion."

In 1906, Roosevelt's attention was attracted to the simplified spelling movement—the president's own spelling was always somewhat idiosyncratic—and he decreed that henceforth the Government Printing Office would adopt some three hundred spelling changes. "Axe" was to become "ax," "woe" would be "wo," "killed" was to be "kilt," "clasped" would become "claspt," "through," "lopped," and "drooped" would be "thro," "lopt," and "droopt." A wave of protest greeted the decree, with *The New York Times* declaring that it would regard these novelties as misprints. The *Baltimore Sun* asked how the president would spell his name. "Will he make it 'Rusevelt' or will he get down to the fact and spell it 'Butt-in-sky?' "

For once, Roosevelt realized that he had gone too far, too fast. When the House of Representatives angrily voted, 142 to 14, to refuse to appropriate funds for printing documents in anything but standard English, including those emanating from the White House, he backed off. To Brander Matthews, a Columbia professor and advocate of simplified spelling, he wrote: "I could not by fighting have kept the new spelling in, and it was evidently worse than useless to go into an undignified contest when I was beaten." In a gesture of defiance, however, he announced he would use the "new spelling" in his own correspondence.

Roosevelt was a godsend for newspaper and magazine cartoonists because they were never at a loss for a subject when he was around. With his flashing teeth, thick glasses, and kinetic energy, he seemed to leap off the drawing board. Over the years, he was the subject of hundreds of cartoons, which helped spread his fame even wider while humanizing him. One cartoon inspired the teddy bear. In November 1902, Roosevelt went on a bear hunt to Mississippi, but game was scarce. Finally, a bedraggled black bear of about 230 pounds was run down by the dogs and roped after it killed one of them. The president was summoned to kill the bear, but he indignantly refused to shoot the animal under such unsporting conditions.

Clifford K. Berryman of the *Washington Post* produced a whimsical cartoon based on the incident, "Drawing the Line in Mississippi." Morris Michtom, a Russian Jewish immigrant who ran a toy shop on Brooklyn, was inspired by the cartoon to create a cuddly stuffed bear for children. He gave it the president's nickname, and the bear became an international phenomenon.*

Informality also marked the president's relations with his Cabinet. Over the years, the McKinley appointees, whom he had kept on as part of his policy of reassuring the nation, were replaced with his own men. Following the death of John Hay, Elihu Root became secretary of state, while William Taft was brought back from the Philippines to become secretary of war. Charles J. Bonaparte, the Baltimore reformer, became first secretary of the navy—a post that Roosevelt all but filled himself—and then attorney general. George B. Cortelyou took over the Treasury Department, and George von Lengerke Meyer, the ambassador to Russia, replaced Cortelyou as Postmaster General. James R. Garfield, the son of the late president, was elevated from chief of the Bureau of Corporations to head of the Interior Department.

---

*Michtom is alleged to have written the president for permission to use his name, and Roosevelt is said to have replied: "I don't think my name will mean much to the bear business, but you're welcome to use it." Schullery, Introduction to *American Bears,* pp. 10–11.

Oscar Straus, a New York businessman-politician, became the first Jew to be appointed to the Cabinet when Roosevelt named him secretary of commerce and labor.* Agriculture Secretary James Wilson, who had ties to the farm vote, was the sole McKinley carry-over to serve throughout Roosevelt's term of office.

The only formal part of the Tuesday and Friday Cabinet meetings was the seating arrangement, in which the chairs were arranged in chronological order of the creation of the departments, and Roosevelt presided at the head of the large table. Each chair had a silver nameplate affixed to its back, giving the occupant a false sense of permanence. Unless the session was called to discuss a specific topic, it opened with a general conversation about governmental affairs. The president then called upon each member to talk about the matters that especially concerned his agency.

Roosevelt allowed the members of his official family wide latitude in the administration of their departments, which sometimes resulted in conflict, as when Leslie M. Shaw, one of his appointees to the Treasury, was outspoken in his fervor for protectionism. In seeking information Roosevelt often bypassed Cabinet members and directly asked bureau chiefs and head clerks. This was praised as a way of cutting red tape, but he was forced to abandon the practice when some department chiefs claimed it was subversive of discipline.

Throughout American history, the function of the Cabinet has been a subject of debate. Should it be a consultative body or merely a staff meeting of departmental administrators? George Washington sought the counsel of his ministers, but later presidents found these officials too specialized and concerned with the operations of their own departments to provide general advice. Nevertheless, they had persisted in trying to make use of the

---

*Before making the appointment, Roosevelt, according to a story that is doubtless apocryphal, sought the advice of Jacob Schiff, the financial baron and pillar of New York's German-Jewish community. Not long afterward, at a banquet sponsored by some prominent Jews in honor of Straus, the president declared that Straus had been named because of his ability, and that the question of the nominee's religion had not even crossed his mind. He turned to Schiff for confirmation. The old man, whose hearing was failing, quickly exclaimed:

"Dot's right, Mr. President! You came to me and said, 'Chake, who is der best Jew I can appoint Segretary of Commerce.' " Harbaugh, p.12.

Cabinet as a consultative body, and Roosevelt followed suit. However, his experience was similar to that of his predecessors, and he turned for advice and information from an ever-widening circle of outside informants.

Perhaps Roosevelt's greatest service as an administrator was his ability to attract to public service a host of young, energetic, and well-educated men whose like had never been seen before in Washington. The bureaucracy was vested with a zeal and excitement it had not known before. Most prominent among these bright young men were Gifford Pinchot, the tall, cadaverous-looking head of the Bureau of Forestry in the Department of Agriculture and a leading conservationist, and James Garfield, who argued that the only way to meet the threat of monopolies was through federal regulation. Lord Bryce remarked that he had never encountered "a more eager, high-minded, and efficient set of public servants. . . ."

Nevertheless, Roosevelt was not above using patronage to advance his policies and political prospects. He had come into office with the reputation of being a firm supporter of the civil service principle, but some of his appointments of Republican wheel-horses caused headshaking among reformers. They were most shocked by his naming James S. Clarkson as surveyor of the Port of New York. Clarkson had been one of John Wanamaker's key aides during Roosevelt's row with Wanamaker during the Harrison years, but as the president privately acknowledged, "In politics, we have to do a great many things that we ought not to do."

White House luncheons were animated affairs—more of a continuation of the president's effort to gather information than a meal. "The President throws off his official harness, which he does very readily, and expresses himself with that freedom which is so natural to him," Oscar Straus reported following a White House lunch. "He was as buoyant and full of spirits as a young college graduate. He has a wonderful fund of humor." Anywhere from two to twenty guests were invited to join the president and his wife at table. The meal itself was usually simple—clear soup, a chop, or grilled fish—in view of the president's tendency to put on weight. Roosevelt had a fondness for game and terrapin, but he could be just as happy with Irish stew, pork and beans, or bread and milk. Quantity was more important to him than quality. He was "an eager and valiant trencherman," said one friend, who had seen him devour a large chicken as well as other food and drink four large glasses of milk at one sitting.

Following lunch, the president returned to his office for more meetings and discussions, to work on papers, and to sign documents and letters he had dictated earlier. He also tried to snatch a few moments of reading and finished at least a book each day. Blessed with the rare gift of absolute concentration, he read whenever he found a free moment and even kept a volume by the White House door to read while waiting for distinguished guests to arrive. When he read a magazine, it was as if he were doing battle with it. He would read a page quickly, tear it out, crumple it up, and continue until the magazine had been devoured.

Roosevelt's reading was eclectic, ranging from Old Norse literature to Lecky's *History of Rationalism in Europe*. He was curious about new writers, and when he found a book that delighted him, he would fire off a letter of appreciation to the author, sometimes accompanied by an invitation to lunch or dinner at the White House. When Kermit discovered the poetry of Edwin Arlington Robinson and sent his father a copy of Robinson's first book, *The Children of the Night*, the president not only read it but reviewed the book favorably in *The Outlook*. Upon learning that the poet was working as a timekeeper on the construction of the New York subway and was having a hard time making ends meet, Roosevelt found him a federal job that would not interfere with his writing.

Roosevelt was the first president since Thomas Jefferson to have an active interest in the graphic arts. When the Detroit industrialist Charles L. Freer offered his great collection of Oriental art to the government, which was reluctant to accept it, the president personally intervened in favor of acceptance. He appointed a fine arts council to advise the federal government on building design, and he is credited with not only being the real father of the National Gallery of Art but supporting the architects that restored Washington to the original L'Enfant plan.

Unhappy with the lack of artistic merit of the nation's coinage, he commissioned his old friend Augustus Saint-Gaudens to design new $10 and $20 gold pieces, which he ranked "close alongside the best and most beautiful of the old Greek coins." There was a barrage of criticism, however, because the coins omitted the motto "In God We Trust," which Roosevelt regarded as blasphemy. Congress, however, quickly passed legislation restoring the motto.* These coins were followed by the buffalo nickel, Lincoln penny, and twenty-five-cent and half-dollar eagle coins, designed by Saint-Gaudens's associates.

All in all, Roosevelt's interests in arts and letters were almost unique among American chief executives. To Louis Einstein, a perceptive observer, he seemed to reincarnate the Renaissance ideal of "the well-rounded life of thought and action." Einstein thought that the president was like Italian princes of the sixteenth century in combining a thirst for learning and adventure—mastering both books and grizzly bears. Although his taste was straitlaced and tended toward the conservative, Roosevelt defended the reputation of Edgar Allan Poe, developed an interest in the paintings of Turner, and in 1913, he wrote a not completely unsympathetic review of the controversial Armory Show, which introduced most Americans to modern art.

When the top of his desk was clear at last, Roosevelt would go for a ride, play tennis, or exercise. If he planned on a ride, he would ask Edith to accompany him. If she could not, he would ask friends, perhaps Leonard Wood, Cabot Lodge, or Jim Garfield. When Edith joined him, they would drive in a carriage out to Park Road, then on the fringes of the city, where Bleistein and Yagenka, their favorite mounts, would be waiting for them. Edith was an excellent horsewoman, and the Secret Service men were hard pressed to keep up with the couple. When they returned home, it was time to dress for dinner.

If Theodore Roosevelt laid the foundation for the modern presidency, then Edith institutionalized the position of first lady. Although the name had first been used a half century before, she helped establish the role of the president's wife as a public, semiofficial one. She was the first to hire a social secretary—Isabelle Hagner, member of an old Washington family—the first to es-

*For Roosevelt's views, see *Letters*, Vol. V, pp. 842–843.

tablish her own office, and the first to include a cameo of herself along with that of the president on the engraved formal invitations sent out by the White House. She also began the White House china collection and that of portraits of first ladies, which were installed in a special gallery. Edith established new standards for official entertaining by holding a series of musical evenings in the newly refurbished East Room, at which Pablo Casals,* Ignace Jan Paderewski, and the entire Philadelphia Orchestra performed.

Even though she was forty-one and responsible for six active children and running the White House, Edith wanted another child. On April 24, 1902, she excitedly informed her sister, Emily Carow, that she was pregnant. "I have the most enormous appetite at present, though I loathe anything sweet, can't touch wine and care but little for tea and coffee! Meat seems to appeal to me!" About two weeks later, there is a cryptic entry in her diary: "Was taken sick in the night." As a biographer put it, the happy exclamation points quickly disappeared from her correspondence. A year later, Edith was again pregnant, but once again miscarried.

Reserved to the point of aloofness, the first lady assiduously avoided the limelight, and was no activist like her niece, Eleanor Roosevelt, who assumed the first lady's role three decades later. But Edith had considerable influence over her husband. "She was the perfection of 'Invisible Government,' " according to Owen Wister. Sometimes, she put a brake upon his natural exuberance with a cautionary "Theodore!!" to which the president would meekly respond: "Why *Ee*-die, I was only . . ." "We all knew that the person who had the long head in politics was Mother," Ethel Roosevelt observed years later.

Edith established her office next door to his study, and he conferred with her several times a day between appointments. She only gave advice when it was sought—usually telling him what to avoid rather than do. "Never when he had his wife's judgment, did [Roosevelt] go wrong or suffer disappointment," according to Mark Sullivan, a prominent journalist. Often, if the president was occupied when officials arrived, Edith would entertain them. As her knitting needles clicked, her sharp eyes and ears registered the nuances of their conversation, for as James

---

*Sixty years later, Casals returned to the White House to play for the Kennedys.

Amos, Roosevelt's valet, remarked, she was a shrewder judge of people than the president. He tended to like almost everyone; Edith viewed the world with a jaundiced eye.

Roosevelt also conferred frequently with his older sister, Mrs. Cowles, seeking counsel on the issues facing him, and valued her advice. "He may have made his own decisions, but talking with her seemed to clarify things for him," observed Eleanor Roosevelt. Edith, however, was unhappy about these contacts and insisted that even the president's sisters be required to make appointments with him. Roosevelt agreed as always—nobody crossed Edith—but Bamie's house at 1733 N Street became the destination of many of his walks about Washington.*

The Roosevelts set new standards for White House entertaining. Trumpets resounded as the president and his wife descended the new grand staircase to greet guests at glittering diplomatic receptions and state dinners. Because of the indifferent quality of the White House kitchen, Edith had major affairs catered by Rauscher, the capital's leading caterer.† The menus included fish, fowl, and meat courses, and four or five wines as well as champagne. Roosevelt himself hardly drank, and Amos, his valet, had standing orders to fill his glasses with ice before the guests came to the table so it would dilute the wine. Even so, he insisted on serving the best champagne, even though the cost of food and drink came out of his own pocket. Roosevelt spent more for entertaining than had any previous chief executive, but shrugged it off, saying he had "a horror of trying to save any money out of his pay."

Family dinners, at which the president liked to entertain visiting celebrities along with his usual round of friends, were far simpler. Usually Roosevelt dominated the conversation, but sometimes he would become fascinated by a guest with an unfamiliar intellectual specialty, one of the means through which he amassed the enor-

---

*Usually the president was accompanied on his walks only by a single Secret Service agent. With his predecessor's fate in mind, he packed a pistol himself, telling one friend, "I should have some chance of shooting the assassin before he could shoot me, if he were near me." He also kept a pistol by his bedside.

†Edith's frugality was evident in the provision that the caterer was to be responsible for both dishwashing and breakage.

mous amount of information he kept at his fingertips. Once, recalled Owen Wister, the featured guests were an Italian expert
on tunnels and an English economist who was urging Britain to
adopt the decimal system of coinage. Roosevelt talked "as if a
decimal and then as if a tunnel was the only thing in the world
he ever loved. He [referred] to the Thames and Severn tunnels . . . to the monetary system of China, or the currency experiments of Frederick II of Hohenstaufen."

"Sometimes an unfamiliar would prove a delightful surprise,"
the novelist said, but it was only when the usual circle was assem-
bled "that Roosevelt let himself go, that the whole company let
itself go, that it became sheer luxury to listen to those distinguished and brilliant men turning their minds loose to the Wright
Brothers' flying machine, the president's interest in judo, government shoptalk and the works of Mark Twain or Sir Walter
Scott." Edith, as well as the other female guests, Mrs. Oliver
Wendell Holmes, Mrs. John LaFarge, Nannie Lodge, Mrs. Winthrop Chanler, took a full part in the repartee. Upon one occasion,
Finley Peter Dunne, the creator of Mr. Dooley, was a guest. "Do
you know, Mr. President," he gravely told Roosevelt, "the appearance of your cabinet is a great disappointment to me. *I don't
believe one of them has ever killed a man.*"

Perhaps the most intellectually brilliant of all the White House
dinners took place on January 12, 1905, when the Roosevelts
entertained Henry James, Henry Adams, Augustus Saint-Gaudens, and John LaFarge. While warding off Adams's barbs, the
president tried to warm up to James, whose convoluted prose
style mystified him as much as his decision to live and work in
England angered him. For his part, James later summed up Roosevelt as "a wonderful little machine . . . destined to be over-
strained perhaps, but not as yet, truly exciting to see. It functions
astonishingly, and is quite exciting to see."

Except upon special occasions, White House evenings ended
at ten o'clock. Edith usually went to bed immediately, while the
president read or wrote in his study, and it was after midnight,
or one o'clock, before he finished work and turned in for the
night.

Fascinated by the vivacious first family, Americans wanted to
know the most intimate details of their lives. Reporters even tried

to pry information out of Quentin about his father's latest esca-
pades. Guardedly, the boy replied that he saw the president occasionally, but added that he "knew nothing about his family life."
Roosevelt tried to shield his family from the limelight and urged
the press to leave the children alone. "I want to feel that there
is a circle drawn about my family," he said. "I ask you to respect
their privacy."

Modern journalism being what it is, he was asking too much.
One of the worst times occurred in 1905 when the press hounded
Ted, who had just enrolled at Harvard, following him to the
classroom door, onto the playing field, and intruding upon his
social life. "You have been having an infernal time through these
cursed newspapers," the president told his son. "The thing to do
is to go on just as you have evidently been doing, attract as little
attention as possible, do not make a fuss about the newspapermen,
camera creatures, and idiots generally, letting it be seen that you
do not like them, and avoid them, but not letting them betray you
into any excessive irritation."

But if Ted wished to avoid public notoriety, Alice cultivated it.
Quickly dubbed Princess Alice by the press, she soon rivaled her
father as a media superstar. Pretty, impulsive, and fun-loving, she
pursued pleasure with the same exuberance he showed in politics.
Six hundred people flocked to the White House for her debut in
January 1902, and although Alice complained that punch rather
than champagne was served, Aunt Corinne noted her niece "had
the time of her life" with "men seven deep around her all the
time." Cousin Franklin was among those in attendance; Eleanor
was still at school in England.

Having had to compete with the other children for attention
at home, Alice used outrageous behavior in public to assert her
own identity. Victorian prudery was giving way to the freer and
easier style of the Edwardians, and it was a happy time for non-
conformity. The newspapers were filled with stories about her—
where she went, what she said, whom she saw, and especially
what she wore. The clothes and cartwheel hats of the period
seemed made to order for her haughty and sensual Gibson Girl
looks, and her favorite blue-gray color, which matched her steely
eyes, was the fashion rage known as "Alice blue." Alice became
a favorite name for babies, and an adoring public sang "Alice
Blue Gown" and "Alice, Where Art Thou?"

Shocking the prissy and prudish, Alice smoked in public, flirted with men, flourished a pet snake named "Emily Spinach,"* was seen betting at the racetrack—and gleefully counting her winnings—and drove her red runabout at "reckless" speeds until the Washington police caught up with her. She consorted with society women and actresses such as Ethel Barrymore, and in Washington her constant companion was the beautiful and scandal-tainted Marguerite Cassini, daughter of the Russian ambassador.†

The president found his daughter both a distraction and a political asset, while Edith, who believed that a lady's name appeared in the press only to announce her birth, marriage, and death, was horrified by Alice's carryings-on. "Sister continues to lead the life of social excitement, which I think is all right for a girl to lead for a year or two," Roosevelt ruefully told Ted, "but . . . I do not regard it as healthy from the standpoint of permanence. I wish she had some pronounced serious taste." Alice and her father argued frequently but the president, who never overcame his lingering guilt about remarrying after her mother's death, never considered that the girl's actions were an attempt to attract his attention. When Owen Wister asked him if he couldn't control Alice, Roosevelt replied: "Listen, I can be President—or—I can attend to Alice."

One evening, not long after Roosevelt had become president, a young farmer from neighboring Syosset arrived at Sagamore and insisted to a Secret Service man that he had an appointment with the president, who wished him to marry Alice. The guard sent him away, but the would-be suitor returned soon after and made a commotion. Hearing the disturbance, Roosevelt came out on the porch, where he was outlined by the light behind him. "There he is now!" the intruder shouted and whipped his horse toward the house, but the guard managed to stop him. A pistol was found on the floor of his buggy. Roosevelt described him to a friend as "a poor, demented creature." But he had a different comment for the family. "Of course he's insane. He wants to marry Alice."

---

*After Edith's very thin sister, Emily Carow

†"It is Anna Karenina!" Roosevelt had exclaimed upon first meeting Marguerite in 1901.

"Why Anna Karenina, Colonel Roosevelt?" the girl asked.

"There is a tragedy in your eyes," was the reply.

Alice, whose own thoughts were turning toward marriage, was being seen more and more in the company of Nicholas Longworth, a balding Republican congressman from Ohio who was fifteen years her senior. A member of a wealthy Cincinnati family, Longworth was a debonair bachelor with a taste for wine and women and the violin—which he played well enough to earn the praise of professionals. He could also perform holding the instrument behind his back and the bow between his knees. At first, Longworth was attracted by the sultry charms of Marguerite Cassini and wooed her with dozens of long-stemmed roses, but after the flirtatious countess refused to marry him, he turned his attentions to Alice.

Both Alice and Longworth accompanied Secretary of War Taft on a trip to the Far East, which one newspaper described as "Alice in Wonderland." Along the way, Nick proposed and Alice accepted, but she still harbored doubts. At a ball held in Tokyo, she expressed them to Lloyd Griscom, a friend:

"Lloyd, do you see that old, bald-headed man scratching his ear over there?" she asked.

"Do you mean Nick Longworth?"

"Can you imagine any young girl marrying a fellow like that?"

"Why, Alice, you couldn't find anyone nicer."

"I know, I know. But this is a case of marriage."

But in her diary, Alice was more passionate. "I love you with everything that is in me Nick, Nick, my Nick," she wrote.

Edith, who had previously observed that Longworth drank, was cool to the match, but the president, often a poor judge of character, was pleased. Although Longworth was considerably older than Alice, he was regarded as a possible steadying influence. He was clever, well off, an administration supporter, had a good career in politics ahead of him, and, as the president noted, had been in Porcellian at Harvard. "Alice is really in love and it is delightful to see how softened she is," observed Edith. But, "I still tremble when I think of her face to face with the practical details of life."

Alice's wedding, on February 17, 1906, was the most glittering social event ever held in the White House. Five hundred friends and relatives crowded into the East Room, and thousands of spectators milled about in the streets outside. At the stroke of noon the bride, lovely in white satin and lace that had belonged to her mother and wearing a diamond necklace that was a gift from the groom, appeared on the arm of her rather grim-looking father. The vows were exchanged in a bower of flowers. As soon as the ceremony was over, Alice stunned family intimates who knew of the strained relations between her and Edith by going to her stepmother, arms outstretched. Edith rose in surprise as the bride placed both hands on her shoulders and affectionately kissed her, whispered into her ear, and kissed her again.

Chapter 20

"I Am No Longer a Political Accident"

"How they are voting for me!" an awed Theodore Roosevelt exclaimed over and over again on Election Night, November 8, 1904, as the returns rolled in. "How they are voting for me!" From the moment the polls closed, it was obvious that he would defeat the Democratic candidate, Judge Alton B. Parker, by a landslide. Edith had invited a small circle of family and friends to the White House to await the outcome, and the event rapidly became a victory celebration. Archie, covered with campaign buttons, raced breathlessly back and forth from the telegraph office in the Executive Wing with fresh batches of returns. Such Democratic strongholds as Buffalo and Rochester were piling up sizable margins for the Republican ticket. Even Missouri had fallen into the Roosevelt column.

"I have the greatest popular majority and the greatest electoral majority ever given a candidate for President," Roosevelt wrote Kermit. "I am stunned by the overwhelming victory we have won. I had no conception that such a thing was possible. I thought it probable we should win, but was quite prepared to be defeated, and of course had not the slightest idea that there was such a tidal wave."

The elated president took his wife aside to proclaim, "My dear, I am no longer a political accident."

This stunning mandate was tribute not only to Roosevelt's popularity but to his skill as a politician. From the moment he entered the White House, as we have seen, he began mapping his strategy to become president in his own right, courting allies, calculating policy moves, building his own forces to oppose the alliance of the Old Guard politicos and Wall Street. Although he had antagonized both anti-imperialists and conservatives, he had won the support of the bulk of the Republican party and the people.

"It is a particular gratification to me to have owed my election not to the politicians . . . not to the financiers . . . but above all to Abraham Lincoln's 'plain people'; to the folk who work hard on the farm, in the shop, or on the railroads, or who own little stores, little businesses," Roosevelt told Owen Wister in one of his "posterity letters." "I would literally, not figuratively, rather cut off my right hand than forfeit by any improper act of mine the trust and regard of these people."

The electoral triumph itself was something of an anticlimax, for the real fight had been over the Republican nomination that he won the year before in the face of resistance from Mark Hanna. Throughout Roosevelt's first years in office, the hulking shadow of the Ohio senator had fallen across his political path. Fretting about his chances for nomination and election, he doubted his hold on the public imagination. On a western trip in 1903, Roosevelt discounted the importance of the huge crowds that had cheered him, telling a companion they were merely coming out of curiosity to see the president. The people forget quickly, he continued, but the great corporations and the political bosses forgot nothing and forgave nothing. "*They* don't want me . . . Hanna and that crowd. . . . They've finished me. . . . I have no machine . . . no faction, no money. All this [waving his hand to indicate the crowd] has no significance."

Hanna had watched Roosevelt's progress in the White House closely, undoubtedly with personal misgivings about that "damned cowboy," but if he had any objections to what he saw, he kept them to himself. The president had frequently consulted with him, and Hanna had supported his efforts to end the coal strike, begin work on the Panama Canal, and establish a De-

partment of Commerce and Labor. Nevertheless, Hanna steadfastly refused to endorse Roosevelt for the Republican nomination
in 1904, arousing speculation that he intended to seek it for himself. Some big businessmen, including the heads of several railroads, urged Hanna to run, but the senator coyly claimed that he
was not a candidate, nor would he become one.

Hanna's hand was forced by Senator Joseph B. Foraker, his
longtime rival for control of Ohio, who used Roosevelt's popularity to advance his own interests. To curry favor with the president and embarrass Hanna, "Fire Alarm Joe" called for the state
Republican convention to endorse Roosevelt for the presidency
when it met in June 1903.* Hanna objected that such a move was
premature, and suggested that any endorsement be postponed
until the following year. Wishing to avoid an open break with the
president, he privately telegraphed Roosevelt that the issue had
been forced upon him and gave assurances that once "you know
all the facts I am sure you will approve my course."

Without waiting for "all the facts," Roosevelt sent a reply—
released to the press—that not only announced his candidacy but
sandbagged Hanna by bringing the secret struggle out into the
open. "Those who favor my administration and my nomination
will favor endorsing both and those who do not will oppose," he
declared. Unwilling to actively oppose the president, Hanna abjectly surrendered. "I shall not oppose the endorsement of your
administration and candidacy by our State Convention," he told
Roosevelt.

To Lodge, Roosevelt explained the reasoning behind this intrigue. "I . . . decided that the time had come to stop shilly-
shallying, and let Hanna know definitely that I did not intend to
assume the position . . . of a supplicant. . . . It simplified things
all around, for in my judgement Hanna was my only formidable
opponent so far as the nomination is concerned." Nine months
later, Hanna, who had been in declining health for some time,
died of typhoid. Fortunately, the two men had already set aside
their differences. Writing to Elihu Root, Roosevelt made a generous assessment of Hanna: "No man had larger traits. . . . He
was a big man in every way. . . . The whole party and the whole

---

*Foraker's nickname derived from his ability as a stump speaker.

country have reason to be grateful for the way . . . after I came into office . . . he resolutely declined to be drawn into the position which a smaller man of meaner cast would have taken."

With Hanna and Matt Quay dead and Tom Platt all but in retirement, the last threat of a veto of Roosevelt's nomination had been removed, and the president's popularity was at its highest. Early in 1904, the Supreme Court upheld the government's position in the Northern Securities case, the Senate ratified the Hay–Bunau-Varilla Treaty, putting the stamp of approval on his greatest foreign policy coup, and Union veterans were hailing his unilateral decision to increase their pensions. Vigorous prosecution of frauds within the post office traceable to previous administrations removed any hint of scandal. In an effort to appease conservatives, Roosevelt also made overtures of friendship to such Wall Street barons as J. P. Morgan, James Stillman of the National City Bank, and Joseph Medill McCormick of International Harvester. Morgan was even invited to dinner at the White House.

The Republican national convention that met in Chicago at the end of June 1904 was a listless affair. There were great swaths of empty seats in the Coliseum, and the cheering that greeted the first mention of Roosevelt's name was little more than perfunctory. The greatest applause came when a large portrait of Mark Hanna was unfurled from the rafters. Roosevelt was nominated unanimously and as a sop to the conservatives, Senator Charles W. Fairbanks of Indiana was named vice-presidential candidate. Fairbanks was chosen with little enthusiasm, not because anyone really wanted him, but as Roosevelt observed, "who in the name of Heaven else is there?"

While the convention droned on, the attention of most Americans was fixed upon the plight of Ion Perdicaris, a presumed American citizen.* He and his son-in-law had been kidnapped several weeks before from his luxurious villa outside Tangier, Morocco, by a brigand named Raisuli, who had a grudge against

---

*Perdicaris was born in Trenton, New Jersey. His father, Gregory, had come to the United States from Greece in the 1820s, and became professor of Latin and Modern Greek at Harvard. He moved to South Carolina, where he married into a prominent Charleston family, the Hanfords. Later, he amassed a fortune in the illuminating-gas business and left his son, Ion Hanford Perdicaris, a fortune of more than $1 million. After dropping out of Harvard, the young man went to live abroad.

the sultan of Morocco. In return for the release of the hostages, Raisuli demanded ransom from the sultan and the release of some of his men, who were being held by the Moroccan authorities. Otherwise, both captives would be killed. The sultan refused these terms and to ensure the safety of an American citizen, Roosevelt promptly ordered that a U.S. Navy squadron already on its way to the Mediterranean be diverted to Tangier.

There was talk of landing marines to seize the customs house at Tangier and other points, and possibly to pursue Raisuli, but primarily to force the sultan to negotiate with the brigand. Fearing that such a landing would lead to an eventual foreign takeover of his country—the last independent nation on the North African coast—the sultan finally agreed on June 22 to Raisuli's terms.

In the meantime, Roosevelt, who had been keeping up with the convention from the White House, was worried by its listlessness. Something had to be done to pump life into the delegates and gear them up for the campaign. Just as the convention was about to adjourn that evening, "Uncle Joe" Cannon, the permanent chairman, instructed the clerk to read a news bulletin just in from Washington which reported that the State Department had cabled a message to the U.S. consul general at Tangier stating: WE WANT PERDICARIS ALIVE OR RAISULI DEAD.

Wild cheering greeted the announcement. The delegates leaped to their feet and shouted approval. Although the cable had been crafted by Secretary Hay with the help of a newsman, Roosevelt got the credit for it. One Kansas delegate, W. A. Elstun, seemed to speak for them all when he told a reporter: "Roosevelt is behind that cable message to that fine old body-snatcher Raisuli. Out in Kansas we believe in keeping the peace but in fighting against wrong." Newspapers praised the president for his forthrightness. But in a section of the cable not read to the delegates, the consul was told that marines would not be landed without consultation with Washington. And, of course, the cable had nothing to do with Raisuli's decision to free his captives because he had already agreed to do so. "It is curious," observed a sardonic Hay, "how a concise impropriety hits the public."

Roosevelt's election in 1904 was guaranteed by the disarray and infighting among the Democrats, who met in St. Louis a few weeks after the Republicans. Brushing aside the radicals and the shopworn William Jennings Bryan, who had twice led them to defeat, they appealed to conservatives by nominating Judge Alton B. Parker of New York. Respectable but colorless, Parker tried to project the image of a "safe and sane" candidate as a counterweight to the Rough Rider. Nevertheless, the *New York Sun,* the mouthpiece of conservatism, weighed Roosevelt's accomplishments and failures and endorsed him in an editorial notable for its brevity: "Theodore! with all thy faults—"

The only issue of the campaign was Theodore Roosevelt. Roosevelt himself was frustrated by the long-standing custom that incumbent presidents did not actively campaign. "I wish I were where I could fight more offensively," he told Lodge. "I always like to do my fighting in the adversary's corner." But his active role was limited to a speech and a lengthy letter to Joe Cannon formally accepting the nomination. "We base our appeal upon what we have done and are doing, upon our record of administration and legislation," he declared. "We intend in the future to carry on the Government in the same way that we have carried it on in the past."

The country was prosperous and at peace, and few people had doubts about the outcome of the election—except for the candidate himself, who was undergoing his usual preelection crisis of nerves. Reports that Parker was raising funds from businessmen rattled him, and he urged George Cortelyou, serving both as secretary of commerce and labor and chairman of the Republican National Committee, to redouble his efforts to obtain funds from Wall Street for dispersal in such heavily contested areas as New York. Cortelyou collected a total of $2.2 million, three quarters of it in corporate gifts, including $150,000 from J. P. Morgan; $148,000 from various insurance companies; $125,000 from the Standard Oil interests; $50,000 from E. H. Harriman, and $100,000 from Henry Clay Frick, the steel baron. Later, Frick angrily complained, "We bought the son of a bitch and then he didn't stay bought!"*

---

*Lincoln Steffens suggested to Roosevelt that rather than offering himself for sale to the moneymen, he should call upon ordinary citizens to contribute small sums to his campaign—a plan that "would make the millions feel that it was their Government . . . and that you and your administration were beholden to the many not to the few." There is no sign that the president gave serious consideration to this revolutionary suggestion. Steffens, *Letters,* Vol. I, pp. 170–171.

The campaign suddenly sprang to life late in October when, first, opposition newspapers and then Judge Parker picked up on the issue that Cortelyou was using his Cabinet post to blackmail corporations into contributing to the Republican campaign chest. Roosevelt denied the charges and consigned his accusers to the Ananias Club of liars. It is generally accepted today that campaign officials did not inform him of the contributions.

Parker's charges came too late to be of much help to him. Roosevelt piled up a victory margin of 7,628,875 votes, or 56.4 percent of the total, to 5,084,442 votes for the Democratic candidate. The margin was even greater in the Electoral College: 336 votes for Roosevelt, the most won by any candidate up to that time, to 140 for Parker. Roosevelt's appeal to ethnic voters paid off, for he ran well among Germans, Swedes, Poles, Italians, and Jews, and he carried every state but the Solid South. He "is the most popular man that has come into public life within recent times," declared the magazine *World's Work*. "The people like his energy, his frankness and his robust ways."

Having scored this remarkable triumph, Roosevelt committed the greatest political blunder of his career. In the exuberance of victory, he announced on Election Night: "On the fourth of March next I shall have served three and a half years, and this three and a half years constitutes my first term." Praising "the wise custom which limits the President to two terms," he added, "under no circumstances will I be a candidate for or accept another nomination." Edith, who was standing nearby, was seen to flinch as he made this rash proclamation. She later told Wister that had she known what the president was going to say she would have done everything in her power to prevent it. Roosevelt may have intended the statement to allay charges of dictatorship bandied about by critics, but it made him a lame duck even before his new term had begun—and would cause substantial trouble in the future.

"Tomorrow I shall come into my office in my own right," Theodore Roosevelt supposedly told a friend on the eve of his inauguration on March 4, 1905. "Then watch out for me." Some observers doubt whether he actually made these remarks, but over the next two years, he asserted his newly won power in a whirlwind of activity on both the domestic and international fronts. Liberated from his pledges to McKinley and Hanna, he rode a rising tide of progressivism and made these the most productive years of his presidency. Federal control of the railroads, meat inspection, and pure food and drug laws were wrung from a reluctant Congress, and with a single stroke of his pen, he snatched up sixteen million acres of land and placed them in trust for future generations of Americans.

Roosevelt also broadened his role in the international arena. On first coming into office, he had accepted the Open Door in China—first promulgated by McKinley and John Hay in 1899 to protect American commercial interests—as the foundation of American policy in East Asia. Yet he also believed that the United States should take a more active role in the area. Although businessmen were still hypnotized by the prospect of China as a great outlet for American surpluses, Roosevelt was thinking, as always, in terms of geopolitical strategy rather than economic advantage.

Even before the inauguration, his attention was fixed on Manchuria, where the Open Door was being slammed shut by Russia. Having wrung special privileges from China in southern Manchuria, including control of Port Arthur, at the tip of the Liaotung Peninsula, the Russians had monopolized trade and established political control of the area. Roosevelt privately fumed against these arrogant and treacherous moves, but there was little the United States could do about them. American interests in Manchuria were relatively small, and there was no chance that the American people would support intervention to protect the Open Door.

Japan, on the other hand, felt her security endangered not only by the Russian advance in Manchuria but also in Korea, which the Japanese regarded as within their sphere of influence. Japan had concluded a treaty with Britain that gave it assurances of British support if any third power intervened in a quarrel with Russia, and now the Japanese challenged the Russian advance into East Asia. In February 1904, they launched a devastating surprise attack upon the Russian fleet at Port Arthur in a forerunner of the attack on Pearl Harbor in 1941. Roosevelt proclaimed America's neutrality and sent notes to the belligerents asking them to respect the territorial integrity of China. While accepting the American proposal in principle, the Russians refused to include Manchuria as part of China proper, making the whole process meaningless. Nevertheless, a major principle had been added to American policy in East Asia. For the next forty years, it was to revolve around the twin goals of the Open Door and the territorial integrity of China.

Over the next year, the Japanese triumphed at Mukden, where Russia lost nearly 100,000 men, and their navy overwhelmed another Russian fleet at Tsushima. Roosevelt and most Americans cheered the surprising land and sea victories of the Japanese, whom they had regarded as protégés ever since Commodore Matthew C. Perry opened the island empire to the West a half century before. Reversing his earlier position of worrying about Japan's designs upon Hawaii, Roosevelt now admired both the efficiency and fighting qualities of the Japanese.

In fact, the president declared he was "thoroughly well pleased" with the Japanese triumphs. "Japan is playing our game," he told his son Ted. Later, he claimed to have served notice on France and Germany that the United States would support Japan if either nation came to the aid of Russia. There is no contemporary evidence that he actually did so, but the suggestion that he was prepared to make the United States a silent partner in the Anglo-Japanese alliance is significant as a revelation of his attitude toward the belligerents.*

Initially, relations between the United States and Russia had been friendly and Roosevelt had expressed approval of Russia's "civilizing" influence in Asia. But the autocratic despotism of the czars, exposés of the Siberian penal system, and the periodic pogroms, or riots, against the Jews all exerted an unfavorable influence on public opinion. Following a particularly savage explosion of anti-Semitic violence in Kishinev during Easter week of 1903, Roosevelt, with an eye on the Jewish vote, ordered the State Department to forward a protest petition from prominent American Jews to the Russian government. He also issued a statement expressing "the deep sympathy felt not only by the admin-

---

*Roosevelt could have issued this warning to the French and German envoys, who did not pass it on. It should also be recalled that in the summer of 1904 he had sent a powerful fleet to European waters, the one which appeared in North African waters at the time of the Raisuli affair.

istration but by all the American people for the unfortunate Jews who have been the victims in the recent appalling massacres and outrages." As expected, the Russians refused to receive the petition, but Roosevelt's efforts impressed Jewish voters.

Much of the president's information on conditions in Russia was received through a clandestine correspondence with Cecil Spring Rice, then serving in the British embassy in St. Petersburg. Bypassing the State Department and the British embassy in Washington, Rice addressed his letters to Edith, who passed them on to the president. Czar Nicholas II "lives quietly . . . in his palace and plays with his baby and will hear nothing but baby talk," "Springy" reported in one letter. "If you come with disagreeable truths, he listens and says nothing. His ideas, if he has any, are to maintain the autocracy undiminished and to continue the war until he has gained 'the mastery of the Pacific.' "

Roosevelt began to have second thoughts about Japan, however. As a new Pacific power, the United States was concerned that a total victory by either Japan or Russia would upset the delicate balance then existing and create a dangerous rival in the Far East. He was particularly concerned about Japanese intentions toward the Philippines. The cardinal point of his foreign policy was the maintenance of a rough balance of power as a guarantee that no nation gained a position of power inimical to the interests of the United States or its friends. The integrity of China and the Open Door could best be maintained, he believed, if Russia and Japan became guarantors of each other's good conduct—a system of balanced antagonisms. Under these circumstances, the time had come to bring an end to the hostilities before the Russians were completely humiliated.

Roosevelt's desire to end the struggle coincided with the belligerents' own battle weariness. Having successfully bloodied the Russians, the Japanese were near exhaustion and anxious for an end to hostilities. On the Russian side, it was clear that if the war continued, the nation would be plunged into revolution. On June 11, 1905, just after the smashing victory at Tsushima, it was officially announced that both Russia and Japan had accepted the president's proposal for peace talks. Behind this brief announcement lay a story of dramatic, secret diplomatic maneuvers.

For several months Roosevelt had been doing everything he

could to end the war, including trying to enlist the help of Kaiser Wilhelm. The wily Wilhelm had his own agenda, however. He professed interest in the Open Door and peace in Asia, but his statements were mere cover for his own imperialist designs. He had encouraged his cousin the czar in his foreign adventures with the hope that the Russians would become mired down in them and therefore no threat to Germany in Europe. Thus, he was cool to Roosevelt's attempts to persuade him to help end the struggle. Only after revolutionary disturbances in Russia threatened to spill over into Germany did he support Roosevelt's peace efforts.

The president was far from being taken in by "Bill the Kaiser's" machinations, however. Even though he may not have fully understood the intricacies of German diplomacy and went out of his way to heap public praise on Wilhelm, he understood just how erratic that diplomacy was. He had no intention of following the lead of a ruler "so jumpy, so little capable of being loyal to his friends." Nevertheless, whatever the relationship between these two larger-than-life personalities, events finally led them to cooperate in seeking peace between Russia and Japan.

The turning point came when the Japanese secretly asked Roosevelt to invite "on his own motion and initiative" the two belligerents to negotiate a peace treaty jointly. Carefully concealing this request, the president, who was acting as his own foreign minister because Secretary Hay was in failing health—he died that summer—labored to bring the Russians to the negotiating table. He played his cards so close to his chest that he wrote secret instructions for his ambassadors in his own hand, and for three weeks the State Department did not even know of the Japanese request.

No more delicate task could have been imagined than trying to bring the Russians and Japanese to the peace table. Upon being asked to urge his government to accept the Japanese offer, the Russian envoy, Count Cassini, replied with "his usual rigmarole to the effect that Russia was fighting the battles of the white race . . . that Russia was too great to admit defeat." One can sense the president's rage at Cassini as he denied his country's need for peace. "Oh Lord! I have been going nearly mad in the effort to get Russia and Japan together," the frustrated Roosevelt told Whitelaw Reid, the American envoy in London.

Mistrusting Cassini and to make certain that his views reached the czar, Roosevelt instructed George von Lengerke Meyer, the American ambassador in St. Petersburg, to see Nicholas personally and convince him to consent to the peace talks. To allow the Russians to save face, Roosevelt assured the czar he was acting on his own initiative; if the Russians agreed to the talks, he would get the Japanese assent. Meyer obtained an audience with the czar with some difficulty, but arrived at the Summer Palace at a strategic moment. Nicholas had just received a letter from the Kaiser, obviously prompted by Roosevelt, adding his voice to those calling for peace. Without consulting any advisers, the czar agreed to accept the American overtures.

Overjoyed at Meyer's cable announcing the Russian agreement, Roosevelt plunged into setting up the machinery for the peace conference. The task was not an easy one. "The more I see of the Czar, the Kaiser and the Mikado the better I am content with democracy, even if we have to include the American newspaper as one of its assets," Roosevelt raged to Lodge. Even the choice of a site caused problems. Europe was immediately ruled out, and Washington in summer was intolerable. Finally, all parties agreed on Portsmouth, New Hampshire. The talks were to begin August 9 at the Portsmouth navy yard, where tight security could be maintained and adequate international communications were available.

On August 5, the president invited both delegations to lunch with him on the presidential yacht *Mayflower,* lying at anchor off Oyster Bay. Protocol and precedence threatened to end the conference before it got under way. Which delegate would sit on the presidential right? Which ruler would be toasted first? Who would enter the dining saloon first? Roosevelt cut the Gordian knot by taking Count Sergei Witte, head of the Russian delegation, and his Japanese opposite number, Baron Jutaro Komura, by the arm and steering them across the threshold while talking all the while so volubly that neither was aware which of the two was ahead of the other. A cold buffet lunch was served from a round table and was eaten standing. With champagne glass in hand, Roosevelt also proposed the only toast to which there would be no reply: "To the welfare and prosperity of the sovereigns and people of the two great nations, whose representatives have met one another on this ship."

The talks opened with Japanese demands for the Russian lease on the Liaotung Peninsula in Manchuria and the railroad running from Harbin to Port Arthur, for evacuation of Russian troops from Manchuria, and for a free hand in Korea. The Russians quickly conceded these points, but the conference bogged down when the Japanese demanded a large monetary indemnity and the strategic island of Sakhalin. The Russians rejected these two demands and threatened to break off negotiations and renew military operations.

Officially, Roosevelt had no part in the negotiations, but he had remained at Oyster Bay to be nearby, and when the conference became deadlocked, he took a behind-the-scenes role. Every possible means was used to persuade the Japanese to forgo an indemnity and to convince the Russians that continuing the war would be futile. "I am having my hair turned gray by dealing with the Russian and Japanese peace negotiations," he informed Kermit. "The Japanese ask too much, but the Russians are ten times worse than the Japs because they are so stupid and won't tell the truth."

A flood of telegrams and cables poured out of the summer White House. "Make clear to His Majesty" the need for peace, Roosevelt told Ambassador Meyer in St. Petersburg. The British ambassador was informed that "in my judgement every true friend of Japan should tell as I have already told it that the civilized world will not support it in continuing the war. . . ." "I hope your government will try to persuade the Czar to make peace," he advised Jules Jusserand, the French envoy. "Dear Baron Kaneko," he wrote a Japanese delegate, "it seems to me that it is to the interest of the great empire of Nippon" to make peace. "Peace can be obtained," the kaiser was informed, if he would only use his influence with the czar.

The conference was still deadlocked and the atmosphere was tense as the delegates met on August 29 for what would likely be the final session. In fact, the Russians had received orders to return home. Throughout that morning, Roosevelt worked in the library at Sagamore, going over correspondence with William Loeb. The telephone rang about noon. The secretary answered it. There was a moment's pause. "What!" he exclaimed. The message was repeated.

"What is it?" asked the president, startled by Loeb's behavior.

"The Associated Press has announced, in an official bulletin from Portsmouth," Loeb reported breathlessly, "that the plenipotentiaries have agreed on all points of difference, and will proceed at once to draft a treaty of peace!"

"This is splendid!" Roosevelt declared. "This is magnificent."

With a smile reaching from ear to ear, he raced upstairs to tell Edith. When he returned, he was still beaming. "It's a mighty good thing for Russia, and a mighty good thing for Japan. And," he added, thumping his chest, "a mighty good thing for *me* too!"

While he celebrated over lunch, the president received a cipher message from Portsmouth providing further information on the meeting. Shortly after the session had opened, Baron Komura announced that his nation would withdraw its demands for an indemnity if the Russians withdrew from Sakhalin. Count Witte flatly refused, but offered to divide the island. For a few seconds, absolute silence had reigned. Witte slumped in his chair, absently tearing up sheets of paper; Komura stared stonily ahead. The decision for peace or war lay with him. Finally Komura broke his silence and accepted the Russian proposal to divide the island. The fiftieth parallel was established as the line of demarcation between the two nations and a treaty was to be signed on September 5, 1905.

Congratulations rained down upon Roosevelt. The rulers of Britain, France, and Germany, as well as Russia and Japan, praised the American president for his peacemaking efforts. When the first family returned to Washington, they were greeted by a cheering crowd of well-wishers. Roosevelt deserved this adulation, for no one had done more to end the war. Had he not interceded, peace would have come much later and many more lives would have been lost. He convinced the Russians and Japanese of the need for the conference. He made them moderate their terms—especially the Japanese. He kept the conference going after no European foreign office thought he could succeed. His appeals brought about the compromises that made peace possible.*

---

*One of these compromises sealed the fate of Korea. The United States had signed a treaty with Korea in 1882 in which it agreed to use its "good offices" if another country

*   *   *

Halfway around the globe, another situation was unfolding in Morocco that was even more dangerous for world peace. Roosevelt broke with the American tradition of aloofness from Europe by intervening in this dispute, and by doing so believed he had helped stave off a devastating general European war by nearly a decade. Ostensibly, the dispute involved rivalries over Morocco, but the real issue was the increasingly dangerous rivalry between Germany and a newly formed Anglo-French alliance.

Fourteen nations, including the United States, had agreed in 1880 to a semblance of the Open Door in Morocco, the last unoccupied territory in North Africa. When the sultan was unable to maintain order in his rich domain, however, it was inevitable that some nation was going to fill the vacuum. In April 1904, France gave the British a free hand in Egypt while Britain recognized French interests in Morocco. This agreement angered the kaiser, who was already alarmed over the threat of encirclement by France and its Russian ally. On March 31, 1905, he visited Tangier and delivered a defiant, saber-rattling speech that attacked French policy in Morocco, insisted that German rights be respected, and called for an international conference to settle the fate of the country. He hoped to inflict a diplomatic humiliation on France and disrupt the Anglo-French alliance. Backed by Britain, the French refused to consider such a meeting, and Europe seemed on the brink of a catastrophe.

Realizing too late that he had gotten in deeper than he intended, the kaiser sought mediation by Roosevelt. The president was reluctant to get involved because the United States had no stake in Morocco, and he was well aware of the difficulties he would have with Congress if he got entangled in the affairs of the Old World. "I do not feel as a Government we should interfere in the Morocco matter," he told Will Taft, who was filling in as acting secretary of state. "We have other fish to fry and no real interest in Morocco. I do not care to take sides between France and Germany. . . ."

---

should treat the Hermit Kingdom "unjustly or oppressively." The Koreans interpreted this to mean that the Americans would protect them from Japanese aggression. But in the Taft-Katsura Agreement of 1905, Roosevelt traded Japanese control over Korea for assurances that Japan had no designs on the Philippines, Hawaii, or American interests in China. In 1910, Japan formally annexed Korea.

Both Jules Jusserand and Speck von Sternburg, diplomats whom he trusted, expressed serious concern that unless prompt action was taken, the crisis could escalate into war between France and Germany. So Roosevelt eventually consented to induce the French to come to the conference table. Skillfully playing one nation against another, he engaged in secret maneuvers that paralleled those that led to the Treaty of Portsmouth. From the Germans he extracted a promise to follow the American lead on any matters in dispute. He convinced the British to encourage their French ally to negotiate. Trading on the French belief that he was prejudiced in their favor—which was essentially true—he persuaded them to join the talks. The conference was set to begin on January 16, 1906, in Algeciras, in southern Spain.

The announcement of American participation in the Algeciras Conference brought a howl of protest from Congress and the press, which had been kept in the dark about the negotiations leading up to the meeting. The United States had no interest in Morocco, critics asserted and called Roosevelt's intervention a case of "international meddling" that would embroil the nation in Europe's quarrels. Algeciras was repeatedly linked to the fracas over the Santo Domingo treaty, then under way, as another example of the president's arbitrary conduct of foreign affairs. Roosevelt could hardly ignore these objections, so the American delegate to the conference, Henry White, the nation's ace diplomatic troubleshooter, was instructed not to intervene but to "safeguard legitimate American interests." Privately, however, the president made it clear that White was to keep on the good side of the Germans while "helping France get what she should have."

The conference followed a tortuous path and at one point was on the brink of failure. With France and Germany stubbornly refusing to come to terms, newspaper headlines proclaimed the imminent threat of war. Once again, Roosevelt intervened to break the deadlock by presenting a compromise. Originated by White, it provided for the recognition of an independent Morocco, authorized France and Spain to police Morocco under a Swiss inspector general, and reaffirmed the Open Door. This plan, in effect, preserved the fiction of international control of Morocco

yet gave France substantive control over it.* German reluctance
to allow France such a large role in the area was overcome by
Roosevelt's alternate use of firmness and flattery in dealing with
the kaiser. Would he fight or yield? Wilhelm opted to retreat
because he was not yet prepared for war. "In this Algeciras
matter," Roosevelt later told a friend, "I stood him on his
head. . . ."

Peace in Europe was preserved until 1914, but Germany, convinced more than ever that she was being encircled as the Anglo-
French alliance was broadened to include Russia, embarked on a
vast naval building program that was a direct challenge to Britain.
As far as the United States was concerned, it was clear that it was
no longer isolated from the world's problems. Roosevelt, although
elated by the outcome of the Algeciras Conference, was under
no illusions how long his handiwork would restrain the rivalries
among the great powers. But he was, at both Algeciras and Portsmouth, a vital force for world peace. He well deserved the Nobel
Peace Prize that he was awarded in 1906 for settling the Russo-
Japanese War—the first American to win a Nobel Prize in any
field.†

Roosevelt's foreign successes enhanced his stature at home as
well as abroad, but domestic policy had the highest priority. "Our
internal problems are of course much more important than our
relations with foreign powers," he told George Otto Trevelyan,
the English historian. "Somehow or other we shall have to work
out methods of controlling the big corporations without paralyzing
the energies of the business community and of preventing any
tyranny on the part of the labor unions while cordially assisting
in every proper effort made by the wage workers to better themselves by combinations."

In dealing with these problems, the president clung to his cautious policy of trying to attain "realizable ideals"—the idea that

---

*In 1912, France and Spain partitioned Morocco between them, and southern Morocco
became a French protectorate.

†Roosevelt turned the $39,734.49 prize over to a foundation to promote industrial
peace in the United States. The money was meant to be the nucleus of a larger endowment,
but no further funds came in, so twelve years later the former president donated the fund,
which had with interest grown to $45,482.83, to World War I relief organizations.

half a loaf was better than none. He advocated new governmental powers to deal with "corporate arrogance" while exercising caution in using those powers. Simultaneously, he exhibited considerable skill in outmaneuvering the "stand pat" leadership of his own party while appropriating programs espoused by the Bryanites to keep "radicalism" at bay. Rather than simply opposing new ideas, he sought compromises that would protect the nation against the threat of radical change. "The only true conservative is the man who resolutely sets his face to the future," he declared.

At a Gridiron dinner in January 1905, the toastmaster noted that inasmuch as both the president and Bryan were in attendance, a debate should be arranged between them.

"What's the use?" called out a member. "They're both on the same side."

Roosevelt's aim was to preserve the status quo by saving the big business interests from the consequences of their own narrowness and stupidity. He was particularly alarmed by the increasing appeal of Eugene V. Debs, hero of the railway strike of 1894 and Socialist presidential candidate, whose support had risen from 100,000 votes in 1900 to 400,000 four years later. Writing to Will Taft, Roosevelt declared: "The dull purblind folly of the very rich men, their greed and arrogance . . . and the corruption of business and politics, have tended to produce a very unhealthy condition and excitement in the popular mind, which shows itself in the great increase in Socialist propaganda."

As part of his policy of emergent progressivism, he turned to the question of railroad rate reform. Since the Civil War, railroad rates had been a highly volatile issue in the Middle West and South, where farmers and other shippers charged they were being discriminated against while large corporations, such as Standard Oil and Armour, received special treatment. Although the Interstate Commerce Commission had been established twenty years before, it had almost no power to interfere with the activities of the railroads. "The crying evil of railroad administration is not high rates, but unfairly discriminating rates, secret rebates and tricks and devices which favor one shipper over another," declared *The New York Times*.

For Roosevelt, the key issue was control of the corporations. Behind all the oratory, propaganda, and maneuvering lay his determination to end corporate abuses. Fair rates were equated with the Square Deal. In his annual message to the lame duck session of Congress in December 1904, he described the attempt to curb corporate abuses at the state level as an "absurdity" and called for allowing the Interstate Commerce Commission to regulate the railroads and prevent price gouging. The commission would not be allowed to actually fix rates but would have the power, when a rate was challenged and found unreasonable, to set a maximum rate that would go into effect immediately, subject to court review.

Progressives were distressed by Roosevelt's failure to revise the protective tariff part of his program. Although he personally favored revision, he did not give the tariff—and most issues linked to the marketplace—a high priority. Moreover, he reasoned that the issue would split the Republican party and stood no chance of passage. But he cannily kept the tariff as an ace in the hole, using the threat of revision to persuade Joe Cannon to support railroad regulation. He was less successful with Senator Nelson Aldrich, the "dictator" of the Senate, and other conservatives with close ties to the railroads.

While the House passed an act that enabled the ICC to declare a freight or passenger rate unjust and to fix a new one, the Senate refused to concur. Before adjourning in March 1905, it substituted a bill that merely called for an investigation. Roosevelt accepted defeat philosophically. In the coming year, he believed, "we can get the issue so clearly drawn that the Senate will have to give in." Nevertheless, he feared that "the big financiers" who opposed regulation might "force the moderates to join with the radicals in radical action, under penalty of not obtaining any at all. . . . I much prefer moderate action, but the ultraconservatives may make it necessary to accept what is radical."

Testimony before the investigating committee that summer confirmed that railroad abuses existed and the country rallied behind Roosevelt and reform. Popular anxiety about the trusts and political corruption was fanned by a well-publicized investigation into the corrupt practices of insurance companies in New York led by an attorney named Charles Evans Hughes. A flurry of articles by crusading journalists also focused public attention on the railroads, and the Senate itself was described as a "Millionaire's Club" and the puppet of privileged wealth. In the meantime, the railroads not only trotted out the familiar arguments against government interference with business, but warned southerners that regulation would mean the end of racially segregated railway cars.

The president himself moved on several fronts to drum up support for railroad reform. Using a Rough Riders' reunion in San Antonio as a pretext, he toured the Midwest and Southwest in April and May and spoke often on the need for railroad rate revision. "Personally, I believe that the Federal Government must take an increasing control over the corporations," he declared in a speech in Chicago, and it should begin with "increased supervision and regulation" of the railroads. In October, he toured the South, emphasizing his mother's southern heritage and praising Confederate leaders he had once condemned as traitors. He also arranged for the release of previous investigative reports that revealed railroad malfeasance and emphasized the need for government regulation.

In the spring of 1905, the Roosevelts looked for a new rural retreat. Sagamore no longer offered refuge, for the president was trailed there by reporters, politicians, and favor-seekers. Upon learning that Edith was seeking a country place, a family friend invited her to visit his farm in Albemarle County in the Blue Ridge foothills of Virginia, about 125 miles from Washington. She discovered an old shack on the place, about a mile off the road and near a stream, and bought it and five surrounding acres on the spot for $195. The house had a single room downstairs and two smaller ones upstairs. Edith named the place Pine Knot and went back to the White House to tell her husband about it.

"It is really a perfectly delightful little place," Roosevelt wrote Kermit after the couple's first weekend visit. "In the morning I fried bacon and eggs, while Mother boiled the kettle for tea and laid the table. Breakfast was most successful. . . . Then we walked about the place . . . admired the pine trees and the oak trees and then Mother lay in the hammock while I cut away some trees to give us a better view from the piazza. . . . It was lovely to sit there in the rocking chairs and hear all the birds by daytime and at night the whippoorwills and the owls and little forest folk. . . ."

When it came time to leave for the White House, Edith happily noted the president seemed less tired and the color had come back to his cheeks. Over the remaining years of Roosevelt's term, they returned, summer and winter, to Pine Knot for "rest and repairs." On rare occasions they invited a favored friend to join them for a weekend of roughing it. One, John Burroughs, the naturalist, thought it was risky for the president to go about unprotected, and raised the question with Roosevelt. "Oh, I go about armed," he replied, slapping his hip pocket. Later, Edith confided that she had arranged for two Secret Service men to come each night at nine, stand guard until morning, and hide in a nearby farmhouse during the day. She didn't let her husband know "because it would irritate him."

In his annual message to the new "progressive" Congress on December 5, 1905, Roosevelt unleashed a program described by the *New York World* as "the most amazing program of centralization that any President of the United States has ever recommended." This package included a pure food and drug law, governmental supervision of insurance companies, an investigation of child labor by the Department of Commerce and Labor, and an employers' liability law for the District of Columbia that was to become a model for the nation. Most of the programs had little chance of passage, but Roosevelt gave the highest priority to legislation authorizing the ICC to establish maximum railroad rates after receiving a complaint. These rates would go into effect after a reasonable time, subject to judicial review.*

Six days later, this legislative gambit was backed up by the announcement that Attorney General William H. Moody had directed the eighty-five U.S. attorneys to institute proceedings against companies offering or receiving rebates. These cases were to be brought under the conspiracy statutes, so those who were convicted faced the prospect of going to jail. Standard Oil and American Tobacco were already under investigation, and within days, federal grand juries in Chicago and Philadelphia returned indictments against several defendants, including the Armour, Swift, and Cudahy meat-packing companies and the Chicago and Alton railway line.

---

*This was a slight modification of his original position in which the new rate would go into effect immediately.

The battle over railway regulation was finally joined in January 1906, when Congressman William P. Hepburn of Iowa introduced a bill embodying Roosevelt's program. It gave the ICC general authority to fix limits on railroad rates, a proposal less than that demanded by many progressives such as Wisconsin's Robert La Follette, who wanted the commission to set rate schedules, but it was regarded as a step in the right direction. Railroad reform was popular in the House, and the measure passed with only seven dissenting votes. In the Senate it ran into firm resistance from members opposed to railroad regulation. This might be partially explained by the fact that members of the House were elected by popular vote, while senators were chosen by state legislatures, which in many cases were dominated by railroad interests.

Senators Nelson Aldrich, Joe Foraker, and Stephen Elkins of West Virginia, all "railroad senators," led the opposition. In an effort to embarrass the president, Elkins, chairman of the Senate Interstate Commerce Committee, refused to sponsor the Senate version of the Hepburn bill, which would have been the normal action in an administration measure, and "Pitchfork Ben" Tillman of South Carolina was designated floor leader for the measure. Tillman was not only a Populist Democrat, but he and the president were anathema to each other; the South Carolinian had even been barred from the White House. The Old Guard confidently expected that Tillman would turn the bill into a radical measure that would lose the support of moderates, including the president. To everyone's surprise, however, Roosevelt and Tillman put aside their animosities to work together.*

For the next sixty days, through March and April 1905, the Senate chamber rang with a debate over railroad regulation. One by one, Republican conservatives—Aldrich, Foraker, Philander Knox, who had been Roosevelt's attorney general, William Allison, even Cabot Lodge—denounced a measure that was a keystone of the legislative program of their own administration. Words like "liar," "unqualified falsehood," and "betrayal" were flung back and forth. Exasperated by opponents within his own party, the president privately called them "a curse" and told Cecil

---

*Tillman had been banned from the White House after a brawl on the Senate floor, and he, in turn, had roundly criticized Roosevelt's dinner with Booker Washington as a disgrace.

Spring Rice he would be happy to lend the Russian government several eminent statesmen if they would guarantee to place them where a bomb was likely to go off.

Lacking six of the forty-six votes needed to kill the bill, Aldrich avoided a frontal attack but ate away at its vitals with crippling amendments. While agreeing to grant the ICC power to fix rates after complaints were brought, the Old Guard tried to make it difficult and costly to bring such complaints. "Broad" versus "narrow" court review of rate cases became the issue. Conservatives favored "broad" review, which would allow the courts to establish a lengthy and costly process that would effectively block rate reform; the moderates wished a swifter, less restrictive review process. La Follette also pushed for a physical evaluation of railroad properties as a basis for rate setting.

The president was no mere bystander during this lengthy debate. He not only worked in bipartisan harmony with a coalition of Republicans and Democrats favoring reform, but tried to influence public opinion with speeches—including a threat to impose stiff inheritance taxes to wipe out family fortunes—and the release of reports, such as one by the Bureau of Corporations on Standard Oil. "The report shows that the Standard Oil Company has benefited enormously up almost to the present moment by secret rates," he declared in transmitting it to Congress. "This benefit amounts to at least three-quarters of a million a year."

But as the deadlock dragged on, it became increasingly apparent that the proregulation coalition did not have the votes to pass the Hepburn bill. Twenty Republicans supported the measure, but it was doubtful if Tillman could produce the twenty-six Democrats needed for a majority. Roosevelt, who wanted a bill, reached out to the conservative bloc for a compromise.

One sign of this was a speech the president gave at a Gridiron dinner on March 17, 1906, in which he denounced crusading journalists as "muckrakers." *Cosmopolitan* magazine had just published "The Treason of the Senate," a series of articles by David Graham Phillips that denounced the general pattern of corruption and conflict of interest that permeated the body. Privately, the president may well have agreed with the articles, but to improve relations with the Old Guard, he attacked Phillips and his fellow muckrakers.

Taking as his text a passage from *The Pilgrim's Progress,* Roosevelt likened these writers to "the Man with the Muckrake, the man who could look no way but downward with the muckrake in his hand; who was offered a celestial crown for his muckrake but who would neither look up nor regard the crown he was offered. . . ." A month later he gave a similar speech at the dedication of the cornerstone of a new House office building in which he warned that this kind of antibusiness journalism could go too far. "The men with the muckrakes are often indispensible to the well being of society," he declared, "but only if they know when to stop raking the muck, and to look upward to the celestial crown above them. . . . "*

With the approval of Aldrich, Senator Allison drafted an amendment giving the courts authority over cases arising from the law but leaving the exact extent of the review rather vague, which left the bill as it was when it had been approved by the House. Weary of the battle, everyone except the extremists on both sides found merit in Allison's proposal and amid outraged cries of "betrayal" from Tillman, Roosevelt decided to accept it. The Senate approved the Hepburn Act on May 18, 1906, by a margin of 71 to 3, with Foraker casting the only Republican vote against it. The act represented a substantial advance in railway regulation. A stronger ICC now had jurisdiction not only over freight and passenger rates but over pipelines; terminal, refrigeration, and storage facilities, and sleeping car and express service. The railroad interests were also required to disgorge the steamship lines and coal mines they had bought up to stifle competition—a requirement that they managed to evade.†

Who had won the battle? Critics charged that the president had been too pragmatic and had settled for too little. By obtaining injunctions, the railroads could delay the application of revised rates; moreover, as La Follette emphasized, the ICC lacked knowledge of the true worth of the roads and would be unable to determine what rates should be. Roosevelt obviously conceded

---

*Rather than being offended, investigative journalists accepted the label and in the eyes of most of the public, "muckraker" became a term of approval.

†In the 1960s, "New Left" historians, with characteristic overstatement and manipulation of data, treated the Hepburn Act as a sham used by the railroads to escape regulation by the states or more stringent federal controls. For a prime example of this view, see Kolko, *The Triumph of Conservatism*.

the logic of the latter point, for the following year he came out for such a provision.

In the final analysis, however, the president had gotten what he had set out to obtain. He called the new law "a fine piece of constructive legislation, and all that has been done tends toward carrying out the principles I have been preaching." Furthermore, his position was bolstered by the fact that in the first test of the scope of the review provision, the courts rejected the "broad" review favored by the conservatives. Within two years the ICC heard almost twice as many complaints as in the previous nineteen years. Roosevelt recognized the limitations of the law and before the end of his term asked for more stringent procedures. A rebellious Congress refused to enact them, but his actions make it clear that the Hepburn Act was intended to be only one step in an ongoing process of reform and regulation.

While the fight over railroad regulation raged, the president was also attempting to force pure food and drug and meat inspection laws through a Congress reluctant to extend federal regulatory powers. Over the years, Dr. Harvey W. Wiley, chief chemist for the Department of Agriculture, had warned about the use of adulterants and chemical preservatives in canned and prepared foods, while muckraking magazines pointed to the dangers of poisonous patent medicines—all to no avail. Nevertheless, the movement for pure food and drugs drew together such diverse elements as farmers, chemists, social reformers, women's groups, doctors, and consumers into a progressive alliance.

Influenced by the public outcry, Roosevelt called for federal regulation of "misbranded and adulterated foods, drinks and drugs" in his annual message in 1905. Ultraconservatives in Congress blocked action on the measure, but Senate opposition collapsed in February 1906, when the American Medical Association warned Nelson Aldrich that its 135,000 members would advise patients to lobby their senators if the pure food bill were not passed. Moving quickly to get rid of this hot potato, the Senate approved the bill and sent it to the House, where it was buried deep in committee.

New life was breathed into the issue by the publication a month later of Upton Sinclair's *The Jungle,* a Socialist tract disguised as

a novel. Sinclair's lurid account of Chicago packing houses revealed in revolting detail the conditions in which beef and pork
were prepared for America's tables. Canned beef was made from
old, diseased cattle, and sausage contained rats and rat dung.
Potted chicken contained no chicken at all, only waste cattle products. Men fell into the processing vats and "all but the bones
[went] out into the world as Durham's Pure Leaf Lard!"

President Roosevelt was one of the most outraged readers of
The Jungle. In Mr. Dooley's words: "Tiddy was toying with a
light breakfast and idly turnin' over th' pages iv the new book
with both hands. Suddenly he rose fr'm the table, an' cryin': 'I'm
pizened,' began throwin' sausages out iv th' window. . . ." Roosevelt didn't need Sinclair's book to alert him to conditions in the
packing houses, however. He vividly remembered the loathsome
"embalmed beef" issued the Rough Riders during the war with
Spain. But he was alarmed by Sinclair's contention that government inspectors were not doing their jobs.

Roosevelt and Agriculture Secretary James Wilson immediately ordered an Agriculture Department investigation of the Bureau of Animal Industry, which ran the inspection system.
Suspecting a cover-up, they appointed Commissioner of Labor
Charles P. Neill and James B. Reynolds, a Washington attorney,
to conduct an independent probe. As expected, the Agriculture
Department report whitewashed the bureau and charged that
Sinclair had misrepresented the situation. Neill and Reynolds, on
the other hand, found that, if anything, the writer had understated
conditions in the packing plants.

With the president's blessing, Senator Albert J. Beveridge of
Indiana, a onetime conservative turned progressive, tacked a comprehensive meat inspection bill onto a House Agricultural Appropriations bill pending in the Senate. This was done because
previous Congresses had undermined the inspection system by
refusing to provide adequate funding. Beveridge proposed to shift
funding from the small appropriation provided by Congress to a
fee for each animal inspected. Packers would also be forced to
label canned meats with the date on which they were processed
and their contents. Moreover, the secretary of agriculture was
empowered to establish sanitary regulations for packing plants
that would be enforced by inspectors. If an operator challenged

an inspector's findings, the secretary had "final and conclusive" authority to make a final ruling.

Wielding the threat of releasing the Neill-Reynolds report, Roosevelt forced the Beveridge amendment through the Senate without a dissenting vote. He hoped this smashing victory would influence the outcome in the House, but the meat packers were just beginning to fight. Once again, as in the fight over the Hepburn Act, the real issue was the extension of federal authority over business. The packers had valuable allies in Speaker Joe Cannon and Congressman James W. Wadsworth of New York, the chairman of the House Agriculture Committee. And time. Summer recess was only six weeks away. If the opponents could stall action on the Beveridge amendment until the deadline had passed, the measure would be dead. By the time Congress reconvened, popular outrage would have dissipated. Wadsworth introduced a substitute plan to draw the teeth of the Beveridge amendment and consigned both measures to his committee.

Roosevelt raged against the substitute. "Each change is for the worse and that in the aggregate they are ruinous, taking away every particle of good from the suggested Beveridge amendment," he told Wadsworth. On June 4, he released the "sickening" Neill-Reynolds report, but its shock value had been reduced because much of it had previously been leaked to the press. The bill approved by Wadsworth's committee six days later dropped the fee system and substituted a $1 million appropriation for federal inspection, barely enough to meet current expenses. The provision for dating canned meats was also dropped, and final authority was shifted from the secretary of agriculture to the federal courts.

Faced with another struggle that threatened to split the Republican party—a rift that the Democrats sought to deepen by unsuccessfully attempting to bring the Beveridge bill to the floor—the president and Joe Cannon agreed to work out a compromise. The details were entrusted to Wisconsin Congressman Henry C. Adams, a member of the Agriculture Committee, who was free of the taint that clung to Wadsworth. The fee system was replaced by a $3 million appropriation for inspection, a provision for dating canned meats was restored, and the question of court review was left ambiguous. Roosevelt accepted the compromise, as he told

Beveridge, because "I . . . do not want to get into an obstinate and wholly pointless fight about utterly trivial matters. . . . I am concerned with getting the result, not with the verbiage."

With the last roadblocks removed, the meat inspection bill was quickly passed and signed into law by the president on June 30, 1906. Under the impact of the meat inspection bill, the pure food and drug law had also been pried loose in the House and it, too, was signed that same day. This was a milestone in consumer protection and formed the basis of all subsequent efforts in this field.

"It has been a great session," Theodore Roosevelt told a friend after Congress had decamped for home. "The railroad rate bill, meat inspection bill & pure food bill, taken together, mark a noteworthy advance in the policy of securing Federal supervision and control of corporations." The session also marked the high-water mark of his years in the White House.

Chapter 21

"To Keep the Left Center Together"

Without public-opinion polls, it was impossible to measure Theodore Roosevelt's standing with the American people as he entered the final two years of his presidency. But there were other ways to estimate it. "If five hundred of the first citizens of New York were to make an affidavit that they had seen the president do a certain thing, and he was to deny it the following day," said one observer, "the five hundred affidavits would be as waste paper in the estimation of the country at large."

Nevertheless, little seemed to go right for Roosevelt in these twilight years. Most of his legislative initiatives were ignored and as the bickering between the White House and Congress increased in intensity, he was nearly overwhelmed by a series of crises and confrontations on both the domestic and international fronts. Part of the difficulty lay in the oozing away of power experienced by all presidents as their terms draw to an end. Roosevelt's problem was magnified by his self-created lame duck status and the congressional rebellion against his aggressive use of executive power. He had bitter and powerful enemies among the Old Guard in Congress, men who grew bolder in their opposition as his term wound down.

In fact, during the 1906 congressional campaign, the president angered conservatives by having discovered "an inherent power" beyond those enumerated in the Constitution—"a power to oversee and secure correct behavior in the management of all the great corporations engaged in interstate business." While "the government ought not to conduct the business of the country," he continued in a speech at Harrisburg, "it ought to regulate it so that it shall be conducted in the interests of the public."

Such talk infuriated businessmen, most notably the Standard Oil "people," as the president called them, and E. H. Harriman, who angrily refused to contribute to the party coffers. Informants told Roosevelt that Harriman had said "he had no interest in the Republican party," and in view of the president's attacks on the corporations, "he preferred the other side to win." Roosevelt also engaged in a running fight with Harriman after the railroad baron publically claimed that the president had personally asked him to raise $250,000 for the 1904 campaign. "Any such statement is a deliberate and wilful untruth—by rights it should be characterized by an even shorter and more ugly word," he declared, probably correctly.

While Roosevelt had been uneasy—as always—about the outcome of the election, he had good reason to be pleased with the results. Four new Republican senators were elected; in the House the party maintained a sizable majority despite some losses, and Hughes defeated William Randolph Hearst for the governorship of New York. Nick Longworth was among those who preserved their seats. "It is very gratifying," the president wrote Alice, "to have . . . won [a] striking victory while the big financiers either stood silent aloof or gave furtive aid to the enemy."

Roosevelt's anger at "the big financiers" prompted him to press forward with his campaign for reform. "To use the terminology of Continental politics," he told Arthur Lee, "I am trying to keep the left center together." Provoked by congressional intransigence and the disdain of big business, he expressed his progressive views more openly. Throughout his political life he had curbed the tendency to take advanced positions, but now felt he had nothing to lose, further widening the chasm that divided the Republican party.

In his annual message to Congress on December 3, 1906, the

president reiterated demands for a graduated income tax and federal supervision of all companies engaged in interstate commerce. Laws barring corporations from making contributions to political parties, limiting the use of injunctions in railroad labor disputes, reducing the hours of railroad workers, and forbidding child labor were also placed on the agenda. "I do wish that the same men who got elected on the issue of standing by me would not at once turn and try to thwart me," he said to those who protested this renewed call for more federal regulation. Always, however, his enemies were eager to exploit any false step for maximum effect—and in his handling of the Brownsville affair he offered them an opening.

Shortly before midnight on August 13, 1906, a group of perhaps ten to twenty armed men shot up the Texas border town of Brownsville, killing one citizen and wounding another. White townspeople accused members of the all-black 25th Infantry Regiment, which had recently been stationed just outside town at Fort Brown, of conducting the raid in revenge for racial slights. In the immediate aftermath of the incident, however, some of the regiment's white officers believed the residents might have staged the raid and planted evidence to implicate the black soldiers. Others, however, believed the men guilty.

None of the blacks admitted complicity, but army investigators were convinced by white witnesses and circumstantial evidence that members of three companies had participated in the shooting and that the others were protecting them in a "conspiracy of silence." Since the men "appear to stand together in a determination to resist the determination of guilt," the army's Inspector-General concluded, "they should stand together when the penalty falls." No one produced convincing evidence, then or later, proving the black soldiers were indeed involved, nor were they given the opportunity to prove their innocence. Nor was the theory that the whites were behind the incident examined. Basing his decision on the army's investigation, Roosevelt ordered all three companies—a total of 167 men—discharged "without honor." Some had upward of fifteen years of unblemished service, and six had won the Medal of Honor in Cuba or the Philippines.

Black Americans who had looked to Roosevelt for a measure

of support and help were keenly disappointed by his action. Booker T. Washington, informed by Roosevelt of his decision in advance, was rebuffed when he offered to provide additional information on the case. "You cannot have any information to give me privately to which I could pay heed, my dear Mr. Washington, because the information on which I act is that which came out of the investigation itself," Roosevelt replied. With that, he departed for Panama in November to see "how the ditch is getting along."

Other black leaders—and some whites—persisted in trying to right what they saw as a miscarriage of justice and urged Secretary of War Taft to suspend the mass dismissal until the president returned home. Taft rescinded the order and wired for permission to rehear the case, but Roosevelt rejected the plea unless there was new evidence. Horrified by any violent defiance of the law, he was unmoved by the protests. As sometimes occurred when his actions were questioned, the self-righteous and vindictive sides of his character came to the fore, and he became even more convinced of the justice of his role in the Brownsville affair. Moreover, in making overtures to southern white voters and relying less on the advice of Booker Washington and other African-Americans, he had become less sensitive to the inequities faced by blacks. Roosevelt was not an unprincipled man; when he had strong feelings about an issue, he remained consistent. In the case of equal rights for black Americans, he vacillated—and so did his policies.

Upon his return to Washington from Panama, the president discovered that the black soldiers had an unlikely champion in the person of Joe Foraker. Undoubtedly the senator saw the affair as an opportunity to advance his chances for the presidency in 1908 among black Republicans at the expense of his fellow Ohioan, Will Taft, but he also seems to have been genuinely convinced that the men were victims of prejudice and injustice. Reacting to criticism, Roosevelt went out of his way in his annual message to denounce lynching and to urge improved educational opportunities for blacks—but refused to rescind the dismissal.

Brownsville had become a political issue, and over the next several weeks debate over it raged in the Senate, with Foraker demanding a full investigation of the justice and legality of the presidential order. Some lawmakers, seeing a way to embarrass Roosevelt, joined Foraker in denouncing his action. Roosevelt responded with a bristling message claiming that the soldiers were guilty of an act "unparalleled for infamy in the history of the United States Army." He also dispatched an assistant attorney general to Texas to conduct another investigation, which confirmed the original findings. Privately, the president, still angry over Foraker's opposition to the Hepburn Act, was convinced that the Ohioan was merely a front man for Wall Street and was using the Brownsville affair as a cover to attack him. Finally, at Foraker's urging, the Senate Military Affairs Committee agreed to investigate the incident.

The animosity between the president and Foraker flared into the open at the Gridiron dinner on January 26, 1907. Turning the pages of the program as he sat at the main table, Roosevelt's eye fell upon a cartoon of Foraker wooing the black vote, which was accompanied by a jingle beginning "All coons look alike to me." As Samuel G. Blythe, the club president, put it, he "showed signs of eruption." Dinner was just being served, but Roosevelt asked to speak immediately rather than later as scheduled. Service was halted and the waiters were shooed from the room—all remarks were off the record—as the president began his speech with a spirited discussion of railroad rate regulation and reform in general.

Then Roosevelt picked up the program and opening it to the Foraker cartoon, read the offending caption. Hurling the booklet to the table, he almost shouted, "Well, all coons do *not* look alike to me!" Fixing his eye upon Senator Foraker, who sat directly in front of him, the president heatedly defended his record on Brownsville and sharply criticized the senator for his interference. Roosevelt had many supporters in the audience who applauded and cheered him. During this presidential tirade, the white-faced Foraker stared fixedly at the president, silently twisting a napkin.

Foraker was not on the program, but Blythe thought it only fair to allow the senator to reply. Quickly living up to his reputation as Fire Alarm Joe, Foraker launched into a defense of the black soldiers and a bitter attack on the president. As a senator from Ohio, Foraker declared, he was beholden to no one except the citizens of his state and was not about to be intimidated or coerced by anyone, not even the president of the United States. The chief executive, he declared, should not enjoy a special position when he was wrong on a matter of policy.

Red-faced and boiling at his public scolding, Roosevelt popped up several times to answer and was barely restrained from seizing the floor. In the hubbub of cheers and shouts that broke out after Foraker had finished, he angrily declared that "some of the men were bloody butchers—they ought to be hung," reported Congressman Champ Clark of Missouri. Only the White House had jurisdiction of the matter, Roosevelt continued. "It is not the business of the House. It is not the business of the Senate. All the talk on that subject is academic. If they pass a resolution to reinstate these men, I will veto it; if they pass it over my veto, I will pay no attention to it. I welcome impeachment."

Obviously such a clash between a president and a prominent senator was hot news, and despite Gridiron etiquette, several papers printed accounts of the affair. The consensus was that Foraker had gotten the better of the exchange, but his moment of glory was brief. A vengeful Roosevelt moved to break his political hold on Ohio by directing that in future all patronage for the state be funneled through Taft. Foraker was not only eliminated from the presidential race in 1908, but he also lost his Senate seat after the Hearst papers revealed he had been on the Standard Oil payroll throughout most of his senatorial career.* Roosevelt himself is said to have inspired a story in the *Cincinnati Times-Star* to the effect that any Republican who voted for Foraker was committing treason to the party.

In the meantime, the Brownsville affair dragged on as the Senate Military Affairs Committee carried out its probe. Perhaps having second thoughts about his position, Roosevelt reduced the heat of his language and agreed to allow the reenlistment of those men willing to swear an affidavit that they had not taken part in the raid. In 1908, a committee majority of five Democrats and four Republicans reaffirmed the guilt of the soldiers, while Foraker and three other Republicans pronounced them innocent. The Senate approved the appointment of a military court of inquiry,

---

*Foraker had received at least $50,000 in Standard Oil money—which he described as legal retainers.

which once again sifted through the evidence and a year later—
after both Roosevelt and Foraker had passed from the scene—
confirmed the original verdict. Fourteen men were allowed to
reenlist, however. No reason was given why they were found
"qualified" and the others were not.*

Throughout the affair, Roosevelt defended his conduct and
claimed that the race of the culprits had nothing to do with his
decision. Had they been white, he said, he would have meted out
similar punishment. Undoubtedly this is true, but blacks were
convinced he had acted like a racist. Confident of his own righteousness, he refused to reexamine the evidence or put aside
preconceived notions. In his anger at his critics, Roosevelt permitted a miscarriage of justice. While it is unfair to measure him
by the standards of a later era and tar him as a racist, he clearly
failed to remain true to his own ideals. In years to come, he would
be less convincing when he launched into his frequent theme of
justice for all. Significantly, he made no mention of the Brownsville affair in his *Autobiography*.

Roosevelt and Congress also wrangled over conservation of the
nation's natural resources. The irrigation and land reclamation
projects of the Newlands Act, passed during his first year in the
White House, were popular in the arid sectors of the West, but
not everyone supported his environmental program. By custom
and tradition, westerners were "boomers" who believed in their
God-given right to exploit the continent's resources. Ranchers,
mine operators, lumbermen, and power companies protested efforts to limit their operations. As the president's closest adviser
on conservation matters, Gifford Pinchot became a lightning rod
for criticism and pointed complaints about high-handed and unrealistic bureaucrats who "sit within their marble halls and theorize and dream about forests conserved."

Nevertheless, the president pushed his conservation program
with renewed vigor following the election of 1904. James Garfield,
an ardent conservationist, was named secretary of the interior in

---

*The Brownsville affair was finally closed in 1972, during the administration of Richard
M. Nixon, when the Pentagon announced that the concept of mass punishment was
contrary to army policy, and the discharges of all the men involved were changed to
honorable.

place of an incumbent whose enthusiasm for the cause was insufficient to satisfy Roosevelt. Pinchot was named in 1905 to head a newly organized Forest Service, given enlarged powers, including the right to order arrests for violation of laws and regulations governing the use of forest, mineral, and grazing rights. Roosevelt was labeled a "dictator," and one newspaper commented only half-jokingly that if he "continued to create reserves there would be little ground left to bury folks on." Some critics charged that land suitable for homesteading had been placed in reserve, but Roosevelt contended that other administration programs had opened over a million acres to homesteading and that the individual homestead allotment had been raised to 320 acres.

Not all the criticism of the Roosevelt-Pinchot policies was unfounded. Their reliance on experts and bureaucratic managers made their programs liable to attack by Congress on grounds that their policies did not represent the views of the people who were affected by them. In essence, it was charged, the president had adopted the position that conservation was too complex an issue to be left to the judgment of ordinary Americans. Particularly galling to smaller operators was the administration's preference for the larger timber growers, cattle grazers, and sheep raisers in assigning leases for land use in the public domain, a policy based on economic efficiency and rationalization.

Under Pinchot's direction, numerous westerners were indicted for land fraud and convicted, including almost the entire Republican party organization in Oregon, among them one of the senators from that state. Charles W. Fulton, the other senator, claimed that the special federal prosecutor, Francis J. Heney, who brought the cases, was conducting a political vendetta, and he demanded Heney's removal. "Mr. Heney cannot hurt the Republican party," Roosevelt replied, "and your wrath should be reserved not for him but for those Republicans who have betrayed the party by betraying the public service and the cause of decent government."

By 1907, Fulton had accumulated enough support to strike back. Using the power of the purse, he attached a rider to a Department of Agriculture appropriations bill that barred the creation of new forest reserves in six western states or additions to existing reserves without the approval of Congress. Taking advantage of the opportunity to humble the president, Congress easily approved the bill, placing Roosevelt in a quandary. If he vetoed the bill, the entire Agriculture Department would come to a standstill; if he signed it, the Forest Service would lose control over millions of acres of public lands considered essential to the national reserve.

But Roosevelt was not to be so easily thwarted. Working under forced draft, Pinchot prepared an executive order snatching some sixteen million acres out of the outstretched hands of would-be exploiters, and the president signed it on March 2, 1907. Two days later, he signed the agriculture appropriations bill—locking the barn after having already stolen Fulton's horse. Critics decried the "midnight proclamations" as another sign of presidential arrogance and encroaching dictatorship, but Roosevelt ignored these attacks. "If I did not act, reserves which I consider very important for the interests of the United States would be wholly in or in part dissipated," he declared. And throwing down the gauntlet to the opposition, he challenged Congress to try to rescind his order.

To make matters worse, the boomers charged that the imperious Pinchot had evaded the intent of Congress by placing 2,500 water power sites in reserve by simply designating them ranger stations. Congress reacted with legislation upsetting the order, but Roosevelt vetoed it. On balance, he won his fight, yet the opposition was able on occasion to have it way. Congress summarily refused to fund a National Country Life Commission, branding it "rural uplift," or support an Inland Waterways Commission, established by the president to prepare a comprehensive plan to improve the nation's waterways.

Undeterred, Roosevelt took his case to the state governors. In May 1908, he called a National Conservation Congress, attended by most of the governors and some five hundred political leaders and experts. "As a people," he declared at the opening session in the East Room of the White House, "we have the right and duty . . . to protect ourselves and our children against the wasteful development of our natural resources." Largely as a result of the conference, forty-one states established their own conservation commissions and programs.

By and large, Roosevelt dealt with environmental issues with statesmanship and administrative skill as well as moral fervor. In his seven and a half years in the presidency, he impressed upon the nation the importance of preserving natural resources for future generations. The concept of conservation was broadened to include not only forests and wildlife, but coal and mineral lands, oil reserves, and power sites. Government land reserves were increased from 45 million acres in 1901 to 195 million acres in 1909. During the Roosevelt years, thirty irrigation projects were started, including some of the nation's largest dams; Grand Canyon and Niagara Falls were among the eighteen protected national monuments, and five new national parks and fifty-one wildlife refuges were established. These were to be Theodore Roosevelt's great legacy to the American people.

"We still continue to enjoy a literally unprecendented prosperity," the president announced to the nation at the end of 1906. But as the new year began, there were flurries on the New York Stock Exchange. While the election of McKinley in 1896 had settled the gold standard question, the need for monetary policy and currency reform remained unmet. There had been persistent talk of establishing a central banking system to manage the currency and offset the contraction of credit in times of crisis, but nothing had come of it. The heated expansion of industrial production throughout the world, the demand for funds to finance the rebuilding of San Francisco, which had been devastated by an earthquake and fire the previous year, and the Russo-Japanese and Boer wars had set off a global demand for capital. Roosevelt himself warned in his annual message that "reckless speculation and disregard of legitimate business methods on the part of the business world can materially mar this prosperity."

In March 1907, the stock market was jolted by a sharp wave of selling that slashed values by $2 billion. Union Pacific stock, for example, dropped twenty-five points in a single day. "I would hate to tell you to whom I think you ought to go for an explanation of all this," E. H. Harriman told a reporter bitterly. Wall Street knew full well where the blame lay—with the erratic antibusiness policies of the president of the United States. Roosevelt was described as "the most effective planter of the weeds of uniformed socialist propaganda," and there was talk of an alliance between

John D. Rockefeller and Harriman to ensure the nomination of a conservative Republican in 1908.

Economics and fiscal policy were never Roosevelt's strong points, and he professed to be puzzled by "this belief in Wall Street that I am a wild-eyed revolutionist. I cannot condone wrong, but I certainly do not intend to do aught save what is beneficial to the man of means who acts squarely and fairly." He correctly assigned the blame for the collapse to the inelasticity of the currency supply and a worldwide speculative boom. The credit shortage caused breaks in foreign stock markets as well as on Wall Street as money became even more difficult to obtain. New York City and cities and towns across the country were unable to float bond issues. Such firms as the Westinghouse Electric Company and the Interborough Metropolitan Company could find no purchasers for their stock and went into receivership.

To allay public concern, Roosevelt met with J. P. Morgan, now semiretired at seventy, but he refused to issue the overly optimistic statement on the state of the economy that the financier urged upon him. He kept an eye on the indexes of business health, however, and ordered George Cortelyou, who had just become secretary of the treasury, to deposit $70 million in customs receipts in the New York banks. These efforts helped the economy weather a mild recession that had settled in at this time.

The strain of running battles with Congress and the financial crisis took its toll upon Roosevelt. "The general impression of this town is that we all feel tired," observed Henry Adams at the end of March. "Three or four persons close to the President have assured me that, for the first time, even he complains of fatigue. . . . The President is trying to find out what effect a dose of hard times and unemployed labor will have on the Republican vote. For the first time he is transparently hesitating." Not long after, Roosevelt uncharacteristically took to his bed, complaining of a toothache.

Nevertheless, the president pressed ahead with militant demands for federal regulation of all companies engaged in interstate commerce and for legislation permitting the Interstate Commerce Commission to evaluate railroad property in determining rates. The Department of Justice filed suit to break up the American Tobacco Company and readied similar action against the Union

Pacific. Standard Oil was convicted in the federal district court in Chicago of 1,462 counts of rebating and was fined a shocking $29.2 million.* In August, Roosevelt went to Provincetown, on Cape Cod, to dedicate a monument to the Pilgrims. With upraised arm and baleful eye, he lashed out at "certain malefactors of great wealth," whom he accused of combining "to bring about as much financial stress as possible, in order to discredit the policy of the government and thereby secure a reversal of that policy." Businessmen muttered among themselves about the president's drinking problem and increasing signs of mental instability.

Unable to sleep, Theodore Roosevelt "sat solemnly in scanty attire" with his daughter Ethel on the porch of Sagamore Hill on an unbearably hot night in June 1907, hoping to catch a cooling breeze off the Sound. Relative calm had returned to the marketplace, and the president was trying to unwind following one of the most strenuous periods of his presidency. Edith and the other children were away, and he and Ethel were alone in the big house. Bright moonlight dappled the lawn, and "as always happens on moonlight nights," the president noted, the male servants—the grooms, the housemen, the coachman—harmonized somewhere off in the distance. Enchanted, both father and daughter listened until the singing faded away.

"There isn't any place in the world like home—like Sagamore Hill, where things are our own, with our own associations and where it is real country," mused Roosevelt. Now he ruminated on the fact that the children were growing up with alarming speed and embarking on lives of their own. Alice had married Nick Longworth; Ethel had just turned sixteen, and was often away at parties in town or country weekends. Ted, in his junior year at Harvard, was preparing to take an extended trip to northern Minnesota; seventeen-year-old Kermit was also going west that summer to rough it with the Thirteenth Cavalry out in Dakota. In the autumn, Archie would be off to Groton, leaving only little Quentin at home.

The nation was also taking a breather from the clash of reform and reaction. Newspapers were filled with stories from Idaho

---

*Within less than a year, the decision would be reversed.

about the trial of "Big Bill" Haywood, head of the radical Western Federation of Miners, on charges of complicity in the murder of former Governor Frank Steunenberg. There was an attempt to unseat Senator Reed Smoot of Utah, on grounds that as a member of the Mormon hierarchy he supported polygamy and therefore could not be loyal to the Constitution. The growing passion for speed was gratified by the British liner *Lusitania,* which set a transatlantic record of five days and forty-four minutes.

Mid-October found the president on a bear hunt deep in the red gum and white oak forests of Louisiana. Writing to Ted, he happily reported bagging a bear and noted that wherever he went in the South he was greeted with cheers—even in Mississippi, where the governor had issued an anti-Roosevelt broadside. "I have been greatly interested in what I have seen in the papers," the president added, in a reference to reports that Ted had made Harvard's second-string football team. "I don't suppose you have much chance to make the first eleven, but I shall be awfully interested to see how you do."

J. P. Morgan was also traveling in those autumn days. With two private railway cars of bishops as his guests, he had gone to Richmond, Virginia, to attend the general convention of the Protestant Episcopal Church, where as a prominent layman he helped settle theological disputes and enjoyed the ceremony and hymn singing. But on October 19, 1907, he received word of a banking panic in New York and confided to Bishop William A. Lawrence that he was returning immediately to the city. "They are in trouble in New York. They do not know what to do, and I don't know what to do, but I am going back."

The crash was precipitated by the efforts of two Wall Street buccaneers, F. Augustus Heinze and Charles W. Morse, to corner the copper market. They failed—and this failure started a run on the trust companies that had financed their speculations. Trust companies had been established to circumvent the New York State banking laws that required banks to maintain 25 percent cash reserves against deposits. To lure depositors, they paid higher than the going rate of interest. When the Clearing House Association required the trusts to have at least 10 percent reserves, they dropped out, depriving depositors of the protection afforded by membership. Yet, in reality, the failure of Heinze and Morse was to the financial panic of 1907 what the assassination of Arch-duke Franz Ferdinand in Sarajevo in 1914 would be to World War I: not a fundamental cause but a precipitating event. The real cause was the strain upon credit by speculation and overex-pansion and the unsound condition of the banking system.

Panic and fear swirled about New York like an evil fog. Unable to meed demands for withdrawals, the Knickerbocker Trust Com-pany collapsed. Prices on the stock exchange tumbled to new lows, and interest on short-term loans rose to an astronomical 125 percent. Like a stone cast into a pond, the panic in New York rippled out into a financial crisis throughout the United States as banks in the South and West were unable to make loans to farmers sending their harvest to market. Moreover, New York's demands on European gold to relieve the crisis created the threat of a similar panic in London, Paris, and Berlin.

Upon his return to Washington from Louisiana on October 23, Roosevelt immediately consulted with Elihu Root, George Meyer, and Robert Bacon, who were close to the business community. Cortelyou had already been dispatched to New York, where Ju-piter Morgan was leading the fight to prevent a general collapse. In the president's name, he issued a statement that "the under-lying conditions which make up our financial and industrial well-being are essentially sound and honest." Cortelyou also arranged for the deposit of $25 million in federal funds in the New York banks. These deposits and funds, pledged by Morgan, Harriman, and Rockefeller, seemed to stem the panic.

And then the storm, which had veered away so tantalizingly, roared back with increased fury. Now the Trust Company of America was under attack. Morgan estimated that another $25 million was needed to keep the company from going under. At the same time, Charles Moore's brokerage house, Moore and Schley, which was heavily in debt to the New York banks, was in danger of insolvency. If the firm failed, Morgan believed it would be impossible to obtain funds to save the Trust Company or to prevent a stock market collapse.

Posing as a public benefactor, Morgan hatched a scheme de-signed to save Moore and Schley while benefiting U.S. Steel, his favorite creation. The stricken brokerage house owned the ma-jority shares in the Tennessee Coal and Iron Company and had pledged them as collateral for its loans. If Moore and Schley failed, these shares would be thrown on the market for what they would bring, deepening the panic. Tennessee Coal had valuable iron and coal holdings in Tennessee, Georgia, and Alabama that would complement those belonging to U.S. Steel. To save the banks and stem the panic—or so it was claimed—Morgan suggested that U.S. Steel buy the shares. Moore and Schley would be able to pay off their creditors while the Tennessee Coal stock would be kept off the market. Before carrying out the deal, however, the directors of U.S. Steel decided to obtain assurances from the president that they would not be prosecuted under the Sherman Anti-Trust Act.

Having secured an appointment to see Roosevelt early on the morning of November 4, Henry C. Frick and Judge Elbert Gary of U.S. Steel sped down to Washington on a special train consisting of a single Pullman car hitched to a fast locomotive, with priority over all other traffic. They wished to obtain presidential acquiescence to the purchase of the Tennessee Coal shares before the stock exchange opened on Monday morning. After being ushered in to see the president as he was having breakfast, Frick and Gary told him about the plight of Moore and Schley—without naming or being asked to name the firm—and explained that among its assets was a majority of the shares in Tennessee Coal. Out of a sense of duty, they said, they were willing to buy the shares even though it would be of "little benefit" to U.S. Steel. Would there be any presidential objections?

"It was necessary for me to decide on the instant," Roosevelt recalled in his *Autobiography*. "To stop a panic it is necessary to restore confidence," and only the Morgan interests retained the confidence of both business and the public. Roosevelt regarded U.S. Steel as a "good," or socially responsible, trust because of its cooperation with the Bureau of Corporations. In fact, the informal understanding that federal officials had developed with steel company executives was the prototype of the formal supervision Roosevelt wanted Congress to adopt over all interstate corporations. He accepted Frick's and Gary's claim that the takeover would be of no great benefit to U.S. Steel and was proposed only in the spirit of public interest.

"I answered Messers. Frick and Gary . . . to the effect that I

did not deem it my duty to interfere, that is to forbid the action,"
he related. In other words, he would not invoke the Sherman Act
to block the purchase. Five minutes before the stock exchange
was to open at 10:00 A.M., Gary telephoned 23 Wall Street from
the White House to pass on the news. The market rallied and the
panic was over. To aid in stabilizing the financial situation, the
Treasury issued $100 million in certificates and $50 million in low-
interest Panama Canal bonds, which were sold to the banks on
credit, and authorized them to issue currency with the bonds as
collateral.

Hostile criticism poured in upon Theodore Roosevelt from
every direction in the wake of the panic. The business world
blamed his reform policies for the disaster; progressives saw betrayal in his response to the cries for help from Wall Street. They
charged that he had allowed himself to be duped by Morgan's
emissaries into scuttling his antitrust policy. Senator La Follette
even claimed the bankers had created the panic for their own
ends. Through acquisition of Tennessee Coal, it was pointed out,
U.S. Steel had for a bargain $45 million gained assets later valued
at about $1 billion, which gave it control of 62 percent of the
nation's steel-making capacity. The president, of course, was certain he had done the right thing. "The result justified my judgement," he contended. "The panic was stopped, public confidence
in the solvency of the threatened institution being at once restored."

While juggling the Brownsville affair, the fight over his conservation efforts, and the financial panic, Roosevelt also faced a
crisis in American relations with Japan. Although the emperor
of Japan had thanked him for his "disinterested and unremitting
efforts" in helping to bring an end to the war with the Russians,
the Japanese were angry at not having won all their demands at
the peace table at Portsmouth. Now they accused the United
States of having thwarted their imperial ambitions for its own
purposes. On the other side, many Americans were wary of the
formidable military power of the Japanese. As Mr. Dooley put
it: "A few years ago I didn't think anny more about a Jap thin
about anny other man that'd been kept in th' oven too long. They
were all alike to me. But today, whin-iver I see wan I turn pale
and take off me hat an' make a low bow."

Trade rivalries in the Far East heightened tension between the two countries, which was fanned by a wave of anti-Japanese hysteria that swept California. On October 11, 1906, the San Francisco Board of Education ordered all Japanese children transferred to a segregated school. The excuse given was that the Japanese children—all ninety-one of them—were crowding whites out of the classrooms. The Japanese regarded the measure as a racial and cultural slur, and it was greeted with widespread anti-American rioting in the cities of Japan.

With the Panama Canal still unfinished and the major elements of the U.S. Navy in the Atlantic, the president was anxious to avoid an open break with Japan. The United States was so ill-prepared for a conflict in the Far East that the Orange Plan, drafted by a joint army-navy board in the event of war with Japan, called for abandoning the Philippines and for withdrawing the handful of cruisers and gunboats in Asian waters to the West Coast. "I am horribly bothered about the Japanese business," Roosevelt told Kermit. "The infernal fools in California, and especially in San Francisco, insult the Japanese recklessly and in the event of war it will be the Nation as a whole which will pay the consequences."

However, he had no authority over the public schools of San Francisco, and could only show his total disagreement with their action through gestures, such as pressuring the California legislature not to enact statewide discriminatory measures and requesting Congress to naturalize Japanese who were already living in the United States. "I shall exert all the power I have under the Constitution to protect the rights of the Japanese people who are here," he assured his friend Kentaro Kaneko.

Roosevelt also assailed the San Franciscans in his annual message of 1906. Not only did he brand the segregation of the Japanese students "a wicked absurdity," but he gave a veiled threat of federal action. These words soothed the Japanese, but the Californians dug in their heels even deeper and southern lawmakers bridled at the thought of federal action against a local school board. The public outcry at his proposal for the naturalization of Japanese already in the United States convinced Roosevelt that he had seriously underestimated the amount of xenophobia on the Pacific coast and, always the realist, he switched tactics.

The entire San Francisco school board was invited to Washington at government expense to discuss the problem with the president. Once they were in the White House, the outcome was foreordained. Under the shadow of the Big Stick, the delegation agreed to rescind its action in return for an end to the influx of Japanese immigrants. Over the next several months, Roosevelt and Secretary of State Root negotiated a "gentlemen's agreement" with Tokyo sharply restricting the immigration of Japanese to the United States, and the segregation order was lifted.

In spite of the peaceful end to the dispute, Roosevelt believed that the Japanese interpreted his sympathetic attitude as a sign of weakness. "I am exceedingly anxious to impress upon the Japanese that I have nothing but the friendliest possible intentions toward them," he told Henry White, "but I am not afraid of them and that the United States will no more submit to bullying than it will bully." Late in the spring of 1907, according to a witness, Roosevelt confided to a meeting of his military and naval advisers that he believed Japan was preparing a hostile move against the United States, probably before the completion of the Panama Canal. The entire American battle fleet should be transferred immediately from the Atlantic to the Pacific.

"Mr. President, do you believe the Japanese would dare attack the West Coast of the United States?" asked one official, excitedly rising to his feet. "Why, Mr. President, the women of the Pacific Coast would drive the Japanese into the sea with their broomsticks."

"Oh, sit down, sit down," replied the exasperated Roosevelt.

Turning to the ranking admiral present, according to this account, the president asked how soon the fleet could get under way around South America for the Pacific. The ships could sail in three weeks, he was told. In actuality, the redeployment did not begin until six months later and was transformed into the epic around-the-world cruise of the Great White Fleet. Originally, it was announced that the fleet would move only from Atlantic waters to San Francisco on a "practice cruise," but everyone knew this audacious flourish of the Big Stick was meant to remind the Japanese that the United States had the world's second largest fleet while theirs ranked fifth. It was also designed to impress the nations of the world with America's power, to test the readiness of the fleet, as well as to create popular support for the administration's naval shipbuilding program, which was tied up in Congress.

Congress immediately protested that it had not been consulted; the eastern states objected to stripping the Atlantic coast of its defenses; and there were those who feared the fleet might be destroyed by a Japanese surprise attack or damaged by the elements. Roosevelt paid no heed to these objections. "Thank Heaven we have the navy in good shape," he wrote Root. "It is high time, however, that it should go on a cruise around the world. In the first place, I think it will have a pacific effect to show that it can be done; and in the next place after talking thoroly over the situation with the naval board I became convinced that it was absolutely necessary for us in time of peace to see just what we could do in the way of putting a big battle fleet in the Pacific and not make the experiment in time of war." Informed by Senator Eugene Hale, chairman of the Senate Naval Affairs Committee, that money would not be appropriated for the voyage, Roosevelt had a ready answer. Funds were available to send the fleet to the Pacific coast, he replied; if Congress wanted the ships back, it could provide the money. That ended the argument.

A triumphant Theodore Roosevelt proudly stood at attention in the pale wintry sunshine of December 16, 1907, as the sixteen battleships of the Great White Fleet steamed majestically out of Hampton Roads. Marine guards snapped to attention, salutes boomed out, and bands crashed into the national anthem as the vessels passed in review before the presidential yacht *Mayflower*. "Did you ever see such a fleet and such a day?" cried the ebullient president, acknowledging the honors with a flash of teeth and a wave of his top hat. "By George, isn't it magnificient!" Ready for "a feast, a frolic or a fight," in the words of Rear Admiral Robley D. Evans, its commander, the armada was off on its unprecedented voyage around the world.

Enthusiastic crowds welcomed the Great White Fleet at ports along the Atlantic and Pacific coasts of South America, and by the time it reached San Francisco, an invitation had been received from Japan for the fleet to come calling. The Japanese government seemed genuinely eager to dispel the bitterness created by past misunderstandings, and as the American bluejackets paraded through the streets of Yokohama, thousands of schoolchildren waved American flags and sang "The Star-Spangled Banner" in English. With tensions between them eased, the two nations concluded an agreement to maintain the status quo in the Pacific, to respect each other's territorial possessions, and to support the Open Door and the territorial integrity of China.

Fourteen months later, on February 22, 1909, Roosevelt was again on hand to greet the Great White Fleet when it returned to Hampton Roads after steaming some 46,000 miles without a serious breakdown. "In my own judgement," he was to say, "the most important service that I rendered to peace was the voyage of the battle fleet round the world." It was a moment of high triumph, because British and German naval authorities had doubted that such a large fleet could be safely sailed around the world. But this triumph was muted by the fact that Roosevelt's imperial years were over. Whatever regrets he may have felt at stepping down, however, he had the consolation of knowing his handpicked candidate would succeed him in the White House.

Chapter 22

"We Had Better Turn to Taft"

One evening midway in Theodore Roosevelt's second term, he invited Will Taft and his wife, Nellie, to the White House. When they had adjourned to the library after dinner, the president leaned back in his leather armchair, closed his eyes, and playfully chanted: "I am the seventh son of a seventh daughter. I have clairvoyant powers. I see a man before me weighing three hundred and fifty pounds. There is something hanging over his head. I cannot make out what it is; it is hanging by a slender thread. At one time it looks like the presidency—then again it looks like the chief justiceship."

"Make it the presidency!" cried the ambitious Nellie.

"Make it the chief justiceship," said Taft.

As a result of his 1904 Election Night statement, Roosevelt had taken himself out of contention for another term and the question of a successor dominated his last two years in office. Would the Republican candidate in 1908 be Taft? Perhaps it would be Elihu Root. Or maybe Charles Evans Hughes, who had been propelled into the limelight by his investigation of the insurance scandals. House Speaker Joe Cannon also envisioned himself as the Republican standard-bearer. Many Americans wanted Roosevelt to remain in office, although he repeatedly denied any interest in a third term. But if the president chose to run again, observers agreed he would win with little trouble despite all his troubles with Congress. "With all his faults," observed a British diplomat, "he is unquestionably today the greatest force in the United States and seems likely to remain so."

As speculation about a third term mounted at the beginning of 1908, presidential secretary William Loeb warned his chief that unless Roosevelt chose a candidate and threw his support behind him, the voters would be convinced that he was going to run again. In effect, Loeb called upon the president to select a candidate to run against himself to prevent convention delegates who were divided among several competing candidates from stampeding to him. This prospect was not simply a matter or Rooseveltian ego but a real possibility; practical politicians find it easier to stick with a known and proven product.

No possible contender offered Roosevelt's blend of progressivism and party stature, but if the president had been able to follow his own instincts, he would have chosen Elihu Root as his successor. Root had the sharpest intellect in the Republican party and had served in the Cabinet with distinction, first as secretary of war and then as secretary of state. "I would walk on my hands and knees from the White House to the Capitol to see Root made president," Roosevelt once told a friend. But Root's health was bad, and he was too closely identified with Wall Street for such progressive times. Charles Evans Hughes, the frosty New York governor, was able and progressive but he, too, had an overriding drawback—Roosevelt not only considered him too independent, but cordially disliked him. The faithful Will Taft seemed most likely to continue his program and received the president's blessing.

"We had better turn to Taft," Roosevelt finally told Loeb. "See Taft . . . and tell him . . . so that he will know my mind."

"I must go over and thank Theodore," Taft told Loeb with considerable emotion when the presidential emissary came to the War Department to inform him of Roosevelt's decision.

As Taft told the president of his willingness to run, Roosevelt gave him a slap on the back. "Yes, Will," he said, "it's the thing to do."

Roosevelt had mixed feelings about leaving Washington. While he believed in a strong presidency, the check upon an incipient dictatorship was the limited term of office. Nevertheless, the actual act of leaving the White House involved considerable personal agony. None of his predecessors had enjoyed the office as much as he had. At the age of fifty, he was still comparatively young and craved action. "I should like to have stayed on in the Presidency, and I make no pretense that I am glad to be relieved of my official duties," he told Cecil Spring Rice during the summer of 1908. "The only reason I did not stay on was because I felt that I ought not to. . . ."

William Howard Taft was fifty-one years old when presidential lightning struck his ample figure. Placid, well-meaning, and amiable, he was a member of a prominent Cincinnati family, a Yale man, and had advanced largely through a series of appointive jobs to become governor general of the Philippines and then secretary of war. He was intelligent if not brilliant, a good administrator, was liked by the Washington press corps, and had served Roosevelt well as a troubleshooter. The two men had been close friends since 1890, and Taft had supported Roosevelt's policies, both at home and abroad. His perpetual smile and a deep chuckle that often exploded into nearly uncontrollable laughter inspired warmth and trust. When the president went on a tour of the West, he told reporters that all would be well in Washington. "I have left Taft sitting on the lid"—a remark that, considering Taft's bulk, caused considerable merriment.

The happiest years of Taft's life had been a period of service as a judge on the Ohio Superior Court. "I love judges, and I love courts," he once said. "They are my ideals, that typify on earth what we shall meet hereafter in heaven under a just God." Twice, Roosevelt offered to appoint him to the Supreme Court as vacancies occurred, and Taft had been sorely tempted to accept. But his wife, Nellie, and older half brother, Charles P. Taft, who financed his career, wished him to remain available for the presidency of 1908. In fact, Nellie distrusted the president; she believed he was using her husband as a cover for promoting a third term for himself. When a family friend asked nine-year-old Charles Taft if his father was going to become a Supreme Court justice, the boy replied, "Nope."

"Why not?" asked the friend.

"Ma wants him to wait and be president."

Taft's passive attitude toward the presidency and the towering ambition of his family should have alerted Roosevelt to his friend's unsuitability for the office, but he ignored these signals. Inasmuch as Taft had been a yes-man, never deviating from the views of his leader, he may have believed that Taft would be easy to control and that he could retain authority over him after departing from the White House. Roosevelt also believed—and led the country to believe—that Taft was cut from the same progressive cloth as himself. Even Taft was convinced of it. "I agree heartily and earnestly with the policies which have come to be known as the Roosevelt policies," he declared. In point of fact, Taft was a conservative by instinct, emotion, and ideology, and espoused a different presidential philosophy than his benefactor. With his reverence for legalisms, he believed a president should be less of an activist and should observe the law more strictly than Roosevelt had done.

While deciding upon a successor, Roosevelt, furious with Congress, business, and the courts for opposing his reform program, made his final year in office a demonstration of presidential activism. There would be no letup in his campaign against "speculation, corruption and fraud." In his annual message of December 1907, a document of some 35,000 words—it prompted one observer to joke that the president had already started on his memoirs—he called for immediate congressional action on a wide spectrum of reforms: income and inheritance taxes, a federal incorporation law, the fixing of railroad rates based upon physical evaluation, federal regulation of railway stocks, currency reform to prevent further financial panics, limitations on injunctions in labor disputes, extension of the eight-hour day and workmen's compensation laws, a postal savings bank, and control of campaign contributions.

Roosevelt had experienced no sudden change of heart. While he recognized there was a new progressive spirit abroad in the land, these reforms were extensions of his long-held and oft-expressed belief that the nation required the adequate regulation of business and guarantees of a Square Deal for the workingman.

"We seek to control law-defying wealth, in the first place to prevent its doing evil, and in the next place to avoid the vindictive and dreadful radicalism which if left uncontrolled it is certain in the end to arouse," Roosevelt explained. But if the president was in a mood for change, Congress was not. The legislators, alternately hostile and indifferent, ignored these proposals and filled their time with investigations of the Brownsville affair, placing the blame for the Panic of 1907, and rolling out the pork barrel.

Faced with congressional inaction and knowing he had so little time left in the White House, Roosevelt sent a blistering Special Message to Capitol Hill on January 31, 1908, designed, as he told Kermit, "to draw the issue as sharply as I well can between the men of predatory wealth and the administration." Unleashing the full fury of his frustration, it was the most radical message of his White House years. Every one of his previous proposals for authority to control corporations and protect the workingman were repeated in even stronger language, along with a new demand for regulation of stock market speculation, which he saw as a form of gambling. This message, Roosevelt said, contained his "deepest and most earnest convictions."

Roosevelt's strongest words were reserved for big business and the federal courts, which had just invalidated a railway workmen's compensation law. Terms like "sinister offenders," "wealthy criminals," "predatory wealth," "rottenness," and "criminal misconduct" abounded in the presidential indictment. Standard Oil and "certain notorious railroad combinations" and their "puppets" in politics, the press, and the law were denounced for opposing reform. Some judges were criticized for supporting these interests to the detriment of the public. "Every measure for honesty in business that has been passed during the last six years has been opposed by these men with every recourse that bitter and unscrupulous craft could suggest and the command of almost unlimited money secure," he declared.

With this tongue-lashing, Roosevelt had come to the parting of the ways with political orthodoxy. William Jennings Bryan, the front-runner for the Democratic nomination, in 1908, wrote to congratulate him for the progressive tone of his message, while a shocked Nicholas Murray Butler told his friend it had made "a very painful impression." In reply, the president regretted that he was not at the beginning of his first term, so he could have "a showdown with my foes both without and within the party." Realizing that the possibility for compromise with the Old Guard no longer existed—and not needing its support for renomination and election—Roosevelt intended to arouse public support for progressive reform and keep it at the top of the national agenda, no matter who occupied the White House.

For the moment, however, the president's main interest was to ensure the nomination of Will Taft, and he threw the weight of the presidential office—and federal patronage—behind his hand-picked candidate. In his biography of Thomas Hart Benton, Roosevelt had castigated Andrew Jackson for using "machine politics" to impose his successor, Martin Van Buren, upon the nation. But he did the same thing while denying doing it. "I appointed no man for the purpose of creating Taft sentiment, but . . . I have appointed men in recognition of the Taft sentiment already in existence," he declared as he seeded the Republican convention with Taft-instructed delegates.

Roosevelt did his work so well that when the convention was gaveled to order on June 16, 1908, Taft already had 563 delegates with only 491 needed to nominate. To make certain there was no slipup, Cabot Lodge was dispatched to Chicago to stave off any last-minute attempt to draft the president. Even so, the 12,000 delegates, alternates, and spectators trooping into the barnlike Coliseum may have worn Taft badges, but their hearts beat for Roosevelt. Alice Longworth, in attendance with her aunt Corrine, wished "in the black depths of my heart that something would happen and Father would be renominated."

On the second day of the convention, her wish appeared on the verge of coming true. Lodge had only to mention the president's name in his keynote address to set off a thunderclap of cheering. Someone in the galleries began to bellow, "Four, four, four years more!" and the chant spread from the balconies to the floor. "Four, four, four years more!" Some delegates marched about the hall waving their coats over their heads; others carried teddy bears aloft. Worried Taft supporters ordered the band to play "The Star-Spangled Banner" in a fruitless effort to quell the demonstration.

Back in Washington, Roosevelt sat at a White House telephone with the receiver glued to his ear during the entire tempestuous forty-nine minutes. Archie Butt, his military aide, observed that he was "in as gay a humor as I have ever seen him." Over at the War Department, where Taft and his wife were following the proceedings by telegraph, Nellie was tense and tight-lipped as the Roosevelt demonstration went on and on. She was convinced the president intended to be renominated. Finally, Lodge brought the cheering to an end by telling the delegates the president's decision not to seek a third term was "final and irrevocable. . . . Anyone who attempts to use his name as a candidate for the presidency impugns both him and his good faith." This, he later said, was "the hardest thing I have ever done in public life."

The rest of the convention was anticlimactic. The speaker who nominated Hughes even forgot to mention the governor's name, while Taft's nomination brought on a contrived demonstration. Nellie sat in her husband's swivel chair, tightly clutching the edge of the desk. "I only want it to last more than forty-nine minutes," she grimly declared. "I want to get even for the scare that Roosevelt cheer . . . gave me yesterday."

"Oh my dear!" Taft clucked in disapproval. "My dear!"

In spite of the efforts of the Taft managers to keep it going, however, the cheering faded out after about twenty-five minutes. On the first ballot, Taft received 702 votes, Hughes got 67, and there were a few scattered votes for favorite sons. Roosevelt was playing tennis when he received the news of Taft's victory and pronounced himself "*dee*-lighted!" For vice-presidential nominee, he and Taft preferred either Senators Jonathan Dolliver or Albert Beveridge, both progressives, but they declined and a conservative nonentity, Congressman James "Sunny Jim" Sherman of New York, was chosen as Taft's running mate.

Two days after Taft's nomination, Roosevelt left Washington for Oyster Bay for his annual vacation with the intention of removing himself from the campaign and to consider his future. What would he do when he left the White House? Where would he go? He had no intention of drifting into a genteel retirement. Having worked all his adult life, he would continue to work. Some people proposed that he succeed Charles W. Eliot, who at seventy-five was rounding out his fortieth year as president of Harvard; others believed he should enter the Senate. Henry Adams thought he might follow the example of Adams's grandfather, John Quincy Adams, who had launched a notable career in Congress after leaving the White House. Senator Philander Knox had a better idea, telling a Washington party: "He should be made a Bishop."

Numerous lucrative offers in journalism were presented him, but always suspicious of the marketplace, Roosevelt rejected them. Instead, he agreed to become a contributing editor of *The Outlook,* a small but influential journal, where he would write a dozen or so articles a year at an annual salary of $12,000. This arrangement would provide him with an office in New York City while leaving him free to work on other literary projects and make public addresses. Roosevelt also laid plans for what he called "my last chance at something in the nature of a great adventure"— big-game hunting in Africa. Edith opposed the trip out of concern for his health, but he would not be dissuaded. Oxford University had invited him to give the Romanes Lecture in 1910, which meant Edith could join him as he visited old friends in Europe and England again.

Experienced hunters were queried about costs, equipment, and the locations where he might be able to "see the great African fauna, and to kill one or two rhino or buffalo and some of the big antelopes, with the chance for a shot at a lion." Kermit agreed to drop out of Harvard for a year to accompany him. Roosevelt was anxious, however, that the safari not be seen as "a game butchering trip" and, to give it a scientific cast, offered to provide wildlife specimens to the Smithsonian Institution. The museum accepted his proposal and agreed to send along several field naturalists and taxidermists—much to Edith's relief, since she had envisioned the family being forced out of Sagamore Hill by an ever-increasing collection of stuffed animals. Andrew Carnegie underwrote a large part of the $75,000 cost of the expedition; Roosevelt would pay his own expenses with a $50,000 contract from *Scribner's* for a series of articles that were later to be published as a book.*

In the meantime, Taft campaigned across the country without

Collier's offered Roosevelt $100,000 for the articles, but he felt that *Scribner's* was a more suitable publication for an ex-president's work.

enthusiasm. For the most part, he endorsed Roosevelt's record and seemed content to be a Roosevelt proxy. Never having run for office before, he regarded campaigning as purgatory. He abhorred playing politics, "buttering people up," and being "exposed to all sorts of criticism and curious inquisitiveness." He hated giving speeches, and most of his major addresses were long, droning affairs delivered with all the passion of a legal opinion. One presentation sent Alice Longworth into such gales of laughter she burst a stitch of an appendectomy incision.

Although Roosevelt made no appearances in Taft's behalf—custom prevented presidents from actively campaigning—he was unable to stand idly on the sidelines, especially after the Democrats nominated Bryan for a third time. He wrote several open letters endorsing Taft and inundated the candidate with avuncular advice. Hit them hard, old man! Avoid talking about delicate subjects like religion. Stop citing court decisions. Don't have your picture taken playing golf. Stay in hotels rather than private homes so more people can see you. Above all, "you big, generous, high-minded fellow," you must *"always"* smile for "your nature shines out so transparently when you smile."

Roosevelt's concern turned out to be misplaced, for Taft defeated Bryan by a popular majority of about 1.3 million votes; the electoral count was 321 to 162.* Triple-chinned and beaming, the president-elect told the crowd gathered in front of his brother Charley's home in Cincinnati that his administration would be a "worthy successor to that of Theodore Roosevelt." Roosevelt himself wrote Taft that he had "won a great personal victory as well as a great victory for your party." In reality, Taft's margin of victory was less than that won by Roosevelt in 1904, and while the Republicans retained control of Congress, they lost seats in both the Senate and House. As always, Roosevelt saw only what he wanted to see.

At exactly what point relations between Roosevelt and his successor began to sour is unknown, but in the interval between the election and Taft's inauguration on March 4, 1909, there were

---

*Following this third defeat, Bryan told the story of a drunk who tried several times to get into a private club and was thrown out after each attempt. After being thrown out the third time, he picked himself off the sidewalk and concluded: "They can't fool me. Those fellows don't want me in there."

signs of trouble. Immediately after the election, the president-elect and Nellie went to Hot Springs, Virginia, for a short rest. The first letter Taft wrote went to Roosevelt. "You and my brother Charley made that possible which in all probability would not have occurred otherwise," he declared. The president was taken aback. To be placed in the same category with Charley Taft and his ever-ready checkbook was hardly what Roosevelt expected. Later, he observed bitterly that it was like saying "Abraham Lincoln and the bond seller Jay Cooke saved the Union."

Over the next few months, reporters became aware of an unmistakable attitude on the part of the Taft family, especially Nellie and Charley, that they didn't owe very much to Roosevelt; "dear Will" would have gotten to the White House without him. Tensions were further exacerbated when Nellie tactlessly let it be known that she planned "sweeping changes" in the operation of the White House and would dispense with some of the Roosevelts' favorite servants, much to the distress of the incumbent first lady. Outwardly, Roosevelt and Taft seemed as friendly as ever, but the vital essence of their relationship was changing—and both knew it.

The formation of Taft's Cabinet brought the struggle out into the open. Before the election, Roosevelt had surveyed his Cabinet to determine whether any of its members wished to stay on in the new administration, and four indicated their desire to do so. "Tell the boys I have been working with," Taft said, "that I want to continue with all of them." Once elected, however, he decided that they would be more loyal to Roosevelt than to him and decided to name his own men.* Though puzzled and hurt, the president accepted this decision with at least surface equanimity. "Ha ha! *You* are making up your Cabinet," Roosevelt chortled on the last day of the year. "*I* in a lighthearted way have spent the morning testing the rifles for my African trip. Life has its compensations."

Roosevelt was determined to enjoy his presidency up to the last possible moment, however. "I am ending my career as President with just the same stiff fighting that has marked it ever since

---

*William Loeb was appointed collector of the port of New York, and Archie Butt was asked to stay on as military aide.

I took the office," he told Arthur Lee. "But I am having a thoroly good time." A high point of this "good time" was a struggle with Congress over the administration's use of the Secret Service to ferret out corruption in government. Regarding Roosevelt as a spent force, Congress passed legislation restricting the service to its assigned duties of protecting the president and the suppression of counterfeiting. Heatedly, Roosevelt said that the lawmakers had taken this step because they "did not themselves wish to be investigated." Congress refused to rescind the ban, but by executive order the president created an investigatory agency within the Department of Justice that eventually became the Federal Bureau of Investigation. Thus, Roosevelt had the final word.

Now it was the time of assessments and bittersweet farewells. One by one, the members of the Cabinet gave dinners in honor of the president and first lady, and the process of packing and shipping the Roosevelt's personal belongings home to Oyster Bay was completed. A sign of the new era was the conversion of the White House stable into a garage for the fleet of automobiles ordered by Taft. The old house itself had grown silent, no longer resounding to the cries of children, the staccato clatter of stilts, or the rattle of roller skates on hardwood floors.

Looking back over his seven and a half years in the presidency, Roosevelt saw his greatest achievements as the settlement of the coal strike, the construction of the Panama Canal, the ending of the Russo-Japanese War, the voyage of the Great White Fleet, "the irrigation business in the West, and finally I think, the toning up of the Government service generally." But his achievements went far beyond these. By the time he left office, observes Elting E. Morison,* he had proposed all the basic reforms that became law during the Taft and Wilson administrations—and some that awaited the coming of the New Deal. Critics might question his effectiveness as a reformer, but he understood that the greatest task facing any political leader is to educate the public.

On March 1, the president invited thirty-one members of his "Tennis Cabinet" and some out-of-town associates for a farewell luncheon. They included such intimate friends as James Garfield

---

*The Letters of Theodore Roosevelt, Vol. VI, p. 922 (note).

and Gifford Pinchot, as well as the French ambassador, Jules Jusserand, and Bill Sewall, his hunting companion from Maine. Toasting his "generals" for the last time, the president thanked them for their able, loyal service and credited whatever success he had achieved in the presidency to them. The words were simple, but moving. Some of the group were not yet ready for the transfer of power and regarded Taft as guilty of usurping Roosevelt's place. Overcome by emotion, they could scarcely abide the thought of his departure and wept openly.

Two days later, Roosevelt spent his last full day in his office, now stripped bare of personal mementos. Outside the White House, the finishing touches were being placed on the stand from which the new president would review the inaugural parade. Bureau chiefs, subordinates, journalists, and congressmen came to wish him well while he stook hands and signed bills, letters, and photographs. Between callers he dashed off a note to "Dear Will," warning against dividing the battle fleet between the Atlantic and Pacific before the Panama Canal was opened. "I have done my work; I am perfectly content; I have nothing to add," he wrote a friend. As Mark Sullivan was leaving, Roosevelt accompanied the newsman to the door and unexpectedly expressed his simmering doubts about Taft. "He's all right," the president said. "He means well and he'll do his best. But he's weak. They'll get around him." To illustrate, Roosevelt pushed his bulk against Sullivan's shoulder. "They'll—they'll lean against him."

In an effort to banish these doubts, Roosevelt had invited the president-elect and his wife to move into the White House a day early, and join him for a family dinner on the eve of the inaugural. "People have attempted to represent that you and I were in some way at odds during the last three months," Taft responded to this unprecedented invitation, "whereas you and I know that there has not been the slightest difference between us. . . ." He signed the note, "With love and affection, my dear Theodore."

Wet snow was falling as the Tafts arrived at the White House early on the evening of March 3, and the flakes sparkled in the light reflected from the old mansion. There were twelve at dinner in the State Dining Room—the Roosevelts, the Tafts, Alice and Nick Longworth, Bamie and Will Cowles, Elihu Root, now senator from New York, and his wife, Archie Butt, and Mabel Boardman, a friend of the Tafts. Edith had ordered all the fireplaces lit to take the chill off the atmosphere, and the president was at his conversational best, but a sense of impending change and departing glory hung over the occasion. Taft later referred to it as "that funeral." By the time dinner was over, the snow had turned into a howling blizzard, and Alice expressed grim satisfaction to herself "at the prospect of a wretched day for inauguration."

"I knew there'd be a blizzard when I went out," Roosevelt brightly greeted his successor the following morning over breakfast. Rather than abating, the storm had increased in intensity during the night, and the inaugural ceremony had been moved indoors to the Senate chamber. "You're wrong," Taft chuckled. "It's my storm. I always said it would be a cold day when I got to be President of the United States."

The solemn ritual marking the transfer of power now played itself out: the arrival of the congressional escort at the White House . . . the appearance of the incoming and outgoing presidents on the portico for a photograph, Taft beaming as usual and Roosevelt looking inscrutable . . . the ride together along hastily cleared Pennsylvania Avenue to the Capitol . . . the nods and waves to the clutch of spectators who had braved the elements— the long parade of various dignitaries to their seats on the Senate floor . . . the rustle of the silk robe of the aged Chief Justice, Melville W. Fuller, as he came forward to swear in the new president . . . Taft's voice booming out the responses over a Bible brought from the Supreme Court. . . . The last words of the oath—"preserve, protect and defend the Constitution of the United States"—had barely faded when the hollow thud of the first cannon of a twenty-one-gun salute sounded out for the new president.

The reign of Theodore Roosevelt was over.

Chapter 23

"My Hat Is in the Ring"

"By Godfrey, that's a wonderful sight!" Theodore Roosevelt exclaimed upon catching a glimpse of the brilliant green foliage of Africa after two weeks at sea. The former president's "eyes were shining with intense excitement," according to a newsman, as the German steamer *Admiral* anchored off Mombasa, British East Africa,* on April 21, 1909. Arab dhows bobbed about the vessel, and a heavy tropical rain beat down on the town. Roosevelt's "hands gripped the rail until the knuckles were bloodless, and his entire sturdy body seemed to be poised, expectant, like a well-trained pointer at work in the field."

With his usual zest for the novel, Colonel Roosevelt, as he now wished to be known, immediately took the Uganda Railway up to base camp of his safari in the Kapiti plains, accompanied by Kermit and the team of Smithsonian naturalists and taxidermists. The railway line ran through a game preserve, and to observe the wildlife better, Roosevelt rode in a special seat built over the cowcatcher of the engine. On both sides, high blond grass punctuated by occasional trees and rock outcroppings stretched away

_____
*Now the Republic of Kenya

to a horizon of purple-blue hills. It seemed to Roosevelt like "a vast zoological garden" teeming with animals while birds hovered so near he could almost reach out and touch them. In the dusk, the train nearly ran over a hyena; giraffes loped across the track, their necks occasionally breaking telegraph lines.

Roosevelt was fascinated by everything he saw, and as the near-poetic opening lines of his record of his travels, *African Game Trails,* attest, he was awed:

> I speak of Africa and golden joys; the joy of wandering through lonely lands; the joy of hunting the mighty and terrible lords of the wilderness, the cunning, the wary, and the grim. . . . But there are no words that can tell the hidden spirit of the wilderness, that can reveal its mystery, its melancholy, and its charm. There is delight in the hardy life of the open, in long rides rifle in hand, in the thrill of the fight with dangerous game. . . . Yet mingled with it is the strong attraction of the silent places, of the large tropic moons, and the splendor of the new stars . . . the awful glory of sunrise and sunset in the wide waste spaces of earth, unworn of man, and changed only by the slow change of the ages through time everlasting.

Awaiting him at the Kapiti station was one of the largest safaris ever fitted out in East Africa. It looked like a military expedition ready for the march. In addition to the Roosevelts, the Smithsonian team, and the white hunters, there were 15 *askaris,* or native soldiers, and 260 porters. Such a large contingent was required because in addition to attending to the ordinary needs of a safari, these people carried the elaborate equipment required to preserve the specimens brought back by the expedition. This included a combined naturalist's laboratory and taxidermist's shop as well as four tons of salt to cure skins. As soon as an animal was shot, it was to be skinned and the bones cleaned and packed away for later reassembly, while the party ate the meat.

One of the most important pieces of baggage as far as Roosevelt was concerned was an aluminum and oilskin case containing a pigskin-bound library of thirty-seven literary classics. The selection of titles included the Bible, Shakespeare's plays, Homer, the *Nibelungenlied,* and *Huckleberry Finn.* "I almost always had some volume with me, either in my saddle pocket or in the cartridge bag which one of my gunbearers carried," he later recalled.

"Often my reading would be done while resting under a tree at noon, perhaps beside the carcass of a beast I killed. . . ."

Breaking camp on April 24, the safari began an eleven-month trek into the African interior. The landscape reminded Roosevelt of the American West. "As my horse shuffled forward under the bright hot sunlight, across the endless flats of gently rolling slopes of brown and withered grass," he wrote, "I might have been on the plains anywhere, from Texas to Montana." Zebra, wildebeest, hartebeest, and several kinds of gazelle, impala, and steenbok were sighted, and he shot a pair of wildebeest early on. Roosevelt was really interested in hunting the five most dangerous types of African game: elephant, rhinoceros, buffalo, leopard, and lion. "If only I can get *my* lion," he told one of his companions, "I shall be happy—even if he is small—but I hope he will have a mane!"

Before the end of the first week of the safari, Roosevelt shot the first of the nine lions he bagged. Late one afternoon, as he and several companions were returning to camp after an unsuccessful hunt, the spoor of a pair of lions was detected in a shallow watercourse. Shouts and rocks produced thrashing and loud grunts from the high grass. Dismounting from his horse, the Colonel waited, high-power Winchester rifle at the ready, for a lion to appear.

"I sprang to one side; and for a second or two we waited, uncertain whether we should see the lions charging out ten yards distant or running away," he later recalled. "Fortunately, they adopted the latter course. Right in front of me, thirty yards off, there appeared, from behind the bushes which had first screened him from my eyes, the tawny, galloping form of a big maneless lion. Crack! the Winchester spoke; and as the soft-nosed bullet ploughed forward through his flank the lion swerved so that I missed him with a second shot; but my third bullet went through his spine and forward into his chest. Down he came. . . ."

Chanting and singing, the native porters marched into camp that evening with the carcass of the great beast slung from a pole. They paraded up and down before the fire where the Bwana Makuba (Big Chief) stood, and, after depositing the trophy at his feet, they all celebrated his triumph with a vigorous dance. "The firelight gleamed and flickered across the grim dead beast," the Colonel wrote, "and the shining eyes and black features of the excited savages, while all around the moon flooded the landscape with her white light."

In all, Roosevelt and his son bagged 512 animals, including 17 lions, 11 elephants, 20 rhinoceroses, 9 giraffes, 47 gazelles, 8 hippopotamuses, 29 zebras, 9 hyenas, and a scattering of such odd creatures as the bongo, the dik-dik, the kudu, the aardwolf, and the klipspringer. Sensitive to charges of indiscriminate slaughter, Roosevelt emphasized that he killed only specimens the scientists desired for their collections or were required for food. Complete animal groups were sought for mounting—male, female, and young. The Roosevelts kept about two dozen trophies for themselves; the rest went to the Smithsonian, the American Museum of Natural History in New York, and the San Francisco Museum.

Almost every evening, while the rest of the party gathered around the campfire after dinner, smoking and trading stories, Roosevelt worked by the light of a flickering lamp at a portable writing table on his *Scribner's* articles. Writing in longhand, he eventually produced fourteen pieces, ranging in length from 5,000 to 15,000 words. Periodically, the safari circled back to Nairobi for supplies, and in Nairobi the completed articles were mailed to the United States, where they had an eager audience. "The people follow the account of your African wanderings as if it were a new Robinson Crusoe," Lodge informed him. The safari was not without its amusing moments. Welcomed to a Catholic mission, the members of the party were met by a group of African children who had been patiently drilled by an American nun to present a phonetic greeting in song:

Ose ka nyu si bai di mo nseli laiti
Wati so pulauli wi eli aditwayi laiti silasi gliremi

Roosevelt was often homesick for his wife and children, however. As the year came to an end, he wrote Edith a letter* that poignantly expressed his longing for her: "Oh, sweetest of all

---

*Edith was so touched by this letter that following his death she did not destroy it with the rest of her correspondence with her husband.

sweet girls, last night I dreamed that I was with you, that our separation was but a dream; and when I waked up it was almost too hard to bear. Well, one must pay for everything; you have made the real happiness of my life; and so it is natural and right that I should constantly [be] more and more lonely without you. . . . Darling I love you so. In a very little over four months I shall see you. . . . How very happy we have been for these last twenty-three years. . . ."

"My main reason for wishing to go to Africa for a year," Roosevelt had told William Allen White before leaving the United States, "is so that I can get where no one can accuse me of running nor do Taft the injustice of accusing him of permitting me to run the job." But within a few months of his departure for the bush, weeks-old newspapers and letters from friends brought him news of a political earthquake under way in Washington that threatened the unity of the Republican party. At the epicenter stood the massive, amiable—and somewhat pathetic—figure of William Howard Taft.

Taft's misfortune was that he succeeded Theodore Roosevelt in the White House. As Henry Pringle observed, Roosevelt out of office had been transformed into an ideal, while Taft was the reality. The country was attuned to the Rough Rider's vigorous tempo, and Taft's pace was deliberate. Taft had honesty; he had integrity. But he lacked Roosevelt's ability to appeal to the average American. He did not have Roosevelt's catlike grace in working both sides of the political street without permanently crossing from one side to another. Nor did he have the explosive vitality that Roosevelt had used to keep mutually antagonistic factions subordinate to his leadership. He was physically and mentally lazy, and Archie Butt reported that Taft once fell asleep at a funeral while sitting in the front row of mourners and had to be nudged awake.

Throughout his time in Roosevelt's Cabinet, Taft had been a follower, not a leader—the reason Roosevelt chose him as a successor. Without his predecessor's sure hand to guide him, he quickly bumbled into political quicksand. In 1908, the Republican platform had pledged tariff revision, which was generally understood to mean a reduction from the high levels established in

1897. Although Roosevelt had avoided tinkering with the tariff because he felt it was a divisive issue, Taft, long an advocate of reform, promptly called a special session of Congress to deal with the matter. It quickly became entangled with progressive efforts to end the dictatorship of the Speaker of the House, Uncle Joe Cannon.

In the final two years of Roosevelt's presidency, progressivism had developed such volatility that the Old Guard could no longer contain it. The Insurgents, as they became known, formed a bloc of fourteen seats in the Senate and thirty in the House, centered on the Midwest. Elected as Republicans, they nevertheless defied party discipline in their determination to put an end to the injustices of monopoly capital. The reactionary Cannon, through his tight control of the committee system and the flow of legislation, typified all that they opposed.

Cannon and the Senate majority leader, Nelson Aldrich, in an effort to block the coup, sought Taft's backing in exchange for a promise to support tariff reform. Taft had his predecessor's example before him. Roosevelt had maintained an uneasy truce with Cannon and, in parting advice to Taft, had told him "to make peace with Uncle Joe." Although Taft detested the poker-playing, whiskey-drinking Cannon, he agreed to support him in the hope of preserving tariff revision. While this action helped Cannon to survive at least temporarily the attempts to trim his power, it made Taft suspect among the progressives.*

The House promptly approved a bill introduced by Congressman Sereno E. Payne of New York that reduced rates on a number of important items. Once it came before the Senate, however, the bill was at the tender mercies of Aldrich, the high priest of protectionism. The resulting Payne-Aldrich Tariff emerged with 847 changes from the House version, most of them upward. "Th' Republican party has been thrue to its promises," observed a caustic Mr. Dooley. "Look at th' free list if ye don't believe it. Practically ivrything nicessary to existence comes in free. Here it is. Curling stones, teeth, sea moss, newspapers, nux vomica, Pulu, canary bird seed. . . . Th' new tariff bill puts these familyar commodyties within th' reach 'iv all."

---

*The following year the Insurgents succeeded in stripping Cannon of his powers.

Led by Senator La Follette, the angry Insurgents attacked Aldrich's changes item by item—exposing the links between the protective tariff and the trusts. Roosevelt had encouraged the progressives, and they expected Taft to do the same. But Taft was no progressive. He believed Roosevelt had gone too far, too fast in pushing the limits of the Constitution in the name of reform. Rather than the nation needing additional reforms, Taft was convinced that only a fine-tuning of those that had been enacted by the Rough Rider was necessary. Taft was so distressed by the strain of having to make decisions on his own that he had difficulty sleeping and wandered about the White House at night in his size 54 pajamas. To add to his misery, Nellie, his wife, suffered a stroke, which affected her speech.

While furious over Aldrich's betrayal, the president was reluctant to interfere in the legislative process, which Roosevelt had often done, or to break with the party leadership. In the end, he not only refused to veto the law as demanded by the Insurgents, but became even more dependent upon Aldrich and Cannon. Having been a yes-man in Roosevelt's Cabinet, he inevitably became one for the Old Guard. In 1910, he unsuccessfuly tried to purge some of the Insurgents in the Republican primaries. The president, said Senator Jonathan Dolliver, is a "ponderous and amiable man completely surrounded by men who know exactly what they want."

Roosevelt was disappointed with his successor's handling of the Payne-Aldrich Tariff, not, as he wrote Lodge, because he favored tariff reduction, but because of the effect of the president's ineptness upon the party. Even more unsettling was Taft's tactless praise of the law in an off-the-cuff speech in Winona, Minnesota, a hotbed of insurgency, calling it "the best tariff bill the Republican party has ever passed." Such remarks outraged midwesterners and undermined the alliance with eastern Republicans. Before the summer of 1909 was over, newspapers all over the Mississippi Valley were saying the Rough Rider could have the party's presidential nomination in 1912 merely for the asking.

Taft's handling of conservation, the keystone of Roosevelt's legacy, aroused further discord. Eyebrows had been raised by the presidential decision to drop James Garfield, the ardently pro-conservation secretary of the interior, but his replacement, Richard A. Ballinger, caused great unease among environmentalists. Ballinger, a onetime reform mayor of Seattle, had served under Garfield as commissioner of the General Land Office, but like most westerners, he was a "boomer" who favored rapid exploitation of the nation's resources. A clash between Ballinger and Gifford Pinchot, still chief of the Forestry Service and regarded as the watchdog of Rooseveltian conservation, was inevitable. Minor cutbacks initiated by the new secretary convinced Pinchot that all conservation programs were now in danger.

Late in the summer of 1909, a young Interior Department investigator named Louis R. Glavis came to Pinchot with a disturbing report. Ballinger, he claimed, was conspiring to turn over valuable coal lands in Alaska to a Morgan-Guggenheim syndicate. Pinchot immediately sent Glavis to the president with these charges. Taft, angry at Glavis and Pinchot for having gone over Ballinger's head and worried about a scandal that would besmirch his administration, accepted Ballinger's explanations and fired Glavis. In an effort to appease Pinchot, Taft assured him of his support for conservation, but instead of snuffing out the flickering flames of discontent, his action kindled a raging prairie fire that eventually consumed his presidency.

Pinchot now courted martyrdom. As long as Roosevelt had been on hand, he had been idealistic if exuberant in his championship of the environment. Now that the restraining hand was removed, he became harsh and intolerant toward Ballinger and orchestrated a campaign by muckraking journalists against the interior secretary. Glavis joined the fray with a *Collier's* article revealing, among other things, that Ballinger had represented the Morgan-Guggenheim interests as an attorney before his old agency in the period between leaving the Land Office and becoming interior secretary.

The president, furious at the leaks emanating from Pinchot's office and on the advice of Elihu Root, finally demanded Pinchot's resignation on January 7, 1910. Taft fully realized this would anger Roosevelt and the Insurgents, but believed the chief forester's calculated insubordination could not be ignored. From a political standpoint, he should have insisted upon Ballinger's resignation, too. The whole affair, noted Charles J. Bonaparte, was marked "by the most notable unbroken succession of colossal blunders known in American politics."

Ten days after Pinchot's dismissal, a native runner brought the news to Roosevelt, who was hunting the rare white rhino* in the Congo. "I cannot believe it," he wrote Pinchot at once. "I do not know any man in public life who has rendered quite the same service you have rendered. . . ." This was followed up by a letter in which he emphasized that it would be "a very ungracious thing for an ex-President to criticize his successor; and yet I cannot as an honest man cease to battle for the principles [for] which you and I and Jim . . . and the rest of our close associates stood." He suggested a meeting with Pinchot when he reached Europe to discuss the situation "before I even in the smallest degree commit myself."

Taft, in an effort to show his continued commitment to conservation, replaced Pinchot with Henry S. Graves, director of the Yale School of Forestry, and after Ballinger's resignation† in 1911, appointed Walter L. Fisher, a confirmed environmentalist, to replace him. Nevertheless, Roosevelt never again felt the same toward Will Taft as he had before Pinchot's dismissal.

Like birds flocking to a telephone wire, dozens of newspaper correspondents descended in mid-march 1910 upon Khartoum in the Anglo-Egyptian Sudan to await Theodore Roosevelt's emergence from the heart of Africa. In an effort to scoop their less enterprising brethren, some chartered launches and raced up the White Nile to head off the steamer carrying the Bwana Makuba and his party. What did he think of the Payne-Aldrich Tariff? the firing of Pinchot? What about Taft's first year in the White House? Would he run for president in 1912?

Tanned and fit in khaki shirt and trousers and a green tie donned specially for the occasion, the Colonel was affability itself. He was eager to talk about the expedition, but the newsmen's efforts

---

*Between them Roosevelt and Kermit accounted for nine of these beasts—five for the Smithsonian group, three for the American Museum of Natural History, and one for the Bronx Zoo. Contrary to legend, TR noted, they were more gray than white.

†Following a lengthy congressional investigation that lasted all through 1910, Ballinger was officially exonerated of corruption and fraud charges, but the proceedings indicated that his hands were not entirely clean. In effect, Taft and Ballinger stood officially acquitted by Congress but guilty in the eyes of a significant segment of the public.

to prod him into commenting on political matters met an un-Rooseveltian reticence. "I have nothing to say and will have nothing to say on American or foreign political questions or on any phase or incident thereof. . . ."

Edith and Ethel joined the travelers in Khartoum, and to Edith's great relief, her husband and son were "in splendid condition." Following a round of sight-seeing, the entire party left a few days later by steamer for Cairo, to retrace Roosevelt's boyhood Nile adventure of thirty years before, and to begin what was to be a triumphant grand tour of Europe.

Upon his arrival in the Egyptian capital, Roosevelt found himself once again the center of controversy. He had come to Africa at a time when European subjugation of the so-called Dark Continent was coming to an end, and African nationalism was beginning to emerge. Before leaving Khartoum, he had addressed a meeting of Egyptian and Sudanese officers. Favorably impressed by Britain's administration of its African protectorates, he had extolled her rule in the upper Nile region, expressed faith in the civilizing effects of western imperialism, and urged restraint in seeking sovereignty. As a result, he told Bamie, the English "hail me as half ally, half teacher and are wild to have me instruct the Egyptians here and their own people at home what the facts are."

On the other hand, Egyptian nationalists, struggling to lift the yoke of British imperialism, were incensed. The Colonel had scheduled a speech at Cairo University, but the authorities advised against it, especially in view of the recent assassination of the pro-Western premier by a Muslim fanatic. Roosevelt being Roosevelt, he insisted upon making an appearance. He not only condemned the murder, but, believing the Egyptians no better suited for self-government than the Filipinos, called for a period of tutelage before they attained nationhood. "God is with the patient," he counseled the students, quoting an Arab proverb, "if they know how to wait." In no mood to show patience, the students paraded in protest the following day.

Controversy accompanied the Colonel across the Mediterranean to Italy, where he was embroiled in "an elegant row" with the Vatican. Before granting Roosevelt an audience with Pope Pius X, the papel secretary of state, Cardinal Merry del Val, insisted that he agree not to visit a group of American Methodists who were proselytizing in Rome. One of these, gifted with the Dickensian name of Ezra Tipple, was particularly obnoxious, having denounced the pope as "the whore of Babylon." Roosevelt had no intention of seeing the Methodists, but informed Merry del Val, a Spaniard whom he described as a "furiously bigoted reactionary," that he would accept no restrictions upon whom he would see or not see.

Through an intermediary, the cardinal then cynically proposed that the former president secretly agree not to see the Methodists, while stating publicly that no such concession had been made. Roosevelt scornfully dismissed this proposition as one of which "a Tammany Boodle alderman would have been ashamed," and he did not see the pope. In the meantime, the Methodists, gloating at this victory over the Vatican, issued a statement hailing Roosevelt's decision, so he decided not to meet with them either. "The only satisfaction I had of the affair," he told Lodge, "was that on the one hand I administered a needed lesson to the Vatican, and on the other I made it understood that I feared the most powerful Protestant Church as little as I feared the Roman Catholics." It was just as well, he added, that he did not intend to run for office again.

Kings and commoners alike were fascinated with the Rough Rider, and large crowds turned out to see if he really carried a Big Stick or wore a brace of pistols. Everywhere, he dined with royalty, delivered speeches, and accepted honors, all with considerable relish, yet with the conviction that the United States was also being honored. "I thoroughly liked and respected almost all the kings and queens I met," Roosevelt observed, but this feeling was mixed with pity for "the tedium, the dull, narrow routine of their lives," which reminded him of the American vice presidency.

Politics intruded upon the tour despite Roosevelt's efforts to resist. Anxious to confer with his mentor, Gifford Pinchot caught up with him in the picturesque seaside town of Porto Maurizio near Genoa, where Edith was visiting her sister, Emily Carow. Pinchot brandished a sheaf of letters from progressive senators and other anti-Taft people containing bitterly worded indictments of the president's policies. "It was a very good talk," Pinchot recalled years later. "I brought him up to date—told him the facts. We discussed them, and he understood them. They left him in a very embarrassing position, but that could not be helped."

Immediately after Pinchot's departure, the Colonel wrote Lodge a lengthy letter expressing deep concern about the political situation back home. While still insisting that he had no intention of prejudging Taft and his administration, he felt that the president and Congress had "on many important points completely twisted around the policies I advocated and acted upon." He added that he would not campaign for conservative Republicans in the November elections and had "the very strongest objection" to running for the presidency in 1912. Roosevelt's resolve to "keep out of things political" for the time being was stiffened by a meeting with Elihu Root, a strong Taft supporter, in London in late May.

Whirlwind trips to Vienna and Budapest followed his stay in Italy, including dinner with the aged emperor Franz Josef. Then, he was off to Paris. At the Sorbonne he gave a lecture—more of a sermon, really—on "Citizenship in a Republic." The French found a certain beguiling innocence in his talk of the need for sound character, homely virtues, and the maintenance of a high birth rate.* Brussels, The Hague, and Copenhagen came next, followed by Christiana (now Oslo), in Norway, where on May 5, 1910, he gave his Nobel Prize speech.

Roosevelt called for limitations on naval armaments and strengthening of the Hague Tribunal, but the most notable element of the speech was the suggestion that a world organization be created to prevent war. "It would be a master stroke," he declared, "if those great Powers honestly bent on peace would form a League of Peace, not only to keep the peace among themselves, but to prevent, by force, if necessary, its being broken by

---

*This speech includes the "Man in the Arena" passage, among the most famous of Roosevelt's remarks:

It is not the critic who counts; not the man who points out how the strong man stumbles, or where the doer of deeds could have done them better. The credit belongs to the man in the arena, whose face is marred by dust and sweat and blood; who strives valiantly . . . who knows the great enthusiasms, the great devotions; who spends himself in a worthy cause; who at the best knows in the end the triumph of high achievement, and who at the worst, if he fails, at least fails while daring greatly, so that his place shall never be with those cold and timid souls who have never known neither victory nor defeat. *Works,* Vol. XIII, pp. 506–529.

others." Both Woodrow Wilson's League of Nations and the United Nations espoused by Franklin Roosevelt were foreshadowed by this speech, although little attention was paid to it at the time.

In Germany, he was warmly received by Kaiser Wilhelm. Andrew Carnegie had asked him to persuade the German ruler to take the initiative in establishing international peace, but Roosevelt expended little effort on it. The kaiser was eager, however, to learn how he was regarded in the United States, and Roosevelt replied: "Well! your Majesty, I don't know whether you will understand the political terminology, but in America we think that if you lived on our side of the water you would carry your ward and turn up at the convention with your delegates behind you— and I cannot say as much for most of your fellow sovereigns!"

The kaiser broke all precedents by inviting Roosevelt to attend the field maneuvers of the German Army—making him the first civilian to be so honored. Photographs were taken of the two men chatting on horseback, and the kaiser sent copies to his guest with his own comments penciled on the back. "When we shake hands, we shake the world," read one inscription. "The Colonel of the Rough Riders lecturing the Chief of the German Army," said another. When the German Foreign Office learned of the kaiser's latest indiscretion, they tried to get the photographs back, but Roosevelt refused to give them up.

Britain's King Edward VII died while Roosevelt was in Norway and President Taft asked him to represent the United States at the state funeral in London on May 20, 1910. "With him and the Kaiser present, it will be a wonder if the poor corpse gets a passing thought," observed Archie Butt. The late king was related to all the crowned heads of Europe, and his funeral was the last gathering of Continental royalty before most of their thrones were consumed in the flames of World War I. Roosevelt and the French foreign minister, Stephen Pichon, were the only commoners among the official mourners, but the Colonel was the center of attention. Everyone wanted to meet him and get his opinion on their problems. "Confound these kings," he declared in amused desperation. "Will they never leave me alone!"

On the evening before the funeral, the official guests were

invited to Buckingham Palace by the new monarch, King George V, for a banquet that reminded Roosevelt of nothing so much as an old-fashioned Irish wake. "Each man as he arrived said some word of perfunctory condolence to the king our host, and then on with the revel!" Roosevelt recalled. The kaiser established himself as arbiter of which members of royalty were worthy of the Colonel's attention. "Roosevelt, my friend," he declared, planting himself squarely in front of Czar Ferdinand I of Bulgaria, with whom the Colonel had been speaking, "I want to introduce you to the King of Spain; *he* is worth talking to!" When the prince consort of the Netherlands attempted to approach Roosevelt, the kaiser shooed him away.*

Pichon, the French minister, sidled up to Roosevelt during the evening, indignantly insisting that the representatives of the two great republics were being slighted. Had not the American ex-president noticed that this coachman wore a mere black coat while the coachmen of the royal mourners wore scarlet? "I told him I had not noticed, but I would not care if ours had been green and yellow," Roosevelt said. "My French, while fluent, is never very clear, and it took me another half hour to get it out of his mind, that I was not protesting because my livery was not green and yellow. . . ."

Promptly at eight the next morning, Roosevelt appeared in formal evening dress for the start of the funeral procession to Windsor Castle—much to the relief of Whitelaw Reid, the American ambassador to London, who feared he might wear his old Rough Rider uniform. Ever alert to slights to the honor of France, Pichon, with whom Roosevelt was to share a carriage, was already fuming about the fact that while the kings and princes were to have glass coaches, they would not. "As I had never heard of a glass coach excepting in connection with Cinderella," Roosevelt

---

*Roosevelt was fond of a story passed on to him by the kaiser. A special train had been made up in Vienna to carry mourners from southeastern Europe to the funeral, and a dispute broke out between Archduke Franz Ferdinand of Austria and the Bulgarian czar over whose private railway car was to have precedence. The archduke triumphed and his car was placed next to the engine with the czar's car next and then the dining car. All went well until dinner, when the archduke passed word to the czar that he would like to walk through his car to the dining salon. The Bulgarian refused permission and the archduke was forced to wait until the train halted at a station, get out, and walk to the dining car. After dinner was over, he had to wait for another station stop so he could pop back into his car. *Letters,* Vol. VII, p. 410.

declared, "I was less impressed with the omission than he was."

Next, Pichon angrily hissed that "*ces Chinois*" were to precede their carriage in the procession and they would be eighth in line behind such nonentities as the king of Portugal. The final straw was the placement of a third party—"*ce Perse,*" a Persian prince—in their carriage. Frightened by the Frenchman's angry imprecations, the unfortunate prince huddled across from his companions, "looking about as unaggressive as a rabbit in a cage with two boa constrictors," according to Roosevelt, who was convulsed with helpless laughter.

Following the royal funeral, the British took the Colonel to their hearts. *Punch* ran a cartoon that seemed to sum up their affection for him. It showed the lions guarding Nelson's column in Trafalgar Square being protected by guards and wearing placards reading: NOT TO BE SHOT. Roosevelt visited old friends, and eminent men called upon him.* Cecil Spring Rice was deeply impressed with his friend's standing among Britain's leaders. "Roosevelt has turned us all upside down," he noted. "He has enjoyed himself hugely and I must say, by the side of our statesmen, looks a little bit taller, bigger and stronger." Cambridge awarded him an honorary degree and during the ceremonies, a large teddy bear was lowered by pulley from the ceiling; when Darwin was similarly honored, he was told, a monkey had been dropped on him.

In a speech at a colorful Guildhall ceremony in which he was made a Freeman of the City of London, Roosevelt harked back to his talks in Khartoum and Cairo in which he had extolled British rule. Now, he said, Britain must decide whether it should be in Egypt at all. He expressed the hope that his audience would "feel that your duty to civilized mankind and your fealty to your own great traditions alike bid you stay." But if the British decided to keep control of Egypt, he added, they should immediately establish and maintain order and deal with the grievances of the local population. This "govern-or-go" speech created a splendid controversy, with staunch imperialists applauding Roosevelt while "Little Englanders" were offended.

---

*He avoided meeting with Winston Churchill, whom he detested. See *Letters,* Vol. VII, p. 87.

At Oxford, where he delivered the Romanes Lecture, the original reason for his European tour, Roosevelt was introduced by Lord Curzon, the chancellor of the university. Speaking in Latin, Curzon called him a "peer of the most august kings, queller of wars, destroyer of monsters wherever found, yet the most human of mankind. . . . Before whose coming comets turned to flight, and all the startled mouths of sevenfold Nile took fright."*

Entitled "Biological Analogies in History," the lecture was a treatise on the fallacy using parallels between the evolution of animal life and the development of human societies and glib attempts to apply Darwinism to social development. While Roosevelt had labored over it for some time and sought expert scientific advice, the lecture did not achieve the success for which he had hoped. "In the way of grading which he have at Oxford," observed the archbishop of York, "we agreed to mark the lecture 'Beta Minus,' but the lecturer 'Alpha Plus.' While we felt the lecture was not a very great contribution to science, we were sure that the lecturer was a very great man."

> Teddy, come home and blow your horn,
> The sheep's in the meadow, the cow's in the corn.
> The boy you left to tend the sheep
> Is under the haystack fast asleep.

With these words, a versifier in *Life,* the humor magazine, heralded the return of Theodore Roosevelt to the United States on June 18, 1910. Whistles shrieked, salutes were fired, and the largest crowd in the history of New York City lined Broadway and Fifth Avenue to give the Colonel a reception that bordered on hysteria. He was at the zenith of his popularity. Upon meeting his old chief, Archie Butt, who represented President Taft at the ceremonies, sensed a change in Roosevelt. "His horizon seemed to be greater, his mental scope more encompassing. . . . He is bigger, broader, capable of greater good or greater evil, I don't know which."†

---

*This was probably a reference to Halley's comet, which had recently come and gone.

†Among those who came out to Roosevelt's ship as it dropped anchor off quarantine were Eleanor and Franklin Roosevelt. FDR was thinking of following in the elder Roosevelt's footsteps by running for the New York State Senate, but as a Democrat, like his father.

"Ugh! I dread getting back to America and having to plunge into the cauldron of politics," Roosevelt had lamented to Lodge before leaving London. The prospect must have seemed even more bleak when he opened a letter from Taft that he had received just before sailing. "It is now a year and three months since I assumed office [the president wrote] and I have had a hard time—I do not know that I have had harder luck than other Presidents, but I do know that thus far I have succeeded far less than others. I have been conscientiously trying to carry out your policies but my method of doing so has not worked smoothly. . . ."

Roosevelt replied that he was "much concerned about some of the things I see and am told; but what I felt it best to do was to say absolutely nothing—and indeed to keep my mind open as I keep my mouth shut!" Similar sentiments were expressed upon his landing in New York; he told newsmen that he planned a sixty-day moratorium on speeches, to "close up like a native oyster. . . ." The reporters smiled to themselves—certain he was incapable of remaining silent for very long. Politely declining an invitation to visit the White House, he returned to Sagamore after attending the wedding of son Ted and Eleanor Alexander at the Fifth Avenue Presbyterian Church.

Try as he might, the Colonel was unable to divorce himself from politics. Only a few days before his return, Pinchot and James Garfield had electrified progressives by predicting the organization of a third party in time for the 1912 election with Roosevelt as its leader. "Back from Elba" was the progressive watchword. Oyster Bay became a mecca for political figures from all over the nation—mostly of the progressive stripe. "I want to tell you that Colonel Roosevelt is the greatest living American and he is in fighting trim," said Senator La Follette after one such visit. Nellie Taft made a remarkably accurate assessment of these pilgrimages to the Rough Rider's shrine. "I suppose you will have to fight Mr. Roosevelt for the nomination, and if you get it he will defeat you," she told her husband.

Roosevelt himself had no intention of splitting the Republican party. Never a political risk-taker, he saw no chance that a new party would win the presidency for him in the face of the power, wealth, and organization of the Republican bosses. A run for the presidency would also be tantamount to admitting that he had been wrong in his choice of Taft as his successor. And it would probably ensure the election of a Democrat in 1912—a calamity worse than keeping Taft in the White House. The best course was to try to bridge the chasm within the party, and he urged progressives to win control of their state Republican organizations rather than create an alternative party.

At the end of June, he and Lodge visited the president at the summer White House in Beverly, Massachusetts. Ostensibly, the meeting—the first between the two in sixteen months—was to discuss the political situation in New York, where Governor Hughes was trying to force a bill establishing direct primaries through a reluctant legislature; in reality, it was an opportunity for Taft and Roosevelt to get a renewed feel of each other.

"Ah, Theodore, it is good to see you!" exclaimed Taft, stretching out both hands in greeting.

"How are you, Mr. President?" replied Roosevelt affably. "This is simply bully."

"See here now, drop the 'Mr. President,' " said Taft, playfully punching his visitor on the shoulder.

"Not at all. You must be Mr. President and I am Theodore. It must be that way.".

For an hour or so they chatted on the porch, first about the situation in New York, and then amid much laughter, Roosevelt gleefully recounted his adventures at King Edward's funeral. He came away from the meeting with hope that the administration might yet be saved. Only a few days earlier he had cautioned Pinchot to be less outspoken in opposition to the president because it was likely that they would have to support him for reelection in 1912. "Taft has passed his nadir," he wrote, and "independently of outside pressure he will try to act with greater firmness, and to look at things more from . . . the interests of the people, and less from the standpoint of a technical lawyer."

Roosevelt wished to settle in at Sagamore to pursue his many interests and take up his editorship at *The Outlook,* but the American people would not be denied the delight of seeing and hearing their hero. On August 23, he embarked on a sixteen-state, three-week speaking tour of the trans-Mississippi West. There was a strong tide running against the divided Republicans in the upcoming November elections, and his aim was to turn the voters' sentiments around. When asked whether he would speak on behalf of the party, he replied: "My speeches on the trip will represent myself entirely, nobody else."

In these speeches, he advanced a doctrine that he called the "New Nationalism," which was intended as a synthesis of his basic political and economic views. The phrase was not original, having been coined by a New York journalist-turned-political-philosopher named Herbert Croly. In a rather arcane book published the previous year, *The Promise of American Life,* Croly, inspired by Roosevelt, criticized the Jeffersonian ideal of limited government and argued that a strong central government of the type envisioned by Alexander Hamilton, but without Hamilton's elitism, would best spread the benefits of the modern industrial state. Thus, Hamiltonian means would be used to achieve Jeffersonian ends. Roosevelt read the book, admired it, and blended its ideas with the reformist views he had repeatedly expressed in his final years as president.

Traveling in a private railway car paid for by *The Outlook,* he swept through the Midwest with the unrestrained fury of a Kansas twister. Farmers, tradesmen, and small-business men waited in the baking prairie sun or endured cloudbursts to greet the Rough Rider. He spoke before state legislatures, in dusty town squares, at picnics, in ball parks, rousing the crowds to a frenzy with militant demands for reform. In Denver, on August 29, he attacked the Supreme Court as a barrier to social justice and advocated checks on the power of the judiciary to declare legislative acts void. If Taft had become more conservative in office, Roosevelt had grown more radical out of power.

Two days later, the Colonel defined the New Nationalism in an eloquent speech at Osawatomie in Kansas, where John Brown had launched his bloody crusade to purge the land of slavery. With words that sent shivers down the spines of nervous conservatives, he called for property rights to be secondary to human rights and for government to be set free of the control of money in politics. "The struggle for liberty has always been, and must always be to take from some one man or class of men the right to enjoy power, or wealth, or position, or immunity, which has not been earned by service," he declared. "The man who wrongly holds that every human right is secondary to his profit must now give way to the advocate of human welfare. . . .

"The New Nationalism puts the national need before sectional or personal advantage," Roosevelt continued, straining to be heard at the farther reaches of the crowd. "This New Nationalism regards the executive power as the steward of the public welfare. It demands of the judiciary that it shall be interested primarily in human welfare rather than property, just as it demands that the representative body shall represent all the people rather than any one class of or section of the people."

Foreshadowing the modern welfare state, he advocated positive action by the national government to advance equality of opportunity, justice, and security for all. Graduated income and inheritance taxes, a revamped financial system, a comprehensive workmen's compensation law, a commission of experts to regulate the tariff, limitations on the political activities of corporations, stringent new conservation laws, and regulation of child labor were all part of his grab bag of reforms.

Westerners cheered the New Nationalism; conservatives condemned it, calling it socialism or anarchism or communism—all the same thing as far as they were concerned. President Taft was so outraged by Roosevelt's attack on the judiciary that upon hearing of it, he flung a golf club across the course. The New Nationalism was hardly radical, however. Roosevelt was not promoting revolution or an end to private property. Instead of trust-busting, a strong federal government would stand as a countervailing power against industrial combinations. Even Lodge and Elihu Root were basically nationalists in that they prescribed a national approach to resolving problems rather than leaving those problems to state governments.

Upon his return home, the Colonel attempted to moderate the effects his supposedly radical Osawatomie speech had had in the East and plunged into a battle to take control of the New York State Republican convention from William Barnes, Jr., the successor to Tom Platt. Party officials suggested that he confer with Taft to show they were united in this fight. When Roosevelt and Taft met in New Haven, where the president was visiting friends, Taft called Barnes a "crook" and volunteered his help.

Roosevelt came away feeling more benign toward the president than he had in some time, but this mood was quickly shattered. Taft's slippery secretary, Charles Norton, told the press—whether with or without his chief's knowledge is unknown—that the Colonel had sought Taft's assistance because he was in trouble in New York. Roosevelt described these comments as "outrageous" and wrote Lodge that the incident demonstrated clearly why any exchange with Taft was impossible. For his part, Taft told Archie Butt that he and Roosevelt had come to the parting of the ways.

Roosevelt's anger was fanned further by Taft's efforts to purge Republican insurgents and by what he regarded as another example of presidential double-dealing. Although the Colonel had kept quiet in public about his disagreements with the president, Taft at least tacitly encouraged Vice President "Sunny Jim" Sherman to oppose Roosevelt when the New York State Republican Committee met in advance of the state convention to choose the temporary chairmanship. The post was more important than it sounded. The temporary chairman would make the keynote speech outlining party policy and, as chairman of the Credentials Committee, would determine which delegates should be seated. One conservative Republican claimed that he had "just talked on the telephone to the President and the President says he wants Sherman chosen." Taft denied this but Roosevelt was defeated, much to the glee of Taft and his secretary, Charles Norton. "We have got him," chortled Norton, "we've got him, we've got him as sure as peas."

Nevertheless, the Colonel had the last laugh. Rather than rubber-stamping Sherman's nomination, the convention, when it met in Saratoga toward the end of September 1910, overwhelmingly chose Roosevelt as its temporary chairman. Roosevelt savored the victory, but instead of breaking openly with Taft after this blatant double cross, he tried to heal the breach between the progressives and the Old Guard. He not only delivered a keynote speech that praised the president's achievements, but he supported a platform that endorsed the Payne-Aldrich Tariff. Moreover, thanks to Roosevelt's backing, Henry L. Stimson,* a protégé of Elihu Root's and a Taft man for 1912, was nominated for governor.

---

*Stimson, who lived until 1950, later served as secretary of war under Taft; secretary of state under Herbert Hoover, and again as secretary of war under Franklin Roosevelt

The former president's efforts as a compromiser failed to pay off, however. Under attack in the East as a radical, he now faced a storm of protest from westerners, angered by what they regarded as a sellout. Newspapers lampooned him as the "man on two horsebacks." In trying to bridge the gap within the party, Roosevelt had failed to grasp that the split between the two wings was irreparable. That autumn he campaigned for moderate Republicans all over the nation—avoiding Dutchess County, where Franklin Roosevelt was running for the state senate as a Democrat in a predominately Republican district—but had no illusions about the outcome of the election.* With Taft as their leader, the Republicans were doomed to defeat.

The outcome of the balloting was as sour as Roosevelt had predicted. Only the western progressives held their own as the voters repudiated the divided Republican party. For the first time in sixteen years, the Democrats won the House of Representatives and took a healthy bite out of the Republican majority in the Senate. They elected more than half the governors, including Woodrow Wilson, the president of Princeton, who won in New Jersey. In New York, Stimson was snowed under and the Democrats captured both houses of the legislature. The anti-Roosevelt *New York World*, which had claimed that a vote for Stimson in 1910 was a vote for Roosevelt in 1912, now happily announced that he was finished politically.

Roosevelt himself seemed to agree. He regarded the Republican debacle as a personal defeat and buried himself at Oyster Bay. "I think that the American people feel a little tired of me, a feeling with which I cordially sympathize," he glumly declared. Progressives criticized him for being too conservative; Taft supporters blamed the loss of the election upon his refusal to strongly support the president. "The one comfort," he told a friend, "is that I think it prevents my having to face the very unpleasant task of deciding whether or not to accept the nomination in 1912."

---

and Harry Truman. He was one of the last links between the era of Theodore Roosevelt and the world shaped by the aftermath of World War II. His intellectual heirs, Dean G. Acheson, Robert Lovett, and John J. McCloy, were the architects of America's Cold War policies.

*Franklin "is a fine fellow," he wrote Bamie, but wished he were a Republican. *Letters*, Vol. VII, p. 110. FDR won the election, claiming that he really represented the Colonel's views while his opponent was a representative of the Republican Old Guard.

*   *   *

Politics abhors a vacuum and as soon as the 1910 race was over, maneuvering began for the presidential election two years later. While the Colonel was, in his wife's words, "safely caged at Sagamore" with his books and horses, the final wedge was being driven between the two wings of the Republican party. The progressives, encouraged by their victories and the defeat of the Taft Regulars, met in Senator La Follette's home in Washington on January 21, 1911, and organized the National Progressive Republican League, dedicated to such democratic reforms as popular election of senators, direct primaries, and initiative, referendum, and recall.

Gifford Pinchot, his brother Amos, and William Allen White, charter members of the new organization, urged Roosevelt to take part in the league, but he warily declined to go further than endorse some of its objectives in *The Outlook*. He was reluctant to become involved with the schismatics, particularly when the league was so obviously intended to promote La Follette's presidential ambitions. Plainly, the Colonel expected Taft to be renominated and then defeated in the general election, leaving him alone on the bloody field to reorganize the party and be its logical presidential nominee in 1916. So, he flirted with both the Insurgents and the Regulars, courting the league while effecting a reconciliation of sorts with Taft.

The president made the first overture by sending Roosevelt a copy of his annual message, requesting his comments and counsel on dealing with the Japanese. Roosevelt read the message "with great interest" and cordially replied to Taft's foreign affairs inquiries. He also supported administration efforts for tariff reciprocity with Canada. When border problems with Mexico grew menacing, he sought permission to raise a division of cavalry if war came. The president promptly acquiesced, but no conflict materialized to provide renewed military glory. In early June, the two men met in Baltimore at the silver jubilee of James Cardinal Gibbons. Newsmen noted they shook hands affably, whispered together, and broke into unrestrained laughter. But the Colonel vigorously denied a report that he had endorsed Taft for renomination. Taking this as a sign, La Follette announced his candidacy for the Republican presidential nomination on June 17, 1911.

While Roosevelt professed to "greatly admire" La Follette, he still thought Taft would be renominated because the Wisconsin senator was viewed as a dangerous radical by large segments of the public. Nevertheless, he turned back efforts by La Follette supporters, who wished him to declare that he would not be a candidate in 1912, in order to keep his options open. The Baltimore meeting was his last contact with Taft and throughout the spring and summer of 1911, he kept up a steady drumfire of criticism of the president and his policies. In a series of articles in *The Outlook,* Roosevelt attacked Taft's proposed arbitration treaties with Britain and France, rebuked the administration for opposing the recall in Arizona, which was seeking statehood, and attacked its position on conservation in Alaska.

"Our life is very happy," Roosevelt wrote Cecil Spring Rice that summer. "I almost feel as if it were a confession of weakness on my part to be as thoroughly contented as I am." He was enjoying his work at *The Outlook* and money was no problem, since the royalties from *African Game Trails* were likely to reach $40,000 in a single year. The family circle at Sagamore was down to three with only Ethel living at home: Kermit was back at Harvard, Archie was at school in Arizona, and Quentin was attending Groton. From California, where Ted was pursuing a business career, came the happy news that Eleanor was pregnant. "We don't care a rap whether it's a boy or girl," the Colonel wrote the mother-to-be, delighted at the prospect of his first grandchild. The baby, a girl, was born on August 16 and named Grace.

This idyll was shattered on October 27, 1911—Roosevelt's fifty-third birthday—when the morning newspapers announced that the Justice Department had filed suit against U.S. Steel for violating the Sherman Anti-Trust Act. Among other counts, the corporation was accused of having attained a monopoly by duping President Theodore Roosevelt into approving its acquisition of the Tennessee Coal and Iron Company during the panic in 1907. Roosevelt, who had been defending his actions in this matter for four years, including a recent appearance before a congressional committee, angrily called the allegations an attack upon his honor. In point of fact, the indolent Taft had not read the documents

before they were filed—a blunder in itself.* The rupture between
the two men was now irrevocable.

Furious with the president for "playing small, mean and foolish
politics," Roosevelt published a reply in his magazine in which
he not only asserted that he had acted in the public interest, but
condemned Taft's "hopeless" policy of attempting to deal with
the trust problem through "a succession of lawsuits" under the
Sherman Act. Rather than viewing the trusts as inherently evil,
he believed they contributed to economic efficiency and should
be carefully regulated rather than dissolved.

Lighting up the sky like a streak of summer lightning, the
Outlook article reopened interest in the Colonel's candidacy for
the presidential nomination. Some businessmen, weary of the
administration's trust-busting efforts, looked favorably upon his
prescription for resolution of the trust problem, and progressives
who were disenchanted with La Follette were ready to jump ship.
Even though he professed to be uninterested in running, nothing
whets the appetite of political operatives for a candidate more
than his unavailability. Politicians, businessmen, progressives, reformers, and Rough Riders were drawn to Sagamore like the
needle of a compass to the pole.

"I am not a candidate. I will never be a candidate," Roosevelt
insisted. Logic told him to bypass 1912, allow Taft to be beaten
by a Democrat, and bide his time until 1916. Nevertheless, by
the beginning of 1912, he was yielding to those who wished him
to run. "Roosevelt for President" clubs were springing up all over
the country and he was assured that the checkbook of George
W. Perkins, a Morgan partner and director of U.S. Steel and
International Harvester, would be open and that the newspapers
of press baron Frank Munsey would be at his service. "In my
opinion, the matter has already passed out of your hands," wrote
his friend Joseph B. Bishop. "Whether you wish to be a candidate
or not does not weigh a particle. The party needs you and will
take you willy-nilly."

What impelled Roosevelt to run against all logic? No one can
plumb the inner thoughts of any man, especially one like the

*The U.S. Steel case dragged on for years, and the Supreme Court finally ruled against
the government in 1920, thereby sustaining Roosevelt's position.

Colonel, who was a master of masking his motives with the gloss of morality. Certainly, his anger at Taft played an important role. "Taft is utterly hopeless," he told a friend. "I think he would be beaten if nominated, but in any event it would be a misfortune to have him in the presidential chair for another term, for he has shown himself an entirely unfit President." Personal ambition, a lust for power, boredom with his current way of life, the wish to implement the reforms outlined at Osawatomie, the U.S. Steel suit, the need to stem the tide of socialism—all these figured in his decision. But the key factor was his keen sense of duty.

Usually it is stated that Roosevelt plunged into the race against Taft more or less on the spur of the moment, his anger flooding its bounds after the U.S. Steel case was filed. In reality, however, he became a candidate in 1912 only reluctantly and at the urging of the reformers. A compulsive political pulse taker, the Colonel knew full well that the odds were against him and that he was risking his place in history by running. He was aware that he would be branded a madman and a power-hungry egotist. But he also realized that he was responsible for much of the progressive movement within the Republican party, and to betray his principles and others who were fighting for those principles would be an abdication of everything for which he had stood.*

In fact, in 1911, when the Colonel told several Republican state chairmen that he did not want to run because it would destroy his reputation, Frank Knox† of Michigan, a former Rough Rider, bluntly accused him of being a coward—probably the only man ever to do so to Roosevelt's face. "Colonel, I never knew you to show the white feather, and you should not do so now." Roosevelt's sense of duty left him no choice but to seek the nomination in 1912. He was responsible for Taft being in the White House, and only he could prevent him from being renominated. To remain on the sidelines would be irresponsible and craven.

Two roadblocks stood in the way of Roosevelt's candidacy: the

---

*For TR's views on his decision to run, see Gable, *Adventure in Reform,* pp. 15–16; Stoddard, *As I Knew Them,* pp. 390–392; *Letters,* Vol. VIII, Appendix II, pp. 1465–1461; and Wood, *Roosevelt as We Knew Him,* pp. 250–254.

†In 1936, Knox was the Republican vice-presidential candidate. Four years later, Franklin Roosevelt appointed him secretary of the navy as part of a unity move prior to America's entry into World War II.

presence of La Follette in the race and the Colonel's desire for a call from the people. The first impediment was removed when the exhausted senator collapsed on February 2, 1912. Although La Follette was soon back on the stump, Pinchot, White, and others wasted no time in switching their allegiance to Roosevelt. The popular call came in a round-robin letter on February 10 from seven Republican governors—all of whom had previously asked the Colonel to run—urging him to enter the race. Two more governors also endorsed the letter after it was written. Alice Longworth observed owlishly that the letter "did not take Father by surprise," but it was as near to a draft as any political figure has ever received.

Formal announcement of the Rough Rider's candidacy was to follow on February 24, but he was unable to restrain himself. Three days before, following a speech at the Ohio constitutional convention in Columbus, a reporter asked him if he had decided to run. "My hat is in the ring!" the Colonel replied, adding another dramatic phrase to the political lexicon. "The fight is on and I am stripped to the buff."

To have any chance of winning the Republican presidential nomination, Roosevelt had to appeal to party moderates, but his campaign began with a fumble from which it never fully recovered. While most of his speech at Columbus was a more or less restrained restatement of the New Nationalism, he also chose to attack judicial interference in efforts to control big corporations and to secure the rights of workers. "The people should have the right to recall that decision if they think it is wrong," he declared. While he did not advocate the recall of Supreme Court decisions—only those of a state's highest court—the statement sent shock waves through the ranks of both moderates and conservatives.

Usually, this speech is attributed to Roosevelt's lengthy absence from public life and the loss of his sure political touch or to an expedient appeal to La Follette's followers. In reality, the decision was a conscious one, based on fidelity to long-held principles. Throughout his presidency, he had been critical of the judiciary for blocking social legislation and was convinced that no comprehensive program of reform could be achieved unless the courts could be curbed. Once again, he had dramatized his basic difference with Taft. Taft's allegiance was to the law; Roosevelt's to justice.

The Columbus speech marked the parting of the ways for the Colonel and many of his oldest friends. Lodge said he was against the constitutional changes advocated by Roosevelt, but rather than oppose his lifelong friend, he would take no part in the campaign. "I have had my mishaps in politics," he sadly wrote "dear Theodore," "but I never thought that any situation could arise which would have made me so miserably unhappy as I have been during the past week." Roosevelt was forbearing. "My dear fellow, you could not do anything that would make me lose my warm personal affection for you." Root also declined to follow his former chief on his latest crusade, and Nick Longworth, who represented Taft's home district in Cincinnati, was in the throes of indecision.*

Roosevelt faced an uphill fight, having to campaign and create a campaign organization at the same time. But once he had plunged into the political waters again, he swam with powerful strokes. With the Taft forces firmly in control of the party organization—especially the "rotten boroughs" of the South—his only hope of winning a share of the delegates was to contest each of the primaries and persuade other states to hold them. He also had to meet charges of betraying Taft, La Follette, and his own promise, made in 1904, not to run for a third term. To deal with the latter, he disingenuously claimed that he meant he would not run for a third *consecutive* term, but the issue continued to haunt him throughout the campaign.

The debate between Roosevelt and Taft was vitriolic and increasingly personal. The Colonel referred to his successor as a "fathead" with "brains less than a guinea pig"; Taft branded his opponent "a demagogue" and "a man who can't tell the truth." One evening, the private railway cars of both candidates stood side by side in the yards at Steubenville, Ohio, and people gathered, hoping for a fight, but each candidate refused to recognize

---

*One of the casualties of the split between Roosevelt and Taft was Archie Butt, who had remained loyal to both men although they were unable to remain loyal to each other. Noticing that his military aide was looking unwell after TR's announcement, the president kindly suggested that he take some leave. Butt went to Europe and booked return passage on the maiden voyage of the *Titanic*. He was among those who were lost with the great ship.

the presence of the other. Public passions ran high and cut across ties of blood, family, and friendship. Newspapers denounced the "revolutionary" schemes of that "madman" Roosevelt and claimed he intended to seize the White House and turn it into a hereditary fiefdom.

In Illinois, in Maryland,* in Pennsylvania, in California, in Nebraska, Minnesota, and the Dakotas—in almost every state where there were primaries—Roosevelt was the clear winner. There was a seesaw battle in Ohio, the president's home state, but Roosevelt eventually won every district and every delegate. He was the victor in nine primaries; La Follette won two and Taft only one. Had the Republican party been responsive to the public will, Roosevelt would have been nominated. In all, he had won 278 delegates to Taft's 48, and nearly double his popular vote, reinforcing his claim to be the choice of the party's rank and file. But as a result of the Old Guard's tight control of the state conventions, advance polls showed that the president had about 550 of the delegates, who were streaming into Chicago, enough for a majority, while Roosevelt had about 100 less. The nomination turned on the ultimate possession of 254 contested seats, and the final decision would rest with the Credentials Committee of the Republican National Committee, which was firmly in the hands of Taft partisans.

"My own belief is that I shall probably not be nominated at Chicago," Roosevelt acknowledged to a British editor. "But they will have to steal the delegates outright in order to prevent my nomination, and if the stealing is flagrant no one can tell what the result will be." Undoubtedly he hoped to force the Old Guard to surrender by demonstrating that the price of renominating Taft would be the breakup of the party, but he had not reckoned that they might prefer a defeat with Taft than a victory with Roosevelt.

"Victory is by no means the most important purpose before us," Taft wrote Boss Barnes. "It should be to retain the party. . . ." Roosevelt's backers charged "fraud," "robbery," and "naked theft," but the committee, as expected, awarded Taft 235 of the disputed seats to only 19 for the Colonel. Unless Roosevelt's

---

*In Maryland, despite the Brownsville incident, black voters provided Roosevelt his margin of victory over Taft.

supporters could overturn these rulings, Taft would be renominated on the first ballot. Breaking with tradition, Roosevelt angrily rushed to Chicago. In full fighting trim, eyes narrowed and jaw snapping, he proclaimed that he felt "like a bull moose!" Soon, the bull moose became the symbol of a fighting cause.

The night before the convention opened, Roosevelt made it clear to a massive rally of supporters that he would not accept the results of a convention that refused to abide by the will of the people. "We who are in this fight are not feeble, and we intend to carry the fight to the end," he declared. The threat of a bolt from the Republican party if Taft won the nomination was in the air. Mr. Dooley thought the convention itself would be "a combination iv th' Chicago fire, Saint Bartholomew's Massacre, th' battle iv the Boyne, th' life iv Jesse James, an' th' night iv the big wind. . . . Iv coorse I'm goin'! I haven't missed a riot in this neighborhood in forty years."

Filibusters, parades, threats of a bolt, protests by beautiful women wrapped in American flags did not prevent the Taft forces from taking control of the convention as soon as it convened on June 18 in the Chicago Coliseum. They rammed through the election of Elihu Root as chairman by a narrow margin, and whatever slim chance Roosevelt had of being nominated evaporated. Implacably, his former friend repeatedly ruled in favor of Taft on all matters that came before him. The galleries jeered him with cries of "Steam Roller! Toot! Toot! Steam Roller!" but Root coolly ignored them. Tempers flared, curses were hurled, and fistfights broke out. Most authorities now agree that Roosevelt was robbed of enough votes to have at least deadlocked the convention, which was more likely to have turned in the end to him rather than Taft, especially if La Follette's delegates had come over to his side.

However, rather than cooperate with Roosevelt, La Follette preferred to see Taft nominated. Rumors of a compromise slate of Charles Evans Hughes and the Wisconsin senator rippled across the convention hall, but both Taft and the Colonel were adamantly opposed to anyone but themselves. Some Roosevelt strategists wanted to bolt the party; other argued it would be impossible to organize a new party on such short notice. Roosevelt nervously paced the room during an all-night debate while Perkins and

Munsey whispered to each other in a corner. "We are frittering away our time. We are frittering away our opportunity," objected Governor Hiram Johnson of California, who supported a bolt. "And what is worse, we are frittering away Theodore Roosevelt." Gradually, a majority was put together in favor of a bolt. Perkins and Munsey ceased whispering and approached Roosevelt. Each placed a hand on his shoulders and one said: "Colonel, we will see you through."

Refusing to endorse a fraud, Roosevelt instructed his delegates to refrain from voting during the roll call of states the next day. Warren G. Harding, a silver-haired small-town editor from Ohio, put Taft's name in nomination, while jeers and catcalls echoed from every part of the hall. In the balloting that followed, the president was renominated with 547 votes as against 107 for the Colonel and 41 for La Follette. But another 344 delegates had refused to vote, indicating that the battle was far from over. Roosevelt's supporters left the hall, leaving the now worthless nomination to Taft. They marched in a body to Orchestra Hall, about a mile away, amid cries of "Thou Shalt Not Steal!" to hold a rump convention. Under a huge portrait of Theodore Roosevelt, they took the first steps toward forming a third entity—the Progressive party—with the Rough Rider as its leader. "If you wish me to make the fight," I will make it," Roosevelt cried, "even if only one state should support me!"

Seven weeks later, on August 5, the Progressive party's founding convention was gaveled to order in the Chicago Coliseum amid bunting and flags left behind by the Republicans. Unlike anything that had ever occurred before in American politics, it was less a meeting of a political party than a gathering of crusaders. Representing every militant reform impulse at large in the nation, the delegates had an earnestness that bordered on fanaticism. Overwhelmingly white* and middle class, they included intellectuals, professionals, dissenting small-business men, professors, clergymen, small-town editors, social workers, farmers, and woman

---

*Northern blacks, attracted by the Progressives' championing of the underdog, tended to support the party even though it ignored the plight of their all-but-disenfranchised brethren in the South.

suffragettes. In the background were the incongruous figures of Bill Flinn, a onetime Republican boss of Pennsylvania, and the magnates Frank Munsey and George Perkins. Seldom has a political party attracted more intelligence, talent, and diverse skills.

Roosevelt himself was under no illusions about his chances for victory in November. "In strict confidence, my feeling is that the Democrats will probably win if they nominate a progressive candidate," he told a friend. He expected that candidate would be Woodrow Wilson, who had had an excellent record as a progressive governor of New Jersey. No third party had ever won a presidential election—new parties replaced old ones—then went on to the White House as the Republicans had replaced the Whigs. Some authorities have suggested that Roosevelt bolted in the heat of the moment and later regretted it. This was hardly likely. Roosevelt continued the fight for the same reason he had entered it. Having been given an overwhelming mandate in the primaries, he could not in good conscience abandon his followers now. The bosses had robbed him of a deserved victory at the Republican convention, not the people. "In loyalty, honor and duty there was nothing for me to do but to heed their call and make the race with all my might, regardless of present or future consequences to myself," he later told a friend. The fight had to go on, not for his sake, but for the future of the Progressive Movement, the Republican party, and the nation.

In the period between the nomination of Taft and the Progressive convention, the Colonel suffered several serious rebuffs. Progressive hopes that the Democrats would nominate a candidate as conservative as Taft, so reformers from both parties could unite behind Roosevelt, were dashed when they chose Wilson on the forty-sixth ballot at a raucous convention in Baltimore. Some Progressive Republicans who had supported Roosevelt against Taft refused to join him under a third-party banner. Senator La Follette also declined to support him. "What a miserable showing some of these Progressive leaders have made," Roosevelt complained. "They represent nothing but mere sound and fury."

Waving a sea of red bandanas and exploding into a Niagara of cheering, some fifteen thousand people greeted Roosevelt when he came to the rostrum on August 7 to accept the party's presidential nomination. California's governor, Hiram Johnson, was to be his running mate. For fifty minutes, the delegates swirled about the hall, brandishing banners and singing "Onward Christian Soldiers" and "The Battle Hymn of the Republic."* When the Colonel tried to halt the demonstration, everyone burst into song:

> Thou wilt not cower in the dust,
> Roosevelt, O Roosevelt!
> Thy gleaming sword shall never rust,
> Roosevelt, O Roosevelt!

Finally, as the din receded, Roosevelt delivered a lengthy "Confession of Faith," which has been described as the most sweeping charter for reform yet presented by a major presidential candidate. Establishing the agenda for the twentieth century, it embraced direct election of U.S. senators; preferential primaries in presidential years; votes for women;† a strict corrupt-practices act; initiative, referendum, and recall; a federal securities commission; the regulation of trusts; reduced tariffs; unemployment insurance; old-age pensions; abolition of child labor, and pure food laws. "Our cause is based on the eternal principles of righteousness; and even though we who now lead may for the time fail, in the end the cause itself shall triumph," Roosevelt thundered. "We stand at Armageddon, and we battle for the Lord!"

Before embarking on what he promised would be a strenuous campaign, Roosevelt returned to Sagamore to recharge his batteries. Edith was not at all happy with his decision to run, nor with the breakup of old friendships, but she kept her innermost feelings to herself. Nicholas Roosevelt, a young cousin, found the Colonel looking "younger by ten years than when I had last seen him. . . . In such wonderful spirits that he behaved like a boy." He professed to be "dumbfounded" by the success of the con-

---

*Edith Roosevelt always remembered the sight of Oscar Straus singing "Onward Christian Soldiers" with "his heart and soul, which amused him immensely after all was over." Straus became the unsuccessful Progressive candidate for governor of New York and the first Jew to run for the office.

†Roosevelt's nomination was seconded by Jane Addams, the founder of the settlement house movement, an unusual event in itself. Women, he assured her, would have a place in the party equal to that of men and he was "without qualification" for woman suffrage. *Letters*, Vol. VII, pp. 594–596.

vention and the spirit of religious fervor that had taken over the movement.

The campaign rapidly developed into a two-man fight between Roosevelt and Wilson as they all but ignored the hapless Taft and Eugene Debs, the Socialist candidate. "Tell us about Taft!" someone in a crowd shouted. "I never discuss dead issues!" the Colonel shot back. Yet he remained clearheaded about his prospects. Wilson, he wrote Arthur Lee, was "an able man and would make a creditable president," and he believed that while he would run well ahead of Taft, the Democratic candidate would win.

Both Roosevelt and Wilson were Progressive, but they were as far apart in personality—Wilson was austere,* almost grim, and a far cry from the ebullient and impulsive Roosevelt—as they were in their commitment to social justice. As a states' rights Democrat of southern background, Wilson regarded such Rooseveltian proposals as woman suffrage and the abolition of child labor as purely state issues and rejected a stronger role for government in human affairs. Under the influence of Louis D. Brandeis, a Boston attorney and longtime foe of monopoly, Wilson, after some hesitation, made the trusts the key issue of his campaign. Emphasizing the predatory nature of monopoly, he favored trust-busting rather than regulation. He called his program the "New Freedom" and attacked Roosevelt's New Nationalism for its acceptance of monopoly with government regulation. Roosevelt branded Wilson's attitude toward big business "rural toryism" and argued that his promises to break up the corporations were simply out of date.

Unfazed by the bleak prospects for victory, the Colonel stumped the country in a private railroad car, the *Mayflower,* and drew such multitudes that the traveling press corps was reminded of John the Baptist. In Providence, ten thousand people packed the railroad station to welcome him. Businesses closed and traffic was halted as 200,000 turned out in Los Angeles. People smiled when it was reported he had clambered over the tender of a

---

*When a scandalmonger offered, first to the Republicans and then the Progressives, a sheaf of letters purportedly linking Wilson to an extramarital affair, the Colonel approved the decision not to touch them. "Those letters would be entirely unconvincing," he said. "Nothing, no evidence could ever make the American people believe that a man like Woodrow Wilson, cast so perfectly as the apothecary's clerk, could ever play Romeo."

speeding transcontinental train into the cab to shake the hands of the crew and to take a turn at the throttle. But the years had left their mark; he was grayer and slower, and sometimes his voice gave out. Yet, as an underdog he now inspired more devotion than ever before.

Three weeks before the election, on October 14, Roosevelt was in Milwaukee to attend a rally. Shortly before eight that evening, he left the Hotel Gilpatric, and stood in an open car to acknowledge the cheering of the crowd. Suddenly, a man in the front rank raised a pistol and fired a shot into his chest from a distance of less than thirty feet. The bullet tore through his overcoat, steel spectacles case, and the folded manuscript of his speech before lodging below his ribs. Reeling backward under the impact, which he later compared to the kick of a mule, the Colonel coughed and put his hand to his mouth to see if he could spit up blood. When he didn't, he assumed the wound was not fatal.

"Stand back! Don't hurt that man!" he shouted as the crowd surged forward to lynch his assailant. He ordered him brought before him, and stared penetratingly at the trembling fellow—John F. Schrank, an anti–third-term fanatic—before turning away.*

Even with a bullet in his body, Roosevelt shrewdly took advantage of the dramatic situation. Although his shirt was soaked with blood, he insisted on being taken to the rally. "It is one thing or another." Unaware that he had been shot, the packed auditorium greeted him enthusiastically, but Roosevelt raised his arm for silence. "I shall ask you to be as quiet as possible," he declared in a low voice. "I don't know whether you fully understand that I have just been shot; but it takes more than that to kill a Bull Moose."

There was a sound like air being sucked out of the hall and cries of "Oh, no! Oh! no!" Reaching into his coat pocket, the Colonel pulled out the copy of his speech, and seeing that the bullet had torn a hole through the fifty folded pages, held it up for all to see. "There is a bullet—there is where the bullet went through—and it probably saved me from it going into my heart.

---

*Schrank claimed he had been inspired by the ghost of William McKinley, who accused his vice president of having assassinated him and asked Schrank to avenge him. Judged insane, Schrank remained in a mental hospital until his death in 1940, the year in which Franklin Roosevelt broke the third-term tradition.

The bullet is in me now, so that I cannot make a very long speech but I will try my best. . . ."

Foolhardy it may have been, but Roosevelt saw this as a supreme moment. Over the next hour and a half, he held the stage, waving off repeated appeals for him to stop and seek medical treatment. He talked of the Progressive Movement, a cause to which he said he was devoted "with my whole heart and soul," and its appeal to all Americans. Unless the movement succeeded, the nation would become divided—"have-nots" against "haves," ethnic group against ethnic group. Finally, he allowed himself to be taken to a hospital, where he joked and talked politics with the examining physicians. They found the wound superficial and credited Roosevelt's excellent physical condition with having prevented him from being more seriously hurt.

Admiration for Roosevelt's courage was widespread, and the other candidates suspended active campaigning while he convalesced from his wound. But cheers were no substitute for votes. He won the support of most progressive Republicans, but reform Democrats, smelling victory for the first time since 1892, remained with Wilson. Wilson piled up 6,301,254 votes to Roosevelt's 4,127,788, while Taft ran a poor third with 3,485,831. Debs polled another 900,000 votes. Although Wilson scored a landslide in the electoral vote, he had polled only about 42 percent of the popular vote.

Roosevelt had fared better than expected. Five states—Pennsylvania, Michigan, Minnesota, South Dakota, and Washington— as well as eleven of California's thirteen electoral votes were his and he lost in nine others by razor-thin majorities. In Kansas, he lost by only 1.77 percent, in Maine by 2.03 percent, and in Iowa by 3.7 percent, while running second in a total of twenty-three states. He had done well in both urban and rural areas, and the only section he had been unable to penetrate was the South. In view of the obstacles to his election—"We had all the money, all the newspapers and all the political machines against us," he said—the returns were a remarkable tribute to the spell he cast upon the American people.

"We have fought the good fight, we have kept the faith, and we have nothing to regret," the Colonel wrote Jim Garfield a few days after the election.

Chapter 24

"The Great Adventure"

Theodore Roosevelt returned to Sagamore with all crusading zeal spent. The rejection of his candidacy in 1912 was the bitterest of his career, and he was certain the gates of political opportunity had slammed shut for good. Winter closed in on the sprawling house on the hilltop with a chill sharper than usual. The place was deserted except for the master, his wife, and daughter. The telephone that had jangled day and night was silent. Roosevelt was a pariah to Republican Regulars—a "wild man," "a destroyer"—the man who had smashed the party and allowed the Democrats to win the presidency. There was no powerful surge to shake his hand when he appeared in public. "When it is evident that a leader's day is past," he gloomily noted, "the one service he can do is step aside and leave the ground clear for the development of a successor."

If a man cannot face the future with certainty, he often looks backward, and the Colonel turned inward to write his memoirs. A few weeks after the election, he was at his desk, working under the gaze of a large portrait of Theodore Roosevelt, Sr., on what he called "Some Chapters of a Possible Autobiography" for *The Outlook*. After serialization, these articles were published in book form as his *Autobiography.** Dipping his pen into an inkwell made of a rhino's foot, he began the journey down the long corridor of the past: "My grandfather on my father's side was almost of purely Dutch blood? . . ."

Soon, he lost himself in the tale of his boyhood and the effort to recapture the strength and charm of his father, his mother's beauty and southern grace, the Victorian solidity of the Twentieth Street house. The shock of the attempted assassination faded and he was back in the New York Assembly climbing the political ladder that led to the White House . . . out in the Bad Lands on the round-up and hunting trail . . . hearing the *whit-whit* of Spanish bullets as they sang past his head in Cuba. . . .

What should he tell? What should he keep to himself? "It is very difficult to strike just the happy mean between being too reticent and not reticent enough!" Roosevelt told Emily Carow. "I find it difficult as both regards my life when I was a child and my political experiences. I can only hope that I am handling it in a proper way." To Cabot Lodge, with whom he had renewed his correspondence, he wrote: "The hardest task I have is to keep my temper, and not speak of certain people, the editors of the *Evening Post,* for instance, anti-imperialists, universal peace and arbitration men and the like as they richly deserve."

The narrative of his childhood is fascinating and he made a spirited defense of his conduct in Panama and in the U.S. Steel case. Understandably, he chose to end the account on a note of triumph, with the return of the Great White Fleet rather than with the dismal defeat of 1912. But the book is almost as notable for what is unmentioned than what it includes. A reader comes away with a sense of passion restrained and judgments not rendered. There is no mention of the unhappy life of his brother Elliott, or of his passionate courtship of Alice Lee and their marriage. It is as if the pretty, fun-loving girl from Chestnut Hill had never existed.

The Colonel was still titular head of the Progressive party, but his preoccupation with his memoirs told more about his attitude

---

*He also worked with Edmund Heller, a naturalist who had accompanied him on the African safari, on *Life-Histories of African Game Animals,* which was published in 1914 in two volumes.

toward its future than anything he might have said. There were already signs of a breakup, with Frank Munsey, the press lord, among the first to desert. Some leaders made overtures to the Republicans. Others turned from the combat against the Democrats and Republicans to fighting among themselves, with the radicals such as the Pinchot brothers, Amos and Gifford, questioning the Progressive credentials of George Perkins, whom they considered the representative of the money trust.

Nevertheless, Roosevelt realized that as the trustee of the hopes of the four million Americans who had affirmed their faith in him at the polls, he could not immediately abandon the party. But he found the effort to preserve unity "very weary work," and complained that "I ought not to be required to do such work. . . . This whole business of leading a new party should be for an ambitious young colonel and not for a retired major general."

With some reluctance, the "retired major general" attended a party conference in Chicago, where he successfully resisted the efforts of radicals to drum Perkins out of the party. "Our purpose," he told the delegates, "is to keep up a continuous campaign for social and industrial justice and for genuine government by the people and for the people." Privately, however, he saw little hope for the party unless the crazy-quilt alliance of southern conservatives, Tammany boodlers, and Bryanites that had elected Woodrow Wilson should fly apart and open the way for Progressive victories in the 1914 off-year elections.

Roosevelt's mind was diverted from these troubles by a lively controversy. Throughout his political career, he was dogged by charges that he was a heavy drinker—how else to explain the constantly ruddy complexion, the strange gusts of laughter and anger—and they had reached flood tide during the 1912 campaign. He had always treated these stories as one of the annoyances of public life, but finally lost patience and brought a libel suit against George J. Newett, the editor of *Iron Ore,* an obscure weekly newspaper in the Upper Peninsula of Michigan, who had written: "Roosevelt lies and curses in a most disgusting way; he gets drunk, too, and that not infrequently, and all his intimates know about it."

The case came to trial in the tiny courthouse in Marquette, Michigan, on May 26, 1913, and for the next five days, former

Cabinet officers, family members, and newspapermen testified to his sobriety. "I do not drink either whiskey or brandy, except as I shall hereafter say," the ex-president related on the witness stand. "I do not drink beer; I sometimes drink light wine. . . . I have never drunk a highball or a cocktail in my life. . . . I may drink a half dozen mint juleps in a year. . . . At home, at dinner, I may partake of a glass of white wine. At a public dinner, I will take a glass or two glasses of champagne. . . ."

Newett was unable to present a single witness who could testify to having seen Roosevelt drunk. He had to admit that he was wrong and apologized abjectly. The Colonel told the court he had not filed suit for money or to destroy a small newspaper, only "so that never again will it be possible for any man, in good faith, to repeat" these charges. In accordance with his wish, nominal damages—six cents—were assessed against the editor.

"I have to go. It's my last chance to be a boy!"

With these words, Theodore Roosevelt embarked on another wilderness expedition—this time to map the unexplored River of Doubt, which flowed through the heart of the Amazonian rain forest—and it was to be the most harrowing physical experience of his life. It all began in June 1913, when the governments of Brazil, Argentina, and Chile invited him to give a series of lectures and, eager for excitement and change to wash away the defeat of 1912, he accepted.

Roosevelt and his wife left New York for Rio de Janeiro on October 4, 1913, and Edith noted that on the long voyage he was more lighthearted than she had seen him in years. "I think he feels like Christian in *Pilgrim's Progress* when the bundle falls from his back," she wrote Bamie, ". . . in this case it was not made of sins but of the Progressive Party." In Brazil, they were joined by Kermit, who was employed by the Anglo-Brazilian Iron Company as a bridge construction supervisor. Postponing his wedding to Belle Willard, whose family owned Washington's Willard Hotel and whose father was currently U.S. ambassador to Spain, Kermit volunteered to accompany his father on an expedition into the heart of the continent just as he had on the African safari.

Following his series of lectures, the Colonel planned, at the suggestion of an explorer friend, to traverse the continent from south to north—up the Paraguay River into the Amazon basin and then to follow the Negro and Orinoco rivers to the sea. The American Museum of Natural History agreed to send along a team of naturalists, as the Smithsonian Institution had done during the African trip, to help collect animal and botanical specimens.

Roosevelt's sun may have been in eclipse at home, but the South American tour rivaled his triumphal procession through Europe. Over a period of six weeks, he traveled from Rio to São Paulo in Brazil; to Montevideo, Uruguay; Buenos Aires, Argentina, and across the Andes to Santiago, Chile. Along the way, the Colonel preached the Progressive party gospel and the necessity of stamping out corruption in government to enthusiastic, if somewhat bemused, audiences.* Early in December, Edith sailed for home and he was ready to begin the journey into the interior. Before he left Rio, however, Brazilian officials proposed a change in his plans for a south-to-north trek.

Four years earlier, Roosevelt was told, Colonel Candido Rondon, a Brazilian explorer, had come upon a large, previously unknown waterway in the Mato Grosso that he had named Rio da Divuda, the River of Doubt, which appeared to be a tributary of the Amazon. Rondon was now planning to trace the course of this mighty river, and Roosevelt and his party were invited to join the expedition. Without hesitation, he accepted and journeyed by steamer up the Paraguay River to join Rondon at the Brazilian frontier.

Early in January 1914, the Roosevelt-Rondon expedition plunged into the jungle. They hacked their way through the tangled underbrush with machetes, waded through marshes up to their knees, then up to their hips, and swam across bayous. "We were drenched with sweat. We were torn by the spines of the innumerable clusters of small pines with thorns like needles," Roosevelt recorded.† "We were bitten by the hosts of fire-ants,

---

*The eagerness of South Americans to welcome Roosevelt belies the repeated charges by his critics that his seizure of the Panama Canal Zone had aroused widespread animosity in the area. In fact, he took an entirely different view toward Argentina, Brazil, and Chile than he did of the unstable republics of the Caribbean basin, and believed that these governments should be co-guarantors of the Monroe Doctrine, transforming it from a unilateral to a multilateral agreement. *Letters,* Vol. VII, p. 756.

†Once again, he had contracted to write a series of articles for *Scribner's,* which were published later in book form as *Through the Brazilian Wilderness.*

and by the mosquitoes, which we scarcely noticed when the fire-ants were found, exactly as all dread of the latter vanished when we were menaced by the big red wasps, of which a dozen stings will disable a man. . . ." Boots and clothes rotted, weapons fouled unless constantly attended to, and their blood was poisoned by fever.

Upon reaching the headwaters of the River of Doubt at the end of February, the expedition split up, with some returning to civilization over the route they had come, while Roosevelt and his son, Colonel Rondon, and a handful of others set out in seven wooden dugout canoes to trace the course of the river with the hope that it would eventually lead to the Amazon. They carried only basic equipment and enough provisions for fifty days, if carefully rationed. "We were about to go into the unknown, and no one could say what it held," the Colonel observed.

The voyage down the river was a harrowing one. "No less than six weeks were spent in slowly and with peril and exhausting labor forcing our way down what seemed a literally endless succession of rapids and cataracts," he reported. "For forty-eight days we saw no human being. In passing the rapids we lost five of the seven canoes with which we started and had to build others. One of the best men lost his life in the rapids. Under the strain, one of the men went completely bad, shirked all his work, stole his comrades' food and when punished by the sergeant murdered the man and fled into the wilderness."

Mile after mile, they drifted down the river and then carried the canoes on long and difficult portages around rapids and cataracts. Roosevelt himself badly injured a leg, the same one he had hurt in the 1902 carriage accident in Pittsfield, when trying to save two canoes that had become caught in the rapids. Abscesses developed, causing serious infection, and weakened by constant exertion and poor diet, he came down with dysentery and malaria. With a temperature of 105 degrees, he was unable to walk and had to be carried in a litter. He begged the others to leave him behind and considered suicide to keep from becoming a burden. In his delirium, he recited the same line of poetry over and over again: "In Xanadu did Kubla Khan a stately pleasure dome decree. . . . In Xanadu did . . ."

Two months after the expedition had set out, what was left of the party reached a village, from which they took a steamer to

Manaus, where they learned they had been given up for dead. Roosevelt, who had lost fifty-seven pounds, was worn out and hardly able to walk. "The Brazilian wilderness stole ten years away of his life," his longtime friend William Roscoe Thayer later observed. But in spite of their privations, Roosevelt and his party had mapped the 1,500 miles of the River of Doubt, a powerful waterway as long as the Rhine or Elbe, which the Brazilians renamed the Rio Roosevelt in his honor. In Brazil, it is popularly known as the Rio Teodoro.*

No crowds, no welcoming committee, no parade greeted the Colonel upon his return to New York from South America on May 19, 1914. *The New York Times* reported that he was "thinner and older looking," yet "none of the old time vivacity of manner was lacking." Waving his cane about, he joked, "You see I still have the big stick." The reporters asked the usual questions—Would he be the Progressive candidate for governor of New York that fall? Would he make another run for the presidency in 1916?—but he declined to discuss politics.

Edith thought the loss of weight became him. He seemed "younger for every pound he has lost—a year a pound, I should say." Arriving by boat at Oyster Bay that evening, he was greeted at the dock by Ethel, who threw herself into her father's arms as he stepped ashore.† "Oh, my darling, my darling," he said as he hugged and kissed his youngest daughter.

Even though he was seriously weakened by his hardships—"I am now an old man," he told Leonard Wood—he immediately became embroiled in controversy. While Roosevelt had been fighting for his life in the Brazilian jungle, William Jennings Bryan, whom Woodrow Wilson had named secretary of state, had with guileless innocence reached an agreement with Colombia that expressed "sincere regret" for the way in which the United States had acquired the Panama Canal Zone and provided for a $25 million payment. The Colonel was incensed when he learned of

---

*In 1992, an expedition that included Tweed Roosevelt, the Colonel's great-grandson, retraced the route covered by the earlier explorers and found the country even wilder than it had been seventy-eight years before.

†The previous year, she had married a Manhattan surgeon, Dr. Richard Derby, and their first child had recently been born.

it. "I regard the proposed Treaty as a crime against the United States, an attack upon the honor of the United States," he raged, and described the agreement as "simply the payment of belated blackmail." The Senate refused to approve the treaty—but Roosevelt never forgave Woodrow Wilson for the insult.*

Gnashing his teeth all the while, Roosevelt was also forced to sit idly by and watch as Wilson stole much of the Progressive platform of 1912, and in a virtuoso performance, rammed it through a reluctant Congress. The banking system was restructured by the establishment of the Federal Reserve System, the tariff was reduced, and trusts and monopolies were regulated by the Federal Trade Commission in line with Roosevelt's program rather than being broken up as Wilson had originally argued. Child-labor and workmen's compensation laws were passed, along with a graduated income tax.

The pleasures of Sagamore did not distract him for long. Two days after his return home he wrote one of his "Personal and Private" letters to General Francis V. Greene, seeking permission to raise a division of cavalry in case of war with Mexico.† And then he was off to Madrid for Kermit's wedding to Belle Willard.

Following the Colonel's return from Spain, the Progressive party tried to draft him to run for governor of New York with the hope that his candidacy would lend credibility to Progressive candidates running all over the nation. "As soon as I got back here I was plunged into politics—and really under very disheartening conditions, for the confusion passes belief," he told a friend. He didn't want to run for governor; he had already been governor. The threat was removed when the doctors determined that he was still suffering from malaria and ordered him to rest.

Roosevelt made his first appearance during the 1914 off-year

---

*In 1921, after Roosevelt's death, the Senate ratified the treaty under pressure from oil interests that were seeking concessions in Colombia.

†In the course of "protecting American lives and property" during the Mexican Revolution, the U.S. Navy seized the Customs House at Veracruz on presidential orders, and in the fighting 126 Mexicans and 19 Americans were killed. War seemed imminent, with Franklin D. Roosevelt, who had followed Uncle Ted in the post of assistant secretary of the navy, being among the most bellicose members of the administration. Wilson, shocked by the bloodletting, sought a way out of the debacle. When Argentina, Brazil, and Chile offered to mediate, he gratefully accepted.

election campaign in Pittsburgh on June 30, where he vigorously attacked both the Democratic and Republican bosses and set forth the Progressive position on current issues. The Bull Moose was alive, and if not well, at least full of fight. No mention was made of an event that had occurred two days earlier in the far-off town of Sarajevo in Bosnia, where a Serbian nationalist had assassinated Archduke Franz Ferdinand, heir to the throne of the Austro-Hungarian Empire. The cataclysm Roosevelt had worked to avert at Portsmouth and Algeciras was about to burst upon the world.

In happy ignorance, most Americans had never even heard of Sarajevo, or Bosnia, or Serbia, and the newspapers dismissed the murder as "another mess in the Balkans." Europe had experienced four decades of peace, and it was expected that the violence would be confined to that obscure corner of the world. But over the next month, nation after nation slipped into war. Conflicting nationalisms and dynastic rivalries had transformed Europe into a house of swords so delicately balanced that the murder of Franz Ferdinand, even though he was almost universally detested, brought the whole edifice tumbling down on the heads of all. Two alliances were pitted against each other: one of Britain, France, Russia, and Serbia, later joined by Italy and Japan; the other of Germany, Austria-Hungary, and Turkey, which became known as the Central Powers.

Violating existing treaties, which they called mere "scraps of paper," the Germans ruthlessly invaded neutral Belgium to attack France. On August 4, as the German Army drove through the prostrate country toward Paris, President Wilson proclaimed the official neutrality of the United States and urged the American people to be "impartial in thought as well as in action." The following day, ex-President Roosevelt supported Wilson's policy of neutrality. This was not the time to criticize the president, he said, and offered to work "hand in hand" with anyone to see the country "through this crisis unharmed." He followed up this statement on August 22 with an article in *The Outlook*. "Only the clearest and most urgent national duty," he wrote, "would ever justify us in deviating from our rule in neutrality and noninterference." In fact, he professed to see the military justification for

Germany's invasion of Belgium and the necessity to maintain Germany as a barrier to Russia.

Privately, Roosevelt was hardly neutral. By heritage, associations, and long-standing suspicion of Germany's imperialistic ambitions, he was for the Allies. In later years, Justice Felix Frankfurter, then a young lawyer, recalled visiting Sagamore on August 4, as the British cabinet was considering a declaration of war. "You've got to go in!" the Colonel told Charles Booth, president of the Cunard Line. "You've got to go in!" He wrote Arthur Lee that if he were president, he would "register a very emphatic protest that would mean something" against the German violation of Belgian neutrality.* "For the last forty-three years Germany has spread out everywhere, and has menaced every nation where she thought it was to advantage to do so. . . ."

Roosevelt might have come out openly for the Allies that fall but for the November elections. Many members of the Progressive party were pacifists or lauded Wilson's policy of nonintervention, so he thought it best to maintain silence on the issue. Moreover, there was another point of friction between them and the party's titular leader. As a political realist, Roosevelt recognized that without some form of cooperation with the moderate Republicans, the party was doomed—even though fusion was anathema to the more radical Progressives.

To ensure the defeat of the old-line Republican and Democratic bosses, William Barnes and Charles Murphy, the Colonel proposed that New York Progressives nominate as their candidate Harvey D. Hinman, an independent Republican who seemed likely to be his party's nominee for governor. Roosevelt overcame the aversion of party leaders to fusion, but the plan ultimately failed when Hinman refused to agree to run as a Progressive if he lost the Republican nomination. In several speeches, Roosevelt forthrightly said that Barnes was working in collusion with his Democratic counterpart and taking a leaf from the Colonel's own book, Barnes immediately filed suit for libel.

Although he was still suffering from the rigors of his Brazilian

---

*"I have never liked Winston Churchill," Roosevelt also told Lee, "but in view of what you tell me as to his admirable conduct and nerve in mobilizing the fleet, I do wish . . . you would extend to him my congratulations on his action." Churchill was serving in the British cabinet as First Lord of the Admiralty. *Letters,* Vol. VII, p. 810.

adventure, Roosevelt dutifully barnstormed across the country in support of Progressive candidates. Yet, he was well aware of the futility of his cause. Poor crowds and the general lack of spirit told him that disaster lay ahead. "I have done everything this fall that everybody has wanted," he told O. K. Davis as he completed a final campaign swing. "This election makes me an absolutely free man. Thereafter I am going to say and do just what I damn please." The outcome was what Roosevelt had expected. Except in California, where Hiram Johnson was reelected governor, the Bull Moosers lost almost everywhere. "I don't think they can much longer be kept as a party," he told Ethel.

With the election—and the necessity to cultivate pacifist elements in the Progressive party—out of the way, Roosevelt took the wraps off his pro-Allied sympathies and threw himself into his last great adventure. In the remaining four years of his life, he campaigned to prepare America to defend itself, then to persuade it to enter the war on the Allied side, and finally to urge prosecution of the war with the fullest rigor. Hermann Hagedorn, his friend, called him "The Bugle That Woke America." He was vilified as a spiteful, bloodthirsty monster and then saw his influence rise to new heights as he stood again on the threshold of the presidency. He struck Woodrow Wilson with even heavier blows than those directed against William Taft. And he was to suffer one of the great tragedies of his life, the loss of a son in battle.

In a series of articles for *The Outlook* and *The New York Times* that were syndicated to papers around the country, Roosevelt lashed Wilson for not having protested the German invasion of Belgium—forgetting his own earlier acceptance of its strategic necessity. Along with Lodge and Leonard Wood, he called for increases in the size of the army and navy so that the United States might avoid a fate similar to that of Belgium. However, he combined preparedness with a call for an international body—a World League for Peace and Righteousness—to adjudicate disputes between nations and to enforce a just peace, as he had suggested in his Nobel Prize speech of 1910.

Secretary of State Bryan, the Colonel declared, was "a professional yodeler, a human trombone," and he described the president as possessing "an astute and shifty mind, a hypocritical ability to deceive plain people, unscrupulousness in handling machine leaders, and no real knowledge or wisdom concerning internal and international affairs." He charged that the administration had virtually acquiesced in Germany's invasion of Belgium by not protesting this gross violation of the Hague Convention. Even if this effort had failed, he continued, it would have been a check on Germany's brutal conduct.

Eventually, Roosevelt convinced himself that it was necessary for the United States to enter the war. But he did not favor war merely to defend the Allies or to restore the balance of power; if there was war, it should be for the defense of the United States and American rights. Although Germany did not plan to attack America immediately, he acknowledged in a letter to J. Medill McCormick of the *Chicago Tribune,* she was bound to challenge American interests in the Caribbean. "In this way, we would be thrown into hostilities with Germany sooner or later and with far less chance of success *than if we joined with the powers which are now fighting her.*" Roosevelt knew that many Americans regarded him as an alarmist, but pressed ahead anyway with his demand for preparedness.

The Army League, the Navy League, and the National Security League—bankrolled by J. P. Morgan, Henry C. Frick, and Samuel Insull, the utilities tycoon, among other wealthy men—lobbied for bigger military appropriations and universal military service. Preparedness was described as insurance against war. A propaganda film, *The Fall of a Nation,* depicted a helpless America invaded by spike-helmeted attackers. For his part, Wilson coldly dismissed all talk of preparedness. "We shall not alter our attitude . . . because some amongst us are nervous and excited"—obviously a reference to Theodore Roosevelt.

The Colonel's loathing for Woodrow Wilson was almost psychopathic in intensity. "Wilson is, I think, a timid man physically," he wrote Cecil Spring Rice, now the British envoy to Washington. "He is certainly a timid man in all that affects sustaining the honor and the national interests of the United States." Wilson was "worse than Jefferson and Madison," and Roosevelt feared the parallel between the current crisis and the War of 1812, when an unprepared nation had gone to war. "Unscrupulous," "shifty," "coldblooded," and "hypocrite" are some of the terms Roosevelt used to describe the president. Someone remarked that Roosevelt had no need for profanity because he could say "Pacifist" or "Woodrow Wilson" or "William Jennings Bryan" in tones "that would make the Recording Angel shudder."

Roosevelt's antipathy to Wilson was based upon more than mere envy and petty jealousy. There were basic temperamental differences between the two men, frustrations arising from the Republican defeat of 1912, the insulting apology to Colombia, dissatisfaction with the policy of "mushy amiability" toward Mexico, and the fact that Wilson had appropriated much of the Progressive party's platform. The overriding reason, however, lay in the fact that throughout his presidency, Roosevelt had hoped for some momentous event whose resolution would enable him to stand with Washington and Lincoln in the pantheon of American heroes. How it galled him that a trick of fate had given Woodrow Wilson—"this schoolmaster"—the golden opportunity to make a mark upon history that had been denied him. Roosevelt was a man of wide emotional range, and he careened between extremes. War and its accompanying hysteria summoned up his spirit of sacrifice and patriotism, yet it stripped him of proportion and fairness. In his bitterness, he began to think about 1916, when, as the head of a reunited Republican party, he would drive Professor Wilson from the White House.

New Yorkers turning the pages of newspapers on the morning of May 1, 1915, found an unusual notice beside the Cunard Line advertisement announcing the sailing of the *Lusitania* that day. Placed by the German embassy in Washington, the notice warned passengers that the waters around the British Isles were a war zone where enemy vessels were subject to attack. Few people canceled their bookings and a Cunard official claimed the majestic liner was "perfectly safe; safer than the trolley cars in New York City." Had he been president when the Germans placed such a notice in the papers, Roosevelt later told a friend, he would have immediately summoned the German ambassador, given him his passport—the symbol of breaking off relations—and hustled him under guard on board the *Lusitania*.

Unrestricted submarine warfare was one of the dubious blessings of twentieth-century technology. Instead of ending quickly, as many had thought it would, the European conflict had bogged down into a bloody war of attrition, and each side was attempting to starve the other into submission. The Royal Navy swept German shipping from the high seas, and Germany struck back with a radical new weapon—the submarine. In February 1915, the Germans announced that the waters around Britain were now considered a war zone and enemy ships found there would be sunk without warning.

International law required warships to halt merchant ships suspected of carrying contraband, verify their identity and cargo, and provide for the safety of passengers and crew. These rules, the Germans complained, pertained to the age of sail. Enemy merchantmen were armed, submarines were fragile, and it would be suicidal for a U-boat captain to surface and give warning. The U.S. government warned Germany that it would be held "to strict accountability" if any American lives were lost. One after another, ships were sunk in the waters around Britain, but Americans continued to travel into the war zone in British vessels.

Shortly before two o'clock on the afternoon of May 7, 1915, Lieutenant Walter Schweiger of the German submarine U-20 was cruising off the southern coast of Ireland when he sighted a large, four-funneled steamer. Schweiger had general orders to sink any British vessel. "Clean bow shot from 700 meters range," he later reported. "Shot hits starboard side right behind bridge. An unusually heavy detonation follows with a very strong explosion cloud. . . . She has the appearance of being about to capsize. Great confusion ensues on board. . . . In the bow appears the name *Lusitania*." In little more than twenty minutes, the ship plunged bow first toward the bottom, taking with her 1,198 men, women, and children, 128 of them American citizens.

Roosevelt was in a courtroom in Syracuse for the trial of the libel charges brought against him by William Barnes when he received a sketchy report of the torpedoing of the *Lusitania*. The tragedy placed him in a dilemma. He knew that the newspapers would ask him for a statement; yet two members of the jury were of German origin and any condemnation of Germany might alienate them. He made no comment at the time, but that night he was awakened by a telephone call from a New York editor who gave him further details of the sinking. "That's murder!" he fairly shouted when he realized the full magnitude of the disaster. "Will I make a statement? Yes, yes, I'll make it now. Just take this," and he dictated:

> This represents not merely piracy, but piracy on a vaster scale of murder than old-time pirates ever practiced. . . . It is warfare against innocent men, women, and children. . . . It seems inconceivable that we can refrain from taking action in this matter, for we owe it not only to humanity but to our own national self-respect.

"Gentlemen," he told his lawyers the next morning, "I am afraid I have made the winning of this case impossible. But I cannot help it if we have lost the case. There is a principle at stake here which is more vital to the American people than my personal welfare is to me." The Colonel turned out to be wrong. Following a spectacular trial in which he was cross-examined by Barnes's attorneys for several days, the jury accepted his description of the political boss as justified.

That evening, May 8, Roosevelt checked with *Metropolitan* magazine,* for which he now wrote, to see what action the president had taken in retaliation for the sinking of the *Lusitania.* "Has he seized the ships?" he asked. When told that Wilson had not ordered a takeover of the German vessels interned in American ports, he exploded in outraged anger. The next day he prepared an editorial for the magazine entitled "Murder on the High Seas," which hotly demanded that Wilson take action to preserve American rights and self-respect. "I do not believe that the firm assertion of our rights means war," he declared, "but in any event, it is well to remember there are things worse than war."

Privately, Wilson was just as angered as the Colonel by the torpedoing of the *Lusitania,* but believed that the American people, although horrified by the ghastly event, were not ready for war. Envisioning America's great moral mission, he declared: "There is such a thing as a man being too proud to fight. There is such a thing as a nation being so right that it does not need to

---

*He had resigned from *The Outlook* at the end of the previous year and had signed a three-year contract with *Metropolitan* at an annual salary of $25,000, double what he had been getting previously.

convince others by force that it is right." These words were greeted with both cheers and denunciations—Roosevelt was contemptuous—and Wilson, realizing the words suggested that he had no backbone, withdrew them.

Five days later, on May 15, Wilson sent the first of a series of protest notes to Berlin. Written by the president himself, the note vigorously upheld the rights of Americans to sail the high seas and demanded an indemnity for the loss of American lives. The German reply, with its emphasis upon the fact that the *Lusitania* had been carrying war matériel, was unacceptable. Wilson followed up with a second note demanding the end of unrestricted submarine warfare. Bryan, who favored a more conciliatory policy, regarded this note as provocative and resigned on June 9, 1915, rather than sign it. Protest note followed protest note until eventually the Germans agreed to pay the indemnity—which was not forthcoming until the 1920s—and ordered U-boat skippers to spare unresisting passenger liners.

Bryan's resignation inspired Roosevelt to pledge his support for any steps the president might take "to uphold the honor and interests of this great Republic," but he was completely out of sympathy with Wilson's note-writing. It reminded him of "a man whose wife's face is slapped by another man, who thinks it over and writes a note telling the other man he must not do it, and when the other man repeats the insult and slaps his wife's face again, writes him another note of protest, and then another and another. . . . " Visiting her father at Sagamore, Alice Longworth reported that Wilson had dispatched a fresh note to Germany. "Did you notice what its serial number was?" he joked. "I fear I have lost track myself, but I am inclined to think it is No. 11,785, Series B."

The *Lusitania* affair transformed Roosevelt into an all-out interventionist. Until the declaration of unrestricted submarine warfare, he had based his attacks on Germany on the violation of Belgian neutrality. "I never wished to take part in the European war until the sinking of the *Lusitania*," he later explained. It also buoyed the cause of preparedness. Anti-German feeling was heightened and increasing numbers of Americans were now convinced that the Colonel was right: The victory of German absolutism would be a menace to the nation and to democratic institutions everywhere. Popular attention was focused upon the military and naval impotence of the United States. Fearing that the Republicans might seize control of the movement, even Woodrow Wilson reluctantly embraced the cause of preparedness. Preparedness, he now declared, was not a partisan cause but a national necessity.

Not everyone was for preparedness, however. Senator La Follette, Jane Addams, Bryan, and many other antimilitarists argued that preparedness was inspired by the munitions makers and Wall Street bankers who had made vast loans to the Allies. It was the first step down the slippery slope to war and would kill reform and curtail civil liberties. Besides, how could the Allies claim they were fighting for democracy when Russia was a member of the coalition? For his part, Roosevelt lashed out at the "professional pacifists," the "mollycoddles," and the "hyphenated-Americans," particularly the German-Americans, who hoped for a German victory. They were, he declared, "not Americans at all but Germans in America."

Under Leonard Wood's direction, summer camps had been established, most notably at Plattsburg, New York, on Lake Champlain, where young businessmen, professionals, and college students underwent military training. Three of Roosevelt's sons, Ted, Archie, and Quentin, enrolled to learn the elements of soldiering. Invited by his old friend, the Colonel came up to the camp in August 1915 to address the trainees. Knowing Roosevelt only too well, Wood looked over the speech and requested that he delete several passages critical of President Wilson. The Colonel was deep into an attack on pacifists, hyphenated-Americans, and those who favored "a policy of supine inaction," when a stray dog wandered up to him and rolled over on its back, eagerly waiting to be petted. "That's a very nice dog and I like him," Roosevelt ad-libbed. "His present attitude is strictly one of neutrality."

Roosevelt found public vindication in preparedness. Since the beginning of the war, Wilson had represented the mood of the majority of the American people—the desire to uphold the nation's honor but to keep out of the war—but now the Colonel's old popularity was reviving and he was looking forward to the presidential campaign of 1916. "Don't imagine that I wouldn't like to be at the White House this minute," he declared that winter.

The dream of a return to the White House was implausible but not impossible. Obviously the odds were against his getting the Republican nomination, but he thought popular sentiment was building for him. "My own judgment is that among the rank and file of the Republican voters . . . there is much more sentiment for me than for any other candidate," he wrote a friend. Now, as the major critic of Wilson's policy of neutrality and the leading advocate of preparedness, he even appealed to some elements of the financial and business communities and of conservative Republicans. What the Republicans needed, he believed, was a candidate progressive enough to appeal to the moderates and who was "right" on foreign policy and preparedness. None of the possibilities being mentioned—Charles Evans Hughes, who was now a justice of the Supreme Court, Elihu Root, and Senator Philander C. Knox—met these qualifications. Hughes was a liberal, but his foreign policy views were unknown, while Root and Knox were too conservative on domestic issues. There was but one possible candidate who filled the bill: Theodore Roosevelt.

Roosevelt's supporters undertook a publicity campaign to drum up support for him, including a four-page spread in the *Saturday Evening Post* headed "Why Roosevelt Would Be Our Best Guarantee of Peace," and sounded out Republican politicians all across the country. In state after state, the answer was the same: The Old Guard was firmly against Roosevelt and while many others favored the Colonel, they were committed to various favorite sons or preferred a "safe" candidate like Hughes. The situation was bleak, but Roosevelt had his own strategy. The Progressive party would be kept together to blackmail the Republicans with the threat of a third-party candidacy, while he concentrated on Americanism and preparedness and avoided a fight with other potential candidates by refusing to enter the primaries.

Admittedly, this was a tricky strategy, but the Colonel believed it would provide time for the politicians to realize that he was the only candidate who could appeal to reformers and conservatives alike. But there was real doubt whether he could bring the Progressives back into the Republican party or that the Republicans would nominate him. Republican professionals believed it was more important to keep control of the party than to win a single election. They were perfectly willing to see another Wilson victory rather than take a chance on Roosevelt. Besides, in Hughes they had a candidate with whom they were comfortable.

On March 9, 1916, while on winter vacation with Edith in the Caribbean, Roosevelt issued a statement that was both a denial of his candidacy and an announcement of his availability. "I will not enter into any fight for the nomination and I will not permit any factional fight to be made on my behalf," he declared. "Indeed I will go further and say that it would be a mistake to nominate me unless the country has in its mood something of the heroic— unless it feels not only devotion to ideals but . . . to realize these ideals in action." Critics claimed this was an invitation to vote for him—and to go to war.

Roosevelt spent the last weeks before the conventions preaching the cause of preparedness, belaboring pacifists and Wilson in the Midwest, the center of isolationism and antiwar sentiment. In Detroit, he gave a rousing speech in favor of universal military service, which was interrupted by a woman who stood in the balcony, American flag in hand, to announce: "I have two sons. I offer them." Amid thunderous applause, Roosevelt replied: "Madam, if every mother in the country would make the same offer, there would be no need for any mother to send her sons to war." Fifty thousand people cheered him in Kansas City, and then he was off to St. Louis, the center of German-American influence, where his attack on pro-German sentiment was hardly likely to endear him to German-American delegates to the Republican convention. Whatever chance he may have had for the nomination—if there ever was one—was killed by this tour.

Both the Republican and Progressive party conventions opened on June 7 in Chicago, an idea broached by Roosevelt supporters to buoy the chances of a joint nomination. The Colonel himself remained at Sagamore, keeping in direct communication with the two conventions by telephone. Efforts were made to arrive at similar platforms to make a coalition easier, but the Republicans, bowing to antiwar sentiment, watered down planks calling for a stronger army and navy and universal military service. The Progressives, placing their devotion to Roosevelt above their pacifist convictions, endorsed strong preparedness planks.

The Republicans were determined, however, to nominate anyone but Roosevelt, while the Bull Moosers would nominate no one else. To break the deadlock, Roosevelt proposed Cabot Lodge—undoubtedly surprised to find himself designated a Progressive—as the joint nominee—but he was rejected by the Bull Moosers. The Progressive delegates were now eager to nominate the Colonel by acclamation, but were kept on a tight leash until the Republicans had acted. Roosevelt, who had already resolved not to run on a third-party ticket, wanted to make sure that the Republican nominee would not be an embarrassment before he pulled out of the race.

Roosevelt's nomination was greeted by a thirty-six-minute demonstration—the longest at the Republican convention—but Hughes pulled well ahead on the first two ballots and appeared likely to be nominated on the next ballot. Bedlam reigned at the Progressive convention across town. Many of the delegates had been willing to accept fusion with the Republicans if the Colonel were the nominee, but they refused to join a coalition on any other terms. Shortly after noon on June 10, just as the Republicans were nominating Hughes, the defiant and cheering Bull Moosers shouted through the choice of Roosevelt as their candidate. The cheers were still echoing through the hall when a telegram arrived from Sagamore Hill: I AM VERY GRATEFUL FOR THE HONOR YOU CONFER ON ME BY NOMINATING ME AS PRESIDENT. I CANNOT ACCEPT IT AT THIS TIME. . . .

"For a moment there was silence," reported William Allen White. "Then there was a roar of rage. It was the cry of a broken heart such as no convention ever had uttered in this land before. Standing there in the box I had tears in my eyes, I am told. I saw hundreds of men tear the Roosevelt picture or the Roosevelt badge from their coats and throw it on the floor."

"Well, the country wasn't in a heroic mood!" Roosevelt wrote Bamie following the conventions. "We are passing through a thick streak of yellow in our national life." Once Hughes pledged himself to preparedness and progressivism, the Colonel endorsed the Republican nominee. Most of the Bull Moosers followed Roosevelt back into the Republican camp, but others opted for Wilson, who was renominated by the Democrats. The Progressive party itself was left dying on the floor of the Chicago auditorium.

With his old fire unabated, Roosevelt campaigned strenuously for Hughes, whom he privately derided as "the Bearded Lady," but the real driving force was his hatred of Wilson. The Wilsonites of 1916 were one with the Tories of 1776 and the Copperheads of 1864. He denounced the president's failure to declare war on Germany following the sinking of the *Lusitania* as "weak" and "cowardly," and derided Wilson's "too proud to fight" declaration. The cautious Hughes, who had stepped down from the Supreme Court, would not go that far, but said he would have been "firmer" in dealing with the Germans. Banking on the fact that a majority of Americans were still opposed to entering the conflict, the Democrats campaigned with the slogan "He kept us out of war."

"I am now being worked to the limit by the Hughes people," the Colonel wrote Corinne during the height of the campaign, "the very people who four months ago were explaining that I had no 'strength.' " Large crowds came out to Sagamore to see him, but as one of the accompanying reporters, Edwin N. Lewis, observed, they were there for Roosevelt, not for Hughes or the Republican party. Lewis also noted that the Colonel, who was approaching his fifty-eighth birthday, appeared much older. "At times in the thick of the excitement, an expression of fatigue flashes across his features. There is a touch of sadness, too, I believe, in his face, as he looks out over these crowds. . . ."

Roosevelt was at his most emotional in his final speech of the campaign at Cooper Union in New York City on November 3. Toward the end, he tossed aside his text and noted that Wilson was resting at Shadow Lawn, the vacation White House:

> There should be shadows now at Shadow Lawn; the shadows of the men, women and children who have risen from the ooze of the ocean bottom and from graves in foreign lands; the shadows of the helpless whom Mr. Wilson did not dare protect lest he might have to face danger; the shadows of babies gasping pitifully as they sank under the waves. . . . Those are the shadows proper for Shadow Lawn; the shadows of deeds that were never done; the shadows of lofty words that were followed by no action; the shadows of the tortured dead.

As Election Day, November 7, 1916, dawned, bookmakers were laying odds of ten to six in favor of Hughes. Roosevelt was less confident of a Republican victory. He believed that Hughes had failed to make "a straight-from-the-shoulder fighting campaign" and was not progressive enough to attract western Bull Moosers. Nevertheless, early returns indicated a Republican sweep, and the pro-Wilson *New York World* conceded defeat. Roosevelt expressed his "profound gratitude as an American proud of his country that the American people had repudiated the man who coined the phrase about this country that it is 'Too proud to fight.' "

By morning everything was topsy-turvy.* Returns from the midwestern and far western states, where progressive and antiwar feelings were strongly concentrated, swung the pendulum back toward the president. Several days were required to sort out the returns, and it was finally determined that Wilson had carried California and its thirteen electoral votes by fewer than four thousand votes. Wilson had 277 electoral votes to 254 for Hughes, and led in the popular vote by 9,129,606 to 8,558,321—a margin of approximately 600,000 votes.

Brooding in the winter grayness of Sagamore following Wilson's narrow victory, Theodore Roosevelt thought perhaps he had made a mistake in 1912 by running against Taft. Had he remained loyal to the Republican party, had he restrained his restless spirit for four years, he would have been the nominee in 1916 and he could have been elected. Now embittered, he told Arthur Lee that "I am completely out of sympathy with the American people." The election was the triumph of pacifism over patriotism— "This is yellow, my friend! plain yellow!"—and he saw himself becoming "an elderly literary gentleman of quiet tastes and an interesting group of grandchildren."

There were still battles to be fought and no matter what the result of the election, the Colonel was constitutionally unable to

---

*There is a story, perhaps apocryphal, that Hughes had gone to bed believing he was elected. When the returns began running toward Wilson, a reporter called the Hughes home, only to be told that the "President-elect" was asleep and could not be disturbed. "Okay," replied the newsman, "when he wakes up tell him he isn't President-elect anymore."

give up the struggle. When Lee suggested he come to England and write a series of magazine articles on the British war effort, he replied that the only way he would return to Europe was at the head of a division of American troops.

Events took an unexpected turn that rendered this a not unrealistic prospect. Shortly after the election, Wilson launched a diplomatic offensive aimed at ending the war, which was climaxed by a speech before Congress on January 22, 1917, that called for "peace without victory." Roosevelt was incensed by what he considered Wilson's failure to blame Germany for starting the war. "Any peace which does not mean victory over those responsible for these outrages, will set back the march of civilization," he declared.

Nine days later, the Germans delivered a body blow to Wilson's peace efforts. In a desperate attempt to starve Britain into surrender, they announced that unrestricted submarine warfare would be resumed against all neutral and belligerent shipping found in the war zone. Each week, a single American ship would be permitted to sail a specified course through the war zone, provided it carried no contraband and was painted with wide red and white stripes. No independent nation could tolerate these conditions. The German High Command fully realized that the United States would be brought into the war on the Allied side, but gambled on winning the conflict before American resources could be mobilized. On February 3, the German ambassador was given his passport and relations between the nations were severed.

Hardly had this shock subsided than British intelligence provided Washington with the decoded text of a secret cable sent by the German foreign minister, Arthur Zimmermann, to the German envoy in Mexico instructing him to offer the Mexicans the return of all territory lost after the Mexican war in exchange for their attacking the United States in case it declared war against Germany. When reporters told Roosevelt of the Zimmermann telegram, he let loose a fusillade of profanity. "I don't apologize . . . this man [Wilson] is enough to make the saints and the angels, yes, and the apostles, swear, and I would not blame them," he said. "My God, why doesn't he do something?"

For another month, Wilson agonized over a declaration of war, holding on to a forlorn hope of peace, and talked of "armed neutrality." The Germans responded by torpedoing three American vessels, with heavy loss of life. In February, the czarist regime in Russia was overthrown and now the Allies could claim to be fighting for democracy. By this time, Roosevelt's scorn for Wilson knew no bounds, but he refrained from a public attack for he was bombarding Newton D. Baker, the secretary of war, with requests for permission to raise a volunteer division when war came. Finally, on the evening of April 2, 1917, the president went before a joint session of Congress to ask for a declaration of war against Germany. "The world," he declared, "must be made safe for democracy."

"The President's great message of April 2 was literally unanswerable," Roosevelt wrote in the *Metropolitan*. "Of course, when a war is on, all minor considerations, including all partisan considerations, vanish at once. All good Americans will back the President with single-minded loyalty. . . ." Having extended this olive branch, the Rough Rider sought a meeting with Wilson in a desperate attempt to get into action, even though he was fifty-nine years old, blind in one eye, and weakened by tropical fever.*

Outwardly cordial, the longtime rivals were wary as they met on April 10 in the Red Room of the White House. "Mr. President, what I have said and thought, and what others may have said and thought, is all dust in a windy street," the Colonel later related that he had assured Wilson. "Nothing could have been pleasanter or more agreeable," reported Joseph P. Tumulty, the president's secretary. The two men met "in the most friendly fashion, told each other anecdotes, and seemed to enjoy what the Colonel was accustomed to call 'a bully time.' "

They also sparred. Roosevelt observed that the president's war message would rank "with the great state papers of Washington and Lincoln" if it were made good. Wilson shot back that his desk was piled high with petitions from former Indian fighters, Texas Rangers, southern "colonels," and other military adventurers. He said the nation would rely upon a conscript army rather

---

*Upon being told by Roosevelt that he did not expect to return alive from France, Elihu Root observed in his old teasing manner that if Wilson were assured of that, permission for him to go overseas would be immediately forthcoming.

than on a volunteer force. Roosevelt agreed to support the draft—
or selective service, as it was to be called—but emphasized that
a volunteer division could be organized and sent overseas quickly,
would show the British and French "what was on the way," and
would be good for morale at home and abroad.

"Yes," Tumulty quoted Wilson saying after Roosevelt had left,
"he is a great big boy. I was . . . charmed by his personality. There
is a sweetness about him that is very compelling. You can't resist
the man."

But there was to be no Roosevelt Division. Wilson rejected his
appeal on grounds that recruitment of such a unit would interfere
with the draft. There was no room in modern warfare for romance
and glory and gallant charges. Perhaps this decision was based
on personal and partisan factors such as a desire for revenge and
the refusal to give the Colonel a leg up for the 1920 presidential
election, but there certainly was logical support for it. Roosevelt's
service in Cuba hardly qualified him to command a division, he
was no team player, and an elite division would drain away officers
and training cadres needed elsewhere.

On the other hand, there were substantial arguments in favor
of the Roosevelt Division. Roosevelt was probably no less competent than some of the generals eventually sent to France; he
offered to go in a subordinate position rather than as a division
commander; he was willing to forgo the list of officers he initially
presented, and all the troops sent to Europe required further
training in France before going into the front lines. Most persua-
sive of all was the argument that the early arrival of American
troops led by Roosevelt would have stimulated Allied morale and
the will of people to keep fighting on.

No one understood this better than Georges Clemenceau, the
French premier. "The name of Roosevelt has legendary force in
this country," he wrote Wilson in a letter that he made public,
and urged the president to "send . . . Roosevelt." The Colonel
took Wilson's rejection hard; no matter what the official reason,
nothing could persuade him that Wilson was not "playing the
dirtiest and smallest politics," and he carefully nursed his grudge.*

---

*Had Wilson given the matter more thought, he might have offered Roosevelt some
seemingly important yet relatively powerless position such as chairman of the Liberty

Unable to fight himself, the old lion sent his cubs to do battle. "It's rather up to us to practice what Father preaches," said Quentin. Ted and Archie were commissioned in the U.S. Army; Kermit, unwilling to wait to fight, became a captain in the British Army and was assigned to a motorized machine-gun unit during the fighting in Mesopotamia (now Iraq). Quentin, who had trouble with his eyes, memorized an eye-examination chart and was accepted for flight training in the fledgling air service. Ethel became a nurse in the same hospital in France where her husband was surgeon, while Ted's wife, Eleanor, was the first female YMCA member to arrive in the fighting area. Shortly afterward, it was announced that the president's son-in-law was joining that organization. "How very nice," remarked the Colonel mischievously. "We are sending our *daughter-in-law* to France in the YMCA!"

One by one they all went off to war—for Edith the parting with Quentin, at nineteen "her baby," was the most difficult—and the night before he left she came into his room after he fell asleep to tuck him into bed. Now only she and her husband were left in the sprawling house at Oyster Bay with its flag bearing five service stars. "They have all gone away from the house on the hill," she wrote sadly.

Complete confusion reigned in Washington as the nation mobilized for war. Lulled by his dreams of peace, Wilson had, despite Roosevelt's warnings, failed to prepare the nation for the conflict. It was assumed that the United States would provide only financial and material assistance to the Allies and perhaps a small naval force, but the exhausted Allies demanded men and more men. Bottlenecks held up industry's conversion to the production of military equipment, and the gears of the massive war machine needed to crush Imperial Germany still ground with confusion and friction.

It was a situation made to order for Roosevelt. Although frequently in pain from fever and the abscesses in his leg and ear, he threw himself into a campaign to stimulate the American war effort. If he could not be with his sons overseas, he could at least

drives, which he undoubtedly would have accepted. Once in service, he would have been kept from attacking the administration during the war.

urge Americans to work harder in behalf of victory. While he found little comfort in his role of "slacker-in-spite-of-myself" and feared being regarded as "merely a scold," he may have performed a greater service at home than had he gone to France. Beginning in September 1917, he toured the country, demanding speedy and vigorous prosecution of the war. He called upon Americans to support the Red Cross and to buy Liberty bonds and bought $60,000 worth himself, which was a substantial part of his capital. When the men of Archie's company needed shoes, he sent them two hundred pairs. In countless speeches and in a syndicated weekly column in the *Kansas City Star,* he extolled an all-out, unified effort to win the war. "This is the people's war," he declared to audiences all over the country.

There was a downside to Roosevelt's activities, however. While his attacks on Wilson for the slowness in preparing American troops for battle and the inadequacy of their equipment served a useful purpose, his theme of "one-hundred per cent American-ism" translated into a rabid condemnation of pacifists, consci-entious objectors, and socialists. "In this war," he proclaimed, "either a man is a good American, and therefore is against Ger-many, and in favor of the allies of America, or he is not an American at all." He opposed the publication of German-language newspapers. "We must have one language, the language of the Declaration of Independence, of Washington's farewell address, and of Lincoln's great speeches. . . ."

Though Roosevelt tried to make it clear that his attacks on German-Americans were limited to those who were disloyal, his audiences were unable to make such distinctions, and his speeches helped unleash a wave of anti-German feeling. On the other hand, he vigorously condemned a race riot that flared in East St. Louis, Illinois, in July 1917. "We have demanded that the Negro submit to the draft and do his share of the fighting exactly as the white man does," he said. "Surely when such is the case, we should give him the same protection by the law, that we give to the white man. . . . Murder is not debatable." President Wilson remained silent.

Once again, Roosevelt found himself in favor with the Amer-ican people, and by 1918 he was regarded as the front runner for the Republican presidential nomination in 1920. When he visited

Alice and Nick Longworth in Washington that winter, he was mobbed by newsmen. The Longworths' home became the rallying point for those Republicans, and some Democrats, who were dissatisfied with Wilson's conduct of the war. When it was suggested the Republican convention would nominate Roosevelt by acclamation, Boss Barnes, who had sued him for libel, interjected: "Acclamation hell! We'll nominate him by assault."

"By George, if they take me, they'll have to take me without a single modification of the things that I have always stood for!" Roosevelt declared. The old progressive fire blazed forth again. No one should be allowed to profit from the war, he declared, and he supported an even higher gross-profits tax than the administration advocated. Similarly, he spoke out against inequities in the draft. Young men with money enough to attend college should not be deferred to leave the poor to fight the war. He also presented a revised version of the Progressive party platform of 1912 geared to the postwar world.

Speaking in March 1918 at the Maine Republican convention, he called for domestic reforms that included old age pensions, sickness and unemployment insurance, public housing projects, the reduction of working hours, aid to the farmer, and the regulation of large corporations. This agenda was far ahead of anything yet announced by Wilson. "I wish to do everything in my power to make the Republican Party the Party of sane, constructive radicalism, just as it was under Lincoln," he informed William Allen White. "If it is not that then of course I have no place in it."

Roosevelt's work in mobilizing the country and his attacks on Wilson's muddling conduct of the war established a common bond with William Taft, who also was critical of his successor, and they resumed their correspondence. Once again it was "Dear Will" and "Dear Theodore." In May 1918, Taft checked into the Blackstone Hotel in Chicago and learned that Roosevelt was in the dining room. Locating the Colonel at a table across the room, he walked over to him. Roosevelt was intent on his meal, but sensing a sudden hush, looked up and saw Taft's smiling face looming over his table. Immediately throwing down his napkin, he rose, hand extended. They shook hands vigorously and eagerly slapped each other on the back. The other guests applauded, and suddenly realizing they had an audience, the two ex-presidents bowed and smiled. Then they sat down and chatted for a half hour.

One of the topics was certainly the war records of their offspring. Roosevelt was proud of his sons, and young Charlie Taft had refused a commission and was serving as a private at the front. But the Colonel's pride was mixed with anxiety. Perhaps because of a premonition of what was to come, he lavished attention on his eight grandchildren. "I can't say how much I have enjoyed them," he wrote Ted's wife, Eleanor, after her children visited him. "Gracie is the most winning little thing I have ever known. . . . Ted's memory was much clearer about the pigs than about me; he greeted me affably but then inquired of a delighted bystander . . . 'What is that man's name?' "

The burden of having his sons in peril and not being able to take part in their adventures caused him to work beyond his endurance and, approaching sixty, he suddenly began to age. "I wake up in the middle of the night," he said, "wondering if the boys are all right, and thinking how I could tell their mother if anything happened." Underneath the usual exuberance, he was plagued by rheumatic fever. He often felt exhausted. "I feel as though I were a hundred years old and had *never* been young," he said at one point. A twilight chill fell across his life: Nannie Lodge, and then Cecil Spring Rice, and finally Henry Adams all died within a few months of each other. In February 1918, he was taken to Roosevelt Hospital in New York for an emergency operation on the abscesses of his leg and thigh. Rumors spread that he was dying, but despite complications he recovered. Uncharacteristically, he remained in the hospital for a month, indicative of the seriousness of his illness. When he emerged he had trouble with his balance, and he had lost the hearing in his left ear. He was now half-deaf as well as blind in one eye.

"Here, spring is now well underway," Roosevelt wrote Quentin in April 1918. "The woods are showing a green foam; the gay yellow of the forsythia has appeared. . . ." On the western front, spring meant a renewal of violence and the Allies were facing disaster. The Bolsheviks took Russia out of the war. The French Army was still shaky after a widespread mutiny the previous year.

Four hundred thousand British troops fell in vain attacks on the German lines, and the survivors were exhausted. But American troops were beginning to take their place at the front.

Archie was severely wounded in the left leg and arm while leading an attack and was awarded the Croix de Guerre. At first, it was feared that his leg would have to be amputated, and then it was found that a nerve in his arm had been severed, but he recovered. "You don't know how proud we are of you, how our hearts go out to you," his father wrote him. Before the summer was over, Ted was wounded and also gassed, winning the Distinguished Service Cross and the Silver Star.

Quentin was much on the minds of his parents. Before going overseas, he had become engaged to Flora Payne Whitney, and they now wanted to get married, but the War Department denied her permission to go to France. Although the Colonel pulled strings in an effort to have this "idiotic ruling" suspended, it was to no avail. "It is wicked; she should have been allowed to go, and then marry Quentin," he told Bamie. "Then even if he were killed she would have known their white hour."

On July 5, Quentin was in his first dogfight. "You get so excited," he wrote his mother, "that you forget everything except getting the other fellow and trying to dodge the tracers, when they first start streaking past you." A few days later he shot down his first German plane. "Of course we are immensely excited by the press reports of Quentin's feat," the Colonel told Ethel. "Whatever now befalls Quentin he has now had his crowded hour. . . ."

Roosevelt was in the library dictating to his secretary, Josephine Stricker, late in the afternoon of July 16 when Phil Thompson, the Oyster Bay stringer for the Associated Press, asked to see him. Thompson was puzzled by a telegram to the *New York Sun:* WATCH SAGAMORE HILL FOR————. That was all. The rest had apparently been censored. The Colonel was immediately filled with dread, and fearing that Edith would hear, closed the door. "Something has happened to one of the boys," he surmised. It could not be Ted or Archie because they were recovering from wounds; nor could it be Kermit, who had recently transferred to the U.S. Army, because he was down with malaria. It had to be Quentin. Lacking information, there was nothing to do but wait.

In the meantime, he asked the newsman to say nothing to his wife.

That evening, the Colonel's iron control of his emotions was put to its severest test. He dressed for dinner as usual, then he and Edith talked and read in the North Room as he tried to maintain normality. Thompson returned the following morning before breakfast, and the look on his face said everything. Roosevelt led him out onto the porch and was told that unofficial dispatches reported that Quentin had been shot down behind the German lines. Separated from his flight, he had been jumped by seven red-splashed Fokkers. For a few minutes, the Colonel paced up and down, trying to calm himself. "But . . . Mrs. Roosevelt!" he cried out. "How am I going to break it to her?"

Returning to the house, he remained inside for a half hour while Thompson waited on the porch for a statement. When the Colonel reappeared, he had a paper in his hand. The statement consisted of a single line: "Quentin's mother and I are very glad that he got to the front and had a chance to render some service to his country, and to show the stuff that was in him before his fate befell him." Three days later, the German government confirmed Quentin's death and reported that he had been buried with full military honors near where he had fallen.

In public, the Roosevelts were self-controlled, but memories of their youngest son stayed with them. "Mother will carry the wound green to her grave," the Colonel wrote Kermit. For a while, Quentin's letters continued to arrive, and once his favorite nephew, Ethel's son Richard, heard a plane overhead and said, "Perhaps that's Uncle Quentin." While Roosevelt was riding on a train, he was seen to put down his book and stare vacantly at the horizon and murmur, "Poor Quinikens!" Hermann Hagedorn observed that with Quentin's death, the "boy in him had died." Paying tribute to his son, Roosevelt wrote: "Only those are fit to live who do not fear to die; and none are fit to die who have shrunk from the joy of life and the duty of life. Both life and death are part of the same Great Adventure. . . ."

With fresh American troops pouring into the front, the tide of battle shifted to the Allies during the summer of 1918, and Roosevelt's own efforts turned to winning the peace. New York Republicans wanted him to run for governor that autumn, but he refused on grounds that he was too absorbed in national and international issues to become involved in a state campaign. But to Corinne he confided: "I have but one fight left in me. . . . I think I should reserve my strength in case I am needed in 1920." Although Roosevelt's store of energy was limited, Wilson's partisan call for a Democratic Congress to give him a free hand in the peace negotiations drew Roosevelt into the 1918 congressional campaign. He aimed to administer a sound defeat to the president and his plans for "peace without victory."

When Wilson announced his Fourteen Points as a basis for peace negotiations, the Colonel attacked them as nothing more than "Fourteen Scraps of Paper" and urged Congress to pass a resolution demanding the unconditional surrender of Germany. Unconditional surrender would "safeguard the world for at least a generation to come from another attempt by Germany to secure world dominion," he declared on the eve of the armistice. As for the postwar world order, he was cautiously affirmative toward Wilson's League of Nations but with reservations that would prevent any international body from overriding such fundamental principles of American nationalism as the Monroe Doctrine. Later, after Roosevelt's death, it was Wilson's messianic refusal to accept similar modifications to the League that contributed to it being rejected by the Senate.

On October 28, the day after his sixtieth birthday, Roosevelt addressed a rabidly Republican rally at Carnegie Hall in which he denounced Wilson in some of the strongest language he had ever used. Despite his body being wracked with pain, he went on for ninety minutes before asking if he had spoken too long. "Go on!" shouted the audience. "Go on!" Wilson was ignoring Republican counsel in the making of the peace, he thundered. "We can pay with the blood of our hearts' dearest, but that is all we are to be allowed. . . . The peace the nation is to get for that price" was to be made by "men of cold heart who did not fight themselves, whose nearest kin are not in danger, who prepared for war not at all, who helped wage the war feebly. . . ."

Plagued with inflammatory rheumatism and with his feet swollen so badly he could not wear shoes, he was ordered to bed a few days later. On November 5, he disobeyed orders and hobbled out to vote. Wilson was repudiated at the polls and the Republicans won both houses of Congress. Roosevelt hailed this as a "stinging rebuke" in a letter to Ted and observed that if the United States had been a parliamentary democracy, Wilson would have been thrown out of office. He was in such pain, however, that he had to dictate the letter to Edith, who added a postscript: "Father is flat on his back with his gout. . . . He is having a horrid suffering time." On November 11, 1918, the day on which Germany surrendered, he again entered Roosevelt Hospital.

For the next seven weeks, the Colonel remained in the hospital, while Edith occupied an adjoining room so she could be with him. At one stage the rheumatism in his left arm was so bad the arm had to be bound to a splint. Nevertheless, he kept up a brisk correspondence with friends and political acquaintances, produced his weekly column for the *Kansas City Star,* and doggedly continued his fight with Wilson. When the president left for the Paris peace conference early in December, the imprecations of the Rough Rider resounded around his head. "Mr. Wilson has no authority whatever to speak for the American people. His leadership has just been emphatically repudiated by them. . . ."

One day while Corinne, who had recently been widowed, was sitting with him, he mentioned his sixtieth birthday. "Well, anyway, no matter what comes, I have kept the promise I made to myself when I was twenty-one." What was that? asked his sister. "I promised myself," he replied as he banged a fist on the edge of the chair, "that I would work *up to the hilt* until I was sixty, and I have done it." Even if he were to be an invalid or even if he died—here he snapped his fingers in dismissal—"what difference would it make?"

Republican leaders beat a path to his room to sound him out for the 1920 presidential nomination. While Roosevelt vowed that he would not actively seek the nomination, he told Joseph Bishop that "if the leaders of the party come to me and say that they are convinced that I am the man the people want and the only man who can be elected, and that they are all for me, I don't see how I could refuse to run." Before releasing the Colonel on Christmas Day, the doctors warned that he might be confined to a wheelchair for the remainder of his life if he did not slow down. "All right!" he replied. "I can work that way, too!"

*   *   *

But Roosevelt's own Great Adventure was rapidly coming to an end. Ethel, Alice, and Archie, who had been invalided out of the service because of his wounds, were at Sagamore to greet him and the Christmas table was set. Everyone tried to be gay, but a sense of hidden sadness hung over the gathering. In the week after being released from the hospital, Roosevelt browsed among his books, gazed at mementos of his career, dictated letters, and worked on his column. Dr. George Faller, the family physician, made two calls a day and found him looking ruddy and as sturdy as ever. But he tired easily, his joints were stiff, and he spent the afternoons resting on a sofa. At night, he slept in Ethel's former room, which had a connecting door with the master bedroom, where Edith slept.

On Friday, January 3, 1919, he prepared a strongly worded message on Americanism for delivery at a benefit concert for the American Defense League, which would be held that Sunday at the Hippodrome in New York. "We have room for but one flag, the American flag," he wrote. "We have room for but one language here and that is the English language, for we intend to see that the crucible turns out our people as Americans . . . and not as dwellers in a polyglot boarding house. . . ."

The following day, Edith telephoned James Amos, Roosevelt's former White House valet who was living in New York City, and asked him to come out to help. The Colonel needed a manservant and would have no one else. Amos arrived that evening and was saddened by his old employer's appearance. "His face bore a tired expression. There was a look of weariness in his eyes. It was perfectly plain that he had suffered deeply." He bathed and dressed the Colonel in clean pajamas and helped him into a chair where he could look out over Long Island Sound.

On Sunday, for the first time since his return from the hospital, Roosevelt remained in the little corner bedroom, resting on the sofa rather than dressing and going downstairs. He received a visit from Flora Whitney, corrected the proofs of a *Metropolitan* article, and finished a *Star* column supporting several reservations to the League of Nations proposal by Wilson. As dusk fell, he "watched the dancing of the waves, and spoke of the happiness of being home, and made little plans for me," Edith wrote Ted.

"I think he had made up his mind that he would have to suffer for some time to come and with his high courage had adjusted himself to bear it." When she rose to go, he put down the book he was reading and told her, "I wonder if you will ever know how I love Sagamore Hill."

That evening, Roosevelt complained of being unable to breathe. The trained nurse employed by the family summoned Dr. Faller. He found nothing amiss and gave his patient some medication to help him sleep. Shortly before midnight, as the Colonel was helped into bed by Amos, he asked, "James, will you please put out the light?" Putting out a small lamp on the dresser, Amos settled down before the fire, while Roosevelt turned on his side and soon fell asleep. At about four o'clock on the morning of January 6, Amos became alarmed by Roosevelt's irregular breathing. He summoned the nurse, but when they returned to the bedroom, the Colonel had stopped breathing altogether.

Edith was awakened and rushed to her husband's bedside. "Theodore, darling!" she called softly, leaning over him. "Theodore, darling!" But there was no reply. He seemed "just asleep, only he could not hear." Dr. Faller was called and pronounced Roosevelt dead of an embolism. He had worked up to the end. On the bedside table was a penciled note to tell Will Hays, the chairman of the Republican National Committee, to go to Washington to "see Senate & House; prevent split on domestic policies."

"The old lion is dead," Archie cabled his brothers.

Epilogue

On the night of April 30, 1889, New York City marked the centennial of the inauguration of George Washington with a lavish dinner at the Metropolitan Opera House. Both the speeches and toasts had been numerous and the guests streamed out eagerly into the crisp evening air. Most immediately headed homeward, but one guest noticed a youthful figure standing on the sidewalk, talking with a circle of friends in a vigorous, eager voice that cracked into a falsetto with excitement. Periodically, he raised an arm and pointed skyward.

"It was young Roosevelt," the passerby noted. "He was introducing some fellows to the stars."

Appendix

Books by Theodore Roosevelt

1882	*The Naval War of 1812*
1885	*Hunting Trips of a Ranchman*
1887	*Thomas Hart Benton*
1888	*Essays on Practical Politics*
1888	*Gouverneur Morris*
1888	*Ranch Life and the Hunting Trail*
1889–1896	*The Winnings of the West,* 4 vols. (vols. 1 and 2, 1889; vol. 3, 1894; vol. 4, 1896)
1891	*New York*
1893	*The Wilderness Hunter*
1893	co-editor with George Bird Grinnell and contributor, *American Big-Game*
1895	co-author with Henry Cabot Lodge, *Hero Tales from American History*
1895	co-editor with George Bird Grinnell and contributor, *Hunting in Many Lands*
1897	*American Ideals*
1897	*Some American Game*
1897	co-editor with George Bird Grinnell and contributor, *Trail and Campfire*
1899	*The Rough Riders*
1900	*Oliver Cromwell*
1900	*The Strenuous Life*
1902	co-author with T. S. Van Dyke, D. G. Elliott, A. J. Stone, *The Deer Family*
1905	*Outdoor Pastimes of an American Hunter*
1907	*Good Hunting*
1909	*Outlook Editorials*
1910	*African and European Addresses*
1910	*African Game Trails*
1910	*American Problems*
1910	*The New Nationalism*

1910	*Presidential Addresses and State Papers and European Addresses,* 8 vols.
1912	*The Conservation of Womanhood and Childhood*
1912	*Realizable Ideals*
1913	*Autobiography*
1913	*History as Literature and Other Essays*
1913	*Progressive Principles*
1914	co-author with Edmund Heller, *Life-Histories of African Game Animals*
1914	*Through the Brazilian Wilderness*
1915	*America and the World War*
1916	*A Book-Lover's Holidays in the Open*
1916	*Fear God and Take Your Own Part*
1917	*The Foes of Our Own Household*
1917	*National Strength and International Duty*
1918	*The Great Adventure*

The *Memorial Edition* (24 vols.) and *National Edition* (20 vols.) of the *Works of Theodore Roosevelt,* edited by Hermann Hagedorn and published by Charles Scribner's Sons in 1923–1926, are the most complete collections of Roosevelt's writings and supersede the Sagamore, Elkhorn, Allegheny, and other multi-volume collections that were published in TR's lifetime. But no collection includes all of Theodore Roosevelt's published books, articles, reviews, essays, and speeches. All of Roosevelt's published writings are listed in the *Theodore Roosevelt Collection: Dictionary Catalogue and Shelf List,* prepared by Gregory C. Wilson, 5 vols., Cambridge, Mass: Harvard University Press, 1970.

Notes

Prologue

The basic sources for this section are the *Washington Evening Star, Baltimore Sun,* and *The New York Times* and *New York World* for March 5, 1905. "Tomorrow I shall," quoted in Mowry, *The Era of Theodore Roosevelt,* p. 197; Morley is quoted in Bishop, *Theodore Roosevelt and His Times,* Vol. I, p. 338; Hay's gift of the Lincoln ring, in Morison, ed., *The Letters of Theodore Roosevelt,* Vol. IV, p. 1131; for a physical description of TR, see Hale, *A Week in the White House,* pp. 14–16; TR's inaugural speech, in *Messages and Papers of the Presidents,* Vol. XIV, pp. 6930–6932; "Everybody," in a letter to George Otto Trevelyan, *Letters,* Vol. IV., p. 1133; for children's activities and Alice Roosevelt's comments, see Longworth, *Crowded Hours,* p. 67; "I never expected," Eleanor Roosevelt, *This Is My Story,* p. 123; for the Roosevelt family on inauguration night, see S. J. Morris, *Edith Kermit Roosevelt,* p. 284; "How I wished," in a letter to Robert Barnwell Roosevelt, *Letters,* Vol. IV, p. 1131.

Chapter 1
"Teedie"

For TR, Sr., and his children, see Theodore Roosevelt, *Autobiography,* pp. 7–14, and Robinson, *My Brother Theodore Roosevelt,* pp. 3–7; "We all craved him," Robinson, *op. cit.,* p. 7; for a description of New York City at the time of TR's birth see Lloyd Morris, *Incredible New York,* pp. 3–15; Wharton quotation is from *The Old Maid,* in *The Edith Wharton Omnibus,* p. 427; for background on TR, Sr., see Miller, *The Roosevelt Chronicles,* pp. 138–142, and Anna Roosevelt Cowles, *Reminiscences;* description of Mittie, *ibid.;* TR's comments on his ancestry, *Autobiography,* pp. 1–2; for the most recent and complete history of the Roosevelt family, see Miller, *op. cit.;* "Old Fifty-seven varieties," E. Morris, *The Rise of Theodore Roosevelt,* p. 22; "I can see him now," quoted in Putnam, *Theodore Roosevelt,* p. 44; "My father was," *Autobiography,* pp. 7–8; for the self-created aspects of TR's character, see Wagenknecht, *The Seven Worlds of Theodore Roosevelt,* pp. 2, 3, 23, and 157; for a biography of Anna Roosevelt Cowles, see Rixey, *Bamie;* also see Miller, *op. cit.,* pp. 149–152; the

570

details of TR's birth, quoted in Martha Stewart Bulloch to Susan Elliott West, October 28, 1858, in TRAC; "All quite well," quoted in Putnam, *op. cit.,* p. 23; description of effects of an asthma attack comes from personal observation; TR's early memory of an attack, *Autobiography,* p. 14; letters beginning "Teedie has a," etc., in Putnam, *op. cit.,* pp. 25–26; "Always afterward," Cowles, *op. cit.;* "in part compensation," Putnam, *op. cit.,* pp. 48–49; for the senior Roosevelt's wartime activities, see the statement of William E. Dodge and letters in Robinson, *op. cit.,* pp. 19–32; "Something occurred," in letter of December 15, 1861, in Theodore Roosevelt Collection; "grind the Southern troops," *ibid.,* p. 19; "He begins to," quoted in Putnam, *Theodore Roosevelt,* p. 23; for running the blockade, see Cowles, *op. cit.,* and Robinson, *op. cit.,* pp. 36–38; "Handsome dandy," in Steffens, *Autobiography,* pp. 349–350; "We used to sit," Robinson, *op. cit.,* p. 2; for Edith Carow's background and early relationship with TR, see S. J. Morris, *Edith Kermit Roosevelt,* pp. 1–2 and pp. 15–27; dragging Livingston's book, Putnam, *op. cit.,* p. 27; for TR's education under Anna Bulloch, see *Autobiography,* p. 12; his early reading, *ibid.,* pp. 16–19; "I was nervous," *ibid.,* p. 29; the seal incident, *ibid.,* pp. 14–15; "Dearest Mamma" letter, *Letters,* Vol. I, p. 3; adult reaction to TR's interest in zoology in S. J. Morris, *op. cit.,* p. 47; "as they always did," *Autobiography,* p. 15; "I will always have," TR, *Theodore Roosevelt's Diaries of Boyhood and Youth,* p. 6; for the senior Roosevelt's post–Civil War charities, see the comments of William Dodge in Robinson, *op. cit.,* pp. 20–21; for family visits to charities, see Cowles, *op. cit.;* "It was verry," TR, *Diaries,* p. 13; "cordially hated," *Autobiography,* p. 14; all diary entries, from TR, *Diaries,* pp. 3–227; letter confirming distances walked by TR, quoted in Putnam, *op. cit.,* p. 63; "On Lake Maggiore," quoted in *ibid.,* pp. 63–64.

Chapter 2
"I'll Make My Body"

"I study English," *Letters,* Vol. I, p. 6; looked like a stork, in E. Morris, *The Rise of Theodore Roosevelt,* p. 60; "sharp, ungreased squeak," Putnam, *Theodore Roosevelt,* p. 78; "a bright, precocious boy," Rockwell, *Rambling Recollections,* p. 261 and p. 315; details of TR's man-to-man chat with his father, in Robinson, *My Brother Theodore Roosevelt,* p. 50; Wood's recollections, in *New York World,* January 21, 1904; "One of my most vivid," Robinson, *op. cit.,* p. 50; "pertest little ape," Hagedorn, *The Boy's Life of Theodore Roosevelt,* pp. 39–40; "I picked up a salamander," Roosevelt, *Diaries of Boyhood and Youth,*; description of Bell, in *Autobiography,* p. 19; "When he does," Robinson, *op. cit.,* p. 56; "an excellent gun," *Autobiography,* p. 20; "One day they read," *ibid.,* p. 19; Moosehead Lake episode, *ibid.,* pp. 29–30; Laughlin comment, in Laughlin interview with Pringle; for background on James Alfred Roosevelt and Robert B. Roosevelt, see Miller, *The Roosevelt Chronicles,* p. 163 and pp. 142–149; also see Bleyer, "The Forgotten Roosevelt," *Newsday Magazine,* October 6, 1985, and Hammond, "Robert Barnwell Roosevelt"; TR's comments on Britain, in *Diaries*, pp. 264–265; comments on Bologna and the Rhone Valley, *ibid.*; "Alexandria!," *ibid.*; "On seeing," *ibid.*, pp. 276–277; "the nicest, coziest," *ibid.*, p. 290; "It was not beautiful," *Diaries*, p. 304; visit to Emerson, in Robinson, *op. cit.*, p. 63; "Healthy, natural," quoted in McCullough, *Mornings on Horseback*, p. 125; "distinctly dangerous," Robinson, *op. cit.*, p. 57; Elliott's limerick, in McCullough, *op. cit.*, p. 125; "I think I have," Robinson, *op. cit.*, p. 67; "He is the most," *ibid.*, pp. 56–57; "This difficulty," *Diaries*, p. 312; "remarkably small," *ibid.;* "We saw . . . various," *ibid.*, p. 313; "What we should call," *ibid.*, p. 317; bribery, *ibid.*, p. 318; comments on Baalbek, *ibid.*, pp. 320–321; "we his children," Robinson, *op. cit.*, p. 64; "wild sort of dance," *Diaries*, p. 317; visits to Athens and Vienna, in *Diaries*, pp. 327–333; "One, a famous," *Autobiography*, p. 23; "The plan is this," *Letters*, Vol. I, pp. 10–11; "Father is," *ibid.*, p. 9; TR's impressions of the German people, in *Autobiography*, p. 23; "Health; good," Robinson, *op. cit.*, p. 84; "Picture to yourself," quoted in Putnam, *op. cit.*, p. 107; "Of course I could," quoted in Morris, *op. cit.*, p. 73; "If you offered rewards," *Letters*, p. 9; "Did you hear," *Letters*, Vol. I, p. 11; "You need not," Vierick article; "Anything less tranquil," Robinson, *op. cit.*, pp. 88–89; "dresses well," *Letters*, Vol. I, p. 23; "the unquenchable gaiety," Parsons, *Perchance Some Day*, p. 29; "Why Elliott," *ibid.*, p. 29; "I worked," *Autobiography*, pp. 23–24; "alert, vigorous," in Cutler statement, TRC; TR falls while skating, in *Letters*, Vol. I, p. 14; "Is it not splendid," *ibid.*, p. 13; TR's physical measurements, in *Diaries*, p. 356; "My only regret," McCullough, *op. cit.*, p. 158; "Take care," *Letters*, Vol. I, p. 33.

Chapter 3
Harvardman

Harvard in Roosevelt's time is described in Wilhelm, *Theodore Roosevelt as an Undergraduate*; "cosy and comfortable," *Letters*, Vol. I, p. 16; sitting beside Quincy, in Harry S. Rand interview in TRC; makeup of TR's class, in McCullough, *Mornings on Horseback*, p. 199; Henry Adams's remarks in *The Education of Henry Adams*, p. 305; "an immature being," *Boston Saturday Evening Gazette*, November 9, 1879; "Is there a university," quoted in Putnam, *Theodore Roosevelt*, p. 131; "It was not considered," in Sherrard Billings interview in TRC; "Now look here," Wilhelm, *op. cit.*, p. 35; Thayer remarks and Saltonstall anecdote, in interviews in TRC; Welling's story, in "My Classmate, Theodore Roosevelt," *American Legion Magazine*, January 1929; "I have avoided being intimate," quoted in Putnam, *op. cit.*, p. 136; joins eating club, in *Letters*, Vol. I, p. 16; "For breakfast we have," *ibid.*, pp. 23–24; "Roosevelt was perfectly," in Rand interview, *op. cit.;* "I want to pet," *Letters*, Vol. I, p. 23; "I have never spent," *ibid.*, p. 19; "I do not think," *ibid.*, p. 18; for background on Eliot, see Bailyn, et al., *Glimpses of the Harvard Past;* "Our new President," Morison, *The Development of Harvard University 1869–1929*, p. 558; "Mr. Eliot, this is," Wilhelm, *op. cit.*, pp. 62–63; TR's visit to Eliot in 1905, in James, *Charles William Eliot*, Vol. II, p. 159; TR's schedule, in *Letters to Anna Cowles,* pp. 15–16; "a little under the weather," *Letters,* Vol. I, p. 26; "a miserable failure," Private Diary, June 17, 1878; Bacon's refusal to go to TR's room, in Mrs. Robert Bacon interview in TRC; "The fellows," *Letters,* Vol. I, p. 26; "As I have taught," Private Diary, January 11, 1880; "danced just as you'd," in Rose Saltonstall interview in TRC; "I have been having," *Letters,* Vol. I, p. 29; sleighing party, *ibid.,* p. 23; "I enjoyed *her,*" *ibid.,* p. 28; frock coat, *ibid.,* p. 17; TR's grades, *ibid.,* pp. 25–26; "The night was dark," quoted in Cutright, *Theodore Roosevelt,* pp. 101–102; "My father had," *Autobiography,* pp. 25–26; "extremely interesting," *Letters,* Vol. I, p. 29; "My respect for," *ibid.,* p. 30; for a discussion of the politics of the Gilded Age and the fight over civil service reform, see Miller, *The Founding Finaglers,* pp. 142–143, 282–286; the most detailed account of the collectorship battle is in McCullough, *op. cit.,* pp. 169–180; "Tell Father," *Letters,* Vol. I, pp. 29–30; "It looks as though," *Letters,* Vol. I, p. 39; "A great weight," quoted in McCullough, *op. cit.,* pp. 179–180; "very uneasy," *Letters,* Vol. I, p. 31; "Xmas," Private Diary, December 25, 1877; "I have never caused," *ibid.,* January 2, 1878; "I have just sat with him," quoted in S. J. Morris, *Edith Kermit Roosevelt,* p. 57; "You would have felt," quoted in McCullough, *op. cit.,* p. 182; "Elliott gave unstintingly," Robinson, *My Brother Theodore Roosevelt*; "Newsboys from the," *ibid.*; "The agony in his face," Elliott Roosevelt memorandum in TRC; "Every now and then," *Letters,* Vol. I, p. 31; "Sometimes when I" and other comments are all from the Private Diary; "My own, sweet sister," *Letters to Anna Cowles,* p. 27; "brave Christian gentleman," Private Diary, September 1, 1878; "I most valued," *ibid.,* May 7, 1878; "I am well satisfied," *ibid.,* August 10, 1878; TR's activities at Oyster Bay are from entries in *ibid.* for June and July 1878; "Afterwards," *ibid.,* August 22, 1878; shooting of dog, *ibid.,* August 24, 1878; blazing away with rifle, *ibid.,* August 26, 1878; "We both of us," in a letter to Bamie, September 20, 1886, quoted in Morris, *op. cit.,* p. 58; "had not been nice," *ibid.,* p. 59; Sewall's comments on TR, in Sewall, *Bill Sewall's Story of T.R.*; "That's where the name," Hagedorn, *The Boy's Life of Theodore Roosevelt,* p. 60; "the kind of Americanism," quoted in Putnam, *op. cit.,* p. 158; "Oh, Father," Private Diary, October 28, 1878; Porcellian and A.D. incident, *ibid.,* October 7, 1878; initiation into Porcellian, *ibid.,* November 2, 1878; "I am delighted," *Letters,* Vol. I, p. 35.

Chapter 4
"Teddykins" and "Sunshine"

"See that girl?" in Mrs. Robert Bacon interview with Henry Pringle; "I loved her," TR, *In Memory of My Darling Wife;* "the very sweet" and "chestnutting," Private Diary, October 19 and 20, 1879; "as sweet," *ibid.,* November 10, 1879; "Win her," *ibid.,* January 25, 1880; for description of Alice, see letter from Corinne Robinson to Pringle, September 18, 1930; "studious, ambitious," in Mrs. Bacon's interview with Pringle; "had no intention," *ibid.;* "You *must* not forget," *Letters,* Vol. I, p. 36; "We have by," in a previously unpublished letter of December 8, 1878; calling him "Thee," in Private Diary, January 26, 1879; "Thank Heaven," *ibid.*, April 18, 1878; "and to do nothing," *ibid.*, December 11, 1879; "distinguished himself," *ibid.*, October 24, 1879; "Snowed heavily," *ibid.*, April 19, 1879; "What a royally," *ibid.*, May 8, 1879; "dear honest," *ibid.*, August 20, 1879; "if she's in a," *Letters*, Vol. I, p. 36; "We little suspected," Laughlin, "Roosevelt at Harvard"; the first version of the boxing match appeared in Wister, *Roosevelt, The Story of a Friendship*, pp. 4–5; Spalding's account, in *Harvard Alumni Bulletin*, December 4, 1931; "row with a mucker," Private Diary, June 19, 1879; Class Day scene, *ibid.*, June 20, 1879, and in Putnam, *Theodore Roosevelt*, pp. 180–182; "the most loveable," Wise, *Recollections*; "As Katahdin expedition," *ibid.*, August 24–September 2, 1879; for the Munsungan Lake expedition, see previously unpublished letter to Martha Bulloch Roosevelt, September 14, 1879; "The horse, harness" "Dickey" burlesque, in Wister, *op. cit.*, pp. 14–15; "Everything went off," Private Diary, November 22, 1879; "like moths," in Pringle interview with Thomas Lee; "Oh the changeableness," in a letter to Bamie, November 11, 1879 (?); "I did not think," in a letter to John Roosevelt, February 25, 1880; "She is the most," Private Diary, December 26, 1879; for Alice's visit to New York, see *ibid.*, December 26, 1879–January 1, 1880; "My sweet little queen," in a previously unpublished letter of January 28, 1880; Corinne on Edith, in S. J. Morris, *Edith Kermit Roosevelt*, p. 62; "It seems like a dream,"in a letter dated February 3, 1880; "There is hardly another," in a letter dated February 1, 1880, TRC; thoughts of studying law, in Private Diary, August 18, 1879; "I am going to try," Thayer, *Theodore Roosevelt*, p. 21; "I had no more desire," *Autobiography*, pp. 26–27; letter to Minot, in *Letters*, p. 43; "When we are alone," Private Diary, February 23, 1880; "Only one gentleman," *Letters*, Vol. I, pp. 42–43; the text of TR's senior essay, in TRC; also see Silverman, *Theodore Roosevelt and Women;* "I have certainly lived," Private Diary, May 5, 1880; for medical checkup and its aftermath, see Hagedorn, *The Boy's Life of Theodore Roosevelt*, pp. 63–64; "He talked to her," Kleeman, *Gracious Lady*, p. 101; "she shall always," Private Diary, July 1, 1880; "Very embarrassing," *Letters*, Vol. I, p. 45; "I hope we have good sport," previously unpublished letter dated August 15, 1880; "The farm people," *Letters*, Vol. I, p. 46; "As soon as we," *ibid.*, pp. 46–47; "Don't you think," in previously unpublished letter dated August 30, 1880, in Longworth Collection; "I have been spending," Private Diary, October 6, 1880; "It was the dearest," Parsons, *Perchance Some Day*, p. 43; "Our intense," Private Diary, October 27, 1880.

Chapter 5
"I Have Become a Political Hack"

"A perfect dream of delight," *Letters*, Vol. I, p. 47; "I tried faithfully," Riis, *Theodore Roosevelt, the Citizen*, pp. 36–37; "The pertinacity in which," Putnam, *Theodore Roosevelt*, p. 219; "never seemed to have," *ibid.;* "Some of the teaching," *Autobiography*, pp. 55–56; "Alice is universally," Private Diary, December

11, 1880; "little inner group," Wharton, *The Age of Innocence,* p. 381; "gentle and lovely," in a letter from Corinne Robinson to Henry Pringle, September 18, 1930; "This feeling," *ibid.;* "I have to read the Bible," in Longworth Collection, April 6, 1881; Wister's account, in Wister, *Roosevelt, The Story of a Friendship,* p. 24; "The men I knew best," *Autobiography,* pp. 56–57; "I insisted on taking part," Shannon, *Beatrice Webb's American Diary,* p. 15; spoke with such "ginger," in Hagedorn, *The Boy's Life of Theodore Roosevelt,* p. 67; "stood stiffly," *Autobiography,* p. 61; "Hurrah! for a summer," Private Diary, May 11, 1881; "We had a beautiful passage," *Letters,* Vol. I, p. 47; "a man lying on the road," quoted in Putnam, *op. cit.,* p. 230; visit to the Bullochs, described in a letter to Mittie Roosevelt in Longworth Collection, June 5, 1881; lions' manes, in *Letters,* Vol. I, p. 48; "has a far keener," *ibid.;* TR's art criticism, in a letter to Martha Bulloch Roosevelt, June 5, 1881; TR's comments on Napoleon, in *Letters,* Vol. I, pp. 51–52; "like fairyland," in a letter to MBR in Longworth Collection, June 21, 1881; "Teddy enjoys," Longworth Collection, July 1, 1881; TR's account of the ascent of the Matterhorn, in *Letters,* Vol. I, pp. 49–50; "I have plenty of information," *ibid.,* p. 50; "I have enjoyed it greatly," quoted in Putnam, *op. cit.,* p. 236; "Am working fairly hard," Private Diary, October 17, 1881; "I am so busy," Longworth Collection, October 14, 1881; "Every brace and bowline," TR, *The Naval War of 1812,* pp. 247–248; post–Civil War condition of U.S. Navy, in Miller, *The U.S. Navy,* pp. 191–197; "perhaps the most incapable," *The Naval War,* p. 405; "to kill," Private Diary, October 6, 1881; conversation between Murray and TR, in Hagedorn interview of Murray in TRC; "Strong Republican," Private Diary, November 4, 1881; TR and Fischer exchange, in Murray statement to Hagedorn; Choate's letter is dated October 31, 1881; "untrammelled and unpledged," *New York Herald,* November 1, 1881; "The canvass is getting on," in Longworth Collection; bounced check, in Pringle, *Theodore Roosevelt,* p. 39; TR's margin of victory, in *New York Times,* November 10, 1881; "I have become," *Letters,* Vol. I, p. 55.

Chapter 6
The Gentleman from New York

Roosevelt's "step across the hotel corridor," in Spinney memorandum in TRC; Hunt's statement, in TRC; making his "gyrations," *New York World,* April 15, 1883; "A number of republicans," TR, Legislative Diary, January 7 and 12, 1882; new "boy at a strange school," *Autobiography,* p. 65; analysis of legislators, *ibid.,* p. 71; "A long deadlock," Legislative Diary, January 2, 1882; "both stupid and monotonous," *ibid.,* January 7, 1882; "He threw each paper," Hudson, *Random Recollections,* pp. 144–145; advice on public speaking, in *Autobiography,* pp. 64–65; "very favorable impression," *New York Evening Post,* January 24, 1882; "Just where I wished," Private Diary, February 14, 1882; "The Chairman is," Legislative Diary, February 14, 1882; "All the men," *Letters,* Vol. I, p. 64; "a very charming," Hunt's statement in TRC; description of TR on the floor of the Assembly, in statement by John Walsh in the Pringle Papers; TR's account of the Manhattan Elevated Railway case, in *Autobiography,* pp. 75–77; *Harper's* review of *The Naval War of 1812,* dated November 1882; *New York Times* review is dated June 5, 1882; for Elliott Roosevelt and Anna Hall, see Miller, *The Roosevelt Chronicles,* pp. 191–193; Hunt's account of the Westbrook affair, in TRC; also see Spinney memorandum in TRC and *Autobiography,* pp. 77–79; letter to Alice, in Longworth Collection, April 5, 1882; attempt to blackmail TR, in Hagedorn statement in TRC; "Big business of the kind," *Autobiography,* p. 78; "My acting became," *Letters,* Vol. I, p. 56; discussion of possibility of buying farm, *ibid.,* pp. 56–57; Schurz's remarks, in *New York Times,* October 28, 1882; "All Hail," *Letters,* Vol. I, p. 58.

Chapter 7
"There Is a Curse on This House!"

"I rose," *Letters,* Vol. III, p. 634; "country Republicans," Private Diary, January 1, 1883; background on Conkling, in Miller, *The Founding Finaglers,* pp. 287–288; Nast cartoon, in Lorant, *The Life and Times of Theodore Roosevelt,* p. 199; Hunt's comment, in his statement in TRC; "Professional agitators," in "Phases of State Legislation," *The Century,* January 1885; "One partial reason," *Autobiography,* p. 92; "The businessmen who spoke," *ibid.,* p. 81; Gompers takes TR on an inspection tour, in Gompers, *Seventy Years of American Life and Labor,* p. 187; "I have always remembered," *Autobiography,* p. 82; "My first visits," *ibid.,* p. 83; "although he felt very doubtful," quoted in Putnam, *Theodore Roosevelt,* p. 303; "It was this case," *Autobiography,* pp. 82–83; "I felt as if my heart," Private Diary, December 31, 1882; "a small pleasant place," Parsons, *Perchance Some Day,* p. 44; "What has the poor," *ibid.,* pp. 44–45; "I can imagine nothing," Private Diary, January 3, 1883; "I weakly yielded," quoted in Putnam, *op. cit.,* p. 284; "the Sodom and Gomorrah," *The Works of Theodore Roosevelt,* Vol. XIV, p. 16; "Young Mr. Roosevelt," *New York Observer,* March 10, 1883; "I came an awful cropper," *Letters,* Vol. III, p. 634; "There is great sense," Hudson, *Random Recollections,* p. 147; "Dr. Wynkoop," in Longworth Collection; "a nightmare," in a letter from TR to Martha Bulloch Roosevelt, September 4, 1883, TRC; "very large," Felsenthal, *Princess Alice,* p. 28; "I have been miserably," Longworth Collection; for details of TR's visit to the Bad Lands, see Hagedorn, *Roosevelt in the Bad Lands,* pp. 3–46; "Darling Wifie" letter is dated September 20, 1883, in Longworth Collection; letter to ALR telling of intention to invest is dated September 23, 1883, in Longworth Collection; TR's contract with Ferris and Merrifield, in Hagedorn, *op. cit.,* Appendix I; "I am a Republican," *Letters,* Vol. I, p. 63; "He's a," *New York Times,* December 27, 1883; "merely the creatures," *Autobiography,* p. 84; "I love you and long for you," *Letters,* Vol. I, p. 64; "Teddy's here," in a letter from CRR to Pringle, September 18, 1930; ALR's last note, in Longworth Collection; Anna Gracie's comments, in a note in Longworth Collection; "I shall never forget," Hunt, *op. cit.;* "Suicidal weather," *New York Times,* February 14, 1884; for the funeral, see *ibid.* and *New York World,* February 17,

1884; "in a dazed," Sewall, *Bill Sewall's Story of T.R.*, p. 11; letter to Schurz, *Letters*, Vol. I, p. 66; "I looked the ground over," Riis, *Theodore Roosevelt, the Citizen*, p. 59; "You could not," Hunt, *op. cit.;* "she was beautiful," *In Memory of My Beloved Wife;* "decidedly mottled," *Letters*, Vol. I, p. 70; "It is now," quoted in *Works*, Vol. XIV, pp. 37–38; "I am by inheritance," in *Boston Herald*, July 20, 1884; "dumbfounded," Thayer, *John Hay*, Vol. I, p. 55; "Blaine's nomination," *Letters*, Vol. I, p. 88.

Chapter 8
"The Wine of Life"

"I heartily enjoy," *Roosevelt-Lodge Correspondence*, Vol. I, p. 7; "The country is growing," *Letters*, Vol. I, p. 74; it "looks," in *Hunting Trips (Works*, Vol. I), p. 11; "Hell with," quoted in Cutright, *Theodore Roosevelt*, p. 146; if had not gone west, in speech at Fargo, North Dakota, September 4, 1910, Vivian, *The Romance of My Life*, pp. 57–77; "one hundred and fifty pounds," Sewall, *Bill Sewall's Story of T.R.*, p. 11; "is now hearty and strong," quoted in Putnam, *Theodore Roosevelt*, p. 530; "We led a free," *Autobiography*, pp. 94–95; the encounter with the drunk, *ibid.*, pp. 124–125, and Hagedorn, *Roosevelt in the Bad Lands*, pp. 151–153; the encounter with "Hell-Roaring" Bill Jones, *ibid.*, pp. 115–116; "Hasten forward" story, *ibid.*, p. 110; "Black care," in *Ranch Life (Works*, Vol. I), p. 329; "You have your child," Sewall, *op. cit.*, p. 47; "Don't talk to me," in Merrifield interview with Hagedorn, TRC; "The poor little mite," *Letters*, Vol. I, p. 100; "No writer until then," Cutright, *op. cit.*, p. 152; "I wish I could make," Wister, *Roosevelt, the Story of a Friendship*, p. 41; "short work of either," TR, *Hunting Trail*, p. 231; TR's account of the grizzly hunt, *ibid.*, pp. 239–240; Schullery's comments, in TR, *American Bears*, pp. 5–9; my account of the Marquis de Morès is based upon Dresden, *The Marquis de Morès;* "By all means bring them," Hagedorn, *op. cit.*, pp. 337–338; Roosevelt's reaction to an expected challenge, in Sewall, *op. cit.*, pp. 27–28; "I can't help writing you," *Letters*, Vol. I, p. 484; "Can these be called," quoted in Putnam, *op. cit.*, p. 550; the article mentioning Davis, in *North American Review*, October 1885; "I answered," quoted in Pringle, *Theodore Roosevelt*, p. 12; looking "pretty gay," *Roosevelt-Lodge*, Vol. I, p. 34; Alice's reaction, in Longworth, *Crowded Hours*, p. 4; "I don't grudge," *Roosevelt-Lodge*, Vol. I, p. 35; "she was passionately," quoted in Felsenthal, *Princess Alice*, p. 34; "I have no constancy," quoted in Putnam, *op. cit.*, p. 557; Stickney's recollections, *ibid.*, pp. 568–569; TR's account of the pursuit and capture of the thieves, in *Ranch Life*, pp. 383–398; ". . . Being by nature," *Letters*, Vol. I, p. 102; "Some days he would write," Hagedorn, *op. cit.*, pp. 397–398; "By right," *Thomas Hart Benton*, p. 173; "too much muscular," *The Nation*, March 29, 1888; "I haven't the least idea," *Letters*, Vol. I, p. 108; "The statement itself is true," quoted in S. J. Morris, *Edith Kermit Roosevelt*, pp. 90–91; "They didn't want it at all," Alice R. Longworth interview with Hagedorn, TRC; "Overstocking may cause," *Ranch Life*, p. 290; "We were throwing," Sewall, *op. cit.*,

p. 92; TR's talk with Sewall, *ibid.*, pp. 93–94; "If I make a good run," *Letters*, Vol. I, p. 111; "He is very bright," quoted in S. J. Morris, *op. cit.*, p. 93; "worse ever than I fear," *Roosevelt-Lodge*, Vol. I, p. 50; "at least I have a better," *Letters*, Vol. I, p. 113.

Chapter 9
"I'm a Literary Feller"

"I love you," quoted in S. J. Morris, *Edith Kermit Roosevelt*, p. 105; "and going to the ranch," *Letters*, Vol. I, p. 117; "I hardly know," in TRC, January 10, 1887; "It almost broke," Cowles, *Reminiscences;* "Remember, darling," Teague, *Mrs. L.*, pp. 12–13; description of Bad Lands following the blizzard, in Hagedorn, *Roosevelt in the Bad Lands*, pp. 432–441; "I am bluer," *Letters*, Vol. I, pp. 126–127; "For the first time," *ibid.*, p. 127; "I intend to," *New York Tribune*, March 28, 1887; Godkin's remark, in *New York Post*, May 13, 1887; "You are not," *Puck*, May 25, 1887; "If I get," *Letters*, Vol. I, p. 131; "like a couple," *Roosevelt-Lodge Correspondence*, Vol. I, p. 55; "too good and cunning," *Letters*, Vol. I, p. 132; "Now, pig!" Hagedorn, *The Roosevelt Family*, p. 16; "Theodore . . . *needs*," quoted by Arthur Schlesinger, Jr., in the foreword to a series of reprints of the American Statesmen series; "entertaining scamp," *Letters*, Vol. I, p. 129; "I am rather," *Roosevelt-Lodge*, p. 57; "the filthy little," TR, *Gouverneur Morris*, p. 289; "In a government," *ibid.*, pp. 322–323; "I don't know whether," *Letters*, Vol. I, p. 131; "Edie is getting along," *ibid.*, p. 132; "He plays more," quoted in Hagedorn, *The Roosevelt Family*, p. 17; "*My* little brother's," *Letters*, Vol. I, p. 133; "ugly duckling," *Washington Post*, February 12, 1974; "stretch each foot," Longworth, *Crowded Hours*, p. 17; "When we approached," *ibid.*, p. 8; "They were, of course," Hagedorn, *The Roosevelt Family*, p. 23; "something I should," Longworth, *op. cit.*, p. 12; "We'd better," quoted in Felsenthal, *Princess Alice*, p. 45; "seventh heaven," Hagedorn, *The Roosevelt Family*, p. 40; Eleanor Roosevelt and swimming, in E. Roosevelt, *This Is My Story*, pp. 49–50; discussion of Edith Roosevelt and her family is based upon S. J. Morris, *Edith Kermit Roosevelt*, and discussions with family members; "We've seen enough," Teague, *op. cit.*, p. 36; TR's hunt in 1887, in Hagedorn, *Roosevelt in the Bad Lands*, pp. 453–454, and E. Morris, *The Rise of Theodore Roosevelt*, pp. 382–383; for Boone & Crockett Club and its work, see Cutright, *Theodore Roosevelt*, pp. 167–177, Trefethen, *An American Crusade for Wildlife*, and Reiger, *American Sportsmen*; "I should like," *Letters*, Vol. I, pp. 135–136; "I realize perfectly," *ibid.*, p. 211; "Writing is horribly," *ibid.*, p. 95; "It seems impossible," TRC, July 1, 1888; "Of course I know," *Letters*, Vol. I, p. 140; "On the 26th," *The Winning of the West*, in *Works*, Vol. VIII, p. 479; Turner's review, in *The Dial*, August 1889; "I'm a literary feller," TRC, October 5, 1888; "It is unfortunate," *Letters*, Vol. I, p. 136; "I am myself," *ibid.*, p. 143; for the campaign of 1888, see Miller, *The Founding Finaglers*, pp. 295–297; "I . . . act as target," *Letters*, Vol. I, p. 149; "I would like above all things," *ibid.*, p. 154; "My real trouble," quoted in E. Morris, *op. cit.*, p. 392; "I had a little talk," *Roosevelt-Lodge,* p. 76.

Chapter 10
"We Stirred Up Things Well"

Halloran's comments, in Halloran, *Romance of the Merit System,* p. 56; "You can guarantee," quoted in Lorant, *The Life and Times of Theodore Roosevelt,* p. 243; "wanted to put an end," quoted in Pringle, *Theodore Roosevelt,* p. 86; TR giving dictation, in Halloran, *op. cit.,* p. 60; for details on the Civil Service Commission, see Hoogenboom, "The Pendleton Act"; "Do let me know," *Letters,* Vol. I, p. 161; "Running his eye quickly," Halloran, *op. cit.,* pp. 57–58; "No republic," *Scribner's,* August 1895; "Plain, commonsense," *The Works of Theodore Roosevelt,* Vol. XIV, p. 111; "He did not get," quoted in Pringle, *op. cit.,* p. 91; "intend to have," *Letters,* Vol. I, p. 163; "about as thorough-paced," *ibid.,* p. 166; "We stirred up," *ibid.,* pp. 166 and 168; "rigidly," *ibid.,* p. 167; "it is to," *ibid.,* pp. 168–169; "the poor wretch," Adams, *Letters of Henry Adams,* Vol. III, p. 175; "I am rapidly," *Letters,* Vol. I, p. 174; "it is four weeks," quoted in S. J. Morris, *Edith Kermit Roosevelt,* p. 120; "Edith can't say," TRC, October 13, 1889; baptism of Kermit, in Hagedorn, *The Roosevelt Family,* p. 19; "homelike and comfortable," *Letters,* Vol. I, p. 208; "What funnily," *ibid.,* p. 200; rats in the White House, in Billings, "Social and Economic Life"; "Nobody got," *Washington Post,* January 2, 1890; for Washington in the 1890s, see Greene, *Washington,* Vol. II, Chapter V, and Billings, *op. cit.;* "There was a vital," Chanler, *Roman Spring,* p. 195; for a profile of Reed, see Tuchman, *The Proud Tower,* Chapter 11; for Adams and his circle, see O'Toole, *The Five of Hearts;* "sympathetic little wife," Adams, *Letters,* Vol. III, p. 398; "looks precisely," quoted in E. Morris, *The Rise of Theodore Roosevelt,* p. 423; "I used to walk," Stoddard, *As I Knew Them,* p. 7; "Half measures simply," TRC, May 2, 1890; "nightmare," *ibid.,* May 23, 1891; "How do you do," Baltimore Report, p. 3; "When I think," S. J. Morris, *op. cit.,* p. 133; "You may tell," quoted in Foulke, *Fighting the Spoilsmen,* pp. 25–36; for the Populist Revolt, see Hicks, *The Populist Revolt;* "The only thing to do," TRC, August 22, 1891; "one of the most beautiful," E. Roosevelt, *This Is My Story,* p. 1; "The repairs of," quoted in S. J. Morris, *op. cit.,* p. 138; "The trouble is," *Letters,* Vol. I, p. 343; "I have used," *ibid.,* p. 282; "for a year or two," *ibid.,* p. 314; "I curled up," Thayer, *John Hay,* Vol. II, p. 333; "I think it represents," quoted in Lorant, *op. cit.,* p. 245; bringing the children to office, in Halloran, *op. cit.,* p. 64; "Then everything," Hagedorn, *op. cit.,* p. 28; "a little like," *Letters,* Vol. VIII, p. 1443; "Poor fellow," TRC, July 29, 1894; "Theodore was more," *ibid.,* August 15, 1894; "He would have been," *ibid.,* August 18, 1894; "The prize was," *Letters,* Vol. I, p. 407; "The last four weeks," *ibid.,* Vol. VIII, p. 1433; "He should never," quoted in S. J. Morris, *op. cit.,* p. 153; "afford to be," *Letters,* Vol. I, p. 439.

Chapter 11
"It Is a Man's Work"

Roosevelt's arrival at Police Headquarters, described by Steffens in his *Autobiography*, pp. 257–258; "My experience," T.R. *Autobiography*, p. 206; for Parkhurst and the Lexow investigation, see Werner, *It Happened in New York*, pp. 36–116; "I have the most," *Letters*, Vol. I, p. 458; "When he asks a question," *New York World*, May 17, 1895; for accounts of TR as police commissioner, see Steffens, *op. cit.*, pp. 255–291, *Autobiography*, Chapter VI, E. Morris, *The Rise of Theodore Roosevelt*, pp. 481–563, and Berman, *Police Administration and Progressive Reform;* "he never thinks," quoted in Kaplan, *Lincoln Steffens*, p. 76; for TR and the press, see Stein, "Theodore Roosevelt and the Press"; for TR and the German anti-Semite, see *Autobiography*, pp. 101–102; "The day is not," *Roosevelt-Lodge Correspondence*, Vol. I, August 31, 1895; "Don't ask me," Steffens, *op. cit.*, pp. 259–260; for TR's views on expansionism, see Beale, *Theodore Roosevelt and the Rise of America to World Power*, pp. 16–52; "For two nickels," Hay, *Letters of John Hay*, Vol. II, pp. 235–236; "Personally I rather," *Letters*, Vol. I, p. 504; letter to the *Harvard Crimson, ibid.*, pp. 505–506; "Every expansion," *The Works of Theodore Roosevelt*, Vol. XIII, p. 336; "I wish our people," *Letters*, Vol. I, p. 522; "little more," *ibid.*, p. 462; the rifle story, in Hagedorn, *The Roosevelt Family*, p. 41; "We always told," *ibid.*, p. 47; Christmas with the Roosevelts, described in TRC, December 26, 1890; "Life was," Chanler, *Roman Spring*, p. 195; "I have always felt," *Letters*, Vol. I, pp. 464–465; "I do not deal," *New York Sun*, June 20, 1895; "I have not one," *Letters*, Vol. I, pp. 502–503; "With proper power," quoted in Chessman, *Theodore Roosevelt*, p. 61; "I cannot shoot," *Letters*, Vol. I, pp. 504–505; "There is one thing," Storer, "How Theodore Roosevelt Was Appointed . . ."; for Hanna and McKinley, see Leech, *In the Days of McKinley;* "as in every way," *Letters*, Vol. I, p. 543; "Upon my word," *ibid.*, p. 522; "McKinley, whose firmness," *ibid.*, p. 543, and *Harper's Weekly*, June 1, 1912; "Not since the Civil War," *ibid.*, p. 550; "as if he were a patent medicine," Pringle, *Theodore Roosevelt*, p. 113; "to the leaders," *Works*, Vol. XIV, p. 258; loaves of bread, *ibid.*, p. 269; "He spoke of you," *Roosevelt-Lodge*, Vol. I, p. 240; "I want peace," Storer, *op. cit.;* "hot-headed," *Letters*, Vol. I, p. 569; "I think I," *ibid.*, p. 568; "I shall write," *Roosevelt-Lodge*, Vol. I, p. 247; "exceedingly polite," *Letters*, Vol. I, p. 572; "Sinbad," E. Morris, *The Rise of Theodore Roosevelt*, p. 560.

Chapter 12
"The Supreme Triumphs of War"

Target practice on the *Iowa*, in *New York Sun*, September 10 and 24, 1897, and *New York Herald*, September 10, 1897; "Oh, Lord!" *Letters*, Vol. I, p. 680; "Until we definitely turn Spain," *ibid.*, pp. 607–608; "As you will see," *ibid.*, p. 604; "Best man," Long, *Journal*, p. 209; "What is the need," *ibid.*; "He broke the record," Harbaugh, *The Life and Times of Theodore Roosevelt*, p. 96; "The machine has worked," *Letters*, Vol. I, p. 799; size of the fleet in 1897, in *Annual Report;* "These battleships are unsafe," quoted in Hagan, ed., *In Peace and War,* p. 296; "We need a large navy," *Works,* Vol. XII, p. 272; Newport speech, in *Works,* Vol. XIII, pp. 182–199; for the conservative reformers, see Josephson, *The President Makers,* Chapters I–III; Long does not make an issue, in *Letters,* Vol. I, p. 623; "I suspect," quoted in E. Morris, *The Rise of Theodore Roosevelt,* p. 572; "the banker, the broker," *Letters,* Vol. I, p. 648; FDR's contact with TR, in *Personal Letters,* Vol. I, p. 110; for Eleanor Roosevelt at Sagamore, see E. Roosevelt, *This Is My Story,* pp. 49–50; Edith's comments, in Rixey, *Bamie,* p. 228; "and he gave me," *ibid.,* p. 637; "The Secretary is away," *ibid.,* p. 655; "largely from the wreck," *ibid.,* p. 642; "to have an entire rest," *ibid.,* p. 642; "do nothing," *ibid.,* p. 664; "Blood on the roadsides," *New York World,* May 17, 1896; the navy's war plan, discussed in Grenville, "American Preparations"; "Of course the President," *Letters,* Vol. I, p. 676; "this was one case," *ibid.,* p. 677; "irresolute," *ibid.,* p. 691; "in a fortunate hour," *Autobiography,* p. 216; "Very unexpectedly," *Letters,* Vol. I, p. 718; the diplomatic maneuvering can be followed in Trask, *The War with Spain;* "began in his usual," Long, *op. cit.,* pp. 212–213; note to Tillinghast, in *Letters,* Vol. I, p. 758; memo to Long, *ibid.,* pp. 759–763; Edith's illness can be followed in S. J. Morris, *Edith Kermit Roosevelt,* pp. 169–170; "You might as well," quoted in Herrick, *The American Naval Revolution,* p. 210; Adler episode, in Beer, *Hanna,* pp. 546–547; "The saddest thing," Long, *op. cit.,* p. 215; "My duty is plain," quoted in Trask, *op. cit.,* p. xiii; for Roosevelt and "accident," see *Letters,* Vol. I, p. 773; "I would give anything," *ibid.,* p. 775; "extremely anxious," *ibid.,* p. 783; "He is so enthusiastic," Long, *op. cit.,* p. 216; "bull in a china shop," *ibid.,* pp. 216–217; Lodge present, in *Autobiography,* p. 216; Dewey credits, *ibid.,* p. 798; letter to Tillinghast, in *Letters,* Vol. I, p. 784; "Everything went well," *ibid.,* p. 790; "crawling back to life," *ibid.,* p. 798; "Hereafter I shall," *Letters,* Vol. II, p. 804; "too busy to exert," Hagedorn, *The Roosevelt Family,* p. 50; "Teddy Roosevelt is capable," *New York Sun,* February 22, 1898; exchange between Wood and McKinley, in *Works,* Vol. XI, p. xvi; "in the name of humanity," *Letters,* Vol. I, pp. 797–798; "Senator Hanna, can we," Brayman, *The President Speaks,* p. 40; "Do you know what that," quoted in Millis, *The Martial Spirit,* p. 130; Edith's visit to the club, in S. J. Morris, *Edith Kermit Roosevelt,* p. 172; "harum-scarum," *Letters,* Vol. I, p. 108; "I believed I could learn," TR, *The Rough Riders,* p. 14; Chanler is quoted in S. J. Morris, *op. cit.,* p. 172; "It was my one chance," quoted in Butt, *Letters of Archie Butt,* p. 146; "If I am to be of any use," *Letters,* Vol. II, p. 808; "We could have," *Rough Riders,* p. 15; "Father went to war," quoted in Hagedorn, *op. cit.,* p. 51.

Chapter 13
"A Body of Cowboy Cavalry"

For the history of the Rough Riders, see TR, *The Rough Riders,* and Jones, *Roosevelt's Rough Riders;* for the war itself, see Trask, *The War with Spain,* Millis, *The Martial Spirit,* and Freidel, *The Splendid Little War;* "exactly as a body," TR, *op. cit.*, p. 32; regimental nicknames, *ibid.*, pp. 36–37; for the regiment's casualties, see Jones, *op. cit.*, p. 6; the wild horse story, *ibid.*, p. 33; for TR and free beer, see *ibid.*, p. 37; "Ted hopes," quoted in S. J. Morris, *Edith Kermit Roosevelt*, p. 175; "This is just a line," *Letters*, Vol. II, p. 832; for naval operations, see Miller, *The U.S. Navy*, pp. 159–162; Roosevelt's little dance, in Jones, *op. cit.*, p. 41; San Antonio farewell, *ibid.*, p. 42; for the trip to Tampa, see TR, *op. cit.*, pp. 38–41; "No head," Private Diary, June 5, 1890; for the "rocking chair" period, see Davis, *The Cuban and Puerto Rican Campaigns*, pp. 45–85; Regular officers scandalized, in *Letters*, Vol. II, p. 835; "an everlasting cad," *ibid.*, Vol. I, p. 358; "energy and enthusiasm," Davis, *op. cit.*, p. 56; "It was a wonderful," *ibid.*, p. 83; for the embarkation of the Fifth Corps, see TR, *op. cit.*, pp. 43–47, and *Letters*, Vol. II, pp. 840–841; "Hello, what can," Freidel, *op. cit.*, p. 68; "sewer," *Letters*, Vol. II, p. 839; Roosevelt worried about delay, *ibid.*, p. 837; "a sapphire sea," *ibid.*, p. 83; "experimented with," quoted in Trask, *op. cit.*, p. 195; "high barren," *Letters*, Vol. II, p. 844; "Shout hurrah," Jones, *op. cit.*, p. 92; for the landing, see TR, *op. cit.*, pp. 51–52, and Jones, *op. cit.*, pp. 97–102; "split the air," *ibid.*, pp. 103–104; preparations for the advance, in Freidel, *op. cit.*, p. 99; "Their frames," TR, *op. cit.*, p. 57.

Chapter 14
"My Crowded Hour"

For the Las Guasimas firefight, see TR, *The Rough Riders*, pp. 60–76, Davis, *The Cuban and Puerto Rican Campaigns*, pp. 134–172, Jones, *Roosevelt's Rough Riders*, pp. 118–140, and Trask, *The War with Spain;* "well, whatever comes," *Letters*, Vol. II, pp. 845–846; "The vultures were wheeling," *Letters*, Vol. II, p. 845; Roosevelt and the military bureaucrat episode, in *Autobiography*, pp. 257–258; "There was no strategy," quoted in Milton, *The Yellow Kids*, p. 323; for the attack on the San Juan Heights, see TR, *op. cit.*, pp. 77–104, Davis, *op. cit.*, pp. 173–223, and Jones, *op. cit.*, pp. 167–194; "I was very glad," TR, *op. cit.*, p. 74; "Above us," Davis, *op. cit.*, pp. 193–194; "It was a very," TR, *op. cit.*, p. 79; "They had no glittering bayonets," Davis, *op. cit.*, pp. 218–220; "The Spanish machine guns!," TR, *op. cit.*, p. 90; "He doubled up," quoted in S. J. Morris, *Edith Kermit Roosevelt*, p. 181; "We are within," *Letters*, Vol. II, p. 846; Rough Rider casualties, *ibid.*, p. 848; "For three days," *ibid.*, p. 851; for naval action off Santiago, see Miller, *The U.S. Navy*, pp. 163–165; "Not since the campaign," *Letters*, Vol. II, p. 849; "He tried to feed them," Stallman, *Stephen Crane*, p. 384; "I have some sick men," *New York Tribune*, September 23, 1898; "You are most kind," *Letters*, Vol. II, p. 860; number of Rough Riders fit for duty, *ibid.*, p. 859; Roosevelt's letter to Shafter and the "round robin," *ibid.*, pp. 864–866; "I feel positively," *New York Herald*, August 16, 1898; "Theodore looks well," Hagedorn, *The Roosevelt Family*, p. 57; "splendid little war," in a letter from Hay to TR, July 27, 1898; the details of the machinations leading up to the gubernatorial nomination, in Appendix II, *Letters*, Vol. II, pp. 1474–1478; Quigg gave his account in a letter to TR, March 19, 1913, quoted in *ibid.*, p. 1474; Roosevelt's account of the meeting with Quigg, in *Autobiog-*

raphy, pp. 280–281; visit of family to Montauk, in Hagedorn, *The Roosevelt Family,* pp. 63–64; for Roosevelt's remarks to the regiment, see Jones, *op. cit.,* pp. 277–280.

Chapter 15
"A Legacy of Work Well Done"

O'Neil's comments, in a letter to TR, October 21, 1898, TRC; "I had a very pleasant," *New York Herald,* September 18, 1898; "We buried past differences," quoted in Harbaugh, *The Life and Times of Theodore Roosevelt,* p. 111; "I do not see how," *Letters,* Vol. II, p. 860; "The matter is," Howe, ed., *Chapman and His Letters,* p. 140; "hardly had been able," *Letters,* Vol. II, p. 880; "idiot variety," *ibid.,* p. 889; if a "man goes," *The Works of Theodore Roosevelt,* Vol. XV, p. 41; "It is something," S. J. Morris, *Edith Kermit Roosevelt,* p. 184; the full details of the affidavit affair, in Chessman, "Theodore Roosevelt's Personal Tax Difficulty"; TR's letter to John Roosevelt, in *Letters,* Vol. II, pp. 799–800; "wearing white flannels," Harbaugh, *op. cit.,* p. 70; Root's comments, in Chessman, *Governor Theodore Roosevelt,* p. 48; "There is great enthusiasm," *Letters,* Vol. II, p. 881; "We cannot avoid facing," *Works,* Vol. XIV, pp. 290–291; to "fix the contest," *Autobiography,* p. 282; Buck Taylor's speech and TR's comment, *ibid.,* p. 127; "No sooner . . . had she," quoted in S. J. Morris, *op. cit.,* p. 188; "has grown mightily," quoted in Gosnell, *Boss Platt,* p. 143; "I have played it," *Letters,* Vol. II, p. 888; "It is absolutely impossible," *Works,* Vol. XV, p. 4; "a most solemn," quoted in S. J. Morris, *op. cit.,* p. 192; "a legacy of work," *ibid.;* "better either," Harbaugh, *op. cit.,* p. 128; "It was necessary to have," *Autobiography,* p. 294; "did not use," *ibid.,* p. 284; "they would have seen," *ibid.,* pp. 297–298; "in appearance," *Letters,* Vol. II, p. 944; "I feel Edie's," S. J. Morris, *op. cit.,* p. 195; "under ferocious attack," Harbaugh, *op. cit.,* p. 305; "I practically went on strike," Longworth, *Crowded Hours,* pp. 26–27; "sheer voodoo," *ibid.,* p. 53; "At that time," *Autobiography,* p. 285; "He has torn down," quoted in Harbaugh, *op. cit.,* p. 122; "I regret to state," *Letters,* Vol. II, p. 1099; "I was hardly prepared," *Autobiography,* p. 308; "at what has been called," *William Barnes Against TR,* Vol. IV, pp. 2368–2375; "I do not believe that it is wise," *Letters,* Vol. II, p. 1008; "clinched his fist," quoted in Harbaugh, *op. cit.,* pp. 117–118; "Well, I suppose," quoted in Chessman, *Governor Theodore Roosevelt,* p. 146; "All together, I am," *Letters,* Vol. II, p. 998; "We cannot avoid," *Works,* Vol. XIII, pp. 319–331; "feeling completely tired," *Letters,* Vol. II, p. 1024; Lee's comments, in *Works,* Vol. X, p. 170; "The more I have," *Letters,* Vol. II, p. 1047.

Chapter 16
"That's an Acceptance Hat"

Bryce is quoted in Thayer, *Theodore Roosevelt,* p. 157; "I have never yet known," *Letters,* Vol. II, p. 1023; Lodge urges, *ibid.,* p. 1124; "I am a comparatively," *ibid.*, p. 1161; "Even to live simply," *ibid.*, p. 1153; much of the following account of TR and the vice presidency is based on Chessman, "Theodore Roosevelt's Campaign Against the Vice-Presidency"; "I don't want him raising hell," quoted in Russell, *The President Makers*, p. 72; for TR's comments on corporations and trusts, see *Works*, Vol. XV, pp. 40–47; "peculiarly intimate relations," *Autobiography*, p. 300; "When I go to war," Chessman, *Governor Theodore Roosevelt*, p. 106; TR's exchange with Odell, in *Autobiography*, p. 300; also see Chessman, *Governor Theodore Roosevelt*, pp. 93–111; "I have always been fond," *Letters*, Vol. II, p. 1141; "All the high monied," *ibid.*; "recent events," *New York Sun*, February 1, 1900; "under no circumstances," *Letters*, Vol. II, p. 1174; "Why Roosevelt," *ibid.*, p. 1337n; "I should be looked upon," *ibid.*; "Teddy has been here," Thayer, *Theodore Roosevelt*, p. 148; "I am expecting," S. J. Morris, *Edith Kermit Roosevelt*, p. 204, and Wood, *Roosevelt As We Knew Him*, pp. 72–73; "an acceptance hat," Olcott, *William McKinley*, Vol. II, pp. 275–276; Bim Bimberg, in *New York Times*, June 12, 1900; for the maneuvering between Platt and TR at the convention, see Russell, *op. cit.*, pp. 83–85, and *Autobiography*, pp. 318–319; "Don't you realize," Pringle, *Theodore Roosevelt*, p. 156; "back sixteen years," *Letters*, Vol. II, p. 1340; "With just a little gasp," *New York World*, June 22, 1900; "Your *duty*," quoted in Leech, *In the Days of McKinley*, p. 542; "I am completely reconciled," *Letters*, Vol. II, p. 1340; "I am as strong as," *ibid.*, p. 1342; "What a thorough-paced," *ibid.*, p. 1408; "The romance of my life," Hagedorn, *Roosevelt in the Bad Lands*, pp. 466–467; also see Vivian, *The Romance of My Life*, pp. 13–32; "He was quiet and dignified," S. J. Morris, *op. cit.*, p. 208; TR inaugural speech, in *Works*, Vol. XV, pp. 77–78; "to see Teddy," Kohlsaat, *From McKinley to Harding*, p. 89; "As I looked at," Longworth, *Crowded Hours*, p. 36; "All in favor," Pringle, *op. cit.*, p. 160; "I shall get fearfully," *ibid.*, p. 14; "I intend studying law," *ibid.*, p. 31; "rather ashamed to say," *ibid.*, pp. 68–69; a "small boisterous person," *ibid.*, p. 74; for the assassination of McKinley, see Olcott, *op. cit.*, pp. 313–333, Leech, *op. cit.*, pp. 586–602, and Lord, *The Good Years*, pp. 41–66; "The President is coming along," *Letters*, Vol. III, p. 139; TR on Mount Marcy, in Murphy, *Theodore Roosevelt's Night Ride*, and *Autobiography*, p. 364; Edith Roosevelt's remarks, in Murphy, *op. cit.*, p. 19; "Mr. Roosevelt, will you," Lord, *op. cit.*, p. 61; "Mr. President, I wish you," *New York Times*, September 16, 1901.

Chapter 17
"I Acted for the Common Well-being. . . ."

Roosevelt's first day in the White House, in Robinson, *My Brother Theodore Roosevelt*, pp. 206–207; "It is a dreadful," *Letters*, Vol. III, p. 150, and Kohlsaat, *From McKinley to Harding*, p. 101; If "you will give," in Pringle, *Theodore Roosevelt*, p. 168; Adams's comments, in Adams's, *Letters of Henry Adams*, Vol. II, p. 469; for TR's views of power, see Blum, *The Progressive Presidents*, pp. 12–22, and *Autobiography*, pp. 364–365; "*How* I wish," quoted in Cashman,

America in the Age of the Titans, p. 61; "Mr. Roosevelt is an entirely," *New York Times*, September 26, 1901; "My view," *Autobiography*, p. 372; "You go into Roosevelt's," quoted in Sullivan, *Our Times*, Vol. III, p. 81; "Every day or two," *ibid.*, Vol. II, p. 394; Steffens's comments, in Steffens, *Autobiography*, p. 503; "Edie says," William Allen White, *Autobiography*, p. 341; "I suppose in a short," Hagedorn, *The Roosevelt Family*, p. 126; "rather vulgar," *ibid.*, p. 128; for the Booker Washington affair, see Gatewood, *Theodore Roosevelt and the Art of Controversy*, Chapter II, and Harlan, *Washington;* "wanted to help," Washington, *My Larger Education*, pp. 168–169; "In the South Atlantic," TRC, October 11, 1901; "as soon as possible," *Letters*, Vol. III, p. 149; "We talked," Washington, *op. cit.*, p. 176; "I had no thought," *Letters*, Vol. III, p. 181; "you need *not*," *ibid.*, p. 195; the "colonel story" appeared in the *Baltimore Herald*, July 3, 1903; "I think that it is," TRC, July 13, 1903; "It really seems hard," quoted in Pringle, *Theodore Roosevelt*, p. 184; "Of all forms," *Autobiography*, p. 439; "the total absence of," *ibid.*, p. 437; first annual message, in *The Works of Theodore Roosevelt*, Vol. XV, pp. 81–138; "to go back on my," *Letters*, Vol. III, pp. 159–160; "prevent violent," *ibid.*, p. 236; "men who seek gain," *Works*, Vol. XV, p. 87; Roosevelt's account of his meeting with Morgan, in Bishop, *Theodore Roosevelt and His Times*, Vol. I, pp. 184–185; for the coal strike, see Greenberg, *Theodore Roosevelt and Labor*, Chapter V; "Is there nothing," *Roosevelt-Lodge Correspondence*, Vol. II, pp. 531–532; "What is the reason," *Letters*, Vol. III, p. 323; for Pittsfield accident, see Lorant, *The Life and Times of Theodore Roosevelt*, pp. 380–382; Edith's comments, in S. J. Morris, *Edith Kermit Roosevelt*, pp. 244–245; "at wit's end," *Letters*, Vol. III, p. 332; "no legal or constitutional," *ibid.*, p. 360; "There would be no," TRC, October 16, 1902; "gross blindness," *Letters*, Vol. III, p. 349; "if it wasn't for," quoted in Pringle, *op. cit.*, p. 190; "a most respectable-looking," *Autobiography*, pp. 489–490; "nearly wild," *Letters*, Vol. III, p. 366; "Suddenly it dawned," *Autobiography*, p. 483; Franklin Roosevelt's comments, quoted in Miller, *F.D.R.*, pp. 38–39; "Occasionally great national crises," *Autobiography*, p. 479; "we shall some day," *Letters*, Vol. IV, p. 1023; second annual message, in *Works*, Vol. XV, pp. 139–168; "I am having a terrific time," *Letters*, Vol. III, p. 406; "I got the bill through," quoted in Pringle, *op. cit.*, p. 240; "It is as right," *Works*, Vol. XV, p. 105; "I do not believe," *Letters*, Vol. III, pp. 272–273.

Chapter 18
". . . The Proper Policing of the World"

For general discussions of TR's foreign policy, see Gould, *The Presidency of Theodore Roosevelt*, Chapter 4, Beale, *Theodore Roosevelt*, and Marks, *Velvet on Iron;* "Whether we desire it," *The Works of Theodore Roosevelt*, Vol. XV, p. 117; "The American people must either," *ibid.*, p. 122; for TR and the navy, see O'Gara, *Theodore Roosevelt and the Rise of the Navy;* for Congress and TR's foreign policy, see Blum, *The Progressive Presidents*, pp. 50–51; "gorged boa constrictor," *Letters*, Vol. IV, p. 734; "More and more," *Works*, Vol. XV, pp. 152–153; "fine figurehead," *Letters,* Vol. VI, p. 1490; "the trouble with our Ambassadors," *Letters,* Vol. IV, p. 1079; "Let them call me," *ibid.,* p. 915; "Will the Prince," *Letters,* Vol. III, p. 230; trying to break through protocol, in Beale, *op. cit.,* pp. 9–11; "For a generation or two," *Letters,* Vol. III, p. 15; Butler's comments, in Beale, *op. cit.,* p. 441; "hammering guns," Balfour, *The Kaiser,* pp. 138–139; "I wish to Heaven," *Letters,* Vol. IV, p. 1159; "unspeakably villainous," in a letter to Douglas Robinson, February 3, 1902, TRC; "If any South American," *Letters,* Vol. III, p. 116; "not guarantee any State," *Works,* Vol. XV, p. 116, and, *Literary Digest,* December 20, 1902; "The striking enforcement," *Letters,* Vol. IV, p. 811; "become convinced that Germany intended," *Letters,* Vol. VIII, pp. 1101–1105; "how I applied," *ibid.,* p. 1107; Thayer's comments, in Thayer, *Theodore Roosevelt,* p. 222; "It is a question of policy," in "The Monroe Doctrine," *Works,* Vol. XIII, pp. 169–181; "I have been hoping," *Letters,* Vol. IV, p. 734; "If a nation shows," *ibid.,* p. 801; "Rather amused at the yell," *ibid.,* pp. 821–822; "I do not much," *ibid.,* p. 1144; "The affairs of the island," *Letters,* Vol. VI, p. 1445; for the Alaska boundary affair, see Collin, *Theodore Roosevelt,* Chapter 7; "To pay them anything," *Letters,* Vol. III, p. 287; TR's letter to Holmes, *ibid.,* p. 529; "settled the last," *Letters,* Vol. VII, p. 28; for the Panama Canal, see McCullough, *The Path Between the Seas,* Bunau-Varilla, *Panama,* and *Diplomatic History of the Panama Canal;* "REVOLUTION IMMINENT" and other cable traffic, in *Diplomatic History,* pp. 345–363; "I took the Canal Zone," *New York Times,* March 24, 1911; comparable to Louisiana Purchase, in *Letters,* Vol. III, p. 685; "the most dangerous," McCullough, *op. cit.,* p. 271; "I was lucky enough," Bunau-Varilla, *op. cit.,* p. 247; "The great bit of work," *Letters,* Vol. III, p. 284; "Why cannot we," *ibid.,* p. 318; "To wait a few months," *ibid.,* p. 595; "bar one of the future," *ibid.,* p. 567; draft of TR's message to Congress, in *Autobiography,* pp. 544–545; fifty-three uprisings, in *Works,* Vol. XV, pp. 207–209; "I have no idea," *Letters,* Vol. III, p. 689; "Instantly I bolted," *ibid.,* p. 644; conversations with Knox and Root, in McCullough, *op. cit.,* p. 383; "There had been innumberable," Wood, *Roosevelt as We Knew Him,* p. 153; "To have acted otherwise," *Autobiography,* p. 539.

Chapter 19
"The Bride at Every Wedding . . ."

Eleanor Roosevelt's account of her wedding, in E. Roosevelt, *This Is My Story,* pp. 124–126; "bully pulpit," in Abbott, *A Review;* "Wherever he goes," A. R. Hale, *Rooseveltian Fact and Fiction,* p. 73; "one of the most wearing," Thayer, *Theodore Roosevelt,* p. 259; "I don't think any family," *Letters,* Vol. V, p. 840; "Distinguished civilized men," Wister, *Roosevelt, The Story of a Friendship,* p. 124; "only . . . President since," Mencken, *Prejudices: Fifth Series,* pp. 184–185; "Now, Jimmy," Thayer, *op. cit.,* p. 266; "the wildest scramble," Hoover, *Forty-two Years in the White House,* p. 28; for the "White House Gang," see Looker, *The White House Gang;* "Hit it!," Amos, *Theodore Roosevelt,* p. 13;

hide-and-seek game from personal conversation; "The other night before," quoted in Busch, *TR: The Story of Theodore Roosevelt*, p. 173; the ride to Warrenton, in *Autobiography*, p. 49; also see Grayson, "Don't Spare the Horses"; Roosevelt's domestic life is reconstructed from Hoover, *op. cit.*, Amos, *op. cit.*, Crook, "The Home Life of Roosevelt," Hagedorn, *The Roosevelt Family*, and W. B. Hale, *A Week in the White House;* "more in the nature," quoted in Wagenknecht, *op. cit.*, p. 26; for the reconstruction of the White House, see Collin, *Theodore Roosevelt*, pp. 27–45, and S. J. Morris, *Edith Kermit Roosevelt*, pp. 245–258; Roosevelt dictating, in Sullivan, *Our Times*, Vol. II, p. 397; "Well, he's a crook," *ibid.;* "A more skillful barber," Brownlow, *A Passion for Politics*, p. 399; Davis's comments, in Davis, *Released for Publication*, p. 128; for the "Dear Maria" affair, see Sullivan, *op. cit.*, Vol. III, pp. 99–128; for the "nature-fakers," see *ibid.*, pp. 146–162; "I ought not," *Letters*, Vol. V, p. 61; for simplified spelling incident, see Sullivan, *op. cit.*, pp. 162–190; "I could not by fighting," *Letters*, Vol. V, p. 527; for Cabinet meeting, see Straus Diary; "a more eager," Pinchot, *Breaking New Ground*, p. 229; "In politics, we have to," quoted in Barry, *Forty Years in Washington*, p. 279; Roosevelt's eating habits are described in Wagenkecht, *op. cit.*, p. 26; "The President throws off," Straus Diary; for Roosevelt and the graphic arts, see Wagenknecht, *op. cit.*, pp. 81–84, and Gatewood, *Theodore Roosevelt and the Art of Controversy*, pp. 213–235; "close alongside," in a letter from TR to L. F. Abbott, November 27, 1907, TRC; Roosevelt as a Renaissance prince, in Einstein, *Roosevelt: His Mind in Action*, pp. v–vi; for Edith Roosevelt as first lady, see S. J. Morris, *op. cit.*, and Anthony, *First Ladies;* "was taken sick," S. J. Morris, *op. cit.*, pp. 237 and 265; "Invisible Government," Wister, *op. cit.*, p. 89; "We all knew," N. Roosevelt, *Theodore Roosevelt*, p. 28; Roosevelt and drinking, in Amos, *op. cit.*, p. 19; he had "a horror," S. J. Morris, *op. cit.*, p. 265; "as if a decimal," Wister, *op. cit.*, pp. 126–128; "Do you know," *ibid.*, p. 114; dinner with Henry James, in Edel, *Henry James: The Master*, pp. 275–276; "I want to feel," quoted in Juergens, *News from the White House*, p. 37; "You have been having," *Letters*, Vol. V, pp. 42–43; for Alice, see Teague, *Mrs. L.*, Felsenthal, *Princess Alice*, Rozek, "The First Daughter of the Land," and Brough, *Princess Alice;* "Sister continues," *Letters*, Vol. III, p. 408; "Listen, I can be," Wister, *op. cit.*, p. 87; "a poor, demented," Brough, *op. cit.*, pp. 155–156; "I love you," quoted in S. J. Morris, *op. cit.*, p. 546; "Alice is really in love," *ibid.;* for the wedding see sources cited above.

Chapter 20
"I Am No Longer a Political Accident"

"How they are voting," quoted in Gould, *The Presidency of Theodore Roosevelt*, p. 127; "I have the greatest," *Letters*, Vol. IV, p. 1024; "It is a particular," *ibid.*, p. 1037; for TR and Hanna, see Josephson, *The President Makers*, pp. 151–154; "*They* don't want," McCaleb, *Theodore Roosevelt*, pp. 198–220; "Those who favor," *New York Tribune*, May 26, 1903; "I . . . decided that the time,"

Letters, Vol. III; "No man had larger," *ibid.,* Vol. IV, p. 703; "who in the name," *ibid.,* p. 749; for the Raisuli-Perdicaris affair, see Hourihan, *Roosevelt and the Sultans,* Chapter VI; for 1904 election, see Sullivan, *Our Times,* Vol. II; "I always like to," *Letters,* Vol. IV, p. 858; "We base our appeal," *ibid.,* p. 921; "The people like," *World's Work,* December 1904; for ER on TR's third-term announcement, see Hagedorn, *The Roosevelt Family,* pp. 272–273, and S. J. Morris, *Edith Kermit Roosevelt,* pp. 280–281; for TR and the Russo-Japanese War, see Esthus, *Theodore Roosevelt and Japan,* Beale, *Theodore Roosevelt,* Chapter V, and Trani, *The Treaty of Portsmouth;* "thoroughly well pleased," *Letters,* Vol. IV, p. 274; for TR and the Jews, see "The Diplomacy of Neutrality: Theodore Roosevelt and the Russian Pogroms of 1903–1906," *Presidential Studies Quarterly,* Winter 1989; "lives quietly," quoted in E. Morris, *The Rise of Theodore Roosevelt,* pp. 285–287; "so jumpy," *Letters,* Vol. IV, p. 1181; "his usual rigmarole," *ibid.,* p. 1222; "Oh Lord!," *ibid.,* p. 1258; "The more I see," *ibid.,* p. 1230; "I am having my hair," *ibid.,* p. 1317; "Make clear to," *ibid.,* p. 1312; "in my judgement," *ibid.;* "I hope your government," *ibid.;* "Dear Baron," *ibid.;* "Peace can be," *ibid.,* p. 1317; "It's a mighty good thing," Hagedorn, *op. cit.,* p. 230; for Moroccan affair, see Beale, *op. cit.,* Chapter VI; "I do not feel as a Government," *Letters,* Vol. IV, p. 1162; "Our internal problems," *ibid.,* p. 1133; "The dull purblind," *Letters,* Vol. V, p. 183; "The crying evil," *New York Times,* March 5, 1905; for the 1904 annual message, see *The Works of Theodore Roosevelt,* pp. 215–266; for the fight over the Hepburn Act, see Josephson, *op. cit.,* pp. 228–236, and Merrill, *Bourbon Leader,* pp. 215–222; "It is really," *Letters,* Vol. IV, p. 1209; "because it would irritate," quoted in S. J. Morris, *op. cit.,* p. 294; for the 1905 annual message, see *Works,* Vol. XV, pp. 270–341; "a curse," quoted in Mowry, *The Era of Theodore Roosevelt,* pp. 200–201; for Roosevelt's attack on the muckrackers, see Juergens, *News from the White House,* pp. 72–79; "a fine piece," in a letter to Kermit in Theodore Roosevelt Papers, June 13, 1906; for the fight over pure food and drug and meat inspection laws, see Davidson and Lytle, "USDA Government Inspected," in *After the Fact,* pp. 232–262; "Each change," *Letters,* Vol. V, p. 291; "I . . . do not want," *ibid.,* pp. 300–301; "It has been," *ibid.,* p. 329; "The railroad rate bill," *ibid.,* p. 328.

Chapter 21
"To Keep the Left Center Together"

"If five hundred," Gould, *The Presidency of Theodore Roosevelt,* p. 225; Harrisburg speech, in *Works,* Vol. XVI, pp. 71–73; "he preferred," *Letters,* Vol. IV, p. 453; "Any such statement," *ibid.,* p. 447; "It is very gratifying," *ibid.,* pp. 488–489; "To use the terminology," *ibid.,* p. 875; the annual message of 1906, in *Works,* Vol. XV; for the Brownsville affair, see Lane, *The Brownsville Affair,* and Weaver, *The Brownsville Raid;* "You cannot have any information," Theodore Roosevelt Papers, November 5, 1906; rejection of Taft's order, in *Letters,* Vol. IV, p. 498; for antilynching and black education statement, see

Works, Vol. XV, pp. 351–355; for Foraker as a front man for Wall Street, see *Letters,* Vol. VI, p. 1026; for the Gridiron episode, see Brayman, *The President Speaks,* pp. 48–55, and Clark, *My Quarter Century,* Vol. I, pp. 446–448; TR and the Cincinnati paper, in Sullivan, *Our Times,* Vol. II, p. 225; for conservation, see Hays, *Conservation and the Gospel of Efficiency,* Pinchot, *Breaking New Ground;* Cutright, *Theodore Roosevelt,* and Gable, "President Theodore Roosevelt's Record on Conservation"; TR's defense of his homesteading policy, in *Letters,* Vol. IV, pp. 681–684; "Mr. Heney cannot hurt," *ibid.,* p. 1177; "We still continue to enjoy," *Works,* Vol. XV, p. 342; for the Panic of 1907, see Allen, *The Lords of Creation,* Chapter IV, and Sobel, *Panic on Wall Street,* Chapter IX; "this belief in Wall Street," *Letters,* Vol. V, p. 631; TR on the causes of the Panic, *ibid.,* pp. 747–748; "The general impression," quoted in S. J. Morris, *Edith Kermit Roosevelt,* p. 320; "certain malefactors," *Works,* Vol. XVI, p. 84; "sat solemnly," *Letters,* Vol. V, p. 696; "There isn't any place," *ibid.,* p. 800; "I have been greatly interested," *ibid.,* p. 821; "the underlying conditions," *ibid.,* p. 823; "It was necessary for me," *Autobiography,* p. 455; "I answered Messers. Frick," *ibid.,* p. 456; "The result justified," *ibid.,* p. 457; for TR and the Japanese crisis, see Esthus, *Theodore Roosevelt and Japan,* Chapters VIII–XII; "I am horribly bothered," *Letters,* Vol. V, p. 475; "I shall exert," *ibid.,* p. 473; "a wicked absurdity," *Works,* Vol. XV, pp. 386–387; "I am exceedingly anxious," TR Papers, July 30, 1907; "Mr. President, do you believe," Van Deman Papers, Part I, pp. 20–21; for the voyage of the Great White Fleet, see Reckner, *Teddy Roosevelt's Great White Fleet;* "Thank Heaven we have," *Letters,* Vol. V, p. 717; for TR's conflict with Hale, see *Autobiography,* p. 568; "In my own judgement," *ibid.,* p. 563.

Chapter 22
"We Had Better Turn to Taft"

"I am the seventh son," Russell, *The President Makers,* p. 87; "with all his faults," Gould, *The Presidency of Theodore Roosevelt,* p. 225; conversations with Loeb and Taft, in Sullivan, *Our Times,* Vol. III, p. 289; "I should like to have," *Letters,* Vol. VI, p. 1242; for the profile of Taft, see Manners, *TR and Will,* Chapter III, and Mowry, *The Era of Theodore Roosevelt,* pp. 231–235; the annual message of 1907, in *Works,* Vol. XV, pp. 410–488; Special Message, in *Letters,* Vol. VI, Appendix III; "deepest and most earnest," in a letter to Kermit, TR Papers, January 31, 1908; "a showdown with my foes," *Letters,* Vol. VI, p. 925; "I appointed no man," quoted in Pringle, *Theodore Roosevelt,* pp. 497–498; for the Republican convention and the campaign of 1908, see Manners, *op. cit.,* and Gardner, *Departing Glory;* "my last chance," *Letters,* Vol. VI, p. 1388; Alice's reaction to Taft's speech is in Longworth, *Crowded Hours,* p. 148; "you big, generous," *Letters,* Vol. V, p. 1231; "won a great personal victory," *ibid.,* p. 1340; "Abraham Lincoln and the bond seller," quoted in Gould, *op. cit.,* p. 390; "Ha ha!," *Letters,* Vol. V, p. 1454; "I am ending my career," *ibid.,* p. 1432; for the Secret Service controversy, see Gatewood, *Theodore Roosevelt and the Art of Controversy,* Chapter VIII; "did not themselves," *Works,* Vol.

XV, pp. 527–528; TR's assessment of his presidency, in *Letters,* Vol. VIII, pp. 1113–1114; "I have done my work," *ibid.,* Vol. VI, p. 1541; "He's all right," Sullivan, op. cit., Vol. IV, pp. 331–332 ; for the family dinner and events leading up to the inauguration, see Manners, *op. cit.,* and Gardner, *op. cit.*

Chapter 23
"My Hat Is in the Ring"

TR's arrival in East Africa, in *New York Times,* April 22, 1909, and Foran, "With Roosevelt in Africa," *Field and Stream,* October 1912; "I speak of Africa," *The Works of Theodore Roosevelt,* Vol. IV, pp. xxiii–xxv; equipment for safari, in *Letters,* Vol. VII, pp. 13–14; article on pigskin-bound library, in *Works,* Vol. XIII, pp. 337–340; "I might have," *ibid.,* Vol. IV, p. 28; "I sprang to one side," *ibid.,* p. 62; "The firelight," *ibid.,* p. 69; musical greeting, *ibid.,* pp. 313–314; "Oh, sweetest of all," S. J. Morris, *Edith Kermit Roosevelt,* pp. 351–352; "My main reason," *Letters,* Vol. VI, p. 1166; Taft falling asleep, in Butt, *Taft and Roosevelt,* Vol. I, p. 85; "to make peace," *ibid.,* p. 296; TR to Lodge on the tariff, in *Letters,* Vol. VII, p. 9; Taft on Pinchot, in Butt, *op. cit.,* pp. 244–245; Bonaparte's comment, in Mowry, *Theodore Roosevelt and the Progressive Movement,* p. 86; "I cannot believe," *Letters,* Vol. VII, p. 45; "before I even," *ibid.,* pp. 50–51; "hail me as," quoted in Manners, *TR and Will,* p. 143; "a Tammany Boodle," *Letters,* Vol. VII, p. 66; "I brought him," quoted in Manners, *op. cit.,* p. 148; "the very strongest," *Letters,* Vol. VII, pp. 69–74; "With him and the Kaiser," Butt, *op. cit.,* p. 348; TR's account of the funeral, in *Letters,* Vol. VII, pp. 409–413; "He has enjoyed," S. J. Morris, *op. cit.,* p. 360; "In the way of grading," quoted in Pringle, *Theodore Roosevelt,* p. 520; "His horizon," Butt, *op. cit.,* p. 396; "Ugh! I do dread," *Lodge-Roosevelt Correspondence,* Vol. II, pp. 379–380; "much concerned," *Letters,* Vol. VII, pp. 88–89; to "close up," *New York Times,* June 19, 1910; "Back from Elba," Mowry, *op. cit.,* p. 118; "I want to tell you," *New York Times,* June 26, 1910; "Ah, Theodore," Butt, *op. cit.,* p. 418; "Taft has passed," *Letters,* Vol. VII, pp. 95–96; Osawatomie speech, in *Works,* Vol. XVII, pp. 5–22; "outrageous," *Letters,* Vol. VII, pp. 135–136; "I think that the American people," Bishop, *Theodore Roosevelt and His Times,* Vol. II, p. 390; "The one comfort," *Letters,* Vol. VII, p. 168; "safely caged," S. J. Morris, *op. cit.,* p. 369; "greatly admire," *Letters,* Vol. VII, p. 293; "that all he wanted," *ibid.,* p. 300; "Our life," *ibid.,* p. 334; "We don't care," *ibid.,* p. 233; "small, mean," *ibid.,* p. 431; TR's reply to the steel case, in *The Outlook,* November 18, 1911; "I am not a candidate," *Letters,* Vol. VII, p. 467; "Whether you wish," in Theodore Roosevelt Papers, December 1, 1911; for TR and the judiciary, see *Works,* Vol. XVII, pp. 74–99; "I have had my mishaps," *Letters,* Vol. VII, p. 515n.; "My dear fellow," *ibid.;* "But they will have to steal," *ibid.,* p. 553; for the conventions, see Mowry, *op. cit.,* pp. 237–267, and Gable, *The Bull Moose Years,* pp. 19–110; "Victory is," quoted in Gable, *op. cit.,* p. 16; "In strict confidence," *Letters,* Vol. VII, p. 568; "What a miserable," in a letter to J. C. O'Laughlin, July 9, 1912, TR

Papers; "an able man," *Letters,* Vol. VII, p. 598; "In loyalty," in Wood, *Roosevelt,* pp. 250–254; for the campaign, see Gable, *op. cit.,* pp. 111–130, and Mowry, *op. cit.,* pp. 268–283; for the Milwaukee speech, see *Works,* Vol. XVII, pp. 320–330; for assessment of the outcome of the election, see Gable, *op. cit.,* pp. 131–132; "We had all the money," *Letters,* Vol. VII, p. 633; "We have fought," *ibid.,* p. 637.

Chapter 24
"The Great Adventure"

"When it is evident," Bishop, *Theodore Roosevelt and His Times,* Vol. II, p. 355; "It is very difficult," *Letters,* Vol. VII, p. 689; "The hardest task," *ibid.,* p. 710; for the plight of the Progressive party, see Gable, *The Bull Moose Years,* pp. 157–181; Mowry, *Theodore Roosevelt and the Progressive Movement,* pp. 284–303; "very weary work," *Letters,* Vol. VII, p. 660; see little hope for the party, *ibid.,* p. 684; for the libel suit, see Lorant, *The Life and Times of Theodore Roosevelt,* pp. 585–586, pp. 588–589; "I think he feels," quoted in S. J. Morris, *Edith Kermit Roosevelt,* p. 398; for the South American adventure, see *Through the Brazilian Wilderness,* Vol. V; "No less than six weeks," *Letters,* Vol. VII, pp. 759–760; "younger for every pound," S. J. Morris, *op. cit.,* pp. 402–403; "I am now an old man," TRC, July 26, 1913; "I regard the proposed," *Letters,* Vol. VII, p. 778; "As soon as I got," *ibid.,* p. 769; for Pittsburgh speech, see *ibid.,* p. 777; TR's support of neutrality, in *New York Times,* August 6, 1914; "You've got to," Hagedorn, *The Roosevelt Family,* pp. 341–342; "register a very," *Letters,* Vol. VII, p. 810; "I have done everything," Davis, *Released for Publication,* p. 441; "I don't think they can," *Letters,* Vol. VIII, p. 831; "professional yodeler," *ibid.,* Vol. VII, p. 790; administration acquiesced in invasion, in *Works,* Vol. XVIII, p. 169; "In this way," quoted in Harbaugh, *The Life and Times of Theodore Roosevelt,* p. 447; "Wilson is, I think," *Letters,* Vol. VIII, p. 841; "That's murder!," Hagedorn, *The Bugle That Woke America,* pp. 68–69; "I do not believe that the firm," *Works,* Vol. XVIII, pp. 377–381; "a man whose wife," quoted in Gardner, *Departing Glory,* p. 335; "Did you notice," Pringle, *Theodore Roosevelt,* p. 584; "I never wished," quoted in Wagenknecht, *op. cit.,* p. 281; "not Americans at all," *Works,* Vol. XX, pp. 324–325; "My own judgement," *Letters,* Vol. VIII, p. 1037; "I will not enter," *New York Times,* March 10, 1916; rejected Lodge, in *Letters,* Vol. VIII, p. 1061; "I AM VERY GRATEFUL," *ibid.,* p. 1062; "For a moment there was silence," W. A. White, *Autobiography,* pp. 526–527; "I am now being worked," *Letters,* Vol. VIII, p. 1120; "At times in the thick," quoted in Gardner, *op. cit.,* p. 357; "There should be shadows," *Works,* Vol. XVIII, pp. 451–452; "a straight-from-the-shoulder," *Letters,* Vol. VIII, p. 1124; "profound gratitude," quoted in Millis, *The Road to War,* pp. 351–352; "I am completely," *Letters,* Vol. VIII, p. 1125; "This is yellow," *ibid.,* p. 1135; "I don't apologize," quoted in Harbaugh, *op. cit.,* p. 466; "The President's great message," *Metropolitan,* April 1917; "Mr. President, what I have said," *Letters,* Vol. VIII, p. 1173; "Yes, he is a great," quoted in Tumulty, *Wilson as I Knew Him,* pp. 285–288; Clemenceau's comments, in *Letters,* Vol. VIII, p. 1201n.; "playing the dirtiest," Wister, *Roosevelt, the Story of a Friendship,* p. 365; "They have all gone," is quoted in S. J. Morris, *op. cit.,* p. 413; "In this war," *Works,* Vol. XIX, p. 31; "We must have one," Bishop, *op. cit.,* Vol. II, p. 364; "We have demanded," *Letters,* Vol. VIII, p. 1211; "I wish to do everything," *Letters,* Vol. VIII, p. 1306; meeting between TR and Taft, described in Manners, *TR and Will,* p. 305; "Here, spring," *Letters,* Vol. VIII, p. 1311; "You don't know how proud," *ibid.,* p. 1301; "It is wicked," *ibid.,* p. 1347; "You get so excited," quoted in S. J. Morris, *op. cit.,* p. 422; "Of course we are immensely," *Letters,* Vol. VIII, p. 1351; "poor Quinikens," Pringle, *op. cit.,* p. 601; "Only those are fit," *Works,* Vol. XIX, p. 243; "I have but one fight," Robinson, *My Brother Theodore Roosevelt;* "We can pay with the blood," *Works,* Vol. XIX, p. 394; "stinging rebuke," *Letters,* Vol. VIII, p. 1390; "Well, anyway," Robinson, *op. cit.,* p. 362; "if the leaders," Bishop, *op. cit.,* Vol. II, p. 469; "We have room for," *Letters,* Vol. VIII, p. 1422; for TR's death, see letters from EKR to Ted and Kermit, January 12, 1919, and Amos, *Theodore Roosevelt,* p. 157.

Bibliography

Manuscript Collections

Theodore Roosevelt Collection, Harvard University Library
Theodore Roosevelt Papers, Library of Congress
Elihu Root Papers, Library of Congress
Alice Roosevelt Longworth Collection, Harvard University Library
Roosevelt Family Papers, Harvard University Library
Anna Roosevelt Cowles, Reminiscences, Typescript in Theodore Roosevelt Collection, Harvard University Library
Oscar S. Straus Diary, Library of Congress
George B. Cortelyou Papers, Library of Congress
Ralph Van Deman Papers, Georgetown University Library
Transcripts of Interviews by Henry Pringle, Theodore Roosevelt Collection, Harvard University Library
William Allen White Papers, Library of Congress
National Archives RG 45, Navy Department Records
National Archives RG 59, State Department Records

Published Correspondence

Adams, Henry. *The Letters of Henry Adams*. 6 vols. Edited by J. C. Levenson et al. Cambridge, Mass.: Harvard University Press, 1982.
Chapman, John Jay. *John Jay Chapman and His Letters*. 2 vols. Edited by Mark A. de Wolfe Howe. Boston: Houghton Mifflin, 1937.
Cowles, Anna Roosevelt. *Letters from Theodore Roosevelt to Anna Roosevelt Cowles*. New York: Scribners, 1921.
Hay, John. *Letters of John Hay*. 3 vols. Edited by Henry Adams. New York: privately printed, 1908.
Lodge, Henry Cabot, ed. *Selections from the Correspondence of Theodore Roosevelt and Henry Cabot Lodge, 1884–1918*. 2 vols. New York: Scribners, 1925.
Roosevelt, Franklin D. *F.D.R. His Personal Letters*. 4 vols. New York: Duell, 1947–1950.

Roosevelt, Theodore. *Letters to Kermit*. Edited by Will Irwin. New York: Scribners, 1946.

————. *The Letters of Theodore Roosevelt*. 8 vols. Edited by Elting E. Morison. Cambridge, Mass.: Harvard University Press, 1951–1954.

————. *Theodore Roosevelt's Letters to His Children*. New York: New American Library, 1964.

Spring Rice, Cecil. *The Letters of Sir Cecil Spring Rice*. Edited by Stephen Gwynn. Boston: Houghton Mifflin, 1929.

Books and Articles

Abbott, Laurence F. *Impressions of Theodore Roosevelt*. New York: Doubleday, Page, 1919.

Abbott, Lyman. "A Review of President Roosevelt's Administration," *The Outlook,* February 27, 1909.

Adams, Henry. *The Education of Henry Adams*. New York: Modern Library, 1931.

Adler, Arthur. *Understanding Human Nature*. Translated by Walter B. Wolfe. New York: Greenberg Publishers, 1927.

Allen, Frederick Lewis. *The Lords of Creation*. Chicago: Quadrangle, 1966.

Amos, James E. *Theodore Roosevelt: Hero to His Valet*. New York: John Day, 1927.

Anthony, Carl Sferrazza. *First Ladies*. New York: Morrow, 1990.

Bailyn, Bernard, et al. *Glimpses of the Harvard Past*. Cambridge, Mass.: Harvard University Press, 1986.

Balfour, Michael. *The Kaiser and His Times*. Boston: Houghton Mifflin, 1964.

Baltzell, E. Digby. *The Protestant Establishment*. New York: Vintage Books, 1966.

Barry, David S. *Forty Years in Washington*. Boston: Little, Brown, 1924.

Beale, Howard K. *Theodore Roosevelt and the Rise of America to World Power*. Baltimore: Johns Hopkins University, 1984.

Beer, Thomas. *Hanna, Crane and the Mauve Decade*. New York: Knopf, 1941.

Berman, Jay S. *Police Administration and Progressive Reform: Theodore Roosevelt as Police Commissioner*. Westport, Conn.: Greenwood Press, 1988.

Billings, Elden E. "Social and Economic Life in Washington in the 1890's," *Records of the Columbia Historical Society*. Washington, D.C.: Columbia Historical Society, 1966–1968.

Bishop, Joseph B. *Theodore Roosevelt and His Times*. 2 vols. New York: Scribners, 1930.

Bleyer, Bill. "The Forgotten Roosevelt," *Newsday Magazine,* October 6, 1985.

Blum, John Morton. *The Progressive Presidents*. New York: Norton, 1980.

————. *The Republican Roosevelt*. Cambridge, Mass.: Harvard University Press, 1977.

Braisted, William R. *The United States Navy in the Pacific, 1897–1909*. Austin, Tex.: University of Texas Press, 1958.

Brayman, Harold. *The President Speaks off the Record*. Princeton, N.J.: Dow Jones, 1976.

Brough, James. *Princess Alice*. Boston: Little, Brown, 1975.

Brown, Charles H. *The Correspondents' War*. New York: Scribners, 1967.

Brownlow, Louis. *A Passion for Politics*. Chicago, Ill.: University of Chicago Press, 1958.

Bunau-Varilla, Philippe. *Panama: The Creation, Destruction and Resurrection*. New York: Robert M. McBride, 1920.

Burton, David H. *Theodore Roosevelt*. New York: Twayne, 1972.

———. *Theodore Roosevelt: Confident Imperialist*. Philadelphia: University of Pennsylvania Press, 1968.

Busch, Noel F. *TR: The Story of Theodore Roosevelt and His Influence on Our Times*. New York: Reynal, 1963.

Butt, Archibald. *The Letters of Archie Butt*. Edited by Lawrence Abbott. Garden City, N.Y.: Doubleday, Page, 1924.

———. *Taft and Roosevelt: The Intimate Letters of Archie Butt*. 2 vols. Garden City, N.Y.: Doubleday, Doran, 1930.

Cadenhead, I. E. *Theodore Roosevelt and the Paradox of Progressivism*. Woodbury, N.Y.: Barron's Educational Series, 1974.

Cashman, Sean D. *America in the Age of the Titans*. New York: New York University Press, 1988.

Challener, Richard D. *Admirals, Generals, and American Foreign Policy*. Princeton, N.J.: Princeton University Press, 1973.

Chanler, Mrs. Winthrop. *Roman Spring: Memoirs*. Boston: Little, Brown, 1934.

Chessman, G. Wallace. *Governor Theodore Roosevelt*. Cambridge, Mass.: Harvard University Press, 1965.

———. *Theodore Roosevelt and the Politics of Power*. Boston: Little, Brown, 1969.

———. "Theodore Roosevelt's Campaign Against the Vice-Presidency," *The Historian*, Spring 1952.

———. "Theodore Roosevelt's Personal Tax Difficulty," *New York History*, January 1953.

Chidsey, Daniel Barr. *The Gentleman from New York: The Life of Roscoe Conkling*. New Haven, Conn.: Yale University Press, 1935.

Clark, Champ. *My Quarter Century of American Politics*. Vol. I. New York: Harper, 1920.

Clymer, Kenton J. *John Hay*. Ann Arbor, Mich.: University of Michigan Press, 1975.

Collin, Richard H. *Theodore Roosevelt: Culture, Diplomacy and Expansionism*. Baton Rouge, La.: Louisiana State University Press, 1985.

———. *Theodore Roosevelt's Caribbean*. Baton Rouge, La.: Louisiana State University Press, 1990.

Commager, Henry S. *The American Mind*. New Haven, Conn.: Yale University Press, 1950.

Cooper, John Milton, Jr. *The Warrior and the Priest*. Cambridge, Mass.: Belknap Press, 1983.

Cosmas, Graham A. *An Army for Empire: The U.S. Army in the Spanish-American War*. Springfield, Mo.: University of Missouri Press, 1971.

Croly, Herbert. *Marcus Alonzo Hanna: His Life and Work*. New York: Macmillan, 1912.

———. *The Promise of American Life*. Cambridge, Mass.: Belknap Press, 1965.

Crook, William H. "The Home Life of Roosevelt," *Saturday Evening Post*, March 11, 1911.

Cutright, Paul Russell. *Theodore Roosevelt: The Making of a Conservationist*. Urbana, Ill.: University of Illinois Press, 1985.

———. *Theodore Roosevelt, the Naturalist*. New York, Harper, 1956.

Dailey, Wallace F. "Theodore Roosevelt in Periodical Literature, 1950–1981," *Theodore Roosevelt Association Journal*, Fall 1982.

Dalton, Kathleen, *The Early Life of Theodore Roosevelt*. Ph.D. dissertation, Johns Hopkins University, 1979.

Davidson, James W., and Mark H. Lytle. "USDA Government Inspected," *After the Fact*. New York: Knopf, 1982.

Davis, Oscar K. *Released for Publication*. Boston: Houghton Mifflin, 1925.

Davis, Richard Harding. *The Cuban and Puerto Rican Campaigns*. New York: Scribners, 1899.

———. *Notes of a War Correspondent*. New York: Scribners, 1911.

Dennett, Tyler. *John Hay*. New York: Dodd, Mead, 1933.

———. *Roosevelt and the Russo-Japanese War*. Garden City, N.Y.: Doubleday, Page, 1925.

Dresden, Donald. *The Marquis de Morès: Emperor of the Bad Lands*. Norman, Okla.: University of Oklahoma Press, 1970.

Dulles, Foster Rhea. *The Imperial Years*. New York: Crowell, 1966.

Dunne, Finley Peter. *Mr. Dooley Says*. New York: Scribners, 1910.

Dyer, Thomas G. *Theodore Roosevelt and the Idea of Race*. Baton Rouge, La.: Louisiana State University Press, 1980.

Edel, Leon. *Henry James: The Master 1901–1916*. Philadelphia: Lippincott, 1972.

Einstein, Louis. *Roosevelt: His Mind in Action*. Boston: Houghton Mifflin, 1930.

Esthus, Richard A. *Double Eagle and Rising Sun: The Russians and Japanese at Portsmouth in 1905*. Durham, N.C.: Duke University Press, 1988.

———. *Theodore Roosevelt and Japan*. Seattle, Wash.: University of Washington, 1966.

———. *Theodore Roosevelt and the International Rivalries*. Waltham, Mass.: Ginn-Blaisdell, 1970.

Felsenthal, Carol. *Princess Alice: The Life and Times of Alice Roosevelt Longworth*. New York: St. Martin's Press, 1988.

Foran, W. Robert. "With Roosevelt in Africa," *Field and Stream*, October 1912.

Foulke, William D. *Fighting the Spoilsmen*. New York: Putnam, 1919.

Freidel, Frank. *The Splendid Little War*. New York: Bramwell House, 1958.

Freidlander, Robert A. "A Reassessment of Roosevelt's Role in the Panamanian Revolution of 1903," *Western Political Quarterly*, April 1961.

Gable, John A. *Adventure in Reform*. Milford, Pa.: Grey Towers Press, 1985.

————. *The Bull Moose Years.* Port Washington, N.Y.: Kennikat Press, 1978.

————. "President Thoedore Roosevelt's Record on Conservation," *Theodore Roosevelt Association Journal,* Fall 1984.

Gardner, Joseph L. *Departing Glory.* New York: Scribners, 1973.

Garraty, John A. *Henry Cabot Lodge: A Biography.* New York: Knopf, 1953.

————. *Right-Hand Man: The Life of George W. Perkins,* New York: Harper, 1960.

Gatewood, Willard B., Jr. *Theodore Roosevelt and the Art of Controversy.* Baton Rouge, La.: Louisiana State University Press, 1970.

Gompers, Samuel. *Seventy Years of American Life and Labor.* 2 vols. New York: Dutton, 1925.

Gosnell, Harold F. *Boss Platt and His New York Machine.* New York: AMS Press, 1969.

Gould, Lewis L. *The Presidency of Theodore Roosevelt.* Lawrence, Kans.: University Press of Kansas, 1991.

————. *The Presidency of William McKinley.* Lawrence, Kans.: University Press of Kansas, 1980.

————. *Reform and Regulation: American Politics from Roosevelt to Wilson.* New York: Knopf, 1986.

Grayson, Gary T. "Don't Spare the Horses: It's Rough to Be Around a Rider When He's President," *American Heritage,* February 1974.

Greenberg, Irving. *Theodore Roosevelt and Labor: 1900–1918.* New York: Garland, 1988.

Greene, Constance. *Washington: Capital City.* Princeton, N.J.: Princeton University Press, 1962.

Grenville, John A. S. "American Preparations for War with Spain," *Journal of American Studies,* Spring 1968.

Hagan, Kenneth J., ed. *In Peace and War.* Westport, Conn.: Greenwood Press, 1984.

————. *This People's Navy: The Making of American Sea Power.* New York: Free Press, 1991.

Hagedorn, Hermann. *The Boy's Life of Theodore Roosevelt.* New York: Harper & Brothers, 1918.

————. *The Bugle That Woke America.* New York: John Day, 1940.

————. *The Roosevelt Family of Sagamore Hill.* New York: Macmillan, 1954.

————. *Roosevelt in the Bad Lands.* Boston: Houghton Mifflin, 1921.

Hale, Annie Riley. *Rooseveltian Fact and Fiction.* New York: Broadway Publishing Co., 1908.

Hale, William B. *A Week in the White House with Theodore Roosevelt.* New York: Putnam, 1908.

Hall, Luella J. *The United States and Morocco 1776–1956.* Metuchen, N.J.: Scarecrow Press, 1971.

Halloran, Matthew. *Romance of the Merit System.* Washington, D.C.: Judd & Detweiler, 1929.

Hammond, Richard R. "Robert Barnwell Roosevelt and the Early Conservation Movement," *Theodore Roosevelt Association Journal,* Summer 1988.

Harbaugh, William H. *The Life and Times of Theodore Roosevelt.* London: Oxford University Press, 1975.

Harlan, Louis R. *Booker T. Washington: The Wizard of Tuskegee, 1901–1915.* New York: Oxford University Press, 1953.

Hart, Albert Bushnell, and Herbert R. Ferleger, eds. *Theodore Roosevelt Cyclopedia.* Westport, Conn.: Meckler Corporation, 1988.

Hatch, Carl E. *The Big Stick and the Congressional Gavel.* New York: Pageant Press, 1967.

Hays, Samuel P. *Conservation and the Gospel of Efficiency.* Cambridge, Mass.: Harvard University Press, 1959.

Herrick, William R., Jr. *The American Naval Revolution.* Baton Rouge, La.: Louisiana State University Press, 1966.

Hicks, John. *The Populist Revolt.* Omaha, Neb.: University of Nebraska Press, 1961.

Hofstader, Richard. *The Age of Reform.* New York: Vintage, 1960.

———. *The American Political Tradition and the Men Who Made It.* New York: Knopf, 1949.

———. *The Paranoid Style in American Politics.* New York: Knopf, 1965.

Hoogenboom, Ari. "The Pendleton Act and the Civil Service," *American Historical Review,* October 1958.

Hoover, Irwin H. *Forty-two Years in the White House.* Boston: Houghton Mifflin, 1934.

Hourihan, William J. *Roosevelt and the Sultans: The United States Navy in the Mediterranean, 1904.* Ph.D. dissertation, University of Massachusetts, 1975.

Hudson, William C. *Random Recollections of an Old Political Reporter.* New York: Cupples & Leon, 1911.

James, Henry. *Charles William Eliot.* 2 vols. Boston: Houghton Mifflin, 1930.

Jessup, Philip C. *Elihu Root.* 2 vols. New York: Dodd, Mead, 1938.

Johnson, Arthur M. "Theodore Roosevelt and the Bureau of Corporations," *Mississippi Valley Historical Review,* March 1959.

Johnson, Gerald W. *The Lunatic Fringe.* Philadelphia: Lippincott, 1957.

Johnston, William D. *TR: Champion of the Strenuous Life.* New York: Theodore Roosevelt Association, 1958.

Jones, Virgil C. *Roosevelt's Rough Riders.* Garden City, N.Y.: Doubleday, 1971.

Josephson, Matthew. *The Politicos.* New York: Harcourt Brace, 1938.

———. *The President Makers.* New York: Frederick Unger, 1964.

Juergens, George. *News from the White House.* Chicago: University of Chicago Press, 1981.

Kaplan, Justin. *Lincoln Steffens.* New York: Simon & Schuster, 1974.

Karsten, Peter. *The Naval Aristocracy.* New York: Free Press, 1972.

Kleeman, Rita H. *Gracious Lady: The Life of Sara Delano Roosevelt.* New York: Appleton-Century, 1935.

Knee, Stuart E. "The Diplomacy of Neutrality: Theodore Roosevelt and the Russian Pogroms of 1903–1906," *Presidential Studies Quarterly,* Winter 1989.

Koening, Louis W. *Bryan: A Political Biography of William Jennings Bryan.* New York: Putnam, 1971.

Kohlsaat, H. H. *From McKinley to Harding.* New York: Scribners, 1923.

Kolko, Gabriel. *The Triumph of Conservatism*. New York: Free Press, 1963.

Lane, Ann J. *The Brownsville Affair*. Port Washington, N.Y.: Kennikat Press, 1971.

Laughlin, T. Laurence. "Roosevelt at Harvard," *Review of Reviews*, LXX(1924).

Leary, John J., Jr. *Talks with T.R.* Boston: Houghton Mifflin, 1920.

Leech, Margaret. *In the Days of McKinley*. New York: Harper, 1959.

Leuchtenberg, William E. *Franklin D. Roosevelt and the New Deal*. New York: Harper, 1963.

Levy, David W. *Herbert Croly of The New Republic*. Princeton, N.J.: Princeton University Press, 1985.

Link, Arthur S. *Woodrow Wilson and the Progressive Era*. New York: Harper, 1954.

Long, John D. *Journal*. Edited by Margaret Long. Ridge, N.H.: Richard D. Smith, 1956.

Longworth, Alice Roosevelt. *Crowded Hours*. New York: Scribners, 1933.

——. *Mrs. L.* Edited by Michael Teague. Garden City, N.Y.: Doubleday, 1981.

Looker, Earl. *The White House Gang*. New York: F. H. Revell, 1929.

Lorant, Stefan. *The Life and Times of Theodore Roosevelt*. Garden City, N.Y.: Doubleday, 1959.

Lord, Walter. *The Good Years*. New York: Harper, 1960.

McCaleb, Walter F. *Theodore Roosevelt*. New York: Boni, 1931.

McCormick, Richard L. *From Realignment to Reform: Political Change in New York, 1893–1910*. Ithaca, N.Y.: Cornell University Press, 1981.

McCullough, David. *Mornings on Horseback*. New York: Simon & Schuster, 1981.

——. *The Path Between the Seas*. New York: Simon & Schuster, 1977.

McDougall, Walt. *This Is the Life*. New York: Knopf, 1926.

Mahan, Alfred Thayer. *The Influence of Seapower Upon History 1660–1783*. New York: Sagamore Press, 1957.

Manners, William. *TR and Will: A Friendship That Split the Republican Party*. New York: Harcourt Brace, 1969.

Marks, Frederick W., III. *Velvet on Iron: The Diplomacy of Theodore Roosevelt*. Lincoln, Neb.: University of Nebraska Press, 1979.

Mencken, H. L. *Prejudices: Fifth Series*. New York: Knopf, 1926.

Merrill, Horace S. *Bourbon Leader: Grover Cleveland and the Democratic Party*. Boston: Little, Brown, 1957.

—— and Marion G. Merrill. *The Republican Command 1897–1913*. Lexington, Ky.: University Press of Kentucky, 1971.

Miller, Nathan. *F.D.R.: An Intimate History*. Garden City, N.Y.: Doubleday, 1983.

——. *The Founding Finaglers: A History of Corruption in America*. New York: McKay, 1976.

——. *The Roosevelt Chronicles*. Garden City, N.Y.: Doubleday, 1979.

——. *Spying for America: The Hidden History of U.S. Intelligence*. New York: Paragon House, 1989.

——. *The U.S. Navy: A History*. New York: Morrow, 1990.

Millis, Walter. *The Martial Spirit*. Boston: Houghton Mifflin, 1931.

————. *Road to War: America 1914–1917*. Boston: Houghton Mifflin, 1935.

Milton, Joyce. *The Yellow Kids*. New York: Harper, 1989.

Morison, Elting E. *Admiral Sims and the Modern American Navy*. Boston: Houghton Mifflin, 1942.

Morison, Samuel Eliot, ed. *The Development of Harvard University 1869–1929*. Cambridge, Mass.: Harvard University Press, 1930.

Morris, Edmund. "A Few Pregnant Days: Theodore Roosevelt and the Venezuelan Crisis of 1902," *Theodore Roosevelt Association Journal*, Winter 1989.

————. *The Rise of Theodore Roosevelt*. New York: Coward, McCann & Geoghegan, 1979.

Morris, Lloyd. *Incredible New York*. New York: Random House, 1951.

Morris, Sylvia J. *Edith Kermit Roosevelt: Portrait of a First Lady*. New York: Coward, McCann & Geoghegan, 1980.

Mowry, George. *The Era of Theodore Roosevelt*. New York: Harper Torchbooks, 1962.

————. *Theodore Roosevelt and the Progressive Movement*. Madison, Wis.: University of Wisconsin Press, 1947.

Murphy, Eloise C. *Theodore Roosevelt's Night Ride to the Presidency*. Blue Mountain Lake, N.Y.: Adirondacks Museum, 1977.

Neu, Charles E. *An Uncertain Friendship: Theodore Roosevelt and Japan, 1906–1909*. Cambridge, Mass.: Harvard University Press, 1967.

Nevins, Allan. *Grover Cleveland: A Study in Courage*. New York: Dodd, Mead, 1932.

Norton, A. A. *Theodore Roosevelt*. Boston: G. K. Hall, 1980.

O'Gara, Gordon V. *Theodore Roosevelt and the Rise of the Modern Navy*. Princeton, N.J.: Princeton University Press, 1943.

Olcott, Charles S. *The Life of William McKinley*. 2 vols. Boston: Houghton Mifflin, 1916.

O'Toole, Patricia. *The Five of Hearts*. New York: Potter, 1990.

Parsons, Frances. *Perchance Some Day*. New York: privately printed, 1951.

Penick, James L., Jr. *Progressive Politics and Conservation*. Chicago: University of Chicago Press, 1968.

Pinchot, Gifford. *Breaking New Ground*. New York: Harcourt Brace, 1947.

Pringle, Henry. *Theodore Roosevelt*. London: Jonathan Cape, 1931.

Putnam, Carlton. *Theodore Roosevelt*. New York: Scribners, 1958.

Reckner, James R. *Teddy Roosevelt's Great White Fleet*. Annapolis, Md.: Naval Institute Press, 1988.

Reiger, John F. *American Sportsmen and the Origin of Conservation*. Revised edition. Norman, Okla.: University of Oklahoma Press, 1975.

Richardson, James D., compiler. *A Compilation of the Messages and Papers of the Presidents*. Vols. X and XI. Washington, D.C.: Bureau of National Literature, 1908, Supplement, 1910.

Riis, Jacob. *Theodore Roosevelt, the Citizen*. New York: The Outlook, 1904.

Rixey, Lillian. *Bamie: Theodore Roosevelt's Remarkable Sister*. New York: David McKay, 1963.

Robinson, Corinne Roosevelt. *My Brother Theodore Roosevelt.* New York: Scribners, 1921.

Rockwell, A. D. *Rambling Recollections.* New York: P. B. Hoeber, 1920.

Roosevelt, Eleanor. *This Is My Story.* New York: Harper, 1937.

Roosevelt, Eleanor Alexander. *Day Before Yesterday: The Reminiscences of Mrs. Theodore Roosevelt, Jr.* Garden City, N.Y.: Doubleday, 1959.

Roosevelt, Kermit. *The Happy-Hunting Grounds.* New York: Scribners, 1920.

Roosevelt, Nicholas. *A Front Row Seat.* Norman, Okla.: University of Oklahoma Press, 1953.

——. *Theodore Roosevelt, The Man as I Knew Him.* New York: Dodd, Mead, 1967.

Roosevelt, Theodore. *American Bears: Selections from the Writing of Theodore Roosevelt.* Introduction by Paul Schullery. Boulder, Colo.: Colorado Associated University Press, 1983.

——. *Autobiography.* New York: Da Capo, 1985.

——. *The Naval War of 1812.* Annapolis, Md.: Naval Institute Press, 1987.

——. *In Memory of My Darling Wife.* New York: privately printed, 1884.

——. *Roosevelt in the Kansas City Star: War-Time Editorials by Theodore Roosevelt.* Boston: Houghton Mifflin, 1921.

——. *The Rough Riders.* New York: New American Library, 1961.

——. *Theodore Roosevelt's Diaries of Boyhood and Youth.* New York: Scribners, 1928.

——. *The Works of Theodore Roosevelt.* National edition. 20 vols. New York: Scribners, 1926.

Roosevelt, Theodore, Jr. *All in the Family.* New York: Putnam, 1929.

Rossiter, Clinton. *The American Presidency.* New York: New American Library, 1962.

Rozek, Stacy A. "The First Daughter of the Land," *Presidential Studies Quarterly,* Winter 1989.

Russell, Francis. *The President Makers.* Boston: Little, Brown, 1976.

Sewall, William. *Bill Sewall's Story of T.R.* New York: Harper & Brothers, 1919.

Shannon, David A., ed. *Beatrice Webb's American Diary.* Madison, Wis.: University of Wisconsin Press, 1963.

Silverman, Elaine L. *Theodore Roosevelt and Women.* Ph.D. dissertation, UCLA, 1973.

Sobel, Robert. *Panic on Wall Street.* New York: Macmillan, 1968.

Spalding, George. Letter to the Editor, *Harvard Alumni Bulletin,* December 4, 1931.

Stallman, R. W. *Crane.* New York: Braziller, 1968.

Steffens, Lincoln. *Autobiography.* New York: Harcourt Brace, 1931.

Stein, Harry H. "Theodore Roosevelt and the Press: Lincoln Steffens," *Mid-America,* April 1972.

Stoddard, Henry L. *As I Knew Them.* New York: Harper & Brothers, 1927.

Storer, Mrs. Bellamy. "How Theodore Roosevelt Was Appointed Assistant Secretary of the Navy," *Harper's Weekly.* June 1, 1912.

Sullivan, Mark. *Our Times, 1900–1925.* 6 vols. New York: Scribners, 1926–1935.

Supreme Court of the State of New York. *William Barnes Against TR.* 4 vols. Walton, N.Y.: The Reporter Co., 1917.

Swanberg, W. A. *Citizen Hearst.* New York: Scribners, 1961.

———. *Pulitzer.* New York: Scribners, 1967.

Tebbel, John, and Sarah M. Watts. *The Press and the Presidency.* New York: Oxford University Press, 1985.

Thayer, William Roscoe. *John Hay.* 2 vols. Boston: Houghton Mifflin, 1916.

———. *Theodore Roosevelt: An Intimate Biography.* Boston: Houghton Mifflin, 1919.

Thelan, David. "Not Classes, But Issues," *The Journal of American History,* September 1969.

Trani, Eugene P. "Cautious Warrior: Theodore Roosevelt and the Diplomacy of Activism." In *Makers of American Diplomacy from Benjamin Franklin to Henry Kissinger.* Edited by Frank Merli and Theodore Wilson. New York: Scribners, 1974.

———. *The Treaty of Portsmouth: An Adventure in American Diplomacy.* Lexington, Ky.: University Press of Kentucky, 1969.

Trask, David F. *The War with Spain in 1898.* New York: Macmillan, 1981.

Trefethen, James B. *An American Crusade for Wildlife.* New York: Winchester Press, 1975.

Tuchman, Barbara. *The Proud Tower.* New York: Macmillan, 1962.

Tumulty, Joseph P. *Wilson as I Knew Him.* Garden City, N.Y.: Doubleday, Page, 1921.

Turk, Richard W. *The Ambiguous Relationship: Theodore Roosevelt and Alfred Thayer Mahan.* Westport, Conn.: Greenwood Press, 1987.

U.S. Civil Service Commission. *Report of Commissioner Roosevelt concerning Political Assessments and the use of Official Influence to Control Elections in the Federal Offices at Baltimore, Md.* Washington: Government Printing Office, 1891. (The Baltimore Report)

U.S. Congress. *Diplomatic History of the Panama Canal.* Senate Document 474. Washington, D.C., 1914.

Vierick, Louis. "Roosevelt's German Days," *Success,* October 1905.

Vivian, James F. *The Romance of My Life: Theodore Roosevelt's Speeches in Dakota.* Fargo, N.D.: Theodore Roosevelt Medora Foundation, 1989.

Wagenknecht, Edward. *The Seven Worlds of Theodore Roosevelt.* New York: Longman, 1958.

Washington, Booker T. *My Larger Education.* Garden City, N.Y.: Doubleday, Page, 1911.

Weaver, John D. *The Brownsville Raid.* New York: Norton, 1970.

Wecter, Dixon. *The Hero in America.* New York: Scribners, 1969.

Werner, M. R. *It Happened in New York.* New York: Coward-McCann, 1957.

Wharton, Edith. *The Edith Wharton Omnibus.* New York: Scribners, 1978.

White, G. Edward. *The Eastern Establishment and the Western Experience.* New Haven, Conn.: Yale University Press, 1968.

White, William Allen. *The Autobiography of William Allen White.* New York: Macmillan, 1946.

Widenor, William C. *Henry Cabot Lodge and the Search for an American Foreign Policy.* Berkeley: University of California, 1980.

Wiebe, Robert H. *The Search for Order, 1877–1920.* New York: Hill & Wang, 1967.

Wilhelm, Donald. *Theodore Roosevelt as an Undergraduate.* Boston: John W. Luce, 1910.

Wise, John S. *Recollections of Thirteen Presidents.* New York: Doubleday, Page, 1906.

Wister, Owen D. *Roosevelt: The Story of a Friendship.* New York: Macmillan, 1930.

Welling, Richard. "My Classmate, Theodore Roosevelt," *American Legion Magazine,* January 1929.

Wood, Fred S., ed. *Roosevelt as We Knew Him: Personal Recollections of 150 Friends.* New York: John C. Winston, 1927.

Index